Plant Molecular Farming

Rajarshi Kumar Gaur • Ramwant Gupta
Dinesh Yadav • Benedicte Riber Albrectsen
Editors

Plant Molecular Farming

Revolutionizing Agriculture Through Innovation and Applications

Springer

Editors
Rajarshi Kumar Gaur
Department of Biotechnology
Deen Dayal Upadhyaya Gorakhpur University
Gorakhpur, Uttar Pradesh, India

Uttar Pradesh Council of Agricultural Research
Uttar Pradesh, India

Dinesh Yadav
Department of Biotechnology
Deen Dayal Upadhyaya Gorakhpur University
Gorakhpur, Uttar Pradesh, India

Ramwant Gupta
Department of Botany
Deen Dayal Upadhyaya Gorakhpur University
Gorakhpur, Uttar Pradesh, India

Benedicte Riber Albrectsen
Department of Plant Physiology
Umeå Plant Science Centre
Umeå University
Umeå, Sweden

ISBN 978-981-95-3822-5 ISBN 978-981-95-3823-2 (eBook)
https://doi.org/10.1007/978-981-95-3823-2

This Springer imprint is published by the registered company Springer Nature Singapore Pte Ltd.
The registered company address is: 152 Beach Road, #21-01/04 Gateway East, Singapore 189721, Singapore

Preface

The use of plant systems for biopharmaceutical drug production offers unique advantages over traditional biotechnological methods and molecular farming. This approach provides high technological, economic, social, and ecological benefits. Plant molecular farming greatly enhances productivity and scaling while saving time and labor requirements. One of the most significant advantages of this approach concerns its diverse range of available production systems. For example, some minor modifications make open field or fully contained systems easy to implement. However, contained production systems do require the use of bioreactors that function as microbiological fermenters. In addition, some of the contained production systems need abundant light to carry out fundamental plant processes like photosynthesis. Fermentation vessels have to be uniformly illuminated, which is difficult to achieve or greatly increase the production costs, so this type of production is placed in locations that harness sunlight as an energy source and where bioreactor design is heavily influenced by algae cultivation techniques.

Recombinant protein quality, quantity, and diversity requirements are growing exponentially, which are principally addressed by recombinant protein engineering technologies and conventional expression system engineering sketched on bacteria or mammalian cell cultures. But most developing countries cannot afford the high prices associated with using such established methods to make medicine. As a result, we have to make both the new drugs and the cheaper forms of those already existing in the market.

We discuss recent advances in plant-based systems for PMF, encompassing improved tissue culture, transformation, and increasing genetic resources. We also outline some remaining challenges and suggest future directions in establishing plant as a robust platform in synthetic biology.

Gorakhpur, Uttar Pradesh, India — Rajarshi Kumar Gaur
Gorakhpur, Uttar Pradesh, India — Ramwant Gupta
Gorakhpur, Uttar Pradesh, India — Dinesh Yadav
Umeå, Sweden — Benedicte Riber Albrectsen

Contents

Editors and Contributors

About the Editors

Rajarshi Kumar Gaur earned PhD in 2005, and is now Professor in the Department of Biotechnology, Deen Dayal Upadhyaya Gorakhpur University, Gorakhpur, Uttar Pradesh, India. He is presently serving as Deputy Director General [Research Project Mangement and Co-ordination] at Uttar Pradesh Council of Agricultural Research, Uttar Pradesh, India. His PhD was on molecular characterization of sugarcane viruses, viz., mosaic, streak mosaic, and yellow luteovirus. He received MASHAV fellowship of Israel government for his post-doctoral studies and joined The Volcani Centre, Israel, and BenGurion University, Negev, Israel. In 2007 he received the Visiting Scientist Fellowship from Swedish Institute Fellowship, Sweden, to work in the Umeå University, Umeå, Sweden. He received Post-Doc. Fellowship from ICGEB, Italy, in 2008.

He has made significant contributions on sugarcane viruses and published 140 national/international papers, authored 30 edited books, and presented nearly about 55 papers in the national and international conferences. He has honored as Fellow of the Linnean Society, Fellow of the Royal Society of Biology, Fellow of the Society of Plant Research, Fellow of the Society of Applied Biology (FSAB), and Fellow of the International Society of Biotechnology (FISBT). He has bagged many other award like Prof. B.M. Johri Memorial Award, Society of Plant Research (SPR); Excellent Teaching Award by Astha Foundation, Meerut; UGC-Research Teacher Award; Young Scientist Award-2012 in Biotechnology by the Society of Plant

Research (SPR), Meerut; Scientific & Applied Research Centre Gold Medal Award-2011 for outstanding contribution in the field of Biotechnology. He has visited several laboratories of the USA, Canada, New Zealand, the UK, Thailand, Sweden, and Italy. Currently, he is handling many national and international grants and international collaborative projects on plant viruses and disease management.

Ramwant Gupta (FSSCE) is working as an Associate Professor in the Department of Botany, Deen Dayal Upadhyaya Gorakhpur University, Gorakhpur, India, since November 2022. Prior to this he worked as an Associate Professor in the Department of Biology, Faculty of Natural Sciences, University of Guyana, Georgetown, South America, from 2020 to 2022. Before this, he served as an Assistant Professor in biology at Fiji National University, Fiji Islands, from 2015 to 2020. He received his PhD in Plant Physiology (Major) and Molecular Biology and Biotechnology (minor) from G B Pant University of Agriculture and Technology, Pantnagar, India, in the year 2010. He completed Postdoctoral studies at Indian Agricultural Research Institute, New Delhi, India, from 2010 to 2014.

He was offered a Post-doctoral fellowship in Plant Sciences by the Agricultural Research Organization (ARO), the State of Israel in the year 2014. He received a foreign fellow award-2020 from the Society for Science of Climate Change and Sustainable Environment, New Delhi, India. He was awarded Excellent Scientist Award-2021 in a Farmers' training program jointly organized by the Indian Council of Agricultural Research and the National Institute of Plant Biotechnology, New Delhi, India. He teaches a wide range of biology courses viz., Plant Physiology, Molecular Biology and Biotechnology, Genetics, Genomics and Bioinformatics, and Industrial Biotechnology, and supervises UG/PG and PhD students enrolled in biology. His research area is photosynthesis and the reproductive biology of plants. He has published over 50 peer-reviewed research papers, 1 book, 4 book chapters, and over 20 conference papers.

Dinesh Yadav is a Professor in the Department of Biotechnology and presently Director of the Research and Development Cell at DDU Gorakhpur University. He has been Dean of the Faculty of Rural Sciences, Head of the Department of Biotechnology, Nodal officer of the IPR cell, and Coordinator of Centre for Genomics and Bioinformatics at DDU Gorakhpur University. He has also served as Associate Professor in the Department of Molecular Biology and Genetic Engineering at G.B. Pant University of Agriculture and Technology, Pantnagar, from 2006 to 2009. He completed his master's degree in Biotechnology from Devi Ahilya University, Indore, in 1996 and PhD from G.B. Pant University of Agriculture and Technology, Pantnagar, in 2002. He has more than 23 years of teaching /research experience.

He has availed DST-BOYSCAST Fellowship at the Australian Centre for Plant Functional Genomics (ACPFG), University of Adelaide, South Australia, from 3 May 2012 to 12 April 2013. He has been awarded for Excellence in Teaching and Research in Higher Education by the University on Teacher's Day 5 September 2021. He has been awarded Dr. Pushpendra Kumar Gupta Vishisht Krishi Vaigyanik Puraskar-2015 in the field of Agricultural Sciences by Uttar Pradesh Academy of Agricultural Sciences (UPAAS), Lucknow, and Young Scientist Award-2008 by Uttarakhand State Council of Science & Technology in the discipline Biotechnology, Biochemistry, and Microbiology. His areas of specialization are Molecular biology, Bioinformatics, Plant Biotechnology, and Enzyme technology. His research group has filed seven patents, two of which have been published. He has published more than 150 research papers including reviews, books, chapters in books, and proceedings in conferences, and more than 300 GenBank accession numbers. A total of 48 indigenously isolated microbial strains (fungal and bacterial) strains have been deposited in the Culture collection centers of India. He has carried out nine projects from funding agencies like DBT, UGC, UP Council of Agricultural Research, Lucknow, and DST as PI/Co-PI and Mentor.

He has guided 16 students to PhD in Biotechnology and 1 registered and pursuing research under his guidance. He has been a mentor for five post-doctoral fellowships, namely, UGC-Dr. D.S. Kothari (twice), DST-Women Scientist-A, DST-Women Scientist-B, and SERB National PDF.

He has delivered more than 100 invited talks/lectures at conferences/symposiums and served as Mentor for the DST-INSPIRE Science internship camp and Inspire awards. Presently he is working on plant-specific transcription factor-DOF (DNA binding with One finger) and Nuclear Factor-Y (NF-Y) for developing biotic and abiotic stress tolerance crops and pectinases groups of enzymes with potential applications in different industries.

Benedicte Riber Albrectsen is Associate Professor at the Department of Plant Physiology, Umeå University, and affiliated with the Umeå Plant Science Centre. She earned her MSc in Denmark, her PhD in Sweden, and completed postdoctoral research in the United States. She was appointed Assistant Professor at Umeå University in 2005 and received her docentur from the Swedish University of Agricultural Sciences (SLU) in 2009.

Her research focuses on specialized plant metabolites - particularly defence compounds . their evolution, biosynthesis, ecological function, and integration with primary metabolism under biotic stress. She has published over 70 peer-reviewed articles, with particular focus on phenolic compounds in Salicaceous species, glucosinolates in Brassicaceae, and solanine in potato.

In addition to her research, she has contributed extensively to teaching and outreach. As study director, she co-developed teaching materials published in The Plant Cell through the COST Action FA1405 "Three-way interactions between plants, microbes and arthropods." She is also active in science communication, with numerous translations and popular science publications.

Contributors

Rajneesh K. Agnihotri Department of Botany, S.L.S. Khandari Campus, Dr. B.R. Ambedkar University, Agra, Uttar Pradesh, India

Ammar Ahmed Department of Biotechnology, Faculty of Science and Humanities, SRM Institute of Science and Technology, Ramapuram, Chennai, Tamil Nadu, India

Mariam Akhtar National Institute for Biotechnology and Genetic Engineering (NIBGE) - Collage PIEAS, Faisalabad, Pakistan

Imran Amin National Institute for Biotechnology and Genetic Engineering (NIBGE) - Collage PIEAS, Faisalabad, Pakistan

Romila Rawat Bisht Department of Botany, Hansraj College, University of Delhi, Delhi, India

Rituraj Borah Applied Microbiology Laboratory, Department of Forestry, North Eastern Regional Institute of Science and Technology, Nirjuli, Arunachal Pradesh, India

Shahzeen Fatma Department of Botany, Deen Dayal Upadhyaya Gorakhpur University, Gorakhpur, India

Rajarshi Kumar Gaur Department of Biotechnology, Deen Dayal Upadhyaya Gorakhpur University, Gorakhpur, Uttar Pradesh, India

Uttar Pradesh Council of Agricultural Research, Uttar Pradesh, India

Ramwant Gupta Department of Botany, Deen Dayal Upadhyaya Gorakhpur University, Gorakhpur, Uttar Pradesh, India

Suman Gusain Division of Biotechnology, CSIR-Institute of Himalayan Bioresource Technology, Palampur, Himachal Pradesh, India

Academy of Scientific and Innovative Research (AcSIR), Ghaziabad, India

Ifrah Imran National Institute for Biotechnology and Genetic Engineering (NIBGE) - Collage PIEAS, Faisalabad, Pakistan

Stefka A. Ivanova Bulgarian Pharmaceutical Science Society, Sofia, Bulgaria

Khushboo Jain Department of Biotechnology, Faculty of Science, MVSC, Janardhan Rai Nagar Rajasthan Vidyapeeth (Deemed to be University), Udaipur, Rajasthan, India

D. C. Joshi ICAR-Vivekananda Institute of Hill Agriculture, Almora, Uttarakhand, India

Rohit Joshi Division of Biotechnology, CSIR-Institute of Himalayan Bioresource Technology, Palampur, Himachal Pradesh, India

Academy of Scientific and Innovative Research (AcSIR), Ghaziabad, India

Shubham Joshi Division of Biotechnology, CSIR-Institute of Himalayan Bioresource Technology, Palampur, Himachal Pradesh, India

Academy of Scientific and Innovative Research (AcSIR), Ghaziabad, India

Madhu Kamle Department of Biochemistry, University of Lucknow, Lucknow, India

Ashok Kumar Department of Botany, Deen Dayal Upadhyaya Gorakhpur University, Gorakhpur, India

Anita Kumari Division of Biotechnology, CSIR-Institute of Himalayan Bioresource Technology, Palampur, Himachal Pradesh, India

Academy of Scientific and Innovative Research (AcSIR), Ghaziabad, India

Khushbu Kumari Division of Biotechnology, CSIR-Institute of Himalayan Bioresource Technology, Palampur, Himachal Pradesh, India

Academy of Scientific and Innovative Research (AcSIR), Ghaziabad, India

Pradeep Kumar Applied Microbiology Laboratory, Department of Botany, University of Lucknow, Lucknow, UP, India

R. Vinoth Kumar Department of Biotechnology, Faculty of Science and Humanities, SRM Institute of Science and Technology, Ramapuram, Chennai, Tamil Nadu, India

Ayushi Malik Nanoparticle Synthesis and Bioinformatics Laboratory, Department of Biotechnology, Vigyan Bhawan, Block – B, New Campus, Mohanlal Sukhadia University, Udaipur, Rajasthan, India

Department of Biotechnology, Nanoparticle Synthesis and Bioinformatics Laboratory, Mohanlal Sukhadia University, Udaipur, Rajasthan, India

Shahid Mansoor National Institute for Biotechnology and Genetic Engineering (NIBGE) - Collage PIEAS, Faisalabad, Pakistan

Avinash Marwal Nanoparticle Synthesis and Bioinformatics Laboratory, Department of Biotechnology, Vigyan Bhawan, Block – B, New Campus, Mohanlal Sukhadia University, Udaipur, Rajasthan, India

Anita Mishra Department of Science (Biotechnology), Biyani Girls College, Jaipur, Rajasthan, India

Arti Mishra Department of Botany, Hansraj College, University of Delhi, Delhi, India

Umeå Plant Science Centre (UPSC), Department of Plant Physiology, Umeå University, Umeå, Sweden

Boris Seryozhev Mladenov University Multiprofile Hospital for Active Treatment and Emergency Medicine "N.I.Pirogov", Sofia, Bulgaria

Ganesa Moorthy Murugan Department of Plant Sciences, School of Biological Sciences, Madurai Kamaraj University, Madurai, Tamil Nadu, India

Rubab Zahra Naqvi National Institute for Biotechnology and Genetic Engineering (NIBGE) - Collage PIEAS, Faisalabad, Pakistan

Jhilmil Nath Division of Biotechnology, CSIR-Institute of Himalayan Bioresource Technology, Palampur, Himachal Pradesh, India

Academy of Scientific and Innovative Research (AcSIR), Ghaziabad, India

Pranhita Nimbalkar Research Scientist, Tojo Vikas Biotech Pvt Ltd, Bengaluru, Karnataka, India

Danka P. Obreshkova Clinic research of Drugs in Medicine, University Hospital "N. I. Pirogov", Sofia, Bulgaria

Ananya Padhy Amity Institute of Environmental Sciences Amity University, Noida, Uttar Pradesh, India

Nikolay Manchev Petrov Virology Laboratory, New Bulgarian University, Sofia, Bulgaria

Sandesh Poudel School of Health and Allied Sciences, Pokhara University, Pokhara, Kaski, Nepal

G. Priya Department of Biotechnology, Faculty of Science and Humanities, SRM Institute of Science and Technology, Ramapuram, Chennai, Tamil Nadu, India

Kailash P. Rajpurohit Department of Biotechnology, Faculty of Science and Humanities, SRM Institute of Science and Technology, Ramapuram, Chennai, Tamil Nadu, India

Varsha Rani Department of Biotechnology, School of Engineering and Technology, Sandip University, Nashik, Maharashtra, India

Rajneesh Rao Department of Botany, Deen Dayal Upadhyaya Gorakhpur University, Gorakhpur, India

Meenakshi Rawat Division of Biotechnology, CSIR-Institute of Himalayan Bioresource Technology, Palampur, Himachal Pradesh, India

Academy of Scientific and Innovative Research (AcSIR), Ghaziabad, India

Rabia Rehman National Institute for Biotechnology and Genetic Engineering (NIBGE) - Collage PIEAS, Faisalabad, Pakistan

Indrajeet Sahani Department of Botany, Deen Dayal Upadhyaya Gorakhpur University, Gorakhpur, Uttar Pradesh, India

Muhammad Waseem Sajjad National Institute for Biotechnology and Genetic Engineering (NIBGE) - Collage PIEAS, Faisalabad, Pakistan

Navnidhi Sharma School of Life Sciences, Coventry University, Coventry, UK

Smriti Shukla Amity Institute of Environmental Sciences Amity University, Noida, Uttar Pradesh, India

Saqib Siddique National Institute for Biotechnology and Genetic Engineering (NIBGE) - Collage PIEAS, Faisalabad, Pakistan

Mansi Singh Applied Microbiology Laboratory, Department of Botany, University of Lucknow, Lucknow, UP, India

Mariya Ivanova Stoyanova Institute of Soil Science, Agrotechnologies and Plant Protection (ISSAPP) "Nikola Pushkarov," Agricultural Academy, Sofia, Bulgaria

Mayank Suthar Nanoparticle Synthesis and Bioinformatics Laboratory, Department of Biotechnology, Vigyan Bhawan, Block – B, New Campus, Mohanlal Sukhadia University, Udaipur, Rajasthan, India

Jebasingh Tennyson Department of Plant Sciences, School of Biological Sciences, Madurai Kamaraj University, Madurai, Tamil Nadu, India

Raj Kumar Thapa Pharmacy Program, Gandaki University, Pokhara, Kaski, Nepal

Research Center, Invention and Innovation Center, Gandaki University, Pokhara, Kaski, Nepal

Jaagriti Tyagi Department of Biotechnology, Dr. MPS Group of Institutions, Agra, Uttar Pradesh, India

Sunita Verma Manas Enclave Residential Society, Indirapuri Street, Bhuwana, Udaipur, Rajasthan, India

Dinesh Yadav Department of Biotechnology, Deen Dayal Upadhyaya Gorakhpur University, Gorakhpur, Uttar Pradesh, India

Kanchan Yadav Department of Biotechnology, Deen Dayal Upadhyaya Gorakhpur University, Gorakhpur, Uttar Pradesh, India

S. Yamini Department of Biotechnology, Faculty of Science and Humanities, SRM Institute of Science and Technology, Ramapuram, Chennai, Tamil Nadu, India

Transient Expression in Plants

Mariam Akhtar, Ifrah Imran, Muhammad Waseem Sajjad, Saqib Siddique, Rabia Rehman, Rubab Zahra Naqvi, Shahid Mansoor, and Imran Amin

Abstract

In the past few years, novel techniques have loomed that permit the employment of plants as manufacturing factories to transiently and frequently express targeted genes of interest with high efficiency, either endogenous or exogenous, in their endemic or modified form. The aforestated strategies permit and expedite targeted gene products for both research and commercial objectives, mitigating the requirement of extensive, tedious, and burdensome stable plant transformation. Moreover, transient expression studies also circumvent the regulatory constraints that are associated with these transformations. Co-expression of proteins for the production of macromolecules and analysis of metabolic pathways can also be effectively achieved using transient expression systems. Three expression systems are employed for transient delivery in plants: biolistic bombardment of leaves with nucleic acids encoding the gene of interest, *Agrobacterium*-mediated transformation of T-DNA fragments, and plant viral vectors. Plant viruses proliferate independently, increasing the possibility of infecting plants in the ecosystem. Needle-less syringes, high-pressure sprays, and vacuum infiltration methods are used to deliver *Agrobacterium* suspension into plant leaves, where the gene of interest is transferred into the nucleus of host plants via T-DNA delivery.

Keywords

Transient transformation · Transient expression systems · Transient assay · Biolistic bombardment · Biolistics · Agroinfiltration · Viral vectors · Protoplast transfection · Gene delivery · Gene silencing · Gene expression · RNAi · VIGS

M. Akhtar · I. Imran · M. W. Sajjad · S. Siddique · R. Rehman · R. Z. Naqvi · S. Mansoor
I. Amin (✉)
National Institute for Biotechnology and Genetic Engineering (NIBGE) - Collage PIEAS, Faisalabad, Pakistan

R. K. Gaur et al. (eds.), *Plant Molecular Farming*,
https://doi.org/10.1007/978-981-95-3823-2_1

1 Introduction

A voluminous gene expression data collected from comparative studies, scrupulously processed using bioinformatics tools, has enabled the analysis of transcriptome networks for predicting specific gene functions and roles in cell processes (Goldenkova-Pavlova et al. 2018). It is important to note that in silico analysis methods are primarily dependent on prognostic classification of gene function based on comparative analysis of homologous genes, but this approach is meager to completely elucidate a specific gene function and the precise mechanism of action (Kabardaeva et al. 2019). Prior to the onset of genome-wide sequencing, direct genetic approaches, which involved the comparison of physiological and phenotypic effects of induced or natural mutations in the target sequence, were used to scrutinize gene function. Now, the heightened accessibility of transcriptomic and genomic information, coupled with dynamic technologies for genetic modification across plant species, has enabled the employment of reverse genetics, from genotype to phenotype, for functional genomics (Marcon et al. 2024; Martín-Alonso et al. 2021). This method involves complementing mutations or altering the gene expression by partial or complete suppression via RNA interference (RNAi) or T-DNA mutations, or by producing transgenic lines overexpressing the target gene. Research suggests that such fluctuations in the gene expression can lead to abnormal phenotypes, making identification more nuanced (Basso et al. 2020; Li et al. 2022a). The development of transgenic plants has given unprecedented experimental validation for many genes' functions. However, the primary challenges in creating transgenic plants are the requirement of significant time and resources and the genotype-dependent efficiency of transformation. It includes selecting transgenic individuals among an immense number of transformants, verifying the incorporation of transferred sequences, confirming single copy integration events into the genome, and estimating the expression level of mRNA and proteins, among other steps (Anjanappa and Gruissem 2021). The prevailing view suggests that transient expression technology in plants offers substantial leverage over stable expression systems, mainly including the elimination of regeneration of transformants, operating independently of position effects related to the integration site of T-DNA, and having no impact on the host genome's stability. Furthermore, this approach considerably quickens research, enabling examination of target gene function in 7–10 days after the introduction of a gene construct into the plant (Tyurin et al. 2020). As a result of these merits, the transient expression system has emerged as a valuable tool for in vivo studies of gene function, facilitating physiological and molecular characterization across diverse plant materials.

2 Gene Delivery Tools in Transient Expression

2.1 Protoplast Transfection

Around six decades have passed since Cocking proposed an inaugural approach for extracting plant protoplasts (Cocking 1960). These are cell wall-free, enclosed by frail membranes, responsive to osmotic pressure, sensitive, totipotent, and serve as adaptable and functional experimental platforms. They scrutinize how stressors and hormones transitorily alter signal transmission and metabolic pathways, provide answers to explicit questions about different cell types, interactions, and movement, and ascertain the subcellular localization of tagged proteins. The progress of modern botany is markedly simplified by the adaptability of plant protoplasts (Kumari 2019).

2.1.1 Methods of Protoplast Isolation

Numerous techniques, including electroporation, PEG-mediated transfection, microinjections, liposome-mediated transfection, and *Agrobacterium*-mediated methods, can be used to introduce macromolecules like recombinant DNA into the protoplast. The PEG-mediated transformation method uses polyethylene glycol (PEG) for binding with the cell membrane, distressing the phospholipid bilayer, and disturbing cell surface potential. Resultantly, this aids the insertion of T-DNA or plasmid into the protoplast (Wang et al. 2022a; Han et al. 2023). Contrastingly, electroporation involves the application of a short, high-voltage pulse to temporarily stimulate plant cell membranes. Small pore channels are transiently formed as a result of this stimulation, thus allowing exogenous DNA molecules to promptly enter the protoplast (Ozyigit 2020). *Agrobacterium* co-transformation involves the introduction of foreign genes into protoplasts by cocultivating them with *Agrobacterium*, resulting in the production of a transgenic plant (Li et al. 2022b). Liposome-mediated transformation requires enclosing DNA within liposomes, which then fuse with protoplasts, facilitating gene transfer (Zhi et al. 2022).

2.1.2 Protoplast-Based Transformation Systems for Transient Expression in Diverse Plant Species

Protoplast-based transformation expression system (PTES) using PEG is an eminently competent technique, extensively employed in disparate applications, such as chromosome transfer via micro protoplast, somatic hybridization, electroporation, DNA or organelle microinjection, nucleocytoplasmic protein trafficking, and gene transactivation (Lacroix and Citovsky 2020a). PTES is quicker as it is transient, more affordable, convenient, and opportune than stable transformation. This method demands high-quality protoplast, precise and subtle transformation protocols (Zhang et al. 2022a). Following transfection, protoplasts are used for subsequent investigations. Notably, PEG-mediated protoplast transfection expression swank high transformation efficiency, making it a favored choice in plant molecular research. The advancement of innovative, prudent, cost-effective, and perceptive reporter gene assays for green fluorescent protein (GFP), firefly luciferase (LUC), and glucuronidase (GUS) in plant cells has intensified the use of protoplast transient expression systems (Bai et al. 2020; Zheng et al. 2021). High-throughput PTES has emerged as a compelling tool for expeditious gene functional analysis and has been well implemented in many plant species, including both monocotyledonous and dicotyledonous crops such as tobacco, rice, cucumber, Arabidopsis, and maize for screening transactivation of various transcription factors, gene silencing cassettes, and efficient genome editing (Trinidad et al. 2021; Jin et al. 2020; Kumari et al. 2021; Pasternak et al. 2021; Coy et al. 2022).

2.1.3 Tailoring Explant Choice: Species Limitation and Optimal Yield

While valid and successful protocols of protoplast isolation exist for various model and crop plants, many horticultural species still pose challenges. Primary source materials are plant tissues, with excellent yield mesophyll protoplast isolated from Arabidopsis, maize, tobacco, rice, and other crop plants like *Panicum virgatum, Medicago sativa, Phaseolus vulgaris, and Hevea brasiliensis* (Talla et al. 2023; Tonnies et al. 2023; Wang et al. 2024a; Xu et al. 2021a; Egesa et al. 2024; Li et al. 2021a; Lai et al. 2020). Species limitation restricts the isolation of adequate and viable protoplasts from mature leaves of many plants. Resultantly, it led to the exploration of in vitro grown seedlings, auspiciously isolating mesophyll protoplast from grape, pepper, tansy, and pineapple (Scintilla et al. 2022; Akter et al. 2024; Ren et al. 2021a). The callus induction, however, is a lengthy process, often taking several months. Alternatively, flower petals have been elected for protoplast isolation in ornamental plants, such as petunia hybrid, Tulipa, and *Rosa rugosa*. Yet this method is impoverished by the limited supply of petals and rapid withering of flowers for uninterrupted protoplast isolation (Naing et al. 2021).

2.1.4 Factors Shaping the Protoplast Isolation Efficiency

The establishment of a stable and reliable transient expression system in protoplasts hinges on two captious factors: obtaining a considerable amount of viable protoplasts and developing eminently adequate and competent transformation techniques. These aspects are influenced by many parameters, such as PEG concentration, pH of the enzymatic solution, and transfection time (Chen et al. 2023a). The enzymolysis solution is a blend of cell wall-degrading lytic enzymes and osmotic stabilizers. For protoplast isolation, sorbitol and mannitol, either in combination or individually, are commonly utilized as osmotic stabilizers. Cell wall digestion during isolation requires the use of lytic enzymes, pectinase, 1,3-glucanase, cellulose, and chitinase. The efficacy of these lytic enzymes on cell wall degradation differs greatly, necessitating the assessment of each enzyme combination (Yao et al. 2016; Zhang et al. 2022b). The period of enzymolysis strongly impacts the protoplast isolation, with extensive periods expanding the proportion of protoplasts formed up to a certain point; after that, further enzymolysis leads to protoplast rupture. Enzyme activity is influenced by the pH of the enzymolysis solution. PEG is contemplated to be a compelling protoplast transfection agent due to its competence to boost cell wall permeability. The PEG concentration affects both cell viability and transformation efficiency. Transfection duration influences efficiency, with extensive periods potentially damaging protoplasts, leading to reduced efficacy (Davis 2020). Also, the ratio of lytic enzymes and osmotic stabilizers plays a significant role in the establishment of a competent protoplast isolation system (Davis 2020; Isaac 2020). After the removal of the cell wall using cellulase, protoplasts are released, maintaining various cellular activities and physiological responses. DNA is introduced into protoplasts using PEG-calcium-mediated transfection, and the cells are incubated for 2–24 h to enable gene expression (Hu et al. 2020a). Some factors greatly increasing protoplast isolation efficiency are dim light or short photoperiods of donor plant, low temperature (15 °C), high humidity (82%), fertilization of donor plant by supplementing with ammonium, young barely expanded leaves, pretreatment of leaves with L-lysine, kinetin, L-arginine, cycloheximide, cadaverine, and putrescine increases protein synthesis by slowing down protoplast lysis, auxiliary calcium application during winters enhance stability, pretreatment with 20 mM cysteine for enhanced cell wall digestion decreased sucrose, increased auxin supply, addition of L-methionine or L-cysteine, and reduction of NH_4NO_3 in cell culture medium enhance yield and viability (Evans and Bravo 2013).

2.1.5 Functional Gene Studies via Protoplast-Based Transient Expression

Transient expression systems based on protoplasts are significant tools for understanding the sophistication of gene function and expression regulation. With the use of rice protoplasts, Nakashima et al. explicated the function and subcellular localization in a transient expression system of *OcNAC6*. It was localized in the nucleus and augmented plant stress tolerance to several biotic and abiotic factors by acting as a transcriptional activator. In maize chloroplast, transcription factor *OsWRKY71* was identified, and transient expression analysis was carried out. The findings revealed that *OsWRKY71* is mostly localized in the nucleus, with trace amounts present in the cytoplasm. By regulating the downstream target genes, this transcription factor assists in increasing cold stress tolerance. An important regulator of starch synthesis is *ZmbZIO91* (Chen et al. 2023a). The gene's function was validated by inserting it into the protoplast, and then its impact on starch synthesis gene expression was assessed. Similarly, a study confirmed *NLP1* and *NIS1* gene activity in significantly inducing cell death in a transient system set up in melon leaves. The evolution of single-cell RNA sequencing has recently transformed the ability to explore the developmental landscape of complex organs. This allows tracking of differentiation pathways of specialized cells and provides insights at single-cell level (Chen et al. 2021). Wang et al. established a protocol for protoplast isolation from *Chirita pumila* using enzymatic digestion and achieved a transformation efficiency of 70%. Arabidopsis *AREB1* homolog, *CpTOE3* gene, was targeted, which binds to acid-responsive elements to stimulate transcription of stress-related genes. The activation of *ABRE*-driven GFP reporter was demonstrated using the *AREB1* gene, thereby validating protoplast's functionality for transient expression studies (Wang et al. 2022b).

2.2 Nano-delivery Systems

2.2.1 Size Range of Nano-carriers for Biomolecule Delivery in Plant Organelles

The difficulty of delivering biomolecules into plants, known as the "biomolecule delivery problem," is mainly attributed to cell walls. The size exclusion limit of approximately 5–20 nm represents the primary obstruction to introducing external biomolecules (Gad et al. 2020). On the other hand, organelle and nuclear membranes have a larger size exclusion limit of about 300–500 nm, but they still present additional obstacles that must be conquered for genetic modifications of plastid or nuclear genome. The comparable dimensions of gold nanoparticles and micrometer-sized plastids used in biolistics make it troublesome to deliver DNA without causing detrimental effects to the organelles. It is suggested that nanomaterials introduced into plants without using force show great potential for delivering biological molecules. Nanomaterials are described as substances with at least one dimension less than 100 nm (1–100 nm, general nanoparticles; 148–200 nm, natural peptides, magnetic nanoparticles; 148–200 nm; 130–140 nm, cationic fluorescence nanoparticles; 15–120 nm, clay nanoparticles), possessing adjustable and distinct chemical and physical properties, which ascend their ability to merge with DNA, RNA, and proteins with precision and exquisite control. Several types of nanomaterials have been used in recent decades, including carbon-based nanotubes (single- or multi-walled), DNA nanoparticles, metal-based nanoparticles (TiO_2, AL_2O_3, and ZnO), and chitosan-complexed carbon nanotubes (Wang et al. 2019; Lv et al. 2020).

2.2.2 Harnessing Various Nanoparticles for Targeted Biomolecule Delivery in Plants

Studies have shown that molecules can internalize into cell walls without mechanical subsidies, such as ultrasound, biolistics, electroporation, or vortexing. It includes layered double hydroxide (LDH) clay nanosheets arbitrating the siRNA molecule delivery via foliar application to *Nicotiana tabacum*, mesoporous silica nanoparticles (MSNs) facilitating DNA plasmid entry upon coculturing with Arabidopsis roots, and DNA origami nanostructures allowing siRNA delivery to *Nicotiana benthamiana* (Zhi et al. 2022; Cheng et al. 2024; Lam and Steinmetz 2018). These substantiate biomolecule internalization and delivery with phenotypic pest resistance, fluorescence microscopy, and mRNA and protein-level quantification. Furthermore, various models and crop plants have been shown to benefit from unassisted transport of siRNA and plasmid DNA using carbon nanotubes (CNTs). CNTs aid the delivery of plasmid DNA *GFP* transgene expression that targets the nuclear genome, with RNA-level quantification. A study documented the delivery of CNT-based plasmid DNA and the transient expression of *YFP* transgene on the basis of fluorescence microscopy imaging in the chloroplast. Studies reveal that CNTs produce transient gene expression, irrespective of their surface chemistry, and shield plasmids from the haul of endonuclease degradation, helping in the transport of diverse biomolecules (Chandrasekaran et al. 2020; Dunbar et al. 2022). In a study, cobalt nanoparticles outperformed Co2+ via enhancing *Agrobacterium*-mediated gene expression in *C. roseus* by enhancing *virB1* gene expression and suppressing ethylene biosynthesis due to reduced plant defense mechanism activation against oxidative stress and heavy metals (Fouad and Hafez 2020).

In-depth design interrogation for nanoparticle-mediated delivery systems must be established in order to advance plant nanobiotechnology, as, other than size, chemical and physical properties allowing adequate delivery into the cell remain ambiguous. Some factors that influence delivery into plant cells are aspect ratio, shape, colloidal stability, tensile strength, charge, and compactness; these factors may not all contribute mutually to internalization (Yan et al. 2022). The lipid envelope exchange penetration (LEEP) model proposes that fortunate membrane penetration depends on the nanoparticle's surface charge, along with its size (Wong et al. 2016). Studies indicating symplastic and apoplastic transport via plasmodesmata are insufficient; for instance, MSNs have been demonstrated to penetrate the cell wall, radially disseminate through plasmodesmata, and travel via xylem (Wang et al. 2022c). Deciphering nanomaterial design and synthesis for governing nanomaterial fate is further complicated by the synergy between the heterogeneous plant biological surface and nanoparticle surface chemistry. A study validated that surface modifications, such as hydrophobicity and nanoparticle size, revamp the way gold nanoparticles are transported to plants (Kumar et al. 2020). Notably, magnetic nanoparticles Fe_2O_3 and Fe_3O_4, with their small size, large surface area, high thermal stability, low sedimentation rate, and low toxicity, are ideal materials for field applications. This magnetofection approach has some drawbacks despite bypassing stable transformation and regeneration. Some species may not be reasonable for transformation with magnetic nanoparticles because of pollen aperture and maternally inherited organelles (mitochondria and chloroplasts) (Lv et al. 2020; Hartono et al. 2023). While maintaining aqueous solubility, nanoparticles may have to traverse many obstacles to carry their cargo, such as compact cell walls made up of glycan networks embedded with aromatics and proteins, hydrophobic and waxy cuticles, and lipid membranes. Furthermore, nothing is well-documented about the spectrum of tissues that nanoparticles can penetrate, including root, seed, leaf, and callus. Thus, for the sake of genetic engineering, interdisciplinary and systemic efforts are required to adjust the properties of nanoparticles (Wang et al. 2019; Lv et al. 2020).

2.3 RNAi and VIGS

Double-stranded RNA (dsRNA)-mediated sequence-specific gene silencing is known as RNAi, also called post-transcriptional gene silencing (PTGS) in plants. RNA-mediated virus resistance and cross-protection are the natural processes strongly associated with PTGS (Ricci et al. 2021). Gene silencing was first observed in transgenic petunia plants that expressed the chalcone synthase (*CHS*) gene, crucial for the production of flavonoid anthocyanin pigments, giving the flower its purple color. Surprisingly, the introduction of *CHS* transgene for overexpression to intensify pigment in flowers yielded an albino phenotype, contrary to what was expected. This unanticipated phenomenon is recognized as co-suppression, where both the native and introduced transgenes were silenced. Later research revealed that it could result from post-transcriptional degradation of sequence-specific transcripts or transcriptional suppression (van der Krol et al. 1990). Transcriptional gene silencing affects methylation of promoters and chromatin remodeling, leading to the silencing of gene transcription. PTGS takes place in the cytoplasm, involves sequence-specific mRNA degradation, and is reestablished every generation. It also parades systemic silencing, spreading throughout the plant, proposing the presence of a signal that travels between regions. Viruses and transgenes primarily provoke PTGS, which involves the production of dsRNA molecules having sequence homology with the silenced genes (Song et al. 2021; Ashfaq et al. 2020).

2.3.1 RNAi Mechanism

RNAi plays a role in targeting consequent target degradation. Unlike exonuclease-mediated mRNA degradation from the terminals, RNAi engages in the formation of double-stranded RNA (dsRNA), which functions as a substrate for dicer, an endonuclease enzyme similar to RNAseIII, which processes dsRNA into 21–23 nucleotide fragments called short interfering RNA (siRNA). These fragments require 2 nucleotide 3′ overhangs and 5′ phosphate ends for functioning suitably. These siRNAs are then associated with the Argonute (*AGO*) family protein, which provides sequence specificity for mRNA targets through base pair complementarity (Hung and Slotkin 2021). After incorporation of AGO with siRNA or miRNA, it forms the RNA-induced silencing complex (RISC), along with the association of other proteins. RISC either cleaves the mRNA or represses translation. The cleaved fragments then serve as a template for another round of dsRNA formation by an RNA-dependent RNA polymerase (RdRP) family protein, initiating another cycle of RNAi to generate secondary siRNAs. This gives rise to a self-expanding cycle of mRNA degradation and siRNA synthesis. Consequently, siRNAs get amplified, leading to further degradation of similar mRNAs, thus lowering the level of protein translation by a process known as post-transcriptional silencing (Wang et al. 2021a).

2.3.2 Why Does VIGS Outperform Other Silencing Methods

Virus-Induced Gene Silencing (VIGS) has numerous superiorities over other methods. VIGS does not need the generation of stable transgenic lines, which hastens the process of functional analysis and makes it cost-effective. VIGS experiments are carried out by inoculation of *Agrobacterium* carrying a suitable vector into the plant, and phenotypes can be scrutinized within 2–6 weeks postinoculation. VIGS is especially beneficial for plants obstinate to transformation and for functional investigation of genes whose silencing results in lethality. In a VIGS experiment, viral DNA or RNA is suitably altered to minimize viral symptoms but retain infectivity. To facilitate the cloning of target genes, the altered VIGS vector needs to have multiple cloning sites (Rössner et al. 2022). More than 40 plant viruses, belonging to 10 different families, have been developed into VIGS vectors thus far (Table 1). This process is comprised of five steps: selecting a suitable vector, selecting and cloning the target gene sequence into a VIGS viral vector, inoculating the vector

Table 1 List of viral vectors used for VIGS

Genome	Family	Virus	Plant	Targeted genes	Method	References
ssDNA	Geminivirdae	Cabbage leaf curl virus (CbLCV)	*Arabidopsis, cabbage, N. benthamiana*	*CH42, PDS, ChlH*	Particle gun bombardment, vacuuum osmosis, agroinfiltration	Xiao et al. (2020)
		Tomato golden mosaic virus (TGMV)	*N. tabacum, N. benthamiana*	*Su*	Particle gun bombardment	Kjemtrup et al. (1998)
		Cotton leaf crumple virus (CLCrV)	*Gossypium*, *Cannabis sativa*	*GFP, ChLI, PDS*	Biolistic bomardment, Agroinfiltration	Tuttle et al. (2012), Tuttle et al. (2008) and Schachtsiek et al. (2019)
		Chili leaf curl virus (ChiLCV)	*N. benthamiana*	*PDS, eGFP*	Agro-inoculation	Kushwaha and Chakraborty (2017)
		Beet curly top virus (BCTV)	*N. benthamiana*	*PDS*	Agroinfiltration	Watanabe et al. (2021)
		Sweet potato leaf curl virus (SPLCV)	Sweet potato	*PDS, CHL*	Agro-soaking	Zhang et al. (2024a)
		Tomato leaf curl virus (ToLCV)	Tomato	*PCNA*	Agro-inoulation	Pandey et al. (2009)
		Bhendi yellow vein mosaic virus (BYMV)	*N. benthamiana*	*GFP, Su, PDS, PCNA, AGO1*	Agroinfiltration	Jeyabharathy et al. (2015)
ssRNA	Comoviridae	Cowpea severe mosaic virus (CPSMV)	*N. benthamiana*, Soy bean	*MYB75*	Agroinfiltration	Zaulda et al. (2022)

(continued)

Table 1 (continued)

Genome	Family	Virus	Plant	Targeted genes	Method	References
ssRNA	Potyviridae	Turnip mosaic virus (TMV)	*C. quinoa, N. benthamiana*	*YC5, GFP, GK*	Mechanical rubbing, Agroinfiltration	Lin et al. (2013)
		Maize dwarf mosaic virus (MDMV)	Maize	*PDS, ChI, lemon white 1*	Mechanical inoculaton	Xie et al. (2021)
		papaya leaf distortion mosaic virus (PLDMV)	Papaya	*PDS, ChlH, CYP83B1, GA, GID1A, GID1B*	Agroinfiltration	Tuo et al. (2021a)
		Potato virus A (PVA)	*N. benthamiana*	*GFP*	Particle bombardment	Gammelgård et al. (2007)
		Watermelon mosaic virus (WMV)	Melon	*PDS, ChlI*	Agroinfiltration	Houhou et al. (2021)
		Telosema mosaic virus	*Passiflora edulis, N. benthamiana*	*PDS, GFP*	Agroinfiltration	Wang et al. (2024b)
		Sugarcane mosaic virus (SCMV)	*N. benthamiana*	*ChlI*	Agroinfiltration	Ali et al. (2017)

ssRNA (+)	Virgaviridae	Tobacco Rattle Virus (TRV)	Cassava, Tomato, *B. cinera*, Cotton, Sunflower, *Areca catechu*, Chrysanthemum, Chinese narcissus, *Crocus sativus L*, Chinese jujube, *Camellia sinensis*, *Rhododendron micranthum*, *A. majus*	*PDS, WRKY3, WRKY7, WRKY10, MAPKK5, CHS, tubulin, OcQR1, RKY15–1, MYB3, ZjCLA, hairy*	Agroinfiltration, vaccum infiltration	Zeng et al. (2019), Xiang et al. (2022), Ding et al. (2024), Mardini et al. (2024), Xingqi et al. (2024), Chen et al. (2023b), Zhou et al. (2021), Kalia et al. (2024b), Zhang et al. (2023), Yang et al. (2023b), Xu et al. (2024) and Tan et al. (2020)
		Tobacco mosaic virus (TMV)	Nicotiana tabacum	*LLP*	Rub inoculation	Labarile et al. (2023)
		Cucumber green mottle mosaic virus (CGMMV)	Cucurbits, *N. benthamiana*, luffa	*PDS, TEN*	Agroinfiltration	Qi et al. (2023), Liu et al. (2020)
		Barley stripe mosaic virus (BSMV)	*Zea mays, N. benthamiana*	*PDS*	Puncture inoculation	Jarugula et al. (2018)
		Pea early browning virus (PEBV)	*Pisum sativum*	*PDS, UNI, KOR1*	Agroinfiltration	Constantin et al. (2004)
		Pepper ringspot virus (PepRSV)	*N. benthamiana*	*PDS*	Agroinfiltration	Tavares-Esashika et al. (2020)

(continued)

Table 1 (continued)

Genome	Family	Virus	Plant	Targeted genes	Method	References
ssRNA (+)	Alphaflexiviridae	Cassava mosaic virus (CMV)	*N. benthamiana*, cassava	*PDS, CHlI*	Agroinfiltration	Tuo et al. (2021b)
		Foxtail mosaic virus (FMV)	*Panicum virgatum*	*ChlD, ChlI, PDS*	Agroinfiltration	Tiedge et al. (2022)
		White clover mosaic virus (WClMV)	Pea, Broad bean	*PDS*	Agroinfiltration	Ido et al. (2012)
		PVX	*N. benthamiana*	*PDS*	Agroinfiltration	Faivre-Rampant et al. (2004)
ssRNA (+)	Bromoviridae	Brome mosaic virus (BMV)	N. benthamiana, Zea mays	*PDS, HSP70*	Agroinfiltration	Ding et al. (2018)
		cucumber mosaic virus (CMV)	Spinach	*PDS, DFR*	Agroinfiltration	Kim et al. (2020)
ssRNA (+)	Tymoviridae	Turnip yellow mosaic virus (TYMV)	*Brasicca rapa*	*PDS*	Agroinfiltration	Yu et al. (2018)
ssRNA (+)	Tombusviridae	Grapevine Algerian latent virus (GALV)	*N. benthamiana*	*Chl*	Agroinfiltration	Park et al. (2016)
		Grapevine leafroll-associated virus-2 (GLRaV-2)	*N. benthamiana*	*PDS*	Agroinfiltration	Kurth et al. (2012)
ssRNA (+)	Solemoviridae	Soybean yellow mottle mosaic virus (SYMMV)	*Pusa Parvati*	*GFP*	Rub inoculation	Sandra et al. (2019)

ssRNA (+)	*Sobemovirus*	*Soybean yellow common mosaic virus* (SYCMV)	soy bean	*PDS, RbcS*	Rub inoculation, agroinfiltration, particle bombardment	Lim et al. (2016)
		Cowpea severe mosaic virus (CPSMV)	*N. benthamiana*, Soy bean	*MYB75*	Agroinfiltration	Zaulda et al. (2022)
dsDNA	Caulimoviridae	Rice tungro bacilliform virus (RTBV)	Rice	*Xa21, PDS, chlH*	Agroinfiltration	Kant and Dasgupta (2017)
ssRNA(+)	Secoviridae	Apple latent spherical virus (ALSV)	Strawberry, Fenugreek, Grapevine, benthamiana, Tomato, Pumpkin	*PDS, PCNA, SU, FT*	Mechanical inoculation, agroinfiltration, biolistic inoculation	Maeda et al. (2020), Yamagishi and Yoshikawa (2022)
		Bean pod mottle virus (BPMV)	Soybean, *Phaselous vulgaris*	*PDS*	Mechanical inoculation	Zhang et al. (2010)
		Tobacco ringspot virus (TRSV)	*Cucumis melo*, *Citrullus lanatus*, *Cucumis sativus*, *N. benthamiana*	*PDS*	Agro-inoculation	Fang et al. (2021)
		Broad bean wilt virus2 (BBWV2)	*Capsicum annuum, N. benthamiana*	*PDS*	Agro-inoculation	Choi et al. (2019)

into plant, assessing the target's silencing efficiency, and documenting the phenotype. The orientation, length, and sequence of the insert all have an impact on the competence of the VIGS vector (Jagram and Dasgupta 2023). Fragments containing 200–350 nucleotides are usually selected to expand silencing adequacy. The majority of VIGS vectors are capable of carrying 150–800 bp segments. Nonetheless, 1500 bp fragments being inserted into tobacco rattle virus (TRV) or PVX-based VIGS vectors have also been reported. The orientation of the insert is the determinant of silencing (Huang et al. 2012).

2.3.3 VIGS Advantages in Functional Genomics

In the past few years, VIGS has been used for the characterization and functional assessment of many genes via silencing that are involved in cell wall biosynthesis, development, growth, seed biology, and inflorescence patterning. The *LaTEN* gene responsible for tendril synthesis has been characterized via CGMMV-VIGS in ridge gourd (Qi et al. 2023). The role of the *ApTT8* gene was characterized for anthocyanin biosynthesis in Agapanthus praecox through TRV-VIGS (Chen et al. 2024). Another example is the silencing of *CrGATA1* through cotyledon-based TRV-VIGS, which led to downregulation of *T30*, *T3R*, and *DAT*, thus decreasing cotyledon's vindoline contents in *C. roseus* (Liu et al. 2024). The role of GhGOLS2 in cotton seed germination was also assessed through seed imbibition VIGS (Si-VIGS) in cotton. VIGS also has the potential for screening genes that are involved in abiotic and biotic stress pathways. Characterization of *DCL* in tomato against *Botrytis cinerea* (Spada et al. 2024), *SYR1* in potato for resistance against *Phytophthora infestans* (Lata et al. 2022), *OsRDR6* in rice against Rice Dwarf virus (Hong et al. 2015), *CYPAE14* in cotton for *Helicoperva armigera* resistance (Ghag 2017), and *Mi-msp5, Mi-msp3,* and *Mi-msp18* for resistance to root-knot nematodes in *Arabidopsis* (Kalia et al. 2024a) are some examples of genes characterized by VIGS for biotic stress tolerance. For abiotic resistance, *OsLG3* in rice (Xie et al. 2019), *SPL9* in *Medicago sativa* (Hanly et al. 2020), *CsATAF1* in cucumber (Wang et al. 2018), *GhGSTU17* in cotton for salt tolerance (Sun et al. 2024), *DREBA4* in tomato for heat stress tolerance (Mao et al. 2020), *CBF* in Bermudagrass for cold tolerance (Huang et al. 2022), and *GhLOG3* in cotton for salinity tolerance (Wang et al. 2021b) have been functionally explored and evaluated.

2.3.4 Delivery Methods for VIGS

Standardizing a competent and appropriate approach for delivering the vectors should be prioritized. Delivery methods include Agro-inoculation, first proposed by Ratcliff et al., involves using needle-less syringes to deliver *Agrobacterium* suspension directly into plants (Vaghchhipawala and Mysore 2008), mechanical rubbing or sap inoculation carried out by rubbing bacterial culture using hands on leaf surface (Tiedge et al. 2022), agrodrench, encompasses drenching the young seedlings with *Agrobacterium* culture for studying genes involved in early root development (Ryu et al. 2004), vacuum puncture infiltration with a 90–100% success rate, seedlings are immersed in *Agrobacterium* suspension upside down inside a vacuum device (Lentz et al. 2018), spray inoculation, using an atomizer or spray gun to spray with a portable compressor (Wang and Jin 2017), fruit agroinjection, injecting fruit stylar apex through syringe (Orzaez et al. 2006). Despite being reliable, VIGS has some limitations; it sometimes results in nonhomogeneous silencing, and the percentage of silencing is variable between experiments and plants. Longer inserts have a negative impact on insert stability. Plant metabolism is harmed by virus vectors. Off-target effects, inability to reach meristem, host specificity of vectors, standardized protocol requirement for each species, and environmental factors like humidity, temperature, and light intensity are some other factors that severely affect silencing (Jagram and Dasgupta 2023).

3 Methods for Obtaining Transient Expression in Plants

GMOs are produced using tissue culture, grafting, and several other techniques that require conventional and laborious transformation protocols (Yang et al. 2023a; Espinosa-Leal et al. 2018; Chen et al. 2022). Novel *in-planta* techniques, which utilize the following methods that allow the introduction of transgenes directly into the plants, have been developed to mitigate these problems (Zale et al. 2009; Hamada et al. 2017).

3.1 Biolistic (Gene Gun) Method

Plant viruses are traditionally infected by mechanically rubbing leaves powdered by carborundum (silicon carbide) or celite (SiO_2) with cultures containing viral nucleic acids (Gaba et al. 2013a). This process is still routinely and effectively used to treat a variety of viruses; however, it is ineffective when woody plants are infected and against phloem-limited viruses (Gal-On et al. 1997). Alternative methods include employing natural invaders such as insects or nematodes, but these are typically time-consuming and technically complex procedures (Gaba et al. 2013b). In contrast to these conventional methods, plant virologists employed biolistic bombardment, also known as particle bombardment, which is a physical method in which small particles, such as gold or tungsten, are coated with DNA and shot into plant tissues at high speed using helium pressure (Gaba et al. 2013b). It is an efficient approach that can be used with many species of plants, especially those not amenable to other transformation methods (e.g., complex grasses such as maize and wheat) (Yanagihara et al. 2024). Micro-particles of this size can penetrate plant cells, allowing entry and uptake of the transgene that can be expressed transiently in the cytoplasm or further integrated into the genome.

This technique was first used in the mid-1980s when a scientist at Cornell University demonstrated that a laboratory-made Gene Gun bombardment device could introduce a virus, tobacco mosaic virus (TMV), into the onion epidermal cells (Klein et al. 1987). In this method, a bombardment device such as a Bio-Rad Helios Gene Gun, HandGun (Gal-On et al. 1995), or HandyGun (Sikorskaite et al. 2010), is utilized to deliver nucleic acid (either DNA, plasmid, or RNA) coated by gold or tungsten particles directly into the live plant cells that may transfer the carried genetic material to their target, which may be epidermal cells, trichomes, or even mesophyll cells (Gal-On et al. 1995; Sikorskaite et al. 2010).

The *β-glucuronidase* gene (*gusA*) in radiata pine species was successfully transferred and expressed utilizing the particle bombardment technique (Rey et al. 1996). In another study, the expression of green fluorescent protein (GFP) was obtained in *Setaria viridis* by using the 0.6 mm gold particles in particle bombardment (Mookkan 2018). Stable transgene integration was performed in the soybean plants when 0.6 μm gold particles coated with 0.8 μg DNA (modified jellyfish GFP (S65T)) were bombarded at 1100 psi at a distance of 6 cm (Khalafalla et al. 2005). A double-barreled particle bombardment system was used to express necrosis-causing genes such as *BAX*, *Avh238*, and *ATR13/Rpp13* in *A. thaliana*, and cell death was observed (Zhao et al. 2020).

3.2 *Agrobacterium*-Mediated Transformation

Agrobacterium tumefaciens, a soil bacterium, naturally introduces part of its DNA into cells and is considered to be one of the most powerful tools for the transfer of genes into the host plants (Scofield et al. 1996; Bendahmane et al. 2000). Naturally, when *Agrobacterium* delivers transfer DNA (T-DNA) into the host, it causes crown gall disease since it contains tumor-inducing genes (Gohlke and Deeken 2014). However, biotechnologists have engineered the Ti-plasmid in such a way that tumor-inducing genes are replaced with the target gene or gene of interest (GOI), which is naturally transferred to the host plant. This mechanism was made possible by the utilization of two different plasmids: one helper plasmid, which carries all the essential genes for the delivery of T-DNA (now containing GOI), and the binary vector, which not only contains the GOI but also the two origins of replication, one for *E. coli* and the other for *Agrobacterium (*Komori et al. 2007*)*. This T-DNA is delivered into the plant host using the homologous recombination process and becomes heritable (Prakash and Zipf 1996). The *Agrobacterium* culture containing the T-DNA is utilized for the inoculation of leaf tissues (agro patch) for transient expression of reporter genes by using the needle or stick, and this process is called agroinfiltration or agro-injection (Johansen and Carrington 2001).

For the transformation process, *Agrobacterium* culture is grown at 28 °C in the selective media containing the antibiotic for which the resistance is conferred by T-DNA until it reaches the optical density (OD) of 1–2 at 600 nm. The culture is then pelleted down and resuspended in media that contains acetosyringone (a chemical that initiates the expression of bacterial genes), and then the agro patch is infiltrated with the culture (Bedoya et al. 2012; González et al. 2010). Studies have shown that different ranges of ODs from 0.2 to 2.0 can be utilized for this process (LiJing et al. 2010). The level and pattern of transient expression can be different for each product being expressed by the T-DNA. In most cases, expression can be observed 3–4 days post-infiltration, and it disappears rapidly after a 5–6-day period (Del Toro et al. 2014). These expression levels depend on the strength of the host plant to resist the silencing, which directly affects the mRNA transcript levels of T-DNA (LiJing et al. 2010).

This approach has been used for transgenes that are to be stably expressed and also transiently. This method works very well in dicotyledon species like *A. thaliana*, *N. tabacum*, and *Solanum lycopersicum* (Wu et al. 2014; Heidari Japelaghi et al. 2018; Van Eck et al. 2019). The *Agrobacterium*-mediated transient transformation had been used in sorghum with GFP as a marker to show its importance as an alternative for protoplasts to test the efficiency of various genome editing tools (Sharma et al. 2020). Similarly, *Agrobacterium*-mediated transient transformation has been employed in the *Marchantia polymorpha* using the *gusA* gene with a GFP marker showing bacterial colonization (Iwakawa et al. 2021). The targeted silencing of the *MdMYB10*, which is an anthocyanin regulator, resulted in altered anthocyanin accumulation in apple plants (Lv et al. 2019), proving that this method can also be utilized for the analysis of gene function in different plants.

3.3 Viral Vectors

Viral vectors are unique in that they make use of the ability of plant viruses to infect cells and move from cell to cell, as well as onto new individuals. Viral vectors give you instant systemic gene expression at each stage in the process. These have been broadly exploited for transient gene delivery in plants since the mid-1980s. They have revolutionized plant research by enabling the transient expression of foreign genes in plants (Habibi-Pirkoohi et al. 2021; Cañizares et al. 2005). Mechanical, spray, or syringe inoculation methods are used for the delivery of viral vectors, which are delivered into host plants by symplastic or apoplastic uptake (Monroy-Borrego and Steinmetz 2022; Hu et al. 2020b; Hull 2009). Major plant species that can be targeted by viral vectors usually include tobacco, tomato, maize, rice, and *A. thaliana*. Plant viruses have long been used for developing recombinant pharmaceutical proteins because of their efficient machinery and enhanced production of recombinant protein in plant tissues (Monroy-Borrego and Steinmetz 2022; Ibrahim et al. 2019; Zaidi and Mansoor 2017). High expression and rapid transmission are also counted among the features that make plant viruses a remarkable choice to be used as vectors (Ibrahim et al. 2019; Zaidi and Mansoor 2017).

There are several kinds of plant viruses, containing either single-stranded (s) or double-stranded (ds) DNA or RNA genomes (Canto 2016). Major plant viruses include Cauliflower Mosaic Virus (CaMV), Geminivirus, Tobacco Rattle Virus (TRV), TMV, and Potato Virus X (PVX)-based vectors. The DNA-based viruses include CaMV and geminiviruses, while RNA-based viruses are TRV, TMV, PVX, and Cowpea Mosaic Virus (CPMV) (Zaidi and Mansoor 2017). CaMV, a virus widely used as a vector, is a dsDNA virus belonging to the *Caulimoviridae* family. Mechanical inoculation method and insect vectors like whiteflies or aphids are used to transmit the vector (Canto 2016; Rizzo and Palukaitis 1990). Geminivirus (family *Geminiviridae*) is a ssDNA virus that, when used as a vector, infects major plant species like cotton, wheat, maize, rice, beans, legumes, tomatoes, tobacco, fruits,

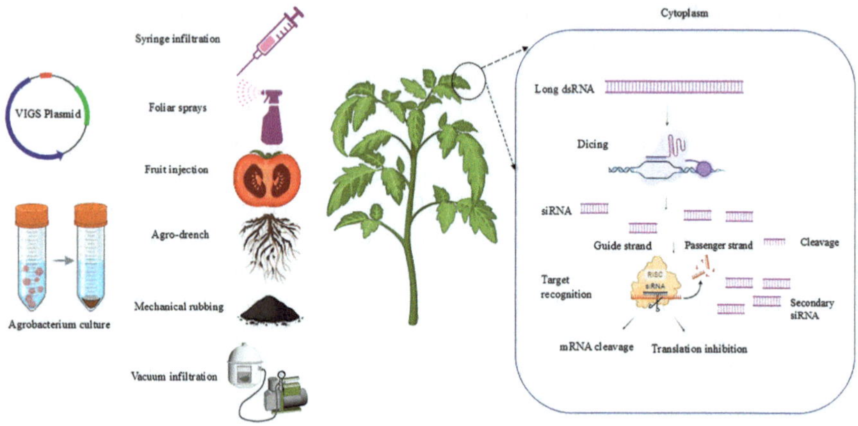

Fig. 1 Mechanism of virus-induced gene silencing in plants. Viral vectors can be introduced into plants via syringe infiltration, foliar sprays, fruit injection, agro-drench, mechanical rubbing, or vacuum infiltration. Inside plant cytoplasm, double-stranded RNA is processed into small interfering RNAs via dicer, which then guides the RISC complex to target and degrade complementary RNA, thereby silencing gene expression via mRNA cleavage or translation inhibition

and weeds. Whitefly and leafhopper are used for the transmission of vectors (Zaidi and Mansoor 2017). DNA viruses produce RNA hairpin structures after transcription, while for RNA viruses, dsRNA acts as an intermediate molecule during viral replication (Kolliopoulou et al. 2017). PVX is a positive-sense strand RNA virus transmitted mechanically to targeted plants. The ability to accommodate large foreign genes in its genome makes it an excellent choice to be used as a vector (Sindarovska and Kuchuk 2021).

In plants, TMV, CPMV, and PVX are majorly used for vaccine production and other pharmaceutical products using a transient gene expression approach. The CPMV system is reported to produce dengue 2 envelope domain III in cowpea plants (Habibi-Pirkoohi et al. 2021). Transient expression of recombinant GFP protein in *N. benthamiana* using a PVX-based system is also reported (Sindarovska and Kuchuk 2021). Combining geminiviral replication and double terminator enhanced the transient expression in plants (Yamamoto et al. 2018) (Fig. 1).

4 Comparative Analysis of Transient Expression Methods

As the most commonly used methods to achieve transient gene delivery are discussed in earlier sections, they come with their pros and cons. There needs to be a detailed comparison of these technologies in terms of their efficiencies, versatility, and environmental impacts.

4.1 Biolistics (Gene Gun) Verus *Agrobacterium*-Mediated Transformation

4.1.1 Advantages and Limitations of the Biolistic (Gene Gun) Method

Biolistics is a species-independent technique; it can be used specifically for monocots that are resistant to *Agrobacterium*, making it an indispensable tool for major cereal crops that have been significantly contributing to world food security. It is capable of transporting large and intricate genetic elements, which, in addition to single gene assemblies, can be used for multigene constructs, RNA, or proteins, among others (Lacroix and Citovsky 2020a; Tian et al. 2019; Lee et al. 2015).

At high velocity, the target tissue or seeds might get physically damaged through bombardment, rendering them unable to thrive and thereby affecting plant health or recovery of delicate tissues. The transformation efficiency of biolistics is often less than that of *Agrobacterium*-mediated methods. It frequently shows more variable gene expression between different transgenic cells or lines because the DNA can be inserted randomly into the genome, so it can give non-repeatable results, which affects the reliability of experiments (Canto et al. 1997; Ozyigit and Yucebilgili Kurtoglu 2020).

4.1.2 Advantages and Limitations of *Agrobacterium*-Mediated Transformation

Agrobacterium is the most valuable biological tool; it enhances gene transfer efficiency in dicotyledonous plants. When *Agrobacterium* delivers recombinant DNA, it does not cause physical damage to plant tissues; the method facilitates good regeneration and more reproducible experimental results. This approach is also fairly inexpensive to implement, needing only simple lab equipment, unlike the more elaborate apparatus needed for biolistics. Agroinfiltration's primary benefit over other delivery systems is that the majority of plant cells within the agro patch receive the T-DNA and express the targeted genes, opening the door to the possibility of large-scale applications. Furthermore, by employing a combination of bacterial cultures carrying distinct binary constructions, it has the potential to express multiple products (Chen et al. 2013; Shreni Agrawal 2022; Rahman et al. 2024).

A primary limitation of *Agrobacterium* is its limited effectiveness in monocot species such as rice and maize due to its reduced natural host range, and certain plants may have intrinsic resistance to *Agrobacterium* infection. Unlike this technique, utilizing bombardment or viral vectors to express multiple products is not easy, unless all distinct products are generated from the same virus in the case of viral vectors (Zhang et al. 2018; Mohammed et al. 2019).

4.2 Advantages and Limitations of Viral Vectors Compared to Other Methods

Rapid and widespread transmission of viral vectors is beneficial for those experiments that need immediate results, such as functional genomic studies. The capacity of viral vectors for high-level transgenic expression is due to the efficacy of vector replication, leading to the amplification of gene products. They are optimized to enable vector-based functional genomics screens or protein expression in planta and hence constitute the preferred choice for rapid gene transfer and comparative redundancy. Combining the effects of multiple techniques, like magnification, has proved to be efficient and affordable for transient gene expression (Habibi-Pirkoohi et al. 2021; Zaidi and Mansoor 2017; Ura et al. 2014).

However, only the genetic constructs of limited size or complexity can be delivered by viral vectors, which restricts their use in diseases for which large transgenes or the insertion of multiple genes are required. A major concern implicated with viral vectors is that the recombinant virus can escape and infect nontarget plants in a field setting, which may cause biosafety issues. Different viral vectors have a limited host range and are only able to infect certain plant species (Ura et al. 2014; Šimčíková et al. 2015; Wang et al. 2021c).

4.3 Comparative Efficiency, Adaptability, and Limitations

Agrobacterium-mediated transformation is highly efficient in dicots, causing minimal tissue damage and enabling reliable gene transfer, making it the preferred method. Biolistics tend to be less efficient, particularly in dicots, but are essential for monocots, which cannot be easily infected by *Agrobacterium*, but it has lower efficiency than viral vectors. Viral vectors offer the highest expression levels, primarily due to the virus's replication within host plants, although their payload capacity and host range are limited (Fan and Sun 2024; Aliu et al. 2024). For species diversity and the types of genetic material one can introduce, biolistics is relatively more universal. The method is compatible with both monocot and dicot species, delivers large DNA constructs or complex gene assembly, and will be a valuable tool for future plant engineering solutions (Ueki et al. 2009, 2013).

Biolistic transgene transfer is subject to tissue damage as well as variable gene expression due to random integration events. This technique is also extremely expensive, not just because of the equipment required to do it, but also the consumable materials as well. The main limitation of *Agrobacterium*, however, is its lower efficiency in monocots and the need for a compatible plant-host relationship. Viral vectors are limited by the risk of environmental release, payload size, and a restricted host range, making them less suitable for certain commercial or large-scale applications (Tyurin et al. 2020; Zaidi and Mansoor 2017; Canto 2016).

4.4 Regulatory and Environmental Challenges

Transient expression systems, particularly when moving from research to commercial applications, must meet regulatory standards ensuring product safety. Biolistics, which involves direct DNA introduction without biological vectors, is considered less environmentally risky. However, unintentional gene flow can occur if the modified material is not adequately contained, particularly during field trials. Biolistics generally faces fewer regulatory hurdles than viral vectors, as it does not involve living organisms capable of replication. Still, regulatory concerns remain, particularly with genetically modified (GM) plants (Canto 2016).

Agrobacterium-mediated transformation poses regulatory and biosafety concerns, especially in field applications, where the bacteria may persist and infect nontarget species. While transient expression methods typically bypass the strict regulations associated with GMOs, engineered *Agrobacterium* is subject to significant regulatory oversight, particularly for stable transgenic plant production. The chances of *Agrobacterium* persisting in the environment and facilitating unintended gene transfer are low, but containment measures, like using greenhouses, can help lessen these risks even more (Tuteja et al. 2012; Konda 2024).

Viral vectors raise significant regulatory concerns due to their ability to spread and infect other plants, leading to fears of horizontal gene transfer and unregulated transgenic spread. Viral vectors are strictly controlled to prevent environmental release, with containment strategies such as nonreplicating viral vectors to prevent transgene dissemination. Employing viral vectors designed to be triplication-deficient or doing research in controlled, enclosed settings can mitigate the environmental risk posed by viral vectors escaping and infecting nontarget plants (Shingote et al. 2024; Okolo et al. 2022).

5 Applications of Transient Expression Systems in Plants

In plants, transient expression systems are used for synthetic as well as recombinant protein production, engineered metabolic pathways, investigation of the function of different types of genes, and molecular farming (Shanmugaraj et al. 2020; Verdú-Navarro et al. 2024).

5.1 Functional Genomics

Functional genomics involves understanding the functions as well as operations of different genes in certain conditions and their products while being present in the locality of their particular genome. (Tyurin et al. 2020). Different transient expression methods in the present century have been utilized worldwide at large levels to downregulate the expression of certain genes in different plants. Some of the prominent methods that have been adopted globally by researchers are virus-induced gene silencing (VIGS), Host-Induced Gene Silencing (HIGS), and Spray-Induced Gene

Silencing (SIGS). These methods have helped a lot in providing detailed insights into the functional role of targeted genes, especially related to plants' defense mechanisms, their growth, and different types of biotic and abiotic stress responses (Tyurin et al. 2020). Different experiments involving the overexpression of a gene are also considered very important in the domain of functional genomics. It helps in examining all types of consequences of raising the level of any certain product, being expressed by a gene (Xia et al. 2020). Multiplex gene analysis is another broad application that has the ability to target many genes at once through the different efforts in the vector design mechanism of temporary expression. Examining the complex gene networks or cascades and their metabolic control approach has been done for many studies in plants like *N. benthamiana* (Stuttmann et al. 2021). Subcellular localization studies have enabled scientists to gain a detailed understanding of the location of a protein in the vicinity of a cell by tagging the desired gene to different types of fluorescent markers, like green fluorescent protein (GFP), to be transiently expressed in a plant (Yuan et al. 2021). Techniques like co-immunoprecipitation (Co-IP) and biomolecular fluorescence complementation (BiFC) have now been used for different transient expression studies and have enabled the researchers to understand the interactions of different types of proteins in different pathways, including transcription control and signal transduction (Wang et al. 2021d). Examples of transient expression in highly valuable crops are listed in Table 2.

5.2 Protein Production for Industrial and Pharmaceutical Use

Transient expression systems are also important tools for the production of different enzymes used in industries, the production of recombinant proteins, and the production of pharmaceutical proteins such as vaccines, therapeutic proteins, and antibodies (Table 3). This ability of transient systems to quickly express and harvest a protein has made this system a highly attractive one for the commercial production of different types of proteins (Burnett and Burnett 2020). Enzymes have very diverse applications and are used in different sectors like food processing, detergent production, biofuel sectors, etc. These enzymes can be transiently produced in plants for the quick testing and checking of the optimized performance of the enzymes. For instance, plants have been utilized to create xylanases and cellulases, crucial in biofuel synthesis, via transient expression (Geem et al. 2021; Benedetti et al. 2020). Different antibodies and other types of therapeutic proteins have been synthesized using plant-based transient expression methods. One well-known example of this is the use of transient expression platforms like pEAQ and MagnICON vectors for the production of monoclonal antibodies in *N. benthamiana* for the treatment of cancer (Nosaki et al. 2021). Due to the multiple types of variants discovered in viruses, the generation of vaccine antigens is now relying on transient expression in plants. Transient expression methods were used in *N. benthamiana* during the pandemic of COVID-19 for the quick generation of candidate vaccine proteins that were utilized in clinical trials (Mamedov et al. 2020). To boost protein production, researchers

Table 2 Examples of transient expression in highly valuable crops

Crop	Gene	Method adopted	Outcome	References
Tomato	*Cry1Ac*	*Agrobacterium*-mediated	Increased insect resistance	Prasanna et al. (2023)
Soybean	*GmFTL3*	Particle bombardment	Induced early flowering	Xu et al. (2021b)
Maize	*ZmLEA3*	Agroinfiltration	Improved drought tolerance	Zu et al. (2020)
Potato	*StMLO*	Viral vector (PVX)	Resistance against powdery mildews	May et al. (2023)
Rice	*OsRac1*	*Agrobacterium*-mediated	Increased resistance against rice blast disease	Zhou et al. (2019)
Wheat	*TaGS5*	Agroinfiltration	Increased size of grain	Ren et al. (2021b)
Cotton	*GhTUB1*	Particle bombardment	Improved fiber strength	Zhang et al. (2024b)
Pepper	*CaPI3*	Agroinfiltration	Increased resistance against bacterial spot disease	Thauland et al. (2019)
Banana	*MaDEF*	Biolistic	Improved morphology of flowers	Affandi et al. (2020)
Cucumber	*CsPR1*	Agroinfiltration	Resistance against fungal pathogens	Yu et al. (2019)
Apple	*MdMYB10*	Viral vector (TMV)	Increased anthocyanin production	Ma et al. (2021)
Grapes	*VvSTS*	*Agrobacterium*-mediated	Enhanced resveratrol production against diseases	(Jiang et al. 2019)
Strawberry	*FaWRKY1*	Agroinfiltration	Increased resistance against bacterial pathogens	Higuera et al. (2019)
Citrus	*CsLOB1*	Biolistic	Improved resistance against citrus canker	Sun et al. (2019)
Tobacco	*NtMYC2*	*Agrobacterium*-mediated	Increased nicotine biosynthesis for pest resistance	Liu et al. (2019)
Sugarcane	*ScbZIP1*	Particle bombardment	Improved salt tolerance	Yuan et al. (2024)
Barley	*HvWRKY38*	Agroinfiltration	Enhanced stress tolerance	Shu et al. (2024)
Carrot	*DcCHI*	Biolistic	Increased production of flavonoids	Baranski and Lukasiewicz (2019)
Lettuce	*LsWRKY75*	*Agrobacterium*-mediated	Enhanced drought tolerance	Chen et al. (2023c)
Pineapple	*AcDFR*	Particle bombardment	Increased production of anthocyanins	Gong et al. (2021)
Spinach	*SoCBF1*	Agroinfiltration	Improved cold tolerance	Jesionowska (2022)

(continued)

Table 2 (continued)

Crop	Gene	Method adopted	Outcome	References
Broccoli	*BoERF1*	Viral vector (PVX)	Increased resistance against bacterial infections	Pretorius (2018)
Onion	*AcACS1*	*Agrobacterium*-mediated	Increased bulb size	Tyurin et al. (2020)
Avocado	*PaERF1*	Particle bombardment	Improved tolerance to bacterial pathogens	Corbineau (2024)
Almond	*PrNSN*	Biolistic	Enhanced nut size	Vishwanath et al. (2024)
Papaya	*CpCHI*	Agroinfiltration	Increased flavonoid production	Tue et al. (2024)
Olive	*OlAPX*	Viral vector (TMV)	Increased drought stress tolerance	Liang et al. (2020)
Mango	*MiAP2*	Particle bombardment	Enhanced resistance against fruit anthracnose	Tayal et al. (2024)
Pine	*PsLEA1*	Agroinfiltration	Improved survival rates under extreme cold	Zhou et al. (2023)
Kiwi	*AcSOD*	Viral vector (PVX)	Enhanced tolerance to oxidative stress	Wen et al. (2023)

Table 3 Success stories in biopharmaceutical production through transient expression

Crop	Gene/Protein involved	Method adopted	Outcome	References
Tobacco	Human Interferon α2b	Agroinfiltration	Successful production of human interferon	Feizi and Baghbankohnehrouz (2021)
	Human Serum Albumin	Agroinfiltration	Production of human serum albumin	Sedaghati et al. (2020)
	Human *EPO*	Biolistic	Erythropoietin production	Bashir et al. (2018)
	Human *MUC1*	Viral vector (PVX)	Production of cancer vaccine	Chabeda et al. (2018)
	Anti-Zika antibodies	Viral vector (TMV)	Development of antibodies for Zika virus treatment	Diamos et al. (2020)
Lettuce	*HBsAg*	Particle bombardment	Production of Hepatitis B vaccine antigens	Sharma et al. (2018)
	Anti-Ebola antibodies	Viral vector (PVX)	Production of antibodies for Ebola treatment	Jugler et al. (2020)
	Human Hyaluronic Acid	Biolistic	Production of hyaluronic acid for cosmetic uses	Serrano-Sevilla et al. (2019)

(continued)

Table 3 (continued)

Crop	Gene/Protein involved	Method adopted	Outcome	References
Potato	*CTB*	Agroinfiltration	Production of cholera toxin B subunit	Kim et al. (2019)
	Recombinant Enzymes	Agroinfiltration	Production of enzymes for mucus breakdown in cystic fibrosis therapy	Alamgir and Alamgir (2018)
Tomato	Human EGF	Viral vector (TMV)	Production of epidermal growth factor	Hanittinan et al. (2020)
	Human Papillomavirus (HPV) L1	Agroinfiltration	Production of HPV vaccine	Massa et al. (2019)
	VEGF protein	Agroinfiltration	Production of vascular endothelial growth factor	Bulaon et al. (2020)
Carrot	Human proinsulin	*Agrobacterium*-mediated	Generation of proinsulin for diabetes treatment	Bhoria et al. (2022)
	Recombinant growth hormone	Agroinfiltration	Production of human growth hormone	Rodriguez-Concepcion and Daròs (2022)
	Therapeutic antiancer peptides	Agroinfiltration	Production of peptides with anticancer properties	Gomes et al. (2019)
Rice	*Hsp70*	Particle bombardment	Heat shock protein production for vaccine adjuvant purposes	Ayan et al. (2022)
	Therapeutic antibodies	Particle bombardment	Generation of antibodies for immunotherapy applications	Dorokhov et al. (2019)
Maize	*ZmFv1*	Agroinfiltration	Production of monoclonal antibodies for cancer treatment	Geis and Goff (2020)
	Human factor IX	Agroinfiltration	Production of the clotting factor for hemophilia treatment	Burnett and Burnett (2020)
Spinach	Human GM-CSF	Biolistic	Granulocyte-macrophage colony-stimulating factor for the immune system	Paradia et al. (2023)
	Recombinant anti-rabies antibodies	Agroinfiltration	Production of antibodies for rabies	Naseri et al. (2019)

(continued)

Table 3 (continued)

Crop	Gene/Protein involved	Method adopted	Outcome	References
Alfalfa	Human *IgG1*	Agroinfiltration	Antibody production for therapeutic applications	Diamos et al. (2020)
	Interleukin-2	Biolistic	Production of IL-2 for immune modulation	Tanveer et al. (2023)
Barley	Hepatitis C Virus (HCV) *E2*	Particle bombardment	Production of HCV antigens for vaccine development	Shakoor et al. (2019)
	Anti-HIV antibodies	Agroinfiltration	Production of HIV-neutralizing antibodies	Singh et al. (2020)
Corn	Recombinant Collagen	Particle bombardment	Production of collagen for tissue engineering	Poudel et al. (2023)
	Recombinant Human Insulin	Agroinfiltration	Production of insulin for diabetes management	Bhoria et al. (2022)
Wheat	Antimalarial Proteins	Viral vector (TMV)	Production of malaria vaccine candidate proteins	Rangasami et al. (2022)
	Human Lactoferrin	Particle bombardment	Production of human lactoferrin for antimicrobial applications	Khan et al. (2020)

have concentrated on improving the effectiveness of transient expression vectors in recent years. The production schedule has been shortened from months to weeks by the pEAQ-HT vector, which has undergone continual refinement, demonstrating elevated production levels of commercial enzymes such as cellulases in *N. benthamiana* (Chuang and Franke 2022). One advantage of using plants to manufacture protein is that they can perform posttranslational alterations, such as glycosylation, in the same way that humans can. As a result, plant-based transient expression technologies are particularly beneficial for producing complex proteins that require proper folding and modification to function (Kermode 2021). Transient systems in plants can be scaled at a large scale without costly infrastructure requirements. The proteins can be synthesized in days or weeks rather than months or years, which also has lower production costs (Anjanappa and Gruissem 2021).

5.3 Metabolic Engineering and Pathway Studies

Metabolic engineering involves the alteration of different metabolic pathways in different organisms to increase the production of various useful substances like nutraceuticals, biofuels, and medicines (Pouvreau et al. 2018). Transient expression enables the insertion of different genes into metabolic pathways, which makes us determine the generation of the desired chemical. This is considered important for the designing of plants to generate different secondary metabolites, including phenylpropanoids, alkaloids, and terpenoids that are intermediates in different metabolic pathways (Rademacher et al. 2019). A modular approach is widely used in synthetic biology, where individual genes or sequences encoding enzymes are assembled into a synthetic metabolic system. In a single experiment, researchers can introduce many pathway genes into plants via transient expression. In a similar vein, temporary expression facilitates rapid testing of synthetic gene circuits, wherein transcription factors, promoters, and other regulatory elements are precisely tailored to achieve control over certain metabolic processes. This work is considered particularly well suited for regulating flow across the different types of intricate metabolic networks. This also guarantees the timely production of our targeted molecules. Transient expression methods have also enabled the quick insertion of many types of gene constructs into plants, which makes them very valuable tools for the optimization of different experimental combinations (Baltes and Voytas 2015; Cravens et al. 2019). Foreign genes have also been expressed in plants through different transient systems, but they do not integrate into the plant genome via any of these temporary expression pathways. The majority of the time, these pathways are used for quick and transient investigations into gene function, protein synthesis, and metabolic engineering. These days, *Agrobacterium*-mediated infiltration or other viral vectors are often used to provide essential genes, enabling rapid and high-level expression. Without permanently altering the plant's DNA, these systems offer a rapid, flexible, and scalable method for assessing synthetic pathways, enhancing metabolic flux, and producing significant proteins or metabolites (Torti et al. 2021).

Transcriptional factors and short DNA sequences called cis-regulatory elements (CREs) work together to regulate the expression of some genes. In plant transient expression systems, CRE studies enable a quick functional assessment of promoter regions to determine their importance in gene regulation. By coupling reporter genes to native or synthetic promoters containing relevant CREs, researchers may determine areas critical for transcriptional control under certain circumstances. With the use of these technologies, it is possible to assess plant regulatory networks quickly and efficiently, leading to a better understanding of how different plant species use other genes, develop, and react to stress (Ijaz et al. 2020).

5.4 Promoter Studies and Reporter Gene Assays

Since transient expression methods in plants enable quick and efficient functional evaluation of promoter sequences, they are often utilized in reporter gene experiments and promoter studies. These techniques simplify the introduction of promoter-reporter gene constructs and assess the promoter's accuracy, efficiency, and inducibility across a range of plant tissues and environments (Lu et al. 2017). Common reporter genes include *GFP, luciferase, and GUS*. Transient assays provide a versatile and very effective environment for quickly testing a wide range of promoter designs. They also give critical insights into the regulatory processes that regulate gene expression and aid in the development of customized promoters for use in plant biotechnology applications (Yuan et al. 2021). The degree of gene expression is significantly influenced by the strength of the promoter. Promoter strength may be quickly assessed through transient expression approaches, which attach promoters to reporter genes like *GFP* and *luciferase* before introducing them into plant tissues. Assessing reporter gene activity provides a clear measure of promoter strength and allows for the comparison of various modifications of the promoter. Stronger promoters can produce high expression of target genes in plant systems, whereas weaker promoters can be used for more focused, controlled expression (Jores et al. 2021). Researchers can study promoters intended to generate genes in particular plant sections, like leaves, roots, or flowers, by using transient expression systems. Researchers can introduce the innovations into entire plants or isolated tissues and evaluate expression patterns by combining these promoters to report genes. In agricultural biotechnology, tissue-specific promoters are helpful because they minimize adverse effects on organs that are not intended targets while directing gene expression to specific tissues. These investigations shed light on the mechanisms guiding plant growth and have the potential to enhance traits including stress tolerance, pest resistance, and nutrient absorption (Miroshnichenko et al. 2020). Induced promoters control genes that activate in response to environmental stimuli like stress, light, temperature, and chemicals. Delivering such promoters into plant tissue transiently stimulates promoter function, and when the reporter gene is exposed to the appropriate inducer, its levels may be evaluated. These promoters show great potential for effectively regulating metabolic pathways and conditionally converting transgenes in plant biology applications where stress response begins at a certain time or under specified conditions (Khan et al. 2023). CRISPR-based reporter assays utilize the CRISPR-Cas systems for promoter targeting and activation or repression in plant tissues using transient expression methods. They claim that the function of regulation of certain promoter elements may be assessed by combining controlled reporter genes with CRISPR activation (CRISPRa) and CRISPR interference (CRISPRi). By functionally increasing the promoter area and regulating gene expression, these studies contribute to our understanding of cis-regulatory regions and their significance in gene expression. It is noteworthy that CRISPR-based reporter assays are an important tool in the study of gene regulation since they allow for the discovery and synthesis of artificial promoters for particular plant gene expression (Mikkelsen et al. 2023). Studies using modified gene reporters in transient expression mechanisms provide a precise assessment of the promoter's expression by analyzing the activity levels of the reporter gene, such as luciferase. Additionally, these assays provide statistical information on promoter strength and shift, which researchers may use to compare various promoter mutants and examine the effects of various regulatory elements. Quantitative data can be obtained in hours or days using luminometers, fluorometers, or enzyme assays of reporter structures inserted into tissues from plants, along with tracking reporter gene activity (Li et al. 2021b).

6 Challenges in Transient Expression Methods

The transient expression systems cannot exceed their limits, as they have technological challenges. These advances are the basis for developing functional genomics, protein production, and metabolic engineering of plants. In this section, the main technological restraints, the mitigation strategies, and the biosafety issues associated with the transient gene expression are presented.

6.1 Technical Challenges

6.1.1 Suboptimal Transformation Efficiency in Monocots

One of the major issues in transient gene expression is the very low efficiency of monocot species like maize, rice, and wheat. On the other hand, most monocots are not very easy to transform through the use of *Agrobacterium*-mediated methods (Xu et al. 2023). Resistance to *Agrobacterium* is the main reason for this low transformation efficiency, and it should be noted that monocots are less compatible than dicots with *Agrobacterium*. DNA is often introduced into monocot cells using the biolistic transformation (Gene Gun) method to achieve this. This method, however, has drawbacks such as nonuniform transgene expression and tissue damage caused by high-speed DNA delivery (Lacroix and Citovsky 2020b). Overall inefficiency in monocots comes in the way of consequent success in the biotechnology sector; thus, the number of high-throughput trials in agriculture is severely limited.

6.1.2 Fluctuating Expression Levels

Another major technical challenge is in consistency in gene expression levels. Different expressions can be due to multiple contributing factors, such as the delivery method of the gene, the plant's physiological state, and the target tissue. Transformation processes often result in random DNA integration, which can lead to uneven gene expression among the transformed cells and experiments. Especially in *Agrobacterium*-mediated transformation, it can be due to the differences in bacterial infection rates and the T-DNA transfer efficiency (Tyurin et al. 2020). This variability is particularly a problem for studies that need a stable and consistent protein produced. In transient systems used to manufacture recombinant proteins or vaccines, fluctuations in expression can lead to unpredictable yields, thus diminishing the efficiency of production pipelines (Meyer 2022). One of the major obstacles that still needs to be overcome in this field is achieving more uniform expression throughout the experimental replicates when scaling the transient systems for larger applications (Belcher et al. 2020).

6.1.3 Tissue Injury

A limitation of biolistic transformation is the injury to the tissue caused by the delivery of the DNA. The injury results from high-speed directed beams of transpiring metal particles (gold or tungsten) transporting the transgene, which, in turn, can diminish cell survival and disrupt plant physiology (Tyurin et al. 2020). Instability, particularly in delicate tissues, such as meristems or embryonic cells, where cell viability is critical for successful regeneration and effective gene expression, is a major concern. Moreover, the stress from the conversion process can trigger the plant's defense mechanisms, which can, in turn, lessen the transgene expression. Activating defense-related genes is a confounding factor in experiments because stress responses, a known problem in biolistic transformation, require careful optimization of delivery parameters (Zhang et al. 2020).

6.2 Overcoming Technical Challenges

Nevertheless, a few paths have been made to deal with the problems. These paths have been developed to improve transformation efficiency, save tissue from damage, and increase the reliability of gene expression in transient systems (Buccitelli and Selbach 2020).

6.2.1 Optimizing Delivery Parameters

In biolistic transformation, altering the size of the particle, the pressure of the helium gas, and the distance between the Gene Gun and plant tissue can be factors that can help to reduce the damage to the tissue and enhance the transformation efficiency. Smaller particle sizes and lower helium pressures will not reduce physical damage but will enable DNA-coated particles to penetrate plant cells in sufficient quantities (Davlekamova et al. 2023). Researchers are also investigating the alternative use of microcarriers, such as biodegradable nanoparticles, which can enter plant cells more gently than traditional metal particles. These nanoparticle-based procedures may open up the possibility of less invasive gene transfer, thus resulting in better transformation and lower tissue damage (Leissing et al. 2022).

6.2.2 Enhancing *Agrobacterium*-Mediated Transformation in Monocots

Changes in procedures for *Agrobacterium*-mediated transformation for monocots aim to alter the strains of bacteria that are generally used in such a process. It has been shown that manipulating *Agrobacterium* strains to enhance T-DNA delivery to monocot cells or including helper plasmids in the process is the fastest breakthrough. Furthermore, certain chemicals, such as acetosyringone, can be used to treat plant tissues, which increases plant cell competence and susceptibility to *Agrobacterium* infection (Agrawal and Rami 2022). Recent progress in plant genetics and genome editing has enabled the design of new genetically modified monocot lines better transformed via *Agrobacterium*. These lines have been designed to knock out the plant's natural defenses, thus actually speeding up T-DNA transfer (Rahman et al. 2024).

6.2.3 Improving Expression Consistency

To decrease the fluctuation in transgene expression, scientists have delved into adding pieces of DNA sequence to ensure that genetic production will be the same across tissues and in multiple instances. Strong plant promoters, such as the 35S promoter of CaMV (CaMV P35S) or the ubiquitin promoter, have effectively stabilized the expression of different plant species (Karpova et al. 2021). Another method of improving the transformation process is to standardize it. Agroinfiltration, which is the uniform distribution of *Agrobacterium* throughout the leaf surface, will make the T-DNA uptake and gene expression uniform. This method has succeeded, resulting in consistent results in functional genomics studies and recombinant protein production (Kaur et al. 2021).

6.3 Environmental Risks and Biosafety Concerns

Given their temporary gene expression, transient expression systems are generally considered safer than stable transformations, so environmental risks and biosafety concerns have been overlooked to some extent, especially in field applications or large-scale production.

6.3.1 Viral Vectors Confinement

An important environmental risk of transient expression systems, particularly those using viral vectors, is the possibility of GM viruses leaking into the environment and infecting nontarget plant species. Viral vectors derived from TMV or TRV can infect plants beyond their original inoculated site and, in some cases, are transmitted from the infected to noninfected plants via mechanical damage or insect vectors (Rodriguez-Concepcion and Daròs 2022). To manage this situation, scientists have been working on nonreplicating viral vectors that become unable to replicate once they deliver the transgene to the target cells.

6.3.2 Preventing Gene Flow

Gene flow is the transfer of genes between different isolated populations via pollen or seed dissemination in field applications, particularly affecting the next generations (Zhang et al. 2020). It is unlikely that transgene sequences will be stably incorporated into the host genome (Smith et al. 2020). In extensive field applications, such as those utilized for the production of recombinant proteins in plants, the danger of gene flow must be controlled. Methods to inhibit gene flow encompass the utilization of male-sterile plants, which fail to generate viable pollen, or the collection of plants before their blooming phase. Furthermore, cultivating modified plants in isolated or regulated conditions might diminish the risk of cross-contamination with adjacent crops or wild plant species (Burnett and Burnett 2020).

6.3.3 Metallic Particle Remains in Soil

Although biolistic transformation is a fascinating procedure, it provokes concern over the fate of the metal particles post-utilization (Lacroix and Citovsky 2020b). Each time we employ this strategy, we are effectively dispersing a minute amount of metal particles into the environment (Leissing et al. 2022). However, what happens when these particles accumulate over time? Although the metals, especially gold and tungsten, are not immediately harmful, their buildup may result in unforeseen ecological consequences. Researchers are investigating environmentally sustainable or biodegradable metallic particles to ensure the continued use of this effective approach without leaving a metallic residue in valuable soil. We need a good balance between technology and sustainability, guaranteeing that our current advances do not result in future environmental challenges.

7 Future Perspectives in Transient Expression Technology

Yet, the area of transient expression technology is dynamic and ever-evolving, with important applications in synthetic biology, plant biotechnology, drug production, and crops. On the other hand, we may be able to expect a lot of new technological advancements that will accelerate efficiency, accuracy, and scalability as researchers push the envelope of what is feasible (Nosaki et al. 2021).

7.1 Potential for CRISPR/Cas Systems in Transient Expression

The development of CRISPR/Cas has fundamentally revolutionized the field of genetic engineering and led to the development of the most precise and effective genome editing tools. CRISPR/Cas systems provide very precise targeted gene editing, functional genomics research, and synthetic biology when combined with transient expression technologies. These systems' unique purity allows for quick testing times and high-efficiency rates, which reduces the risk of off-target consequences (Uranga and Daròs 2023). While CRISPR/Cas techniques do not generate immortal cells, they may be used to quickly and effectively knock down or activate specific genes. Researchers can also modify the genotype of plant cells by introducing CRISPR elements, such as a family of nucleases and short hairpin-targeting RNAs (Cas). This method works well for figuring out how genes affect development, metabolism, and stress responses. Short-term CRISPR expression in plants can be used to detect abnormalities in the activity of genes related to stress, development, and metabolism. The modified cell provides a model for investigating the breakdown of cytokinetic dynein (Zhang et al. 2021).

7.1.1 Gene Activation and Repression (CRISPRa/CRISPRi)

CRISPR-based techniques, including CRISPR activation (CRISPRa) and interference (CRISPRi), are being used more often in transient expression systems. CRISPRa can boost target gene expression, whereas CRISPRi can limit gene activity without altering the DNA sequence. These technologies allow researchers to precisely control the expression of genes and evaluate the consequences of temporarily and controllably switching on or off particular genes. The rapid distribution of CRISPRa/CRISPRi components offers a convenient method for quickly altering metabolic pathways or investigating the regulatory networks driving plant growth (Gentzel et al. 2020).

7.1.2 Base and Prime Editing

Modern CRISPR genetic modification techniques, such as base and prime editing systems, allow researchers to alter the DNA sequence at a specific location without disrupting each strand. When these methods are used with temporary expression systems, researchers may swiftly make precisely customized modifications or fix point mutations. This is, of course, ideal for figuring out how single-nucleotide polymorphisms affect gene activity or modify agricultural genes in plants. Because the changes are only temporary, researchers can be more flexible because they may test and fine-tune the effects of various copies of the chromosome before integrating them into the genome permanently (Li et al. 2021c).

7.1.3 Multiplex Gene Editing

Multiplex editing is a possible feature of CRISPR/Cas systems that might allow for the simultaneous alteration of many genes. Multiplex editing and transient expression work together to enable parallel studies on several targets, offering a high-throughput approach to studying complex features, regulatory networks, and gene-gene interactions. This technique is very useful for unlocking genes associated with pest resistance, drought tolerance, and metabolic engineering (Abdelrahman et al. 2021).

7.1.4 Development of Cas Variants

In the future, creating and optimizing Cas variations specifically meant for plant systems will be a top priority. Enhancing the effectiveness of CRISPR-based transient expression would require producing shorter, more potent Cas proteins that are highly selective and immunogenic. With virtually little chance of off-targets, these enhanced variants will be able to fix gene editing errors, making CRISPR/Cas the most often used in short-term applications (Capdeville et al. 2023).

7.2 Synthetic Biology Approaches for Customized Gene Expression

With modular genetic components and transient expression systems, scientists can create almost any gene in synthetic biology and recombine it for effective and controlled expression in plant cells. This technique works especially well for building metabolic pathways, synthesizing gene circuits, and producing promoters with targeted regulatory effects (Yaschenko et al. 2022). In biology, via recombination and testing, an organism incorporates a circuit system, like pieces of clothing. Transient expression has the advantage of real-time function demonstration and is the fastest and most effective way to introduce novel synthetic circuits into plant cells. This kind of approach helps show how gene regulatory networks, which control metabolic fluxes and protein synthesis, react to their environment. To get a precise and unambiguous result, the various combinations of transcription factors, synthetic promoters, and regulatory RNAs will be investigated (Vazquez-Vilar et al. 2023). Promoters that control gene expression in response to chemical or developmental cues are known as synthetic promoters. Testing and optimization of synthetic promoters are accelerated in a transient expression system. The activity, selectivity, and strength of promoters may be evaluated in a range of tissues and settings by using promoter-reporter constructs. Plant metabolic engineering may be enhanced, and gene expression can be directed with the use of customized promoters made using synthetic biology (Kummari et al. 2020). To bind and balance gene activity, synthetic transcription factors (sTFs) are engineered to be attached to a single isolated DNA strand. When sTFs are judiciously released through transient expression, targeted gene delivery may be used to either restore or repress the gene, as well as measure gene expression in specific tissues. With this kind of fine-tuning, scientists may design entirely new genetic circuits that can be switched on or off with fresh inputs, or they can develop gene control networks that mimic natural signaling pathways. The transitional stage enables quick and low-cost sTF engineering, which is then followed by regulatory adjustment (Cai et al. 2020). Advanced technologies produce high-value chemicals, medications, and biofuels using biological processes. To progressively restore the routes and other related portions, scientists are joining disparate DNA fragments. Using a viral vector, transient expression is the quick creation and screening of route variants. This is how you develop the metabolic engineering strategies that work the best for you and result in the best possible path (Xia et al. 2022).

7.3 Prospects for Commercialization and Scalable Systems

Organizations need to provide a platform that can manage enormous volumes of plant biomass while preserving high levels of gene expression if they want to proceed with transient expression systems. Infrastructure expenditures are made by commercial facilities to provide the ideal climate through the use of technological tools like vertical farming, bioreactors, and automated and controlled environments to target plant tissue on a massive scale. These technologies make use of the light, temperature, and nutrient channels that plants are naturally endowed with, which leads to their extreme proliferativeness and the continuous production of transient genes (Zagorskaya and Deineko 2021). The rigorous Good Manufacturing Practice (GMP) regulations necessitate the creation of such facilities, especially in the pharmaceutical and biomanufacturing industries. It is possible to produce medicines such as therapeutic proteins, antibodies, and vaccinations using the transient expression approach. But we have to scale the process, and regulation is the most crucial element. To contextualize this issue, we shall integrate plant-based expression systems with GMP regulations in this essay. By putting these rules into practice, you can be confident that all the requirements for safety and quality are met during the production process, which will ensure the project's commercial approval (Tusé et al. 2020). New plant-based solutions should be included in regulatory guidelines for the development of transient expression products. Biologics and biosimilars made from plants may require fewer approval steps when regulatory bodies get a deeper comprehension of transient expression systems. Collaboration between companies, regulatory authorities, and scientists is necessary to develop rules that facilitate the commercialization of transient expression technologies (Zhang et al. 2020). To facilitate the temporary expression process, automated and AI-powered platforms will be integrated with and assist future commercialization activities. Automated techniques for protein extraction, gene transfer, and plant tissue culture can reduce labor costs and boost labor efficiency. AI algorithms can regulate vector genes, purification techniques, and environmental factors, which increases the efficiency and replication of temporary expression systems (Ahmed et al. 2024).

8 Conclusion

Today, innovative methods have been established that enable plants to serve as production facilities for an efficient transient expression system of specific genes, whether foreign or native, in their modified or original form. These approaches allow for the targeted isolation of gene products for commercial and scientific applications without needing protracted, stagnant, and likely arduous stable plant transformations, bypassing some of the associated regulatory restraints. Here, we have provided primary approaches for expressing foreign nucleic acid sequences in plants. The choice of a transient expression system depends on the species of plant, the experimental goals, and regulatory and environmental factors. While biolistics is versatile and applicable to both monocots and dicots, it often causes tissue damage and is less efficient. *Agrobacterium* is highly efficient for monocots and causes minimal tissue damage, though it is less effective in dicots. Viral vectors, limited by host range, payload capacity, and regulatory costs, are ideal for rapid, systemic expression but require careful consideration. Balancing these factors is key to achieving the desired experimental results, and the choice of method depends on the specific application of each system. Protoplast isolation, nanomaterial-based delivery systems, and VIGS methods have been used in research and practical settings to generate a considerable quantity of specific products, some with medical and pharmaceutical applications. The spectrum of these products includes modified and recombinant proteins that can be isolated through affinity binding via their tagged epitopes for commercial applications. They also encompass the expression of viral vectors along with peptides on their surface for antibody generation, vaccine production, and alteration of plant metabolic pathways by targeted silencing of endogenous genes using VIGS, a method to examine functional genomics. VIGS and related methods provide insights into molecular mechanisms for species with or without mutant libraries. The isolation of plant protoplasts is a complicated procedure affected by various factors. Future research will focus on exploring and enhancing regeneration techniques via protoplast isolation. VIGS and related methods provide insights into molecular mechanisms for species with or without mutant libraries. With ongoing technological progress, innovations in these fields will create new opportunities for plant research, biotechnological applications, and genetic improvements.

References

Abdelrahman M, Wei Z, Rohila JS, Zhao K (2021) Multiplex genome-editing technologies for revolutionizing plant biology and crop improvement. Front Plant Sci 12:721203

Affandi, Nurhafizah Dahniar, Nur Hikmah Mostaffa, and Aisyafaznim Al-Idrus (2020) 'Interactomics: Development of an efficient and improved Agrobacterium tumefaciens-mediated transformation method for transient expression of heterologous protein in recalcitrant plant tissues in planta', Research Square. https://doi.org/10.21203/rs.2.22256/v1

Agrawal S, Rami E (2022) A review: Agrobacterium-mediated gene transformation to increase plant productivity. J Phytopharmacol 11:111

Ahmed N et al (2024) Advancing horizons in vegetable cultivation: a journey from ageold practices to high-tech greenhouse cultivation—a review. Front Plant Sci 15:1357153

Akter N, Shim J, Lee S (2024) Optimized protocol for isolation and transient expression of placenta-originated protoplast in pepper (Capsicum annuum L.). In Vitro Cell Dev Biol Plant 60(2):161–167

Alamgir A, Alamgir A (2018) Biotechnology, in vitro production of natural bioactive compounds, herbal preparation, and disease management (treatment and prevention). In: Therapeutic use of medicinal plants and their extracts: volume 2: phytochemistry and bioactive compounds, pp 585–664

Ali S et al (2017) Sugarcane mosaic virus-based gene silencing in Nicotiana benthamiana. Iran J Biotechnol 15(4):260

Aliu E et al (2024) Enhancing Agrobacterium-mediated plant transformation efficiency through improved ternary vector systems and auxotrophic strains. Front Plant Sci 15:1429353

Anjanappa RB, Gruissem W (2021) Current progress and challenges in crop genetic transformation. J Plant Physiol 261:153411

Ashfaq MA et al (2020) Post-transcriptional gene silencing: basic concepts and applications. J Biosci 45:128

Ayan A, Meriç S, Gümüş T, Atak Ç (2022) Next generation of transgenic plants: From farming to pharming. In: Genetically modified plants and beyond, p 49

Bai L et al (2020) Development of an efficient protoplast isolation and transfection system for castor bean (Ricinus communis L.). Plant Cell Tissue Organ Cult (PCTOC) 143:457–464

Baltes NJ, Voytas DF (2015) Enabling plant synthetic biology through genome engineering. Trends Biotechnol 33(2):120–131

Baranski R, Lukasiewicz A (2019) Genetic engineering of carrot. In: The carrot genome, pp 149–186

Bashir KMI, Kim M-S, Stahl U, Cho M-G (2018) Agrobacterium-mediated genetic transformation of Dictyosphaerium pulchellum for the expression of erythropoietin. J Appl Phycol 30:3503–3518

Basso MF et al (2020) Insights into genetic and molecular elements for transgenic crop development. Front Plant Sci 11:509

Bedoya LC, Martínez F, Orzáez D, Darròs J-A (2012) Visual tracking of plant virus infection and movement using a reporter MYB transcription factor that activates anthocyanin biosynthesis. Plant Physiol 158(3):1130–1138

Belcher MS et al (2020) Design of orthogonal regulatory systems for modulating gene expression in plants. Nat Chem Biol 16(8):857–865

Bendahmane A, Querci M, Kanyuka K, Baulcombe DC (2000) Agrobacterium transient expression system as a tool for the isolation of disease resistance genes: application to the Rx2 locus in potato. Plant J 21(1):73–81

Benedetti M et al (2020) Expression of a hyperthermophilic cellobiohydrolase in transgenic Nicotiana tabacum by protein storage vacuole targeting. Plants 9(12):1799

Bhoria S et al (2022) Current advances and future prospects in production of recombinant insulin and other proteins to treat diabetes mellitus. Biotechnol Lett 44(5):643–669

Buccitelli C, Selbach MJNRG (2020) mRNAs, proteins and the emerging principles of gene expression control. Nat Rev Genet 21(10):630–644

Bulaon CJI et al (2020) Rapid transient expression of functional human vascular endothelial growth factor in Nicotiana benthamiana and characterization of its biological activity. Biotechnol Rep 27:e00514

Burnett MJ, Burnett AC (2020) Therapeutic recombinant protein production in plants: challenges and opportunities. Plants People Planet 2(2):121–132

Cai Y-M et al (2020) Rational design of minimal synthetic promoters for plants. Nucleic Acids Res 48(21):11845–11856

Cañizares MC, Nicholson L, Lomonossoff GP (2005) Use of viral vectors for vaccine production in plants. Immunol Cell Biol 83(3):263–270

Canto T (2016) Transient expression systems in plants: potentialities and constraints. In: Advanced technologies for protein complex production and characterization, pp 287–301

Canto T et al (1997) Characterization of cucumber mosaic virus. Virology 237(2):237–248

Capdeville N, Schindele P, Puchta H (2023) Getting better all the time—recent progress in the development of CRISPR/Cas-based tools for plant genome engineering. Curr Opin Biotechnol 79:102854

Chabeda A et al (2018) Therapeutic vaccines for high-risk HPV-associated diseases. Papillomavirus Res 5:46–58

Chandrasekaran R, Rajiv P, Abd-Elsalam KA (2020) Carbon nanotubes: plant gene delivery and genome editing. In: Carbon nanomaterials for agri-food and environmental applications. Elsevier, pp 279–296

Chen Q et al (2013) Agroinfiltration as an effective and scalable strategy of gene delivery for production of pharmaceutical proteins. Adv Tech Biol Med 1(1):103

Chen J et al (2021) Comparative transient expression analyses on two conserved effectors of Colletotrichum orbiculare reveal their distinct cell death-inducing activities between Nicotiana benthamiana and melon. Mol Plant Pathol 22(8):1006–1013

Chen Z, Debernardi JM, Dubcovsky J, Gallavotti A (2022) Recent advances in crop transformation technologies. Nat Plants 8(12):1343–1351

Chen K et al (2023a) Isolation, purification, and application of protoplasts and transient expression systems in plants. Int J Mol Sci 24(23):16892

Chen Q et al (2023b) Physiological response of CmWRKY15-1 to chrysanthemum white rust based on TRV-VIGS. Front Plant Sci 14:1140596

Chen L et al (2023c) GA signaling protein LsRGL1 interacts with the abscisic acid signaling-related gene LsWRKY70 to affect the bolting of leaf lettuce. Hortic Res 10(5):uhad054

Chen G et al (2024) Development and application of virus-induced gene silencing (VIGS) for studying ApTT8 gene function in Agapanthus praecox ssp. orientalis. Sci Hortic 324:112595

Cheng X et al (2024) Nanoparticle LDH enhances RNAi efficiency of dsRNA in piercing-sucking pests by promoting dsRNA stability and transport in plants. J Nanobiotechnol 22(1):544

Choi B et al (2019) A plant virus-based vector system for gene function studies in pepper. Plant Physiol 181(3):867–880

Chuang L, Franke J (2022) Rapid combinatorial coexpression of biosynthetic genes by transient expression in the plant host Nicotiana benthamiana. In: Engineering natural product biosynthesis: methods and protocols. Springer, pp 395–420

Cocking EC (1960) A method for the isolation of plant protoplasts and vacuoles. Nature 187(4741):962–963

Constantin GD et al (2004) Virus-induced gene silencing as a tool for functional genomics in a legume species. Plant J 40(4):622–631

Corbineau F (2024) Ethylene, a signaling compound involved in seed germination and dormancy. Plants 13(19):2674

Coy MR, Abbitt SE, Frank MJ (2022) Protoplast isolation and transfection in maize. In: Protoplast technology: methods and protocols. Springer, pp 91–104

Cravens A, Payne J, Smolke CD (2019) Synthetic biology strategies for microbial biosynthesis of plant natural products. Nat Commun 10(1):2142

Davis B (2020) Factors influencing protoplast isolation. In: Fungal protoplasts. CRC Press, pp 45–71

Davlekamova A et al (2023) Optimization of biolistic transformation parameters for nicotiana tabacum. Appl Biochem Microbiol 59(3):368–372

Del Toro F, Tenllado F, Chung BN, Canto T (2014) A procedure for the transient expression of genes by agroinfiltration above the permissive threshold to study temperature-sensitive processes in plant–pathogen interactions. Mol Plant Pathol 15(8):848–857

Diamos AG et al (2020) High level production of monoclonal antibodies using an optimized plant expression system. Front Bioeng Biotechnol 7:472

Ding XS et al (2018) An improved brome mosaic virus silencing vector: greater insert stability and more extensive VIGS. Plant Physiol 176(1):496–510

Ding R et al (2024) Molecular traits of MAPK kinases and the regulatory mechanism of GhMAPKK5 alleviating drought/salt stress in cotton. Plant Physiol:kiae415

Dorokhov YL, Sheshukova EV, Komarova TV (2019) Plant platform for therapeutic monoclonal antibody production. In: Plant biotechnology: progress in genomic era, pp 543–581

Dunbar T, Tsakirpaloglou N, Septiningsih EM, Thomson MJ (2022) Carbon nanotube-mediated plasmid DNA delivery in rice leaves and seeds. Int J Mol Sci 23(8):4081

Egesa AO, Vallejos CE, Begcy K (2024) Cell size differences affect photosynthetic capacity in a Mesoamerican and an Andean genotype of Phaseolus vulgaris L. Front Plant Sci 15:1422814

Espinosa-Leal CA, Puente-Garza CA, García-Lara S (2018) In vitro plant tissue culture: means for production of biological active compounds. Planta 248:1–18

Evans DA, Bravo JE (2013) Plant protoplast isolation and culture. Plant Protoplasts Int Rev Cytol Suppl 16:33–53

Faivre-Rampant O et al (2004) Potato virus X-induced gene silencing in leaves and tubers of potato. Plant Physiol 134(4):1308–1316

Fan X, Sun H (2024) Exploring Agrobacterium-mediated genetic transformation methods and its applications in Lilium. Plant Methods 20(1):120

Fang L et al (2021) A tobacco ringspot virus-based vector system for gene and microRNA function studies in cucurbits. Plant Physiol 186(2):853–864

Feizi A, Baghbankohnehrouz B (2021) Overexpression of the recombinant human interferon-beta (rhIFN-β) gene in tobacco chloroplasts. Biotechnologia 102(4):367

Fouad AS, Hafez RM (2020) Effects of cobalt ions and cobalt nanoparticles on transient expression of gus gene in catharanthus roseus suspension cultures. J Radiat Res Appl Sci 13(1):765–775

Gaba V, Lapidot M, Gal-On A (2013a) HandGun-mediated inoculation of plants with viral pathogens for mechanistic studies. In: Sudowe S, Reske-Kunz AB (eds) Biolistic DNA delivery: methods and protocols. Humana Press, Totowa, pp 53–62

Gaba V, Lapidot M, Gal-On A (2013b) HandGun-mediated inoculation of plants with viral pathogens for mechanistic studies. In: Biolistic DNA delivery: methods and protocols, pp 53–62

Gad MA, Li M-j, Ahmed FK, Almoammar H (2020) Nanomaterials for gene delivery and editing in plants: challenges and future perspective. In: Multifunctional hybrid nanomaterials for sustainable Agri-food and ecosystems. Elsevier, pp 135–153

Gal-On A et al (1995) Particle bombardment drastically increases the infectivity of cloned DNA of zucchini yellow mosaic potyvirus. J Gen Virol 76(12):3223–3227

Gal-On A et al (1997) Simple hand-held devices for the efficient infection of plants with viral-encoding constructs by particle bombardment. J Virol Methods 64(1):103–110

Gammelgård E, Mohan M, Valkonen JP (2007) Potyvirus-induced gene silencing: the dynamic process of systemic silencing and silencing suppression. J Gen Virol 88(8):2337–2346

Geem KR et al (2021) Production of gloeophyllum trabeum endoglucanase Cel12A in nicotiana benthamiana for cellulose degradation. Front Plant Sci 12:696199

Geis FK, Goff SP (2020) Silencing and transcriptional regulation of endogenous retroviruses: an overview. Viruses 12(8):884

Gentzel IN et al (2020) A CRISPR/dCas9 toolkit for functional analysis of maize genes. Plant Methods 16:1–9

Ghag SB (2017) Host induced gene silencing, an emerging science to engineer crop resistance against harmful plant pathogens. Physiol Mol Plant Pathol 100:242–254

Gohlke J, Deeken R (2014) Plant responses to agrobacterium tumefaciens and crown gall development. Front Plant Sci 5:155

Goldenkova-Pavlova IV et al (2018) Computational and experimental tools to monitor the changes in translation efficiency of plant mRNA on a genome-wide scale: advantages, limitations, and solutions. Int J Mol Sci 20(1):33

Gomes C, Oliveira F, Vieira SI, Duque AS (2019) Prospects for the production of recombinant therapeutic proteins and peptides in plants: special focus on Angiotensin I-Converting Enzyme Inhibitory (ACEI) peptides. In: Genetic engineering-A glimpse of techniques and applications. IntechOpen

Gong Z, Cheng M, Botella JR (2021) Non-GM genome editing approaches in crops. Front Genome Ed 3:817279

González I et al (2010) Cucumber mosaic virus 2b protein subcellular targets and interactions: their significance to RNA silencing suppressor activity. Mol Plant-Microbe Interact 23(3):294–303

Habibi-Pirkoohi M, Shahriari AG, Ghodoum Parizipour MH (2021) Transient gene expression: an approach for recombinant vaccine production. J Med Microbiol Infect Dis 9(1):46–54

Hamada H et al (2017) An in planta biolistic method for stable wheat transformation. Sci Rep 7(1):11443

Han X et al (2023) Protoplast isolation and transient transformation system for Ginkgo biloba L. Front Plant Sci 14:1145754

Hanittinan O et al (2020) Expression optimization, purification and in vitro characterization of human epidermal growth factor produced in Nicotiana benthamiana. Biotechnol Rep 28:e00524

Hanly A et al (2020) Characterization of the role of SPL9 in drought stress tolerance in Medicago sativa. Int J Mol Sci 21(17):6003

Hartono F, Widiyanti P, Zaidan AH (2023) Iron sand-based and ionic-based magnetic nanoparticle physical characteristics comparison for biomedical application. In: AIP conference proceedings. AIP Publishing

Heidari Japelaghi R et al (2018) High-efficiency agrobacterium-mediated transformation of tobacco (Nicotiana tabacum). J Plant Mol Breed 6(2):38–50

Higuera JJ et al (2019) The strawberry FaWRKY1 transcription factor negatively regulates resistance to Colletotrichum acutatum in fruit upon infection. Front Plant Sci 10:480

Hong W et al (2015) OsRDR6 plays role in host defense against double-stranded RNA virus, Rice Dwarf Phytoreovirus. Sci Rep 5(1):11324

Houhou, Fakhreddine, Verónica Aragonés, Anamarija Butković, et al. 2021. 'A Virus-Induced Gene-Silencing Vector Based on a Mild Isolate of Watermelon Mosaic Virus for Reverse Genetic Analyses in Cucurbits', Research Square. https://doi.org/10.21203/rs.3.rs-222617/v1

Hu Y et al (2020a) Optimization of isolation and transfection conditions of maize endosperm protoplasts. Plant Methods 16:1–15

Hu P et al (2020b) Nanoparticle charge and size control foliar delivery efficiency to plant cells and organelles. ACS Nano 14(7):7970–7986

Huang C, Qian Y, Li Z, Zhou X (2012) Virus-induced gene silencing and its application in plant functional genomics. Sci China Life Sci 55(2):99–108

Huang X et al (2022) CdWRKY2-mediated sucrose biosynthesis and CBF-signalling pathways coordinately contribute to cold tolerance in bermudagrass. Plant Biotechnol J 20(4):660–675

Hull R (2009) Mechanical inoculation of plant viruses. Curr Protoc Microbiol Chapter 16:Unit 16B.6

Hung YH, Slotkin RK (2021) The initiation of RNA interference (RNAi) in plants. Curr Opin Plant Biol 61:102014

Ibrahim A, Odon V, Kormelink R (2019) Plant viruses in plant molecular pharming: toward the use of enveloped viruses. Front Plant Sci 10

Ido Y, Nakahara KS, Uyeda I (2012) White clover mosaic virus-induced gene silencing in pea. J Gen Plant Pathol 78:127–132

Ijaz U et al (2020) Plant Cis-regulatory elements: methods of identification and applications. Asian J Agric Biol. https://doi.org/10.35495/ajab.2019.08.352
Isaac S (2020) Metabolic properties of protoplasts. In: Fungal protoplasts. CRC Press, pp 171–187
Iwakawa H et al (2021) Agrobacterium-mediated transient transformation of Marchantia liverworts. Plant Cell Physiol 62(11):1718–1727
Jagram N, Dasgupta I (2023) Principles and practice of virus induced gene silencing for functional genomics in plants. Virus Genes 59(2):173–187
Jarugula S, Willie K, Stewart LR (2018) Barley stripe mosaic virus (BSMV) as a virus-induced gene silencing vector in maize seedlings. Virus Genes 54:616–620
Jesionowska M (2022) Investigating interactions between Vitis riparia open stomata 1 (OST1) and inducer of CBF expression (ICE) proteins. University of Guelph
Jeyabharathy C, Shakila H, Usha R (2015) Development of a VIGS vector based on the β-satellite DNA associated with bhendi yellow vein mosaic virus. Virus Res 195:73–78
Jiang J et al (2019) VvWRKY8 represses stilbene synthase genes through direct interaction with VvMYB14 to control resveratrol biosynthesis in grapevine. J Exp Bot 70(2):715–729
Jin DM et al (2020) Development of a rapid selection system for salt-resistant mutants of Nicotiana benthamiana through protoplast culture after gamma irradiation. Plants 9(12):1720
Johansen LK, Carrington JC (2001) Silencing on the spot. Induction and suppression of RNA silencing in the Agrobacterium-mediated transient expression system. Plant Physiol 126(3):930–938
Jores T et al (2021) Synthetic promoter designs enabled by a comprehensive analysis of plant core promoters. Nat Plants 7(6):842–855
Jugler C, Joensuu J, Chen Q (2020) Hydrophobin-protein a fusion protein produced in plants efficiently purified an anti-west nile virus monoclonal antibody from plant extracts via aqueous two-phase separation. Int J Mol Sci 21(6):2140
Kabardaeva K et al (2019) Fine tuning of translation: a complex web of mechanisms and its relevance to plant functional genomics and biotechnology. Russ J Plant Physiol 66:835–849
Kalia RK, Pathak R, Mawar R (2024a) Regulation of plant-microbe interaction through RNAi. In: Plant holobiome engineering for climate-smart agriculture, pp 419–439
Kalia D et al (2024b) Tobacco rattle virus-based virus-induced gene silencing (VIGS) as an aid for functional genomics in Saffron (Crocus sativus L.). Physiol Mol Biol Plants 30(5):749–755
Kant R, Dasgupta I (2017) Phenotyping of VIGS-mediated gene silencing in rice using a vector derived from a DNA virus. Plant Cell Rep 36:1159–1170
Karpova O et al (2021) AtDREB2A gene expression under control of the inducible promoter and virus 5′-untranslated regions improves tolerance to salinity in Nicotiana tabacum. Int J Biol Biomed Eng 15:260–274
Kaur M et al (2021) Agroinfiltration mediated scalable transient gene expression in genome edited crop plants. Int J Mol Sci 22(19):10882
Kermode AR (2021) Regulation of plant gene expression at the posttranslational level: applications to genetic engineering. In: Mechanisms of plant growth and improved productivity modern approaches. CRC Press, pp 317–390
Khalafalla MM et al (2005) Optimization of particle bombardment conditions by monitoring of transient sGFP (S65T) expression in transformed soybean. Breed Sci 55(3):257–263
Khan MS, Joyia FA, Mustafa G (2020) Seeds as economical production platform for recombinant proteins. Protein Pept Lett 27(2):89–104
Khan A et al (2023) Plant synthetic promoters: advancement and prospective. Agriculture 13(2):298
Kim T-G, Lan TT, Lee J-Y (2019) Immunogenicity of fusion protein of cholera toxin B subunit-Porphyromonas gingivalis 53-kDa minor Fimbrial protein produced in Nicotiana benthamiana. Biotechnol Bioprocess Eng 24:954–963
Kim H, Onodera Y, Masuta C (2020) Application of cucumber mosaic virus to efficient induction and long-term maintenance of virus-induced gene silencing in spinach. Plant Biotechnol 37(1):83–88
Kjemtrup S et al (1998) Gene silencing from plant DNA carried by a geminivirus. Plant J 14(1):91–100

Klein TM, Wolf ED, Wu R, Sanford JC (1987) High-velocity microprojectiles for delivering nucleic acids into living cells. Nature 327(6117):70–73

Kolliopoulou A, Taning CNT, Smagghe G, Swevers L (2017) Viral delivery of dsRNA for control of insect agricultural pests and vectors of human disease: prospects and challenges. Front Physiol 8:8

Komori T et al (2007) Current status of binary vectors and superbinary vectors. Plant Physiol 145(4):1155–1160

Konda AK (2024) Modern techniques for testing of transgenic plants and biosafety issues. In: Training manual, p 76

Kumar S, Jha I, Mogha NK, Venkatesu P (2020) Biocompatibility of surface-modified gold nanoparticles towards red blood cells and haemoglobin. Appl Surf Sci 512:145573

Kumari I (2019) A simple procedure for isolation, culture of protoplast and plant regeneration. In: In vitro plant breeding towards novel agronomic traits: biotic and abiotic stress tolerance, pp 263–273

Kumari R et al (2021) Cucumber RDR1s and cucumber mosaic virus suppressor protein 2b association directs host defence in cucumber plants. Mol Plant Pathol 22(11):1317–1331

Kummari D et al (2020) An update and perspectives on the use of promoters in plant genetic engineering. J Biosci 45:1–24

Kurth EG et al (2012) Virus-derived gene expression and RNA interference vector for grapevine. J Virol 86(11):6002–6009

Kushwaha NK, Chakraborty S (2017) Chilli leaf curl virus-based vector for phloem-specific silencing of endogenous genes and overexpression of foreign genes. Appl Microbiol Biotechnol 101(5):2121–2129

Labarile R, Mincuzzi A, Spanò R, Mascia T (2023) Virus-induced silencing of a sequence coding for Loricrin-like protein in Phytophthora infestans upon infection of a recombinant vector based on tobacco mosaic virus. Horticulturae 9(3):360

Lacroix B, Citovsky V (2020a) Biolistic approach for transient gene expression studies in plants. In: Biolistic DNA delivery in plants: methods and protocols, pp 125–139

Lacroix B, Citovsky V (2020b) Biolistic approach for transient gene expression studies in plants. Methods Mol Biol 2124:125–139

Lai Q, Wang Y, Zhou Q, Zhao Z (2020) Isolation and purification of mesophyll protoplasts from Ginkgo biloba L. Cytologia 85(1):27–32

Lam P, Steinmetz NF (2018) Plant viral and bacteriophage delivery of nucleic acid therapeutics. Wiley Interdiscip Rev Nanomed Nanobiotechnol 10(1):e1487

Lata H et al (2022) RNA interference (RNAi) mechanism and application in vegetable crops. J Hortic Sci Biotechnol 97(2):160–170

Lee W-S, Hammond-Kosack KE, Kanyuka K (2015) In planta transient expression systems for monocots. In: Recent advancements in gene expression and enabling technologies in crop plants, pp 391–422

Leissing F, Reinstädler A, Thieron H, Panstruga R (2022) Gene gun-mediated transient gene expression for functional studies in plant immunity. In: Effector-triggered immunity: methods and protocols. Springer, pp 63–77

Lentz EM et al (2018) Cassava geminivirus agroclones for virus-induced gene silencing in cassava leaves and roots. Plant Methods 14:1–9

Li X et al (2021a) HbLFG1, a rubber tree (Hevea brasiliensis) lifeguard protein, can facilitate powdery mildew infection by suppressing plant immunity. Phytopathology 111(9):1648–1659

Li Y et al (2021b) A high-efficiency Agrobacterium-mediated transient expression system in the leaves of Artemisia annua L. Plant Methods 17:1–12

Li Y, Li W, Li J (2021c) The CRISPR/Cas9 revolution continues: from base editing to prime editing in plant science. J Genet Genomics 48(8):661–670

Li X et al (2022a) Application progress of plant-mediated RNAi in pest control. Front Bioeng Biotechnol 10:963026

Li L et al (2022b) Characterization of Agrobacterium-mediated co-transformation events in rice using green and red fluorescent proteins. Mol Biol Rep 49(10):9613–9622

Liang C et al (2020) Resistance analysis of cherry rootstock 'CDR-1'(Prunus mahaleb) to crown gall. BMC Plant Biol. https://doi.org/10.1186/s12870-020-02673-0
LiJing L et al (2010) An efficient system to detect protein ubiquitination by agroinfiltration in Nicotiana benthamiana. Plant J. https://doi.org/10.1111/j.1365-313X.2009.04109.x
Lim S et al (2016) Development of a new vector using soybean yellow common mosaic virus for gene function study or heterologous protein expression in soybeans. J Virol Methods 228:1–9
Lin Y-Y et al (2013) Improving initial infectivity of the Turnip mosaic virus (TuMV) infectious clone by an mini binary vector via agro-infiltration. Bot Stud 54:1–10
Liu H, Kotova TI, Timko MP (2019) Increased leaf nicotine content by targeting transcription factor gene expression in commercial flue-cured tobacco (Nicotiana tabacum L.). Genes 10(11):930
Liu M et al (2020) A cucumber green mottle mosaic virus vector for virus-induced gene silencing in cucurbit plants. Plant Methods 16:1–13
Liu Y et al (2024) A cotyledon-based virus-induced gene silencing (cotyledon-VIGS) approach to study specialized metabolism in medicinal plants. Plant Methods 20(1):26
Lu J et al (2017) An efficient transient expression system for gene function analysis in rose. Plant Methods 13:1–12
Lv Y et al (2019) The infiltration efficiency of agrobacterium-mediated transient transformation in four apple cultivars. Sci Hortic 256:108597
Lv Z, Jiang R, Chen J, Chen W (2020) Nanoparticle-mediated gene transformation strategies for plant genetic engineering. Plant J 104(4):880–891
Ma H et al (2021) The long noncoding RNA MdLNC499 bridges MdWRKY1 and MdERF109 function to regulate early-stage light-induced anthocyanin accumulation in apple fruit. Plant Cell 33(10):3309–3330
Maeda K et al (2020) Virus-induced flowering by apple latent spherical virus vector: effective use to accelerate breeding of grapevine. Viruses 12(1):70
Mamedov T et al (2020) Engineering, production and characterization of Spike and Nucleocapsid structural proteins of SARS–CoV-2 in Nicotiana benthamiana as vaccine candidates against COVID-19. bioRxiv:2020.12. 29.424779
Mao L et al (2020) Characterization of the DREBA4-type transcription factor (SlDREBA4), which contributes to heat tolerance in tomatoes. Front Plant Sci 11:554520
Marcon C, Win YN, Du X, Hochholdinger F (2024) BonnMu: a resource for functional genomics in maize (Zea mays L.). Cold Spring Harb Protoc 2025:pdb.top108465
Mardini M et al (2024) Advancing virus-induced gene silencing in sunflower: key factors of VIGS spreading and a novel simple protocol. Plant Methods 20(1):122
Martín-Alonso S, Frutos-Beltrán E, Menéndez-Arias L (2021) Reverse transcriptase: from transcriptomics to genome editing. Trends Biotechnol 39(2):194–210
Massa S et al (2019) Bioproduction of a therapeutic vaccine against human papillomavirus in tomato hairy root cultures. Front Plant Sci 10:452
May D, Paldi K, Altpeter F (2023) Targeted mutagenesis with sequence-specific nucleases for accelerated improvement of polyploid crops: progress, challenges, and prospects. Plant Genome 16(2):e20298
Meyer P (2022) Stabilities and instabilities in transgene expression. In: Transgenic plant research. Routledge, pp 263–275
Mikkelsen NS et al (2023) Enrichment of transgene integrations by transient CRISPR activation of a silent reporter gene. Mol Ther Methods Clin Dev 29:1–16
Miroshnichenko D et al (2020) Evaluation of plant-derived promoters for constitutive and tissue-specific gene expression in potato. Plants 9(11):1520
Mohammed S, Abd Samad A, Rahmat Z (2019) Agrobacterium-mediated transformation of rice: constraints and possible solutions. Rice Sci 26(3):133–146
Monroy-Borrego AG, Steinmetz NF (2022) Three methods for inoculation of viral vectors into plants. Front Plant Sci 13:963756
Mookkan M (2018) Particle bombardment – mediated gene transfer and GFP transient expression in Seteria viridis. Plant Signal Behav 13(4):e1441657

Naing AH, Adedeji OS, Kim CK (2021) Protoplast technology in ornamental plants: current progress and potential applications on genetic improvement. Sci Hortic 283:110043

Naseri Z, Khezri G, Davarpanah SJ, Ofoghi H (2019) Virus-based vectors: a new approach for the production of recombinant proteins. J Appl Biotechnol Rep 6(1):6–14

Nosaki S, Hoshikawa K, Ezura H, Miura K (2021) Transient protein expression systems in plants and their applications. Plant Biotechnol 38(3):297–304

Okolo JC, Adejumo BA, Amedu JE (2022) Genetically modified crops in Africa: Perspectives and safety concerns. In: Food security and safety volume 2: african perspectives, pp 83–114

Orzaez D, Mirabel S, Wieland WH, Granell A (2006) Agroinjection of tomato fruits. A tool for rapid functional analysis of transgenes directly in fruit. Plant Physiol 140(1):3–11

Ozyigit II (2020) Gene transfer to plants by electroporation: methods and applications. Mol Biol Rep 47(4):3195–3210

Ozyigit II, Yucebilgili Kurtoglu K (2020) Particle bombardment technology and its applications in plants. Mol Biol Rep 47(12):9831–9847

Pandey P, Choudhury NR, Mukherjee SK (2009) A geminiviral amplicon (VA) derived from tomato leaf curl virus (ToLCV) can replicate in a wide variety of plant species and also acts as a VIGS vector. Virol J 6:1–13

Paradia PK, Bhavale R, Agnihotri T, Jain A (2023) A review on edible vaccines and biopharmaceutical products from plants. Curr Pharm Biotechnol 24(4):495–509

Park S-H et al (2016) Development of virus-induced gene expression and silencing vector derived from grapevine Algerian latent virus. Plant Pathol J 32(4):371

Pasternak T, Paponov IA, Kondratenko S (2021) Optimizing protocols for Arabidopsis shoot and root protoplast cultivation. Plants 10(2):375

Poudel SB, Bhattarai G, Kwon T-H, Lee J-C (2023) Biopotentials of collagen scaffold impregnated with plant-cell-derived epidermal growth factor in defective bone healing. Materials 16(9):3335

Pouvreau B, Vanhercke T, Singh S (2018) From plant metabolic engineering to plant synthetic biology: the evolution of the design/build/test/learn cycle. Plant Sci 273:3–12

Prakash C, Zipf A (1996) Workshop on transgenic plants: biology and applications. Plant Mol Biol Rep 14(2):189

Prasanna GL, Gopal K, Ramaiah M, Gowthami L (2023) Methods of gene transfer, sources of transgenes with gene deployment tactics and their role in developing insect resistant transgenic plants: a review. In: Latest trends in agriculture entomology Integrated Publication, p 45

Pretorius L-S (2018) Investigation of virus-induced defence modulations during plant-virus interactions

Qi X et al (2023) Establishment of virus-induced gene silencing (VIGS) system in Luffa acutangula using Phytoene desaturase (PDS) and tendril synthesis related gene (TEN). Plant Methods 19(1):94

Rademacher T et al (2019) Plant cell packs: a scalable platform for recombinant protein production and metabolic engineering. Plant Biotechnol J 17(8):1560–1566

Rahman SU et al (2024) Agrobacterium-mediated transformation for the development of transgenic crops; present and future prospects. Mol Biotechnol 66(8):1836–1852

Rangasami S et al (2022) Application of Plantibodies, the plant-MADE vaccines. Int J Environ Clim Change 12(10):171–180

Ren R et al (2021a) Highly efficient leaf base protoplast isolation and transient expression systems for orchids and other important monocot crops. Front Plant Sci 12:626015

Ren X et al (2021b) Alternative splicing of TaGS3 differentially regulates grain weight and size in bread wheat. Int J Mol Sci 22(21):11692

Rey M et al (1996) Factors affecting transient gene expression in cultured radiata pine cotyledons following particle bombardment. Physiol Plant 96(4):630–636

Ricci A et al (2021) Host-induced gene silencing and spray-induced gene silencing for crop protection against viruses. In: RNAi for plant improvement and protection. CABI, Wallingford UK, pp 72–85

Rizzo TM, Palukaitis P (1990) Construction of full-length cDNA clones of cucumber mosaic virus RNAs 1, 2 and 3: generation of infectious RNA transcripts. Mol Gen Genet MGG 222:249–256

Rodriguez-Concepcion M, Daròs J-A (2022) Transient expression systems to rewire plant carotenoid metabolism. Curr Opin Plant Biol 66:102190

Rössner C, Lotz D, Becker A (2022) VIGS goes viral: how VIGS transforms our understanding of plant science. Annu Rev Plant Biol 73:703–728

Ryu CM, Anand A, Kang L, Mysore KS (2004) Agrodrench: a novel and effective agroinoculation method for virus-induced gene silencing in roots and diverse Solanaceous species. Plant J 40(2):322–331

Sandra N, Jailani AAK, Jain RK, Mandal B (2019) Development of Soybean yellow mottle mosaic virus-based expression vector for heterologous protein expression in French bean. Mol Biotechnol 61:181–190

Schachtsiek J et al (2019) Virus-induced gene silencing (VIGS) in Cannabis sativa L. Plant Methods 15:1–9

Scintilla S et al (2022) Regeneration of non-chimeric plants from DNA-free edited grapevine protoplasts. Front Plant Sci 13:1078931

Scofield SR et al (1996) Molecular basis of gene-for-gene specificity in bacterial speck disease of tomato. Science 274(5295):2063–2065

Sedaghati B, Haddad R, Bandehpour M (2020) Transient expression of human serum albumin (HSA) in tobacco leaves. Mol Biol Rep 47(9):7169–7177

Serrano-Sevilla I et al (2019) Natural polysaccharides for siRNA delivery: nanocarriers based on chitosan, hyaluronic acid, and their derivatives. Molecules 24(14):2570

Shakoor S et al (2019) Role of oral vaccines as an edible tool to prevent infectious diseases. Acta Virol 63(3):245–252

Shanmugaraj B, Bulaon CJI, Phoolcharoen W (2020) Plant molecular farming: a viable platform for recombinant biopharmaceutical production. Plants 9(7):842

Sharma R, Mody N, Vyas S, Vyas SP (2018) Plant-based edible vaccines: an innovative concept to immunization. In: Advanced molecular plant breeding: meeting the challenge of food security. CRC Press

Sharma R et al (2020) Agrobacterium-mediated transient transformation of sorghum leaves for accelerating functional genomics and genome editing studies. BMC Res Notes 13(1):116

Shingote PR et al (2024) Virus–Vector Interactions and Transmission. In: Molecular dynamics of plant stress and its management. Springer, pp 403–425

Shreni Agrawal ER (2022) A review: agrobacterium-mediated gene transformation to increase plant productivity. J Phytopharmacol 11:111

Shu P, Li Y, Sheng J, Shen L (2024) Tomato SlMAPK3 modulates cold resistance by regulating the synthesis of Raffinose and the expression of SlWRKY46. J Agric Food Chem 72(10):5185–5196

Sikorskaite S et al (2010) HandyGun: an improved custom-designed, non-vacuum gene gun suitable for virus inoculation. J Virol Methods 165(2):320–324

Šimčíková M, Prather KL, Prazeres DM, Monteiro GA (2015) Towards effective non-viral gene delivery vector. Biotechnol Genet Eng Rev 31(1–2):82–107

Sindarovska Y, Kuchuk M (2021) Long-term potato virus X (PVX)-based transient expression of recombinant GFP protein in Nicotiana benthamiana culture in vitro. Plants (Basel) 10(10)

Singh AA et al (2020) Plant-based production of highly potent anti-HIV antibodies with engineered posttranslational modifications. Sci Rep 10(1):6201

Smith AL et al (2020) Global gene flow releases invasive plants from environmental constraints on genetic diversity. Proc Natl Acad Sci U S A 117(8):4218–4227

Song ZT, Liu JX, Han JJ (2021) Chromatin remodeling factors regulate environmental stress responses in plants. J Integr Plant Biol 63(3):438–450

Spada M, Pugliesi C, Fambrini M, Pecchia S (2024) Challenges and opportunities arising from Host–Botrytis cinerea interactions to outline novel and sustainable control strategies: the key role of RNA interference. Int J Mol Sci 25(12):6798

Stuttmann J et al (2021) Highly efficient multiplex editing: one-shot generation of 8× Nicotiana benthamiana and 12× Arabidopsis mutants. Plant J 106(1):8–22

Sun L, et al (2019) Citrus genetic engineering for disease resistance: past, present and future. Int J Mol Sci 20(21):5256
Sun Y et al (2024) GhUBC10-2 mediates GhGSTU17 degradation to regulate salt tolerance in cotton (Gossypium hirsutum). Plant Cell Environ 47(5):1606–1624
Talla SK et al (2023) Redox basis of photosynthesis inhibition at supra-optimal bicarbonate in mesophyll protoplasts of Arabidopsis thaliana. J Plant Physiol 287:154047
Tan Y, Bukys A, Molnár A, Hudson A (2020) Rapid, high efficiency virus-mediated mutant complementation and gene silencing in Antirrhinum. Plant Methods 16:1–12
Tanveer S et al (2023) Production of proinflammatory cytokines by expressing Newcastle disease vaccine candidates in corn. J King Saud Univ Sci 35(3):102537
Tavares-Esashika ML et al (2020) Characterization of an infectious clone of pepper ringspot virus and its use as a viral vector. Arch Virol 165:367–375
Tayal R et al (2024) Developing bioengineered fruits and vegetables for postharvest quality improvement using heterologous, overexpression, and combinatorial engineering. In: Bioengineered fruit and vegetables. Apple Academic Press, pp 55–74
Thauland TJ et al (2019) Case study: mechanism for increased follicular helper T cell development in activated PI3K delta syndrome. Front Immunol 10:753
Tian B et al (2019) Biolistic transformation of wheat. In: Transgenic plants: methods and protocols, pp 117–130
Tiedge K et al (2022) Foxtail mosaic virus-induced gene silencing (VIGS) in switchgrass (Panicum virgatum L.). Plant Methods 18(1):71
Tonnies J et al (2023) Scalable transfection of maize mesophyll protoplasts. JoVE (J Vis Exp) 196:e64991
Torti S et al (2021) Transient reprogramming of crop plants for agronomic performance. Nat Plants 7(2):159–171
Trinidad JL, Longkumer T, Kohli A (2021) Rice protoplast isolation and transfection for transient gene expression analysis. In: Rice genome engineering and gene editing: methods and protocols, pp 313–324
Tue NH et al (2024) Partitioning recombinant chitinase from Nicotiana benthamiana by an aqueous two-phase system based on polyethylene glycol and phosphate salts. Int J Biol Macromol 269:131924
Tuo D et al (2021a) An efficient papaya leaf distortion mosaic potyvirus vector for virus-induced gene silencing in papaya. Horticult Res 8:8
Tuo D et al (2021b) A cassava common mosaic virus vector for virus-induced gene silencing in cassava. Plant Methods 17:1–12
Tusé D, Nandi S, McDonald KA, Buyel JF (2020) The emergency response capacity of plant-based biopharmaceutical manufacturing-what it is and what it could be. Front Plant Sci 11:594019
Tuteja N et al (2012) Recent advances in development of marker-free transgenic plants: regulation and biosafety concern. J Biosci 37:167–197
Tuttle JR et al (2008) Geminivirus-mediated gene silencing from Cotton leaf crumple virus is enhanced by low temperature in cotton. Plant Physiol 148(1):41–50
Tuttle JR, Haigler CH, Robertson D (2012) Method: low-cost delivery of the cotton leaf crumple virus-induced gene silencing system. Plant Methods 8:1–8
Tyurin AA, Suhorukova AV, Kabardaeva KV, Goldenkova-Pavlova IV (2020) Transient gene expression is an effective experimental tool for the research into the fine mechanisms of plant gene function: advantages, limitations, and solutions. Plants 9(9):1187
Ueki S et al (2009) Functional transient genetic transformation of Arabidopsis leaves by biolistic bombardment. Nat Protoc 4(1):71–77
Ueki S, Magori S, Lacroix B, Citovsky V (2013) Transient gene expression in epidermal cells of plant leaves by biolistic DNA delivery. In: Biolistic DNA delivery: methods and protocols, pp 17–26
Ura T, Okuda K, Shimada M (2014) Developments in viral vector-based vaccines. Vaccine 2(3):624–641

Uranga M, Daròs JA (2023) Tools and targets: the dual role of plant viruses in CRISPR–Cas genome editing. Plant Genome 16(2):e20220
Vaghchhipawala ZE, Mysore KS (2008) Agroinoculation: a simple procedure for systemic infection of plants with viruses. In: Plant virology protocols: from viral sequence to protein function, pp 555–562
van der Krol AR et al (1990) Flavonoid genes in petunia: addition of a limited number of gene copies may lead to a suppression of gene expression. Plant Cell 2(4):291–299
Van Eck J, Keen P, Tjahjadi M (2019) Agrobacterium tumefaciens-mediated transformation of tomato. In: Transgenic plants: methods and protocols, pp 225–234
Vazquez-Vilar M, Selma S, Orzaez D (2023) The design of synthetic gene circuits in plants: new components, old challenges. J Exp Bot 74(13):3791–3805
Verdú-Navarro F, Moreno-Cid J, Weiss J, Egea-Cortines M (2024) Genetic engineering of plants for vaccine, recombinant protein, and drugs production for health security. In: Genetic engineering of crop plants for food and health security: volume 2. Springer, pp 97–117
Vishwanath AA et al (2024) Mitochondrial Ca2+ efflux controls neuronal metabolism and long-term memory across species. bioRxiv:2024.02.01.578153
Wang M, Jin H (2017) Spray-induced gene silencing: a powerful innovative strategy for crop protection. Trends Microbiol 25(1):4–6
Wang J et al (2018) CsATAF1 positively regulates drought stress tolerance by an ABA-dependent pathway and by promoting ROS scavenging in cucumber. Plant Cell Physiol 59(5):930–945
Wang JW et al (2019) Nanoparticle-mediated genetic engineering of plants. Mol Plant 12(8):1037–1040
Wang Q et al (2021a) Mechanism of siRNA production by a plant Dicer-RNA complex in dicing-competent conformation. Science 374(6571):1152–1157
Wang R et al (2021b) Identification of GhLOG gene family revealed that GhLOG3 is involved in regulating salinity tolerance in cotton (Gossypium hirsutum L.). Plant Physiol Biochem 166:328–340
Wang Y et al (2021c) Is viral vector gene delivery more effective using biomaterials? Adv Healthc Mater 10(1):2001238
Wang L, Yu G, Macho AP, Lozano-Durán R (2021d) Split-luciferase complementation imaging assay to study protein-protein interactions in Nicotiana benthamiana. Bio-Protocol 11(23):e4237–e4237
Wang Y et al (2022a) A highly efficient mesophyll protoplast isolation and PEG-mediated transient expression system in eggplant. Sci Hortic 304:111303
Wang J et al (2022b) An efficient and universal protoplast isolation protocol suitable for transient gene expression analysis and single-cell RNA sequencing. Int J Mol Sci 23(7):3419
Wang D et al (2022c) Nano-enabled pesticides for sustainable agriculture and global food security. Nat Nanotechnol 17(4):347–360
Wang L, Zhang J, Xu X (2024a) A comparison of DNA-methylation during protoplast culture of Ponkan mandarin (Citrus reticulata Blanco) and tobacco (Nicotiana tabacum L.). Plants 13(20):2878
Wang X et al (2024b) A telosma mosaic virus-based vector for foreign gene expression and virus-induced gene silencing in the perennial woody vine, passion fruit (Passiflora edulis). bioRxiv:2024.08. 13.607796
Watanabe K et al (2021) Construction of monopartite geminivirus-based virus-induced gene silencing (VIGS) vectors using a two-component strategy. J Gen Plant Pathol 87(6):366–376
Wen Y et al (2023) VdGAL4 modulates microsclerotium formation, conidial morphology, and germination to promote virulence in Verticillium dahliae. Microbiol Spectr 11(1):e03515-22
Wong MH et al (2016) Lipid exchange envelope penetration (LEEP) of nanoparticles for plant engineering: a universal localization mechanism. Nano Lett 16(2):1161–1172
Wu H-Y et al (2014) AGROBEST: an efficient agrobacterium-mediated transient expression method for versatile gene function analyses in Arabidopsis seedlings. Plant Methods 10:1–16
Xia P et al (2020) An attempt to establish an agrobacterium-mediated transient expression system in medicinal plants. Protoplasma 257:1497–1505

Xia Y-H et al (2022) Release of moth pheromone compounds from Nicotiana benthamiana upon transient expression of heterologous biosynthetic genes. BMC Biol 20(1):80
Xiang J et al (2022) An efficient and novel method to screen Botrytis cinerea resistance genes based on TRV-induced gene silencing with lily petal discs. Physiol Mol Plant Pathol 122:101923
Xiao Z et al (2020) An efficient virus-induced gene silencing (VIGS) system for functional genomics in Brassicas using a cabbage leaf curl virus (CaLCuV)-based vector. Planta 252:1–12
Xie X et al (2019) The roles of environmental factors in regulation of oxidative stress in plant. Biomed Res Int 2019(1):9732325
Xie W et al (2021) Simultaneous gene expression and multi-gene silencing in Zea mays using maize dwarf mosaic virus. BMC Plant Biol 21(1):208
Xingqi Y et al (2024) Construction and optimization of the TRV-mediated VIGS system in Areca catechu embryoids. Sci Hortic 338:113621
Xu M et al (2021a) Stochastic gene expression drives mesophyll protoplast regeneration. Sci Adv 7(33):eabg8466
Xu K et al (2021b) Fine-tuning florigen increases field yield through improving photosynthesis in soybean. Front Plant Sci 12:710754
Xu H et al (2023) An agrobacterium-mediated transient expression method for functional assay of genes promoting disease in monocots. Int J Mol Sci 24(8):7636
Xu Y et al (2024) Development and application of the TRV-induced gene-silencing system in different Rhododendron species. Plant Cell, Tissue and Organ Culture (PCTOC) 157(3):61
Yamagishi N, Yoshikawa N (2022) Efficient virus-induced gene silencing system in pumpkin (Cucurbita maxima) using apple latent spherical virus vector. J Virol Methods 301:114456
Yamamoto T et al (2018) Improvement of the transient expression system for production of recombinant proteins in plants. Sci Rep 8(1):4755
Yan Y et al (2022) Nanotechnology strategies for plant genetic engineering. Adv Mater 34(7):2106945
Yanagihara C, Tsukamoto H, Ishida Y, Komari T (2024) Improved biolistic transformation and genome editing in wheat by using trehalose for high osmotic treatment. Plant Biotechnol. https://doi.org/10.5511/plantbiotechnology.24.0328a
Yang L et al (2023a) Heritable transgene-free genome editing in plants by grafting of wild-type shoots to transgenic donor rootstocks. Nat Biotechnol 41(7):958–967
Yang W et al (2023b) Virus-induced gene silencing in the tea plant (Camellia sinensis). Plants 12(17):3162
Yao L et al (2016) Protoplast isolation and development of a transient expression system for sweet cherry (Prunus avium L.). Sci Hortic 209:14–21
Yaschenko AE et al (2022) Deciphering the molecular basis of tissue-specific gene expression in plants: can synthetic biology help? Curr Opin Plant Biol 68:102241
Yu J et al (2018) Efficient virus-induced gene silencing in Brassica rapa using a turnip yellow mosaic virus vector. Biol Plant 62:826–834
Yu G et al (2019) Mildew resistance locus O genes CsMLO1 and CsMLO2 are negative modulators of the Cucumis sativus defense response to Corynespora cassiicola. Int J Mol Sci 20(19):4793
Yuan G et al (2021) Expanding the application of a UV-visible reporter for transient gene expression and stable transformation in plants. Horticult Res 8:8
Yuan W et al (2024) NIa-Pro of sugarcane mosaic virus targets Corn Cysteine Protease 1 (CCP1) to undermine salicylic acid-mediated defense in maize. PLoS Pathog 20(3):e1012086
Zagorskaya A, Deineko E (2021) Plant-expression systems: a new stage in production of biopharmaceutical preparations. Russ J Plant Physiol 68:17–30
Zaidi SS-e-A, Mansoor S (2017) Viral vectors for plant genome engineering. Front Plant Sci 8:539
Zale JM, Agarwal S, Loar S, Steber CM (2009) Evidence for stable transformation of wheat by floral dip in Agrobacterium tumefaciens. Plant Cell Rep 28(6):903–913
Zaulda FA et al (2022) A cowpea severe mosaic virus-based vector simplifies virus-induced gene silencing and foreign protein expression in soybean. Plant Methods 18(1):116
Zeng H et al (2019) Agrobacterium-mediated gene transient overexpression and tobacco rattle virus (TRV)-based gene silencing in cassava. Int J Mol Sci 20(16):3976

Zhang C, Bradshaw JD, Whitham SA, Hill JH (2010) The development of an efficient multipurpose bean pod mottle virus viral vector set for foreign gene expression and RNA silencing. Plant Physiol 153(1):52–65

Zhang X-F et al (2018) A new mechanistic model for viral cross protection and superinfection exclusion. Front Plant Sci 9:40

Zhang Y et al (2020) A highly efficient agrobacterium-mediated method for transient gene expression and functional studies in multiple plant species. Plant Commun 1(5):100028

Zhang D, Zhang Z, Unver T, Zhang B (2021) CRISPR/Cas: a powerful tool for gene function study and crop improvement. J Adv Res 29:207–221

Zhang L et al (2022a) Optimization of an efficient protoplast transformation system for transient expression analysis using leaves of Torenia fournieri. Plants 11(16):2106

Zhang D et al (2022b) An integrated physiology, cytology, and proteomics analysis reveals a network of sugarcane protoplast responses to enzymolysis. Front Plant Sci 13:1066073

Zhang Y et al (2023) Virus-induced gene silencing (VIGS) in Chinese jujube. Plants 12(11):2115

Zhang Y, Yang X, Huang L, Deng S (2024a) A highly efficient Agrobacterium-mediated infectious system for Sweet potato leaf curl virus and a deltasatellite-based VIGS vector. Phytopathol Res 6(1):23

Zhang H et al (2024b) A brassinosteroid signal regulated beta-tubulin GhTUB17 play a role in cotton fiber elongation by influencing microtubule organization. Ind Crops Prod 221:119309

Zhao H, Shao G, Gao W, Gu B (2020) The application of double-barreled particle bombardment for transient gene expression in Arabidopsis. Chin Bull Bot 55(2):182

Zheng L et al (2021) An improved and efficient method of agrobacterium syringe infiltration for transient transformation and its application in the elucidation of gene function in poplar. BMC Plant Biol 21:1–19

Zhi H et al (2022) The promising nanovectors for gene delivery in plant genome engineering. Int J Mol Sci 23(15):8501

Zhou Z et al (2019) Importance of OsRac1 and RAI1 in signalling of nucleotide-binding site leucine-rich repeat protein-mediated resistance to rice blast disease. New Phytol 223(2):828–838

Zhou P et al (2021) Virus-induced gene silencing (VIGS) in Chinese narcissus and its use in functional analysis of NtMYB3. Hortic Plant J 7(6):565–572

Zhou C, Niu S, El-Kassaby YA, Li W (2023) Genome-wide identification of late embryogenesis abundant protein family and their key regulatory network in Pinus tabuliformis cold acclimation. Tree Physiol 43(11):1964–1985

Zu X et al (2020) Mutation of IDR1 enhances drought tolerance by reducing ROS production and activating ROS scavenging in rice. bioRxiv:2020.08.24.264556

Plant-Derived Proteins and Their Interactions with Microorganisms: Interactions, Applications, and Innovations

Jaagriti Tyagi, Arti Mishra, Ananya Padhy, Romila Rawat Bisht, Smriti Shukla, and Rajneesh K. Agnihotri

Abstract

Plant-derived proteins have emerged as sustainable and nutritionally valuable alternatives to animal-based proteins, gaining increasing attention for their roles in health, food security, and industrial applications. A growing area of research focuses on the intricate interactions between plant proteins and microorganisms, which significantly influence the structural, functional, and bioactive properties of these proteins. Microbial fermentation, enzymatic processing, and co-culturing techniques can enhance the digestibility, amino acid profile, and nutritional quality of plant proteins while reducing anti-nutritional factors such as phytates and tannins. Various microorganisms, including lactic acid bacteria, *Bacillus* spp., and fungi like *Aspergillus* and *Rhizopus*, play pivotal roles in these transformations. These interactions also lead to the generation of bioactive peptides with antioxidant, antimicrobial, and immunomodulatory properties, adding functional

J. Tyagi (✉)
Department of Biotechnology, Dr. MPS Group of Institutions, Agra, Uttar Pradesh, India

A. Mishra
Department of Botany, Hansraj College, University of Delhi, Delhi, India

Umeå Plant Science Centre (UPSC), Department of Plant Physiology, Umeå University, Umeå, Sweden

A. Padhy · S. Shukla
Amity Institute of Environmental Sciences Amity University, Noida, Uttar Pradesh, India

R. R. Bisht
Department of Botany, Hansraj College, University of Delhi, Delhi, India

R. K. Agnihotri
Department of Botany, S.L.S. Khandari Campus, Dr. B.R. Ambedkar University, Agra, Uttar Pradesh, India

R. K. Gaur et al. (eds.), *Plant Molecular Farming*,
https://doi.org/10.1007/978-981-95-3823-2_2

value to plant proteins in food and pharmaceutical formulations. In agriculture, plant-microbe interactions influence protein biosynthesis, stress resistance, and nutrient cycling, contributing to plant health and productivity. Technological innovations, such as synthetic biology and genome-editing tools like CRISPR-Cas9, are being applied to engineer both microbial strains and plant genomes, optimizing protein expression and microbial synergy for targeted outcomes. These advances pave the way for novel applications in the development of high-protein functional foods, plant-based meat alternatives, biostimulants, and therapeutic compounds. This chapter highlights the dynamic interplay between plant proteins and microbes, showcasing the vast potential of their interactions in food technology, agriculture, and biotechnology. Continued interdisciplinary research is essential to fully harness these biological relationships for sustainable development and innovation in protein-based solutions.

Keywords

Plant-derived proteins · Microorganisms · CRISPR-CAS · Genetic engineering · Fermentation

1 Introduction

Plant-derived proteins are protein sources from plant foods like soybeans, peas, wheat, rice, and chickpeas. Plant-derived proteins are gaining grip as people become interested in consuming alternatives to animal-based proteins (like meat, milk, and eggs) on health, environmental, and ethical grounds. Increasingly, individuals are opting to cut meat out of their diets. Some individuals eschew meat altogether, while others, "flexitarians," simply consume smaller amounts of it. As such, the need for plant proteins for food has grown. These plant proteins are usually processed into protein concentrates (more than 60% protein) and isolates (more than 85% protein) to supplement foods such as protein bars, meat alternatives, and milk alternatives. Plant-derived proteins are valuable components in our diets. Proteins are required by the body to repair and construct muscles and tissues and assist in numerous bodily functions, such as constructing hormones, enzymes, and blood constituents. Although most individuals in the past generally derived protein from animal products, proteins from plants are today considered to be a vital alternative. One of the strengths of plant-derived proteins is that they are perceived as being more sustainable. A significant amount of land, water, and energy is required in animal farming, in addition to supporting greenhouse gas emissions. In return, plant proteins could be farmed using minimal resources and ultimately are possibly a cleaner choice, environmentally speaking. With this said, both cultivating and processing plant proteins present problems, such as production wastes and usage of water. Soy protein was among the first plant-derived proteins to ever be used in foods, as well as healthy in that it lowers cholesterol. Wheat gluten has also been used as a plant protein for a long time, especially when baking. The newer plant proteins (pea, rice, and chickpea) are only now being used, and it will take scientists a while to learn about the primary proteins in these sources and how best to use these proteins together with existing proteins. There are advantages to plant-derived proteins, but they are not without their concerns. One of the concerns is that river plant protein typically does not have all of the amino acids needed, which it does not have. For example, most plant proteins do not have enough lysine or methionine. To fix this problem, food makers began blending various plant proteins together. This allows food makers to give the complete product all the amino acids needed by our body. Plant proteins are the future of a sustainable world. With an increasing population and decreasing natural resource security, plant proteins will sustain human health and feed an increasing population with sustainable protein. Plant proteins are plant-based and thus use less water, land, energy, etc., compared to animal proteins. The food sector is investing in enhancing plant-derived proteins to enhance taste, nutrition, functionality, and availability in food systems. With more research and technological advancements on plant proteins, they can be a great, healthy, and sustainable choice to assist in supporting populations globally (Aimutis 2022).

2 Importance of Microorganisms in Food and Health

Microorganisms are extremely small organisms of both animals and plants, and they are significant to us in the sustenance and well-being they provide. They are crucial to the food systems, where fermentation is used in making yogurt, cheese, bread, and sweets. During fermentation, these microbes change the taste, smell, and texture of food. They also kill harmful germs and boost the food's nutritional value by producing vitamins, including different types of vitamin B. The main types of helpful microorganisms in food are bacteria, yeasts, and molds. While some microbes can cause diseases, many are beneficial and help keep food safe and enjoyable. In yogurt and cheese, depending on the type, these enzymes act on complex substances and break them up into simpler substances, thus easing pressure from the digestive system and imparting a characteristic flavor. Such changes certainly make the eating experience enjoyable and beneficial for nutrient absorption. Within our bodies, particularly in the stomach and intestines, these microorganisms constitute the gut microbiota. This group includes many different microbes that live in our digestive system. They help break down food, produce important vitamins like B12 and K, and prevent harmful germs from growing. Good microbes compete with the bad ones for space and food, which helps keep our body safe. A healthy balance of these microbes is essential for good digestion and overall health. They also assist in absorbing minerals like calcium and iron from our food. Some even help control body weight and protect us from certain diseases (Bernal 2024). Eating fermented foods regularly helps improve this balance by adding more good microbes to the gut. So even though we cannot see them, microorganisms greatly influence our food, health, and how our bodies function. (Sionek et al. 2023). When the balance is upset, a state called dysbiosis can result in a spectrum of health concerns, including inflammatory bowel illnesses, obesity, and even mental health problems like anxiety and depression. Particularly early in life, microorganisms are also crucial in the development and control of the immune system. The immune system learns to differentiate between benign substances and harmful intruders by exposing the body to a great range of germs. This lowers the risk of autoimmune diseases and allergies, hence emphasizing microbial diversity as a crucial component of general health. Some microorganisms also naturally battle harmful germs, therefore acting as natural preservatives that enhance food safety and lessen the need for synthetic additives, which many consumers want to avoid. Apart from this, researchers are researching the use of certain helpful microorganisms in producing functional foods with extra health advantages in addition to basic nutrition. Among these are possible reductions in chronic disease risk, immune system strengthening, and gut health enhancement. More is being learned about how these microbes promote our health and how they might be employed to make food more nutritious and safer through ongoing study. Modern food science and public health campaigns depend on them, as they comprise a fundamental component (Silva and Ribeiro 2024).

3 Scope and Objectives of the Chapter

Present-day food science focuses its attention on green plant proteins and on the microorganisms working behind the scenes to turn raw ingredients into safe and delicious meals and on the finer points of making food safer and healthier for the person and the planet. Researchers record a big jump in demand for plant protein because many folks see it as a kinder, cleaner stand-in for meat; they praise its strong nutritional profile, lower land-and-water cost, and constant stream of new ways to grow and use it. At the same time, experts name weaknesses—spotty amino acid mixes and rough processing—that still ask for creative fixes. The article then turns to microorganisms, noting how bacteria and yeasts naturally guard food against spoilage, shape flavor, add vitamins, and even help train the immune system. A wide cast of microbes, it argues, keeps our bodies in balance and out of the hospital, so forward-minded food makers now look to this diversity when crafting next-generation functional foods. All these subjects show how progress in knowledge of plant proteins and microorganisms is crucial for creating healthier, more sustainable food systems. The objectives of the text are to inform about current developments, address challenges in these fields, and encourage further research and innovation to support a growing global population with healthier, more environmentally friendly food options.

4 Overview of Plant-Derived Proteins

4.1 Major Sources of Plant-Derived Proteins

Plant proteins are very popular today, as they are healthier and more environmentally friendly than animal proteins. As they are considered fit and sustainably positive, plant proteins are of particular interest. Compared to the same animal protein, they have fiber, vitamins, and minerals, and a lower cholesterol and fat content. Wheat, rice, oats, barley, and corn are also considered to be excellent protein and energy sources. Pseudo-cereals use quinoa, amaranth, and buckwheat, which are not actually grains but are highest in protein and all the necessary amino acids. Many proteinaceous food origins have beneficial fats, like almonds, peanuts, cashews, and seeds such as hemp, sunflower, flax, chia, pumpkin, and sesame. To maintain a healthy body and support the organization of food, balanced consumption of these aids is necessary. Plant proteins exist in a very large number of plants that offer the body the essential nutrients needed to develop, heal, and remain strong. It is essential for anyone interested in maintaining a healthy and balanced diet that does not overly depend on animals to know the primary sources of plant proteins (Thakur et al. 2024). Legumes are one of the richest sources of plant proteins. Legumes are comprised of foods such as beans, lentils, chickpeas, and peas. Legumes are not only full of protein and fiber but also possess essential vitamins and minerals. Soybeans are particularly well known since they contain all nine essential amino acids; hence, they are a complete protein (Willett et al. 2019). This does not happen much in plant foods, so soy is particularly valuable in vegetarian diets. Legumes are also highly versatile and can be incorporated in numerous kinds of meals, ranging from soups and salads to meat alternatives. Nuts and seeds are also rich in plant protein. Some examples are almonds, walnuts, chia seeds, flaxseeds, and sunflower seeds. They are rich in protein as well as in healthy fats, particularly omega-3 fatty acids that contribute to the well-being of the heart. They are also rich in other nutrients such as magnesium, zinc, and vitamin E. A mixed intake of nuts and seeds may provide individuals with sufficient protein and other nutrients, along with a change in taste and texture to foods. Whole grains like oats, brown rice, barley, and quinoa are another significant source of plant protein. Although grains are primarily famous for supplying carbohydrates, they also supply a little but useful amount of protein. Quinoa is particularly useful because, like soybeans, it is a complete protein. Whole grains also supply fiber and B vitamins, which aid in energy and digestion. With a variety of whole grains included in one's diet, meals become balanced and nutritious. Vegetables are not at their best when it comes to protein, but they still have a vital role. Broccoli, sesame seed protein isolate, Brussels sprouts, and kale are all sources of protein, as well as fiber, vitamins, and antioxidants. These nutrients aid in numerous bodily functions and are beneficial against disease. Consumed in great quantity, vegetables do make an effective contribution toward meeting daily protein requirements. Pseudo-cereals, including amaranth and buckwheat, resemble grains but belong to a different family of plants. Both are gluten-free and are a good source of protein and very rich in lysine, an amino acid lacking in many other grains. As they do not have gluten, they can be consumed well by individuals who have a gluten allergy or sensitivity. They can be easily used in baked foods, porridge, etc. Soy foods such as tofu, tempeh, and soy milk are consumed extensively in vegetarian diets due to their high protein and nutrient content. Tofu and tempeh are processed from whole soybeans and, therefore, retain a greater part of their nutritional value. The foods are also highly concentrated with isoflavones, plant chemicals that have the potential to decrease the risk of heart disease and other disorders. They are simple to prepare and can be employed in both savory and dessert recipes. Plant-based protein isolates are also a useful source of protein. They are specially treated products derived from peas, rice, hemp, and other plant sources. They are employed in protein powders, energy bars, and vegetable meat substitutes. These isolates come in a concentrated protein form and are thus ideal for athletes, bodybuilders, or anyone who desires an instant and convenient boost of protein. A newer and more exciting plant-based protein source is algae and aquatic vegetation, such as sesame seed protein Spirulina and chlorella. These small creatures are high in protein, vitamins, and minerals. They are also environmentally friendly since they can be cultivated with less input than terrestrial crops. Algae are beginning to show up more in health supplements and might become an even larger component of our diets in the future (Ferrari et al. 2022).

4.2 Nutritional Composition and Functional Properties

More people now see plant proteins as healthy, earth-friendly stand-ins for meat and dairy. This shift is especially helpful for anyone new to plant eating, since many plant proteins behave in the kitchen much like their animal cousins. Plant proteins are mainly chains of amino acids, and those tiny building blocks fuel countless processes inside the body. Animal foods offer a ready-made, balanced batch of all the essential amino acids, yet many plants fall short in one or two of them. Bread and maize, for instance, usually deliver too little lysine, while beans and peas tend to skimp on methionine. Combining beans with grains—or yogurt with nuts—brings those missing pieces together so the body can use every essential amino acid. Soy stands out because its profile is almost identical to that of meat; its globulin-rich beans contain about 35–40% protein and offer almost every nutrient the body craves. Pea protein and yellow split peas, after a simple milling and water-removal step, are the source that holds about 23–31% protein, offers easy digestibility, and delivers an amino acid mix that most plants barely match, save for a slight shortfall in methionine and cysteine. Hemp protein, milled from hemp seeds, sits around 20–25% protein and supplies edestin plus albumin, two globulins so digestible they rival egg white, although the lysine level is still lower. Beyond amino acids, plant proteins usually bring along fiber, vitamins, minerals, and phytochemicals that former food chemists used to lump together under the vague title of "goodness." Those extras tie to tangible benefits like smoother digestion, lower cholesterol numbers, and a mild boost to the body's antioxidant defenses. Whole grains and legumes are prime sources of B vitamins and iron; nuts and seeds deliver heart-friendly fats and vitamin E. On top of broad nutrient value, plant proteins also display functional traits—such as gelation, emulsification, and water-binding—that food makers prize when reformulating baked goods, dairy alternatives, and meat analogues. These proteins can dissolve in liquid, soak up water and oil, whip up foam, hold gels, and keep their structure. Each ability shapes food's feel, shelf life, and even the taste that shows up on your tongue. Solubility shows how well a protein disappears in liquid, giving soups or sauces their silky finish. Soy protein isolate tops the chart; it vanishes easily, making it a favorite for drinks and plant-based meats. Pea protein is similar, though factories often tweak it with extra steps so it dissolves perfectly. Plant proteins also grab hold of water and oil, locking moisture in and stopping dry bites. That quality matters in bread, muffins, or meat analogues, where a plump, tender mouthful is the goal. Emulsification is the knack of blending water and oil until they stay friends instead of sliding apart. Think creamy dressings or dairy-free milk-they depend on that steadiness. Soy and pea proteins serve as tiny bridges, clinging to both liquids and keeping them together. Foaming is another characteristic that is beneficial in creating things such as whipped toppings or mousses. Eggs are commonly used to foam, but certain plant proteins, particularly legume-derived proteins, could create stable foams when applied correctly (Roy et al. 2020). Gelation is the process by which proteins create a jelly-like texture, and this is crucial in foods such as tofu and plant milk substitutes. Plant proteins form gels when they are heated or when they interact with other components. For instance, soy protein can produce strong gels that contribute to the firm texture of tofu. Plant proteins gel as well as they do, depending on such factors as the pH of the food, salt content, and other ingredients that are blended with them. Such functional properties demonstrate that plant-derived proteins are not only healthy but also extremely handy to use in preparing a broad range of foods that are visually appealing and have a good taste (Day et al. 2022).

4.3 Challenges: Digestibility, Allergenicity, and Anti-nutritional Factors

Digestibility is a big issue that plant proteins have been carrying with them since time immemorial—much is learned against them when compared to animal proteins, which arises from plant proteins having a more complex structure and one difficult to digest by humans. Further to this, plants have some anti-nutritional factors such as lectins, tannins, and phytates that restrict direct access to nutrients for absorption. These nutrient inhibitors then appreciably lessen the absorption process for some minerals such as iron, calcium, and zinc. For example, from among these inhibitors, phytates found in legumes and whole grains act by binding with minerals, thereby limiting their bioavailability. Another problem is allergenicity: some individuals are allergic to common plant protein sources such as soy, peanuts, or wheat. This can restrict their choices and make switching to safe plant-based diets more difficult. Sometimes plant proteins are incomplete, meaning that they lack all essential amino acids in balance, which forces an overload of combining certain plant sources to achieve a complete protein profile. Appropriate food processing methods, such as soaking, fermenting, and cooking, may help reduce anti-nutritional factors and improve digestibility and would make for excellent plant-based food (Opazo-Navarrete et al. 2025). Plant-derived proteins are becoming more and more related for their ecological sustainability and health benefits. They do, however, offer some dietary difficulties, especially in terms of their lower digestibility than animal proteins. Anti-nutritional elements such as trypsin inhibitors, phytates, and lectins explain most of this lower digestibility. These substances inside the digestive system disrupt protein digestion and nutrient absorption. Particularly, phytates and lectins can chelate vital minerals like iron, zinc, and calcium, therefore generating insoluble complexes that the body poorly absorbs. By inhibiting the activity of trypsin, a vital enzyme essential for protein digestion, trypsin inhibitors therefore worsen protein use. Such interactions greatly lower the bioavailability of important nutrients, therefore raising the risk of deficiencies, especially in groups that depend mostly on plant-based diets.

For example, higher phytate levels in whole grains and legumes are linked to more common cases of iron and zinc shortages among vegetarians. Furthermore, lectins have been proven to destroy the intestinal lining, possibly disrupting digestive capacity and nutrient absorption. Still, several processing techniques, such as cooking, fermenting, soaking, and sprouting, substantially lower the concentrations of these anti-nutritional components and enhance protein quality. Ensuring balanced nutrition through careful meal preparation and proper food processing is critical to reduce nutrient deficits and maximize health results as the consumption of plant-derived proteins keeps growing (Aschemann-Witzel et al. 2021). As the popularity of plant-based diets continues to rise, allergenicity in plant-based protein, that is, the potential of certain proteins found in plants to cause allergic reactions in sensitive people, becomes an important issue. Mild symptoms like itching and swelling to more extreme instances like anaphylaxis characterize these allergic responses. The molecular makeup of plant proteins, their stability during digestion, and their resemblance to other recognized allergens all affect their allergenic potential, which can cause cross-reactivity; that is, a person allergic to one form of protein might also react to another with a similar structure. Common plant-based allergies include soy, peanuts, tree nuts, wheat, and legumes like lentils and chickpeas. Various processing techniques, such as heat treatment, enzymatic hydrolysis, fermentation, and even genetic modification, are employed to modify or degrade allergenic protein structures to lower their allergenic qualities. These approaches can lower the identification of the proteins by the immune system, therefore lowering the risk of triggering an allergy. Still, these same processes could also affect the digestibility, nutritional value, and functional characteristics of the proteins, such as texture, taste, or protein quality. High heat, for instance, may harm other nutrients or lower the amount of necessary amino acids; enzymatic treatments could influence the protein's behavior in food products. Thus, producing hypoallergenic plant-derived proteins calls for deft balancing of lowering allergenicity while preserving adequate nutritional value and purpose.

Although these methods raise issues about safety, cost, and public acceptance, scientists are also investigating genetic modification and selective breeding to eliminate or change certain allergic elements of plant proteins. Particularly among those searching for a sustainable and healthy diet, as plant-based foods become more prevalent, it is critical to enhance testing methods for allergenicity, guarantee correct labeling, and keep research on safe and healthy protein sources fit for all consumers, including those with allergies (Günal-Köroğlu et al. 2025). Natural chemicals found in plant foods, especially in seeds, grains, and legumes, are anti-nutritional elements known as anti-nutritional factors. These factors can hinder the absorption of critical nutrients, including proteins, vitamins, and minerals, as well as make the digestion of food by the body more challenging. Phytates, tannins, and lectins are the three most often occurring factors. Although they protect plants from diseases and infections, they lower the nutritional value of food to humans and animals. Cooking, soaking, or fermenting are examples of food preparation that will usually lower their factor content and enhance the bioavailability of the related nutrients. To maximize the health advantage of plant diets, one must find and reduce factors (Langyan et al. 2022). These factors are endogenous chemicals found in plant foods that interfere with the digestion, absorption, and utilization of nutrients and, consequently, create a major hindrance to the nutritional value of plant proteins.

Important factors are protease inhibitors, trypsin, and pepsin inhibitors, which lower the digestibility of proteins by inhibiting the activities of digestive enzymes involved in degrading proteins to absorbable amino acids. These inhibitors are often present in legumes such as soybeans and may cause decreased protein usage if not well processed. Phytates, or phytic acid, are another prominent factor present in seeds, grains, and legumes. They possess a strong chelating power for vital minerals like calcium, iron, and zinc, producing insoluble complexes that the human body cannot absorb, possibly causing mineral deficiencies, particularly among people who have diets that are very much plant-based. Lectins, found in grains and legumes, can bind to the gut lining and disrupt nutrient uptake, occasionally damaging intestinal cells, which can compromise digestion and enhance intestinal permeability. Moreover, tannins—polyphenolic structures in some fruits, cereals, and legumes—are capable of binding strongly to proteins and minerals, lowering their bioavailability and the proteins' digestibility. Oxalates present in sesame seed protein isolatenach, nuts, and beets are capable of binding with calcium to produce insoluble calcium oxalate, which decreases absorption of calcium and may contribute to the development of kidney stones in susceptible individuals. Saponins present in legumes like chickpeas and soybeans can disrupt cell membranes in the gastrointestinal tract, producing gastrointestinal distress and decreasing nutrient absorption. While most of these factors benefit plants—by fending off pests and microorganisms—they also can cause problems for human health if taken excessively.

Thankfully, several processes such as soaking, sprouting, fermentation, and heating can lessen the quantity of these anti-nutrients greatly, increasing not only digestibility but also the nutrient content in foods from plant sources. It is necessary to understand and manage ANFs to maximize the health benefits of plant-based foods and to ensure that people are getting enough nutrients, particularly as plant-based consumption keeps growing in popularity globally (Ishwarya Shankaran and Kumari 2024). Some indigenous healthy foods are soy, lentils, and chickpeas, and all three of these have little micronutrient naturally occurring agents, which are referred to as anti-nutrients. Anti-nutrients are naturally occurring substances found in food that are harmless but tend to interfere with the body's ability to assist in the absorption of some nutrients in the foods we eat (i.e., vitamins and minerals). These foods are still very good for humans, and one can get loads of nutrients from the foods, especially if other foods are eaten in a variety. Phytoestrogens act like estrogenic compounds produced by plants that could hinder hormonal balance if consumed in massive amounts over time. Oligosaccharides in legumes cause gastric discomfort such as gaseousness and bloating because they are not basically digested in the human gastrointestinal tract. These compounds may reduce the overall safety and nutritional quality of plant-derived proteins (Milana et al. 2025).

5 Microbial Interaction with Plant Proteins

5.1 Microbial Metabolism of Plant Proteins

Microbial metabolism of plant-derived proteins includes lactic acid bacteria, which degrade plant proteins into smaller, easier-to-digest components in a natural process known as microbial metabolism of plant-derived proteins. These include forms of Lactococcus, Lactobacillus, Streptococcus, and Leuconostoc. These beneficial bacteria employ enzymes, special tools, to carry out their tasks. Two key enzymes, PepN and PepXP, break down big plant proteins into smaller fragments known as peptides and amino acids. These little pieces can enhance the flavor of the meal as well as make it easier for our bodies to take in. Once amino acids are generated, other enzymes such as BCAT, ArAT, and CBL assist in converting them into beneficial taste components, including aldehydes, acids, and alcohols. These chemicals provide food with their wonderful flavor and aroma. For instance, they may produce nutty, fruity, or cheesy tastes, therefore enhancing the pleasure of plant-based meals. The kind of plant protein employed, however, will determine how well this procedure works. For example, whereas chickpea protein is more difficult for lactic acid bacteria to digest, soy protein is usually simpler. Soy has a structure that lactic acid bacteria enzymes may easily operate on, whereas chickpea protein may include components that inhibit this process. A tiny amount of glucose, a kind of sugar, can be included to enable lactic acid bacteria to work more effectively. Glucose helps lactic acid bacteria grow faster and produce more enzymes, therefore allowing them to break down more protein. This improves food taste as well as simplifies digestion. For food scientists, knowledge of how lactic acid bacteria digest various plant proteins is really priceless. It lets them pick the finest elements and bacteria needed to produce high-quality, delicious, plant-based cuisine. This is especially crucial now as more individuals seek solutions to animal products like milk, cheese, and meat. Overall, microbial metabolism by lactic acid bacteria is a simple and natural way to make plant proteins more useful and enjoyable in our diets by using lactic acid bacteria in the proper way to improve the taste, nutrition, and texture of plant-based foods without the need for artificial additives (Engels et al. 2022).

5.2 Fermentation: A Natural Partnership

People still rely on fermentation, and it still sits at the heart of modern kitchens. By turning sugars into acids or alcohol, microbes crowd out spoilage germs that would ruin food. The process also breaks down tough compounds, making meals easier to digest, raising good probiotics, and lowering some anti-nutrients. That is why everyday staples like yogurt, kimchi, sourdough bread, and tempeh are everywhere. Fermentation, old and still brilliant, keeps food safe, tasty, and surprisingly shelf stable. Fermentation helps preserve food by creating an environment that inhibits harmful microbes. It also improves digestibility, boosts beneficial probiotics, and can reduce anti-nutrients. Common fermented foods include yogurt, kimchi, bread, etc. (Caffrey et al. 2025). Recent studies show that fermentation does far more than keep food from spoiling; it also boosts our health and eases strain on the planet.

In addition to giving pickles, bread, and cheese their signature flavor and extra shelf life, the process raises the number of friendly probiotics in every bite, microbes that help keep digestion running smoothly. Because those same probiotics have been tied to stronger immune defenses, staples like yogurt and sauerkraut quickly earn their place on any health-minded shopping list. Beyond kitchen shelves, fresh breakthroughs in fermentation science are unlocking new, planet-friendly routes for food production. By mixing old starters with carefully selected modern strains and using low-tech bioreactors, scientists and small makers are crafting nutrient-rich foods that cut waste, conserve water, and shrink energy footprints. That promise makes fermentation an important ally as the world grapples with food shortages, fragile ecosystems, and rising climate risk. As research presses onward, the age-old craft looks set to deliver both healthier people and a more resilient food system (Zhang et al. 2025). Vegetable proteins are fermented in a process whereby beneficial microbes like lactic acid bacteria transform complex plant proteins into simpler forms, thereby increasing their nutritional value, digestibility, and palatability. In the production of vegetable milk replacements like cheeses and yogurts, where taste and texture are of great importance, this process is crucial. Lactic acid bacteria strains commonly found in ferments—*Lactobacillus*, *Streptococcus*, and *Leuconostoc*—make enzymes that gently break plant proteins into free amino acids and tiny peptides, boosting flavor and nutrition. Furthermore, fermentation can reduce anti-nutritional elements and allergens present in plant proteins, therefore making those proteins more fit for consumption. With beneficial consequences for producing better fermentation performance in plant-derived foods, using lactic acid bacteria strains found in fermented vegetable foods such as olives, kimchi, and tempeh could possibly offer more proteolytic action against plant protein than traditional strains derived from milk. The maximum efficiency of processes and attainment of high-grade plant-based food goods depend on optimal use of these microbes and plant protein interactions (Harper et al. 2022).

5.3 Influence of Microbial Enzymes on Protein Structure and Function

Microbial enzymes are small proteins that help shape larger proteins by speeding up chemical reactions, and they work as quick, flexible catalysts in cells and factories alike. Thanks to modern protein engineering, scientists can tweak these enzymes so they plug better into specific jobs, gaining extra kick, steadiness, and focus. Approaches such as rational design and directed evolution guide researchers as they swap amino acids, keeping in mind the close bond between structure and job. In rational design, a scientist makes well-planned edits based on existing knowledge, whereas directed evolution hurls mutations at a gene and then sorts through winners as if sitting on a lab bench version of nature. Both paths birth enzymes that thrive under heat, shrug off poison, or even take on a brand-new reaction.

Because of these breakthroughs, engineered microbial enzymes now power food plants, drug labs, biofuel start-ups, and wash-up crews for polluted water, proving that tiny tools can address tall problems in a greener, inventive way (Réty 2023). By changing the morphology and features of plant-derived proteins, microbial enzymes therefore improve their use in several food products. Microbes like bacteria and fungi create these enzymes, which catalyze certain processes modifying protein structure and resulting in enhanced solubility, digestibility, and functional qualities such as emulsification and foaming. Proteases, for example, enhance the bioavailability of proteins by lowering their possible allergenicity by breaking them down into smaller peptides and amino acids. Besides raising the nutritional value of plant proteins, this enzymatic hydrolysis helps to improve their sensory qualities, therefore increasing their acceptability. Furthermore, enzymes such as transglutaminase help protein molecules crosslink, which leads to more robust gel networks useful in the preparation of dairy replacements and meat analogues. Another enzymatic process, deamination, changes amino acid side chains, raising the net charge of proteins and so improving their functional qualities and solubility. Moreover, the use of enzyme cocktails mixing several kinds of enzymes has been demonstrated to synergistically enhance the functional qualities and extraction of plant proteins from several sources, including soy, pea, and chickpea. Highly specific, these enzymatic changes rely on elements including the kind of enzyme, substrate, and processing circumstances. By customizing these factors, one can create plant-based protein additives with preferred characteristics for culinary uses. The overall quality and versatility of plant-derived proteins in the food sector can be improved by the strategic use of microbial enzymes, which offer a sustainable and effective solution (Gouseti et al. 2023).

6 Fermentation of Plant-Based Proteins

6.1 Traditional Fermented Plant Food

Older fermented plant staples like tempeh and natto rely on solid-state fermentation, where fungi such as Rhizopus or helpful bacteria like *Bacillus subtilis* take center stage. People have turned to this technique for centuries, knowing it boosts both nutrition and function in plant proteins. As the microbes work, they unravel tough protein chains and neutralize anti-nutrients like phytic acid, tannins, and enzyme blockers that normally block nutrient uptake. Fermentation also adjusts the amino acid composition, raising the levels of vital building blocks—often short in legume-like lysine and methionine. The net result is an upgrade in protein quality across the plate. Moreover, the fermentation process induces biochemical and structural modifications that generate distinct textures and complex flavor profiles, transforming plain legumes into highly palatable products. With the increasing global shift toward plant-based diets, traditional fermentation methods provide an efficient and sustainable approach to enhance the nutritional quality, bioavailability, and sensory attributes of plant-derived proteins (Canoy et al. 2024).

Fermented traditional plant foods include a wide range of products prepared by lactic acid fermentation, a naturally occurring food preservation technique that improves the nutrient content, safety, and storage stability of plant ingredients. These foods form an essential part of the global diet pattern and are prepared with diverse substrates, including cereals, legumes, vegetables, and fruits. Consider soybeans as an example, where fermentation is used for the production of natto, miso, and tempeh in Asia, or maize and teff through the production of ogi, kenkey, and injera in Africa. During these fermentative processes, they increase digestibility through carbohydrate metabolism and reduction in anti-nutritional factors, as well as increase bioavailability of minerals, and enrich the food in beneficial microflora, in particular that of lactic acid bacteria; they also improve the sensory attributes of the fermented plant foods and are involved in health actions such as promoting gut health and immunity (Owusu-Kwarteng et al. 2022).

The conventional traditions mostly rely on spontaneous fermentation or back-slopping—i.e., a sample of a fermented batch that was successful in the past is used as an inoculum to start a fresh batch, thus maintaining the stability and resistance of the microbial community. This technique also preserves the characteristic and unique taste and texture of these products. The fermentation itself can also result in the production of bioactive metabolites, such as vitamins, peptides, and antioxidants, which are additional contributors to the health-beneficial effects of these foods. With an increased interest in plant-based diets, there are exciting prospects of learning and adapting traditional fermentation methods to create sustainable, nutrient-dense, and culturally relevant foods to meet the demands of contemporary diets (Horlacher et al. 2023). According to recent studies, while fermentation works significantly in enhancing the nutritional value of plant foods by increasing the digestive capability of proteins, it also renders the essential nutrients such as iron, zinc, and B vitamins more bioavailable. The microorganisms can decompose complex proteins and curb anti-nutritional factors such as phytates and tannins during fermentation, thus preventing nutrient absorption. The transformation tends to make the fermented plant food of nutritional value comparable to that of animal protein. Hence, foods such as tempeh, natto, and fermented legumes are good and sustainable substitutes in plant-based diets. Their better digestibility, improved nutritional status, and environmentally sustainable mode of preparation make these foods worthy of inclusion in healthy and environmentally responsive dietary choices (Cakmak et al. 2025).

6.2 Modern Fermentation Techniques and Microbial Strains

Currently, fermentation processes for plant proteins are using more advanced technologies and microbial strains to better the nutritional and sensory characteristics of plant foods through solid-state fermentation and submerged fermentation utilizing microorganisms, including, for example, lactic acid bacteria species, fungi, and yeasts, which fractionate and hydrolyze plant proteins to improve digestibility and bioavailability. For instance, strains of lactic acid bacteria such as *Lactobacillus plantarum* and *Lactobacillus acidophilus* are popular in food ferments because they churn out useful bioactive peptides, curb harmful germs, extend shelf life, and slow down rancid fat oxidation. Fungal species such as Rhizopus oligosporus and Aspergillus oryzae, commonly employed in the production of soybean tempeh, are utilized for their pronounced proteolytic activity, which contributes functionally to the development of desirable texture and flavor characteristics in fermented food matrices.

Moreover, innovative technologies such as ultrasound, microwave, pulsed electric field, ohmic heating, cold plasma, and high hydrostatic pressure are now being implemented into fermentation technologies to further improve protein digestibility, improve microbial activity, and improve physicochemical properties of plant foods. In addition to resolving limitations of historical fermentation processes, these new technologies facilitate the production of plant-based analogues of food that have positive sensory properties and nutritional content in response to consumer desire for a more environmentally sustainable product, while also being beneficial to consumer health (Farid et al. 2024).

6.3 Impact on Amino Acid Profile, Digestibility, and Allergenicity

Plant-derived proteins generally exhibit an imbalanced amino acid profile compared to animal-based proteins, often being deficient in essential amino acids such as lysine and methionine; consequently, the biological value of most plant proteins ranges approximately between 0.4 and 0.9. Furthermore, the presence of antinutritional factors such as trypsin inhibitors, tannins, and phytates interferes with proteolytic enzyme activity, thereby reducing protein digestibility and limiting amino acid absorption into the bloodstream. Enzymes struggle even more because the tough cell walls of plants trap nutrients inside, blocking easy access. Fortunately, simple kitchen or farm steps—fermenting, soaking, or thorough cooking—open the cell walls and neutralize anti-nutrients, boosting both digestibility and overall protein quality (Anyiam et al. 2025). In vitro protein digestibility, amino acid profile, and some functional properties address the nutritional value of sesame seed protein isolate, mainly looking at how easily the body can digest it, what kind of amino acids it consists of, and if it can cause an allergy. The study indicates that sesame seed protein isolate is very digestible when in vitro protein digestibility was measured at 89.57%, or about as easily broken down and utilized by the body as flaxseed protein isolate. This is significantly greater than the digestibility of proteins in raw soybean flour and conophor protein concentrate. One big reason it is so easy to digest is that during processing, things like polysaccharides (which are not proteins) get removed. This process facilitates the action of digestive enzymes, allowing them to more effectively break down proteins. When considering the amino acid content, glutamic acid and aspartic acid dominate in the highest concentrations, important for a variety of body functions such as metabolism and brain function. The study, however, also determines that sulfur-containing amino acids in the form of methionine and cysteine are present in lower concentrations, and therefore these are the "limiting" amino acids in sesame seed protein isolate. What this means is that the sesame seed protein isolate, being a solid protein source overall, may not be as complete as animal proteins or some other plant proteins when it comes to these essential amino acids. The study observes that sesame seed protein isolate is a useful plant-derived protein source with acceptable digestibility and useful functional properties but pays some attention to its lesser amounts of some amino acids if it is to be considered a primary source of protein in meals. This makes sesame seed protein isolate an acceptable choice for supplementing protein in vegetarian or vegan meals, particularly when blended with other protein foods to compensate for the lacking amino acids (Fasuan et al. 2018).

6.4 Sensory and Functional Enhancements Through Fermentation

Fermentation is increasingly recognized as a key process in making plant-based foods, offering the advantage of improving both their taste and functional qualities. By subjecting plant proteins to different fermentation processes—e.g., alcoholic, lactic acid, acetic acid, propionic acid, and butyric acid fermentation—plant proteins can be greatly enhanced in digestibility, nutritional quality, and palatability, thereby becoming more desirable and acceptable for human consumption. These processes not only reduce the ecological footprint of food production by conserving land-and-water resources but also meet ethical standards through the offer of renewable substitutes for animal products.

The addition of probiotics through fermentation enhances gut health and makes a general contribution to well-being. Further, fermentation fuels diversity in plant-based foods, making them more acceptable and appealing to consumers. Issues, including the formation of unwanted by-products, risks to health, and enhanced costs linked with fermentation, must be overcome. Case studies among innovative plant-based food firms show the technical uses and commercial viability of fermentation technologies. Future technological innovation and research are critical to upscaling fermentation processes as well as further integrating them into sustainable agricultural systems, ultimately improving the sustainability of food and health impacts (Abbaspour 2024).

7 Genetically Engineered Microbes and CRISPR Technology for Functional Protein Enhancement

The history of using microorganisms to produce food is ancient and ironic, dating back thousands of years to the time when ancient societies first learned about fermentation processes. These microscopic creatures have been used by cultures all over the world to produce necessary foods and drinks, including cheese, vinegar, yogurt, and beer (Ganugi et al. 2021; Gholami-Shabani et al. 2023; Praveen and Brogi 2025). In the past, these fermentation processes were conducted without a thorough comprehension of the underlying biological systems.

The nineteenth century witnessed a more systematic approach to employing microbes in food production after researchers like Louis Pasteur discovered that they were the agents causing fermentation. The genetic alteration of microorganisms has become a revolutionary tactic in the food sector due to the development of contemporary biotechnology and advances in molecular biology (Ahmed 2024; Lazim and Nazir 2025; Agrawal 2024).

Genetic engineering is the technique that alters the genetic makeup of microorganisms, such as fungi and bacteria, to improve their capacity to produce particular food ingredients or to accelerate current manufacturing procedures. There are stringent regulations surrounding the utilization of genetically modified microorganisms (GMMs). But researchers and companies have been investigating genetically modified *Pseudomonas syringae* (GMMs) for almost 30 years without any significant mishaps or environmental issues since the first field test for assessing GMMs was conducted on University of California experimental plots in 1987 (Dourado et al. 2017; Chemla et al. 2025). These initiatives could soon suggest that microbiome engineering is a reliable approach to boosting agricultural sustainability. In context with this, we examine current advances in strain engineering and synthetic biology to render microbiome engineering more suitable for agricultural uses and talk about strategies to lessen environmental effects. In a single cell or a whole organism, the CRISPR-Cas system enables quick, site-specific genome modification (Anton and Bultmann 2017; Manghwar et al. 2019). It executes genome-wide screens, modulates epigenomes, regulates transcription, and scans chromosomes. Gene editing based on CRISPR/Cas9 enables the simplest, most versatile, and most precise method for altering the genome of plants. It relies on two essential molecules: a guide RNA (gRNA) and a Cas9 endonuclease. The two short RNA molecules that make up the gRNA are transactivating crRNA (tracrRNA), which serves as a binding scaffold for the Cas9 endonuclease, and CRISPR RNA (crRNA), which is a 20-nucleotide sequence complementary to the target DNA. A certain protospacer-adjacent motif (PAM) must be present right next to the target site for Cas9 to recognize it. 5′-NGG-3′ is the canonical PAM linked to the most popular Cas9 from *Streptococcus pyogenes* (SpCas9).

Because of its many uses in genome editing, the CRISPR-Cas technology is now being employed more and more to produce modified crops. Research teams have created a number of agricultural systems with CRISPR-Cas that are resistant to disease, drought, salt, and heat (Santosh Kumar et al. 2020; Angon et al. 2023; Erdoğan et al. 2023). CRISPR-Cas is helping to improve crop quality metrics like appearance, accessibility, nutritional ingredients, and other desirable characteristics, as well as develop climate-resistant crops. One beneficial aspect includes the enhancement of carotenoids, which are thoroughly distributed and essential for photosynthetic organisms (Sun et al. 2020; Sakure 2022). However, humans require carotenoids in their diet, particularly as precursors to vitamin A and β-carotene, even though they are unable to produce them themselves. Vitamin A deficiency (VAD) is one of the most severe worldwide health issues, resulting in a variety of symptoms such as xerophthalmia, night blindness, pediatric blindness, and an increased risk of morbidity and mortality, particularly in young children.

In order to combat VAD, carotenoid-enriched staple crops, often referred to as "golden crops," were introduced through genetic engineering or traditional breeding methods (Shetty et al. 2018; Mueller and Flachs 2022). Rapid, DNA/transgene-free, targeted multiplex genetic modification of organisms is now a reality to produce "golden" staple crops because of the CRISPR-based genome-editing tool, which appears to be the most effective and popular genetic engineering technique. To combat VAD, several genome-edited golden crops, or carotenoid biofortified crops, have been produced via CRISPR-Cas-based methods. For example, the 5.2-kb carotenogenesis cassette knock-in, which contains the genes for maize PSY and CrtI, has produced the Golden Rice cultivar Kitaake. The endosperm of this cultivar has 7.9 μg/g dry weight (DW) β-carotene (Zheng et al. 2021).

Another vital component of the human diet, vitamin E (tocopherol), is a potent lipid-soluble antioxidant, and its deficiency is associated with a number of human disorders, including cardiovascular disease and some types of cancer. *Hordeum vulgare* homogentisate phytyltransferase (HvHPT) and *Hordeum vulgare* homogentisate geranylgeranyltransferase (HvHGGT) were specifically overexpressed using CRISPR-Cas9 technology, which led to an apparent rise in tocopherol and tocotrienol content (Zeng et al. 2020).

Another important component in plants is iron; metabolic processes, including the electron transport chain and oxygen transport, rely on iron. The most common diseases in humans are caused by issues associated with iron metabolism, which include a variety of conditions with different clinical manifestations, from neurological diseases to anemia. Biofortification of crops and diets may fulfill human iron needs (Lal 2020; Roemhild et al. 2021). The latter form of biofortification is caused by CRISPR-Cas9. According to reports, this pathway interferes with the Inositol pentakisphosphate 2-kinase 1 (IPK1) gene, which results in wheat iron biofortification (Ibrahim et al. 2022).

Through precise genome alterations, CRISPR/Cas9-mediated gene editing has been utilized to minimize allergens in peanut types and increase the percentage of oleic acid in oil (Neelakandan et al. 2022; Han et al. 2023). Nevertheless, despite these developments, issues with off-target effects, unforeseen repercussions, and the broad use of genome editing in peanuts still exist. By addressing its several potential features, such as phytate reduction, fresh seed dormancy, aflatoxin resistance, and abiotic stress tolerance, peanuts could benefit greatly from advancements in CRISPR-based genome editing.

The soybean (G. max (L.) Merr.), a crucial crop for oil and protein, is one of the first crop species that the CRISPR/Cas9 technology targets for genetic modification. The construction of the first soybean CRISPR collection that targets over 100 candidate genes is significant (Chilcoat et al. 2017; Bao et al. 2019). Using improved methods, a number of mutant soybean lines were also produced. The community's efforts to gather mutants of every soybean gene will be crucial for soybean genetics and biotechnology. Chickpea has mostly been modified to generate transgenic plants that express a chimeric gene that codes for *Bacillus thuringiensis'* insecticidal crystal protein Cry1Aabc, which confers resistance against pod borer (*Heliothis armigera*) (Das et al. 2017). The development of gene-edited chickpeas to improve production, seed size, and quality is being spearheaded by the International Crops Research Institute for the Semi-Arid Tropics (ICRISAT) (Bhatnagar-Mathur and Sharma 2018). In the first instance of CRISPR/Cas9-mediated editing of chickpea protoplasts, researchers reported creating knockouts of the genes for Reveille 7 (RVE7) and 4-coumarate ligase (4CL), both of which are linked to chickpea drought tolerance (Badhan et al. 2021).

A novel approach for introducing herbicide resistance in chickpeas and other legumes by gene editing has been made possible by the recent finding that the Ala251Thr mutation in the psbA chloroplast gene is what causes resistance to the photosystem II inhibitor herbicide metribuzin. In another investigation, numerous candidate genes were identified as early knockout targets for increasing the carotenoid content of chickpeas through the discovery and expression analysis of candidate genes involved in carotenoid production (Rezaei et al. 2016).

Due to their nonsynchronous maturity, lack of better varieties, and vulnerability to biotic and abiotic stressors, mung beans have a very poor production (~0.5 t ha^{-1}). Molecular breeding techniques can offer quick, effective, and economical methods for crop improvement; however, conventional and mutant breeding methods were previously used with success. By targeting a symbiosis receptor-like kinase gene, CRISPR/Cas9-mediated gene editing was recently exploited to impair symbiotic nitrogen fixation in cowpea (*Vigna unguiculata*) (Ji et al. 2019). The plant system's effectiveness with CRISPR/Cas9 implies that gene editing can be used on other species, such as mung beans. Among the first targets for mung bean gene editing would be disease resistance and quality attributes. Creating mung bean cultivars that can adapt to shifting weather conditions would aid in the global growth of mung bean production (Bhowmik et al. 2021; Rasheed et al. 2022).

For the faba bean (*Vicia faba*) crop, no CRISPR/Cas9 research has been published so far. Tannins in the seed coat and vicine and convicine (V–C) in the cotyledons are two important anti-nutritional components of faba beans. Finding the potential genes involved in the V–C biosynthesis pathway has been the subject of numerous attempts. According to a recent study, faba bean flowers with no tannin and white color are caused by mutations in Transparent Testa Glabra 1 (TVfTG1), which codes for a WD40 transcription factor, and Transparent Testa 8 (VfTT8, Vf_Mt1g072320), which codes for a bHLH domain transcription factor (Gutierrez and Torres 2019). Additionally, by boosting the amounts of the legumin subunit in comparison to vicilin, CRISPR/Cas9 offers novel ways to improve the sulfur-containing amino acids methionine [Met] and cysteine [Cys] (Singh et al. 2024).

CRISPR-Cas has also shown potential in pea plants, where genes responsible for "off-flavors" are prime targets for CRISPR/Cas9 gene editing to enhance flavor quality. But distinct "off-flavors," which are frequently characterized as green, beany, hay-like, metallic, and astringent, are the primary reason why pea products are still not commonly used despite the high nutritional content of pea seeds. A combination of volatile organic compounds (VOCs), nonvolatile saponins, alcohols, ketones, aldehydes, methoxypyrazines, and others is responsible for these unwanted flavors. The genes causing the "off-flavors" in pea products are clear targets for CRISPR/Cas9 gene editing in order to enhance their flavor quality. VOCs are tiny organic compounds that are mostly created by the lipoxygenase (LOX) pathway during the production, transportation, and storage of peas. Polyunsaturated fatty acids are converted by LOX enzymes into hydroperoxides (9-HPO or 13-HPO), which are further altered by aldehyde dehydrogenases and hydroperoxide lyases to produce a variety of volatile organic compounds (VOCs) (Bhowmik et al. 2021, 2023). CRISPR/Cas9-targeted LOX mutations may be able to help peas lose unwanted off-flavors associated with volatile organic compounds.

The evolution of new plant varieties relies heavily on genetic diversity, and in order to improve lentils genetically, new alleles that go beyond the current pool of suited germplasm must be introduced. New sources of genes and alleles must be found and created in lentil genetic resources in order to achieve additional advancements in increasing yield and meal quality, as well as cultivar stability. When used in combination with conventional breeding methods, targeted genome editing utilizing CRISPR/Cas9 genome-editing technology can help overcome some of the limitations for boosting lentil yield and meal quality (Rasheed et al. 2022; Roy et al. 2023). Agrobacterium-mediated genetic transformation is one of several genetic transformation techniques that have been tried. The functions of genes related to tendril production (tendril-less, tl) and seed size (pea subtilase, SBT1.1), as well as other traits like pod breaking and herbicide tolerance, have been better understood by means of random mutagenesis approaches. To enhance lentils, gene editing may offer a simpler, cost-effective, and more accurate method of gene replacement.

There is a need for novel technologies that can increase plants' stress responses and prepare them for the environment. Plant breeding has advanced significantly since the advent of genome-editing techniques. Sequence-definite nucleases are used by genome-editing technologies to introduce particular known changes in the genome. The versatility and efficiency of the CRISPR-Cas genome-editing method have been widely recognized. CRISPR-Cas is aiding in enhancing agricultural quality metrics like appearance, palatability, nutritional components, and other desired qualities, as well as creating climate-ready crops.

To effectively alter genes for improved abiotic stress tolerance in a variety of leguminous crops, including beans, peas, and lentils, CRISPR-Cas9 and/or CRISPR-Cas12a were employed (Razzaq et al. 2022; Sandhu et al. 2025). Unexpected biotic and abiotic challenges brought on by recent climate change have a detrimental impact on crop production in a number of ways, including yield and quality, endangering the sustainability of agriculture. Thus, new cultivars that can withstand both biotic and abiotic stress are crucial candidates for GE technology breeding.

Cellular mechanisms regulating plants' response to drought have been interpreted by inclusive molecular searches. By regulating stomatal closing and gene expression to limit water loss through transpiration, abscisic acid (ABA) plays a critical role in affecting how plants respond to drought. ABA signaling depends on the transcription factor known as ABA-responsive element (ABRE)-binding proteins, which is a basic leucine zipper (bZIP). While AREB1 loss-of-function increased drought susceptibility, AREB1 overexpression (ABF2) enhanced drought tolerance in soybean, rice, and Arabidopsis (Hossain et al. 2010; li et al. 2024). Furthermore, during drought, AREB1 regulates a large number of genes downstream of the ABA signaling pathway and has a role in ABA production, osmotic stress response, and ABA-mediated antioxidant signaling.

Using a CRISPR-Cas9 multiplexing technique, a tomato gene encoding hybrid proline-rich protein 1 (HyPRP1), a negative regulator of salt stress responses, was edited (Kapoor et al. 2019). This led to the precise removal of its functional domains, the proline-rich domain (PRD) and the eight cysteine motif (8CM). We then showed that the greatest degree of salt tolerance in tomatoes was conferred by removing the PRD domain of HyPRP1. Employing genome-editing technologies, the role of OsmiR535 in salt stress tolerance was investigated, and it was proposed that rice's ability to withstand salinity may be enhanced by OsmiR535 deletion by CRISPR/Cas9.

Rice that has had its genes producing the abscisic acid receptor (PYL1/4/6) knocked out using CRISPR/Cas9 has demonstrated a significant tolerance to high temperatures (Miao et al. 2018). Setting goals to raise heat stress resistance and creating heat-tolerant cultivars are essential to addressing problems like climate change and global warming. Given that PHYB has been found to be a thermosensor, phytochrome (PHY) may be a crucial target in this regard. Following the CRISPR/Cas9 gene deletion, it became apparent that a mutation in the tomato SlAGAMOUSLIKE 6 (SlAGL6) gene, which codes for MADS-box, was the cause of the seedless phenotype (Klap et al. 2017). Thus, SlAGL6 mutations made tomato plants more vulnerable to heat stress.

Employed a multiplex CRISPR-Cas9 technique to precisely alter quinoa's transporter genes and essential nutrient production in order to greatly improve its nutritional profile for athlete recovery diets. Quinoa, a nutrient upgrade made possible by CRISPR-Cas9, is a promising option for performance nutrition and postexercise recovery because it can restore essential amino acids like lysine, antioxidant vitamins C and E, and zinc—all of which are crucial for immune system function and muscle repair. In order to make quinoa an even better whole-food choice for athlete nutrition, we predicted that stacking these modifications would result in quinoa seeds with noticeably greater levels of lysine, zinc, vitamin C, and vitamin E (Marques et al. 2020; Anuradha et al. 2023). Furthermore, we expected that the targeted metabolic reprogramming would be confirmed by transcriptome profiling of edited vs. wild-type seeds, which would show coordinated overexpression of nutrient biosynthesis pathways in the edited plants.

8 Microbial Synthesis of Proteins and Amino Acids

Since many plant proteins lack one or more essential amino acids, establishing a complete amino acid profile is a major nutritional problem in the creation of plant-based food items. Microbial biosynthesis provides a long-term way to make up for these inadequacies. In order to create critical amino acids like lysine, methionine, and tryptophan—which are frequently lacking in plant-derived proteins like rice, corn, or legumes—engineered strains of *Corynebacterium glutamicum*, *Escherichia coli*, and *Bacillus subtilis* are used (Farid et al. 2024; Dhiman et al. 2025). To improve the protein quality and bioefficacy of plant-based meat replacements, dairy substitutes, and nutritional supplements, these microbially produced amino acids are then added. This strategy encourages the creation of plant-based foods that are ethically sourced, full of nutrients, and sustainable in terms of the environment.

9 Production of Proteolytic Enzymes to Improve Bioavailability

Because of their complex structure and anti-nutritional components, plant proteins frequently have worse digestibility than animal proteins. Plant proteins are hydrolyzed into smaller peptides and free amino acids by microbial proteolytic enzymes, such as those generated by *Aspergillus oryzae*, *Bacillus subtilis*, and Rhizopus species (López-Fernández et al. 2020; Morilla et al. 2023). The bioavailability and absorption of vital nutrients in plant-based products are enhanced by this enzymatic treatment. Proteases can improve plant proteins' solubility, texture, and emulsification in addition to their nutritional content, which makes them more appropriate for creating premium meat and dairy substitutes. Therefore, using microbial proteases is essential to maximizing the nutritional value and palatable qualities of plant-based food innovations.

10 Co-fermentation Strategies

To improve the nutritional, functional, and sensory qualities of plant-based meals, co-fermentation—the use of many microbial species in a single fermentation process—is being used more and more. For example, plant-based yogurts and cheeses can have their flavor, texture, and shelf life greatly enhanced by co-fermentation with lactic acid bacteria and yeasts (Erem and Kilic-Akyilmaz 2024; Yang et al. 2024). Co-fermentation improves the availability of vitamins, amino acids, and bioactive substances while lowering anti-nutritional factors in plant substrates high in protein, such as soy or peas. Furthermore, mixing fermentative and proteolytic microorganisms can improve the digestibility of proteins and yield beneficial postbiotic chemicals. One flexible biotechnological technique for converting plant-based ingredients into more nutrient-dense and palatable products is co-fermentation.

11 Methods for Assessing Plant-Based Protein Quality (PDCAAS and DIAAS)

Protein quality is typically evaluated based on amino acid composition and digestibility. Two widely recognized methods are:

By comparing a test protein's amino acid profile to a reference pattern and adjusting it for overall digestibility, the FAO/WHO endorses the standard Protein Digestibility Corrected Amino Acid Score (PDCAAS) technique for assessing protein quality. A scale from 0 to 1.0 (or 0–100%) is used to assign scores, with numbers greater than 1 being truncated. Despite being commonly used, PDCAAS has drawbacks, such as using fecal digestibility rather than ileal digestibility and overestimating protein quality due to truncation. The FAO has approved the Digestible Indispensable Amino Acid Score (DIAAS), a more contemporary technique that uses ileal digestibility data of particular essential amino acids to overcome the constraints of PDCAAS. It offers a more precise assessment of the actual nutritional worth of proteins, particularly those derived from plants that can have limited amino acids or reduced digestibility. Because DIAAS values are not truncated, it is easier to distinguish between proteins of high and low quality (Wolfe et al. 2016; Mathai et al. 2017).

12 Proteomics and Metabolomics Tools

Tools like proteomics and metabolomics are essential for thoroughly characterizing plant-derived proteins. Proteins found in plant materials may be identified, measured, and structurally analyzed through the use of proteomics, which also provides information on the amino acid sequences, posttranslational modifications, and functional characteristics of these proteins. Methods like liquid chromatography-tandem mass spectrometry (LC-MS/MS), two-dimensional gel electrophoresis (2-DE), and

MALDI-TOF MS are frequently employed to profile proteins and bioactive peptides, especially those produced by enzymatic hydrolysis or processing (Mouzo et al. 2018). In contrast, the study of small molecule metabolites that are influenced by or involved in protein metabolism is an aspect of metabolomics. Metabolomics uses techniques such as nuclear magnetic resonance (NMR), gas chromatography-mass spectrometry (GC-MS), and liquid chromatography-mass spectrometry (LC-MS) to evaluate the bioavailability, metabolic pathways, and amino acid content of molecules produced from proteins. Proteomics and metabolomics work together to provide us a systems-level understanding of plant proteins, which helps us create plant-based diets that are both functionally and nutritionally optimal (Darie-Ion et al. 2022).

13 Plant-Based Protein for Industrial Applications

Microbial biotechnology is useful for enhancing the functional and nutritional properties of plant-derived proteins. One prominent example is the fermentation of soy protein with *Lactobacillus plantarum* or *Bacillus subtilis*, which has been demonstrated to improve digestibility, lower anti-nutritional factors like trypsin inhibitors, and raise the concentration of free essential amino acids (Park et al. 2012; Chen et al. 2023). Another example is the co-fermentation of pea protein with lactic acid bacteria and fungal strains, which improves its solubility, sensory qualities, and bioavailability (Ben-Harb et al. 2019; El Youssef et al. 2020). In recent years, there has been a significant increase in the demand for plant-derived proteins worldwide due to a confluence of ethical, environmental, and health concerns. The use of plant-derived protein products, including meat substitutes, dairy substitutes, and protein supplements made from soy, peas, rice, oats, and lentils, has been steadily increasing, according to market trends (Petrusán et al. 2016; Lakkakula et al. 2017; Prasad et al. 2025).

The global market for plant-derived proteins is expected to reach USD 20 billion by 2027, with a compound annual growth rate (CAGR) of more than 7%, according to industry sources (Boukid et al. 2023a, b; Talwar et al. 2024). A major factor driving this increase is the trend of consumers choosing clean-label and sustainable foods. The popularity of plant-based diets has also been pushed by the rising incidence of dietary restrictions and lifestyle-related illnesses, such as lactose intolerance, gluten sensitivity, and veganism.

A number of characteristics, including flavor, texture, nutritional profile, transparency of labeling, and perceived health advantages, affect consumer adoption. The difference between conventional animal-based goods and their substitutes has decreased as a result of improvements in the sensory and functional aspects of plant-based products brought about by developments in food technology and protein processing. Plant-based protein consumption has also been normalized and promoted in large part by promotional strategies including health claims, environmental sustainability communications, and celebrity endorsements (Negowetti 2020; Tziva et al. 2023). Nevertheless, there are still issues with price competition, customer doubts about protein sufficiency, and certain sensory restrictions. In summary, the market for plant-derived proteins is expanding quickly due to high customer demand and innovation-driven momentum. Enhancing product quality, price, and consumer education must continue to be the major priorities for long-term growth to assure broader acceptability across a range of demographic groups.

14 Challenges and Future Perspectives

The widespread embrace and long-term viability of plant-derived proteins are nevertheless hampered by several issues, despite their rising popularity. One of the main problems is that many plant proteins have an incomplete amino acid profile (Gorissen et al. 2018; Dimina et al. 2022). They frequently lack critical amino acids like methionine or lysine, which reduces their nutritional equivalence to animal proteins. Furthermore, even though techniques like fermentation and enzymatic treatment are being investigated to improve protein quality, digestibility and bioavailability continue to be major issues. Consumer acceptability and perception are additionally affected by sensory constraints, such as dissimilar colors, gritty textures, or bad tastes. Technological and scalability obstacles exist from the perspective of production, especially when it comes to the extraction and processing of high-quality proteins from various plant sources while preserving functionality. The availability and consistency of raw materials are further complicated by supply chain limits, environmental issues, and agricultural restrictions.

Protein engineering, precise fermentation, creative breeding, and combining different plant sources to produce nutritionally balanced profiles are the key to the future of plant-based protein (Colgrave et al. 2021; Boukid et al. 2023a, b; Abbaspour 2024). The development of next-generation plant proteins with improved functionality, flavor, and health advantages will heavily rely on emerging disciplines, including proteomics, metabolomics, and synthetic biology. Furthermore, increased funding for consumer education, legislative assistance, and clear labeling can promote acceptance and trust. Plant-based protein production might become more sustainable with the use of circular economic concepts, such as the use of agricultural by-products and waste valorization. Plant-derived proteins are positioned to play a key role in the worldwide shift toward healthier, more sustainable food systems with sustained innovation and strategic cooperation between academics, businesses, and policymakers.

15 Conclusion

Significant progress has been achieved in addressing the worldwide need for environmentally friendly and healthy protein substitutes by means of plant-based protein biotechnology. Plant proteins are becoming more and more entrenched as competitive alternatives to animal-based proteins because of developments in crop improvement, protein extraction, and functional augmentation (Ismail et al. 2020;

Akharume et al. 2021). But problems, including lacking amino acid profiles, worse digestibility, and sensory restrictions, still exist.

Promoting genetically modified crops with higher protein content, using precision fermentation to improve functionality, and combining omics technologies (genomics, proteomics, and metabolomics) for thorough protein characterization are the main areas of focus for this field moving forward. Furthermore, the development of blended protein systems—which combine many plant sources or incorporate microbial proteins—and advancements in green processing technology present encouraging opportunities to enhance the sustainability and quality of products (Lappi et al. 2022; Jareonsin et al. 2024). Plant-based protein biotechnology has the potential to be a key factor in creating nutrient-dense, environmentally friendly, and internationally accessible food systems in the future with sustained multidisciplinary research and industrial cooperation.

References

Abbaspour N (2024) Fermentation's pivotal role in shaping the future of plant-based foods: an integrative review of fermentation processes and their impact on sensory and health benefits. Appl Food Res 4:100468

Agrawal R (2024) Emerging Technologies for Improvement of microbial strains using Omic sciences and measuring methods for maximum product formation. In: Textbook of industrial microbiology, pp 13–39

Ahmed S (2024) Genetic modification of microorganisms: applications in industry and medicine. Front Biotechnol Genet 1(1):72–87

Aimutis WR (2022) Plant-derived proteins: the good, bad, and ugly. Annu Rev Food Sci Technol 13(1):1–17

Akharume FU, Aluko RE, Adedeji AA (2021) Modification of plant proteins for improved functionality: a review. Compr Rev Food Sci Food Saf 20(1):198–224

Angon PB, Mondal S, Akter S, Sakil MA, Jalil MA (2023) Roles of CRISPR to mitigate drought and salinity stresses on plants. Plant Stress 8:100169

Anton T, Bultmann S (2017) Site-specific recruitment of epigenetic factors with a modular CRISPR/Cas system. Nucleus 8(3):279–286

Anuradha, Kumari M, Zinta G, Chauhan R, Kumar A, Singh S, Singh S (2023) Genetic resources and breeding approaches for improvement of amaranth (Amaranthus spp.) and quinoa (Chenopodium quinoa). Front Nutr 10:1129723

Anyiam PN, Phongthai S, Sai-Ut S, Kingwascharapong P, Jung YH, Zhang W, Rawdkuen S (2025) Nutritional components and digestibility profiles of some potential plant-based protein sources. Foods 14(10):1769

Aschemann-Witzel J, Gantriis RF, Fraga P, Perez-Cueto FJ (2021) Plant-based food and protein trend from a business perspective: markets, consumers, and the challenges and opportunities in the future. Crit Rev Food Sci Nutr 61(18):3119–3128

Badhan S, Ball AS, Mantri N (2021) First report of CRISPR/Cas9 mediated DNA-free editing of 4CL and RVE7 genes in chickpea protoplasts. Int J Mol Sci 22(1):396

Bao A, Chen H, Chen L, Chen S, Hao Q, Guo W, Qiu D, Shan Z, Yang Z, Yuan S, Zhang C (2019) CRISPR/Cas9-mediated targeted mutagenesis of GmSPL9 genes alters plant architecture in soybean. BMC Plant Biol 19:1–12

Ben-Harb S, Saint-Eve A, Panouillé M, Souchon I, Bonnarme P, Dugat-Bony E, Irlinger F (2019) Design of microbial consortia for the fermentation of pea-protein-enriched emulsions. Int J Food Microbiol 293:124–136

Bernal P (2024) How are microbes helping end hunger? Microb Biotechnol 17(3):e14432
Bhatnagar-Mathur P, Sharma KK (2018) Genetic modification of grain legumes research Institute for the Semi-Arid Tropics (ICRISAT), India. In: Achieving sustainable cultivation of grain legumes volume 1. Burleigh Dodds Science Publishing, pp 107–152
Bhowmik P, Konkin D, Polowick P, Hodgins CL, Subedi M, Xiang D, Yu B, Patterson N, Rajagopalan N, Babic V, Ro DK (2021) CRISPR/Cas9 gene editing in legume crops: opportunities and challenges. Legume Science 3(3):e96
Bhowmik P, Yan W, Hodgins C, Polley B, Warkentin T, Nickerson M, Ro DK, Marsolais F, Domoney C, Shariati-Ievari S, Aliani M (2023) CRISPR/Cas9-mediated lipoxygenase gene-editing in yellow pea leads to major changes in fatty acid and flavor profiles. Front Plant Sci 14:1246905
Boukid F, Hassoun A, Zouari A, Tülbek MÇ, Mefleh M, Aït-Kaddour A, Castellari M (2023a) Fermentation for designing innovative plant-based meat and dairy alternatives. Foods 12(5):1005
Boukid F, Kumari S, Khan ZS (2023b) Plant protein-based foods, Trend from a Business Perspective: market, consumers' challenges, and opportunities in future. In: Novel plant protein processing, pp 267–282
Caffrey EB, Perelman D, Ward CP, Sonnenburg ED, Gardner CD, Sonnenburg JL (2025) Unpacking food fermentation: clinically relevant tools for fermented food identification and consumption, vol 16. Adv Nutr, p 100412
Cakmak H, Düven G, Subasi BG, Buyukkestelli HI, Bravo-Núñez Á, Ozyurt VH, Sogut E, Simsek S (2025) Plant-based fermented foods: general concepts, improvement of the quality, metabolites, and health-related functions. Food Biosci 69:106960
Canoy TS, Wiedenbein ES, Bredie WL, Meyer AS, Wösten HA, Nielsen DS (2024) Solid-state fermented plant foods as new protein sources. Ann Rev Food Sci Technol 15:189
Chemla Y, Sweeney CJ, Wozniak CA, Voigt CA (2025) Design and regulation of engineered bacteria for environmental release. Nat Microbiol 10(2):281–300
Chen Q, Liu B, Liu G, Shi H, Wang J (2023) Effect of Bacillus subtilis and lactobacillus plantarum on solid-state fermentation of soybean meal. J Sci Food Agric 103(12):6070–6079
Chilcoat D, Liu ZB, Sander J (2017) Use of CRISPR/Cas9 for crop improvement in maize and soybean. Prog Mol Biol Transl Sci 149:27–46
Colgrave ML, Dominik S, Tobin AB, Stockmann R, Simon C, Howitt CA, Belobrajdic DP, Paull C, Vanhercke T (2021) Perspectives on future protein production. J Agric Food Chem 69(50):15076–15083
Darie-Ion L, Whitham D, Jayathirtha M, Rai Y, Neagu AN, Darie CC, Petre BA (2022) Applications of MALDI-MS/MS-based proteomics in biomedical research. Molecules 27(19):6196
Das A, Datta S, Thakur S, Shukla A, Ansari J, Sujayanand GK, Chaturvedi SK, Kumar PA, Singh NP (2017) Expression of a chimeric gene encoding insecticidal crystal protein Cry1Aabc of Bacillus thuringiensis in chickpea (Cicer arietinum L.) confers resistance to gram pod borer (Helicoverpa armigera Hubner.). Front Plant Sci 8:1423
Day L, Cakebread JA, Loveday SM (2022) Food proteins from animals and plants: differences in the nutritional and functional properties. Trends Food Sci Technol 119:428–442
Dhiman S, Kaur S, Thakur B, Singh P, Tripathi M (2025) Nutritional enhancement of plant-based fermented foods: microbial innovations for a sustainable future. Fermentation 11(6):346
Dimina L, Rémond D, Huneau JF, Mariotti F (2022) Combining plant proteins to achieve amino acid profiles adapted to various nutritional objectives—an exploratory analysis using linear programming. Front Nutr 8:809685
Dourado MN, Leite TF, Barroso PAV, Araújo WL (2017) Genetically modified organisms in the tropics: challenges and perspectives. In: Diversity and benefits of microorganisms from the tropics, pp 403–430
El Youssef C, Bonnarme P, Fraud S, Péron AC, Helinck S, Landaud S (2020) Sensory improvement of a pea protein-based product using microbial co-cultures of lactic acid bacteria and yeasts. Foods 9(3):349

Engels W, Siu J, van Schalkwijk S, Wesselink W, Jacobs S, Bachmann H (2022) Metabolic conversions by lactic acid bacteria during plant protein fermentations. Foods 11(7):1005
Erdoğan İ, Cevher-Keskin B, Bilir Ö, Hong Y, Tör M (2023) Recent developments in CRISPR/Cas9 genome-editing technology related to plant disease resistance and abiotic stress tolerance. Biology 12(7):1037
Erem E, Kilic-Akyilmaz M (2024) The role of fermentation with lactic acid bacteria in quality and health effects of plant-based dairy analogues. Compr Rev Food Sci Food Saf 23(4):e13402
Farid MS, Anjum R, Yang Y, Tu M, Zhang T, Pan D, Sun Y, Wu Z (2024) Recent trends in fermented plant-based analogues and products, bioactive peptides, and novel technologies-assisted fermentation. Trends Food Sci Technol 149:104529
Fasuan TO, Gbadamosi SO, Omobuwajo TO (2018) Characterization of protein isolate from Sesamum indicum seed: in vitro protein digestibility, amino acid profile, and some functional properties. Food Sci Nutr 6(6):1715–1723
Ferrari L, Panaite SA, Bertazzo A, Visioli F (2022) Animal-and plant-based protein sources: a scoping review of human health outcomes and environmental impact. Nutrients 14(23):5115
Ganugi P, Martinelli E, Lucini L (2021) Microbial biostimulants as a sustainable approach to improve the functional quality in plant-based foods: a review. Curr Opin Food Sci 41:217–223
Gholami-Shabani M, Shams-Ghahfarokhi M, Razzaghi-Abyaneh M (2023) Food microbiology: application of microorganisms in food industry. In: Health risks of food additives-recent developments and trends in food sector. IntechOpen
Gorissen SH, Crombag JJ, Senden JM, Waterval WH, Bierau J, Verdijk LB, van Loon LJ (2018) Protein content and amino acid composition of commercially available plant-based protein isolates. Amino Acids 50:1685–1695
Gouseti O, Larsen ME, Amin A, Bakalis S, Petersen IL, Lametsch R, Jensen PE (2023) Applications of enzyme technology to enhance transition to plant proteins: a review. Foods 12(13):2518
Günal-Köroğlu D, Karabulut G, Ozkan G, Yılmaz H, Gültekin-Subaşı B, Capanoglu E (2025) Allergenicity of alternative proteins: reduction mechanisms and processing strategies. J Agric Food Chem 73(13):7522–7546
Gutierrez N, Torres AM (2019) Characterization and diagnostic marker for TTG1 regulating tannin and anthocyanin biosynthesis in faba bean. Sci Rep 9(1):16174
Han HW, Yu ST, Wang ZW, Yang Z, Jiang CJ, Wang XZ, Sun XS, Wang CT (2023) In planta genetic transformation to produce CRISPRed high-oleic peanut. Plant Growth Regul 101(2):443–451
Harper AR, Dobson RC, Morris VK, Moggré GJ (2022) Fermentation of plant-based dairy alternatives by lactic acid bacteria. Microb Biotechnol 15(5):1404–1421
Horlacher N, Oey I, Agyei D (2023) Learning from tradition: health-promoting potential of traditional lactic acid fermentation to drive innovation in fermented plant-based dairy alternatives. Fermentation 9(5):452
Hossain MA, Cho JI, Han M, Ahn CH, Jeon JS, An G, Park PB (2010) The ABRE-binding bZIP transcription factor OsABF2 is a positive regulator of abiotic stress and ABA signaling in rice. J Plant Physiol 167(17):1512–1520
Ibrahim S, Saleem B, Rehman N, Zafar SA, Naeem MK, Khan MR (2022) CRISPR/Cas9 mediated disruption of Inositol Pentakisphosphate 2-Kinase 1 (TaIPK1) reduces phytic acid and improves iron and zinc accumulation in wheat grains. J Adv Res 37:33–41
Ishwarya Shankaran P, Kumari P (2024) Nutritional analysis of plant-based meat: current advances and future potential. Appl Sci 14(10):4154
Ismail BP, Senaratne-Lenagala L, Stube A, Brackenridge A (2020) Protein demand: review of plant and animal proteins used in alternative protein product development and production. Anim Front 10(4):53–63
Jareonsin S, Pumas C, Jaitiang D, Uttarotai T (2024) Green fusion proteins: An approach to sustainable nutrition blending plant and algae-based proteins for a circular food system. Future Foods 10:100415
Ji J, Zhang C, Sun Z, Wang L, Duanmu D, Fan Q (2019) Genome editing in cowpea Vigna unguiculata using CRISPR-Cas9. Int J Mol Sci 20(10):2471

Kapoor R, Kumar G, Arya P, Jaswal R, Jain P, Singh K, Sharma TR (2019) Genome-wide analysis and expression profiling of rice hybrid proline-rich proteins in response to biotic and abiotic stresses, and hormone treatment. Plants 8(9):343
Klap C, Yeshayahou E, Bolger AM, Arazi T, Gupta SK, Shabtai S, Usadel B, Salts Y, Barg R (2017) Tomato facultative parthenocarpy results from Sl AGAMOUS-LIKE 6 loss of function. Plant Biotechnol J 15(5):634–647
Lakkakula P, Olson F, Ripplinger D (2017) Pea and lentil market analysis
Lal A (2020) Iron in health and disease: an update. Indian J Pediatr 87(1):58–65
Langyan S, Yadava P, Khan FN, Dar ZA, Singh R, Kumar A (2022) Sustaining protein nutrition through plant-based foods. Front Nutr 8:772573
Lappi J, Silventoinen-Veijalainen P, Vanhatalo S, Rosa-Sibakov N, Sozer N (2022) The nutritional quality of animal-alternative processed foods based on plant or microbial proteins and the role of the food matrix. Trends Food Sci Technol 129:144–154
Lazim NAM, Nazir NM (2025) Biotechnology and genetic engineering: unlocking new possibilities. In: Evolving landscapes of research and development: trends, challenges, and opportunities. IGI Global Scientific Publishing, pp 109–122
Li C, Chen Y, Hu Q, Yang X, Zhao Y, Lin Y, Yuan J, Gu J, Li Y, He J, Wang D (2024) PSEUDO-RESPONSE REGULATOR 3b and transcription factor ABF3 modulate abscisic acid-dependent drought stress response in soybean. Plant Physiol 195(4):3053–3071
López-Fernández J, Benaiges MD, Valero F (2020) Rhizopus oryzae lipase, a promising industrial enzyme: biochemical characteristics, production and biocatalytic applications. Catalysts 10(11):1277
Manghwar H, Lindsey K, Zhang X, Jin S (2019) CRISPR/Cas system: recent advances and future prospects for genome editing. Trends Plant Sci 24(12):1102–1125
Marques RLL, Norrevang AF, Ache P, Moog M, Visintainer D, Wendt T, Osterberg JT, Dockter C, Jorgensen ME, Salvador AT, Hedrich R (2020) Prospects for the accelerated improvement of the resilient crop quinoa. J Exp Bot. https://doi.org/10.1093/jxb/eraa285
Mathai JK, Liu Y, Stein HH (2017) Values for digestible indispensable amino acid scores (DIAAS) for some dairy and plant proteins may better describe protein quality than values calculated using the concept for protein digestibility-corrected amino acid scores (PDCAAS). Br J Nutr 117(4):490–499
Miao C, Xiao L, Hua K, Zou C, Zhao Y, Bressan RA, Zhu JK (2018) Mutations in a subfamily of abscisic acid receptor genes promote rice growth and productivity. Proc Natl Acad Sci 115(23):6058–6063
Milana M, van Asselt ED, van der Fels-Klerx IH (2025) A review of the toxicological effects and allergenic potential of emerging alternative protein sources. Compr Rev Food Sci Food Saf 24(1):e70123
Morilla EA, Taddia A, Sortino M, Tubio G (2023) Mixed cultures of Aspergillus niger and Rhizopus oryzae using lignocellulosic substrates to improve hydrolytic enzyme production. Bioenergy Res 16(4):2285–2296
Mouzo D, Bernal J, López-Pedrouso M, Franco D, Zapata C (2018) Advances in the biology of seed and vegetative storage proteins based on two-dimensional electrophoresis coupled to mass spectrometry. Molecules 23(10):2462
Mueller NG, Flachs A (2022) Domestication, crop breeding, and genetic modification are fundamentally different processes: implications for seed sovereignty and agrobiodiversity. Agric Hum Values 39(1):455–472
Neelakandan AK, Wright DA, Traore SM, Chen X, Spalding MH, He G (2022) CRISPR/Cas9 based site-specific modification of FAD2 cis-regulatory motifs in peanut (Arachis hypogaea L). Front Genet 13:849961
Negowetti NE (2020) A planetary health approach to the labeling of plant-based meat. Food Drug LJ 75:142
Opazo-Navarrete M, Burgos-Díaz C, Bravo-Reyes C, Gajardo-Poblete I, Chacón-Fuentes M, Reyes JE, Mojica L (2025) Comprehensive review of plant protein digestibility: challenges, assessment methods, and improvement strategies. Appl Sci 15(7):3538

Owusu-Kwarteng J, Agyei D, Akabanda F, Atuna RA, Amagloh FK (2022) Plant-based alkaline fermented foods as sustainable sources of nutrients and health-promoting bioactive compounds. Front Sustain Food Syst 6:885328
Park MJ, General T, Lee SP (2012) Physicochemical properties of roasted soybean flour bioconverted by solid-state fermentation using Bacillus subtilis and lactobacillus plantarum. Prev Nutr Food Sci 17(1):36
Petrusán JI, Rawel H, Huschek G (2016) Protein-rich vegetal sources and trends in human nutrition: a review. Curr Top Pept Protein Res 17:1–19
Prasad S, Gupta E, Yadav S, Babulal KS, Mishra S (2025) Plant-based food industry: overview and trends. In: The future of plant protein: innovations, challenges, and opportunities, pp 73–97
Praveen M, Brogi S (2025) Microbial fermentation in food and beverage industries: innovations, challenges, and opportunities. Foods 14(1):114
Rasheed A, Barqawi AA, Mahmood A, Nawaz M, Shah AN, Bay DH, Alahdal MA, Hassan MU, Qari SH (2022) CRISPR/Cas9 is a powerful tool for precise genome editing of legume crops: a review. Mol Biol Rep 49(6):5595–5609
Razzaq MK, Akhter M, Ahmad RM, Cheema KL, Hina A, Karikari B, Raza G, Xing G, Gai J, Khurshid M (2022) CRISPR-Cas9 based stress tolerance: new hope for abiotic stress tolerance in chickpea (Cicer arietinum). Mol Biol Rep 49(9):8977–8985
Réty S (2023) The structure and properties of microbial enzymes. Microorganisms 12(1):45
Rezaei MK, Deokar A, Tar'An B (2016) Identification and expression analysis of candidate genes involved in carotenoid biosynthesis in chickpea seeds. Front Plant Sci 7:1867
Roemhild K, von Maltzahn F, Weiskirchen R, Knüchel R, von Stillfried S, Lammers T (2021) Iron metabolism: pathophysiology and pharmacology. Trends Pharmacol Sci 42(8):640–656
Roy D, Ye A, Moughan PJ, Singh H (2020) Composition, structure, and digestive dynamics of milk from different species—a review. Front Nutr 7:577759
Roy A, Sahu PK, Das C, Bhattacharyya S, Raina A, Mondal S (2023) Conventional and new-breeding technologies for improving disease resistance in lentil (Lens culinaris Medik). Front Plant Sci 13:1001682
Sakure AA (2022) Enhancement of nutritional, pharmaceutical and industrial value of crops through genetic modification with carotenoid pathway genes. In: Biotechnology and crop improvement. CRC Press, pp 63–77
Sandhu R, Bangarwa SK, Attri M, Tiwari S, Kohli S, Fayaz S, Chaudhary N (2025) Effects of biotic stresses and their mitigation strategies in legumes: a review. Legum Res Int J 48(2). https://doi.org/10.18805/LR-5160
Santosh Kumar VV, Verma RK, Yadav SK, Yadav P, Watts A, Rao MV, Chinnusamy V (2020) CRISPR-Cas9 mediated genome editing of drought and salt tolerance (OsDST) gene in indica mega rice cultivar MTU1010. Physiol Mol Biol Plants 26:1099–1110
Shetty MJ, Chandan K, Krishna HC, Aparna GS (2018) Genetically modified crops: an overview. J Pharmacogn Phytochem 7(1):2405–2410
Silva CC, Ribeiro SC (2024) Microorganisms and their importance in the food industry: safety, quality and health properties. Foods 13(10):1452
Singh S, Mishra V, Riyazuddin R, Chugh V, Upadhyay SK (2024) CRISPR/Cas9-mediated genome editing for trait improvement and stress tolerance in Leguminosae (legume family). Plant Breed. https://doi.org/10.1111/pbr.13234
Sionek B, Szydłowska A, Küçükgöz K, Kołożyn-Krajewska D (2023) Traditional and new microorganisms in lactic acid fermentation of food. Fermentation 9(12):1019
Sun B, Jiang M, Zheng H, Jian Y, Huang WL, Yuan Q, Zheng AH, Chen Q, Zhang YT, Lin YX, Wang Y (2020) Color-related chlorophyll and carotenoid concentrations of Chinese kale can be altered through CRISPR/Cas9 targeted editing of the carotenoid isomerase gene BoaCRTISO. Hortic Res 7:161
Talwar R, Freymond M, Beesabathuni K, Lingala S (2024) Current and future market opportunities for alternative proteins in low-and middle-income countries. Curr Dev Nutr 8:102035

Thakur S, Pandey AK, Verma K, Shrivastava A, Singh N (2024) Plant-based protein as an alternative to animal proteins: a review of sources, extraction methods and applications. Int J Food Sci Technol 59(1):488–497

Tziva M, Kalfagianni A, Negro S, Hekkert M (2023) Plant-based protein products in the news: mind the gap between innovation and public discourses. PLOS Sustain Transform 2(1):e0000044

Willett W, Rockström J, Loken B, Springmann M, Lang T, Vermeulen S, Garnett T, Tilman D, DeClerck F, Wood A, Jonell M (2019) Food in the Anthropocene: the EAT–lancet commission on healthy diets from sustainable food systems. Lancet 393(10170):447–492

Wolfe RR, Rutherfurd SM, Kim IY, Moughan PJ (2016) Protein quality as determined by the digestible indispensable amino acid score: evaluation of factors underlying the calculation. Nutr Rev 74(9):584–599

Yang X, Hong J, Wang L, Cai C, Mo H, Wang J, Fang X, Liao Z (2024) Effect of lactic acid bacteria fermentation on plant-based products. Fermentation 10(1):48

Zeng Z, Han N, Liu C, Buerte B, Zhou C, Chen J, Wang M, Zhang Y, Tang Y, Zhu M, Wang J (2020) Functional dissection of HGGT and HPT in barley vitamin E biosynthesis via CRISPR/Cas9-enabled genome editing. Ann Bot 126(5):929–942

Zhang Y, Zhu X, Wang N, Liu X, Wang L, Ning K (2025) Synergy of traditional practices and modern technology: advancing the understanding and applications of microbial resources and processes in fermented foods. Trends Food Sci Technol 157:104891

Zheng X, Kuijer HN, Al-Babili S (2021) Carotenoid biofortification of crops in the CRISPR era. Trends Biotechnol 39(9):857–860

Phytochemicals as Biopharmaceuticals: Unlocking Mechanisms for Disease Prevention and Treatment

S. Yamini, Ammar Ahmed, Kailash P. Rajpurohit, G. Priya, and R. Vinoth Kumar

Abstract

Phytochemicals are naturally occurring bioactive compounds found in plants with diverse therapeutic properties that contribute to the prevention and treatment of various diseases. These include antioxidants, anti-inflammatory agents, antimicrobial agents and cancer-fighting substances, which exert their effects through complex interactions with molecular and cellular pathways. Major categories of phytochemicals, such as flavonoids, alkaloids, terpenoids, and polyphenols, have demonstrated significant efficacy in modulating immune function, reducing oxidative stress and inhibiting the proliferation of pathogenic organisms. Research highlights the potential of these compounds in addressing major health concerns, including cardiovascular diseases, diabetes, neurodegenerative disorders, and infections. Phytochemicals such as flavonoids, curcumin, and resveratrol have shown promise in reducing inflammation, improving cellular function, and promoting neuroprotection by modulating signaling pathways like NF-κB, MAPK, and Nrf2. Their ability to influence critical enzymes and receptors has expanded their application beyond traditional medicine to modern pharmacology. However, challenges remain, including low bioavailability, instability, and inconsistencies in therapeutic outcomes due to plant variability. Advances in drug delivery systems, such as nanoparticles and liposomes, along with improvements in extraction and formulation methods, offer solutions to enhance the effectiveness of phytochemicals. Additionally, ongoing research on the synergistic effects of plant-derived compounds and their role in integrative medicine

Authors Yamini S and Ammar Ahmed have equally contributed to this chapter.

S. Yamini · A. Ahmed · K. P. Rajpurohit · G. Priya (✉) · R. V. Kumar (✉)
Department of Biotechnology, Faculty of Science and Humanities, SRM Institute of Science and Technology, Ramapuram, Chennai, Tamil Nadu, India
e-mail: priyag3@srmist.edu.in; vinothkr4@srmist.edu.in

R. K. Gaur et al. (eds.), *Plant Molecular Farming*,
https://doi.org/10.1007/978-981-95-3823-2_3

presents new opportunities for novel drug development. This chapter focuses on different phytochemicals that represent a rich and promising source of therapeutic agents, with the potential to complement or replace conventional treatments. Their continued study and application in clinical aspects hold promise for advancing global healthcare and providing sustainable, effective, and safe therapeutic alternatives.

Keywords

Phytochemicals · Anti-oxidants · Anti-inflammatory · Neuroprotection · Bioavailability · Drug delivery systems

1 Introduction

Phytochemicals derived from plants exhibit a broad spectrum of bioactive properties with significant medicinal applications. Found abundantly in various dietary sources, over a thousand phytochemicals have been identified to date. Phytochemicals such as carotenoids, polyphenols, isoprenoids, phytosterols, saponins, dietary fibers, and specific polysaccharides demonstrate potent antibacterial, antiviral, and antioxidant activities (Patra et al. 2021; Varghese et al. 2023). Their therapeutic appeal lies in their lower toxicity compared to conventional treatments, while retaining the ability to modulate gene expression and enhance immune responses through antioxidant mechanisms (Venkatalakshmi et al. 2016). Diets rich in plant-based components provide essential minerals and phytochemicals that contribute to overall health improvement and disease prevention. To extract phytochemicals from plant materials, diverse methods have been employed, encompassing both traditional approaches such as maceration, percolation, decoction, reflux extraction, and Soxhlet extraction, and modern techniques like pulsed electric field extraction (PEF), supercritical fluid extraction (SFE), enzyme-assisted extraction (EAE), and ultrasound-assisted extraction (UAE) (Kumar et al. 2023; Pawase et al. 2024; Quitério et al. 2022). These compounds are categorized based on their structural and functional attributes (Borges et al. 2015). While phytochemicals can enhance cellular and organ function, careful consideration of dosage is crucial to mitigate potential adverse effects. Optimizing the utilization of phytochemicals is pivotal for advancing innovative therapeutic strategies and promoting public health sustainability.

2 Classification of Phytochemicals

Phytochemicals, such as isoflavonoids, exhibit diverse biochemical and physiological activities, including notable antioxidant and antiestrogenic properties. These compounds are broadly classified into primary and secondary metabolites, with secondary metabolites, such as phenolics, demonstrating significant implications for human health (Koche et al. 2016). Figure 1 highlights some major classes of

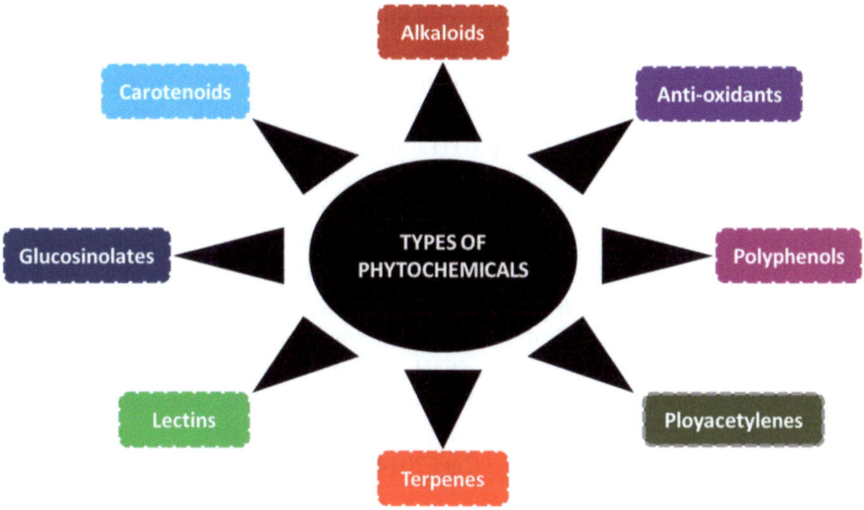

Fig. 1 Classification of the most widely used phytochemicals as biopharmaceuticals

phytochemicals that are extensively studied for their therapeutic potential (Goswami et al. 2024). Phenolic compounds, which include flavonoids, phenolic acids, and polyphenols, are abundant in plants and serve as crucial defense regulators (Rahman et al. 2022). Flavonoids, in particular, are renowned for their antioxidant properties and exhibit a wide array of biological activities, such as anti-microbial and anti-inflammatory effects. Additionally, alkaloids, distinguished by their heterocyclic nitrogen structures and characteristic bitter taste, play vital roles in plant defense against pathogens and herbivores. These compounds also possess significant pharmacological properties, including antihypertensive and antimalarial effects (Rajput et al. 2021). Phytochemicals are integral to both plant survival and human health, showcasing a broad spectrum of bioactivities that underpin their importance in medicine, nutrition, and disease prevention.

2.1 Alkaloids

Alkaloids, a class of secondary metabolites derived from plants, fungi, and marine sponges, are renowned for their continuous contributions to therapeutic interventions against chronic diseases such as cancer, diabetes, and neurodegenerative disorders. For instance, alkaloid derivatives (such as galanthamine and huperzine A) are currently employed in neurodegenerative diseases. These disorders are polygenic and multifactorial, with etiologies linked to inflammation, oxidative stress, and protein aggregation (Kong et al. 2021). Consequently, natural-product-based alkaloids with polypharmacological properties hold promise for drug development and nutraceutical applications in managing neurodegeneration. Rescinnamine, isolated from *Rauvolfia reflexa*, acts as a dual cholinesterase inhibitor (Fadaeinasab et al. 2015). Additionally, monoterpene indole alkaloids, containing a monoterpenoid unit fused with an indole moiety, exhibit selective BuChE inhibition. Compounds such as angustidine, nauclefine, and angustine from *Nauclea officinalis* have IC50 values ranging from 1.03 to 7.70 μM, with kinetic studies revealing mixed-mode inhibition (Ki = 6.12 μM) (Liew et al. 2015).

The urgent need for novel antibiotics targeting multidrug-resistant bacterial and fungal pathogens is well-documented (Bhattarai et al. 2020). Infectious diseases remain the second leading global cause of mortality, claiming 17 million lives annually (Mulani et al. 2019). Notably, the ESKAPE (*Enterococcus faecium, Staphylococcus aureus, Klebsiella pneumoniae, Acinetobacter baumannii, Pseudomonas aeruginosa,* and *Enterobacter* spp.) pathogens account for over 700,000 deaths each year and are of critical concern (Rice 2008). Despite some successes (such as plazomicin developed by Achaogen), financial and regulatory challenges hinder the post-approval viability of novel antibiotics, posing significant threats to global health (Nathan and Cars 2014).

2.2 Polyphenols

Polyphenols (PPs) are a diverse class of phytochemicals characterized by phenolic rings with two or more hydroxyl groups. Known for its potent antioxidant properties, PPs exhibit numerous therapeutic effects and potential health benefits, as evidenced by both in vitro and in vivo studies, as well as clinical investigations. Due to their free-radical scavenging and anti-inflammatory capabilities, PPs hold promise in addressing various metabolic disorders. Dietary PPs have been shown to alleviate symptoms associated with metabolic syndrome (MtS), including obesity, dyslipidemia, atherosclerosis, hyperglycemia, liver toxicity, hypertension, accelerated aging, neurodegenerative disorders, and certain cancers (Gasmi et al. 2022).

Among their biological activities, the cancer chemopreventive properties of dietary PPs have been extensively documented in in vitro studies and using animal models. These effects are largely attributed to their anti-oxidant and anti-inflammatory actions. Tea polyphenols, particularly epigallocatechin-3-gallate (EGCG) and theaflavins (TFs), are among the most widely studied for their potential in cancer prevention and treatment. Zhang et al. (2015) demonstrated that EGCG induced apoptosis in hepatocellular carcinoma (HCCLM6) cells by reducing mitochondrial membrane potential and promoting G0/G1 phase cell cycle arrest. This was associated with decreased expression of Bcl-2 and NF-κB, without affecting noncancerous liver cells (HL-7702). Additionally, the chemopreventive effects of EGCG and TFs were observed in N-nitrosodiethylamine-induced tongue and liver cancer in mice, where they modulated cellular proliferation, apoptosis, and CD44 expression (Sur et al. 2016). Furthermore, TFs have shown significant chemopreventive activity against prostate cancer, thereby highlighting their therapeutic potential.

2.3 Lectins

Lectins are naturally occurring proteins known for their ability to selectively bind to carbohydrates and glycoconjugates via reversible interactions at specific binding sites. These proteins are widely distributed in nature, and exhibit diverse structures and carbohydrate-binding specificities, making them valuable in anti-viral applications. Lectins can detect carbohydrate interactions on cellular surfaces and viral envelopes through hemagglutination assays, and they have shown significant potential in treating viral infections (Nabi-Afjadi et al. 2022).

Additionally, plant-derived lectins have demonstrated utility in cancer diagnostics. Wisteria floribunda agglutinin (WFA), a legume lectin, binds with high affinity to terminal N-acetylgalactosamine residues, and is particularly effective in detecting the cancer biomarker L1 cell adhesion molecule (L1CAM), used for diagnosing intrahepatic cholangiocarcinoma (CC) (Matsuda et al. 2013). While WFA shows low specificity to mucin 1 (MUC1), another glycosylated biomarker, combining both these markers have significantly enhanced the diagnostic accuracy for CC (Matsuda et al. 2010).

2.4 Terpenes

Terpenes, characterized by the general chemical formula $(C_5H_8)_n$, are composed of isoprene units. While most terpenes consist of even numbers of intact isoprene units, exceptions such as C19 diterpenoids do exist. When terpenes incorporate functional groups like alcohols, aldehydes, or ketones, they are classified as terpenoids. A widely adopted classification system categorizes terpenoids based on the number of isoprene units in their structure (Del Prado-Audelo et al. 2021). Monoterpenoids, composed of two isoprene units $(C_{10}H_{16})$, exhibit diverse structural arrangements, including acyclic, monocyclic, and bicyclic forms. Sesquiterpenoids, containing three isoprene units $(C_{15}H_{24})$, often feature simple to complex mono- and polycyclic ring systems. Triterpenoids $(C_{30}H_{48})$ display over 40 different carbon skeletons, while tetraterpenoids $(C_{40}H_{64})$, also known as carotenes, are noted for their large structures. Currently, approximately 1000 monoterpenes, 7000 sesquiterpenes, and 3000 diterpenes have been identified, with the list continuously expanding.

The anti-inflammatory properties of terpenes have been extensively evaluated to treat inflammatory diseases using various in vivo models. Terpenes such as α-phellandrene, 1,8-cineole, borneol, and terpinen-4-ol have demonstrated significant inhibition of pro-inflammatory mediators in lung injury models induced by lipopolysaccharides (LPS) and carrageenan (Caceres et al. 2017). Specifically, 1,8-cineole has been shown to reduce cytokines, including IL-1β, IL-4, IL-6, IL-13, IL-17A, and TNF-α in asthma models sensitized with ovalbumin (OVA) or house dust mite allergens, as well as in short-term cigarette smoke exposure in mice (Kennedy-Feitosa et al. 2016). Its inhibitory effect on prostaglandin E2 (PGE2) were also observed in an ex vivo study using blood monocytes from patients with bronchial asthma (Juergens et al. 1998). Additionally, linalool and humulene, has demonstrated to reduce pro-inflammatory mediators in OVA-sensitized asthma mouse models (Kim et al. 2019a, b).

Terpenes have also been investigated in experimental colitis models. Borneol, terpinen-4-ol, and β-caryophyllene significantly decreased levels of IL-1β, IL-6, and NLRP3 inflammasome components in colitis models induced by 2,4,6-trinitrobenzene sulfonic acid (TNBS) or dextran sulfate sodium (DSS) (Cho et al. 2007). These findings underscore the therapeutic potential of terpenes in inflammatory diseases.

2.5 Polyacetylenes

Polyacetylene (PAs) phytochemicals are increasingly recognized for their potential role in the chemoprotective effects of consuming apiaceous vegetables. Emerging evidences indicate that PAs influence carcinogenesis by modulating diverse signaling pathways crucial for inflammation, apoptosis, and cell cycle regulation. Epidemiological studies have suggested a correlation between dietary intake of PA-rich vegetables and reduced risks of inflammation and cancer. PA supplementation has demonstrated effects on cell growth, gene expression, and immunological responses, and also has been shown to decrease tumor incidence in animal models. The cancer chemopreventive effects of dietary PAs involve mechanisms such as modulation of inflammatory cytokines, the NF-κB pathway, antioxidant response elements, unfolded protein response (UPR) pathways, growth factor signaling, cell cycle regulation, and apoptosis (Alfurayhi et al. 2023).

Carrot-derived PAs, notably falcarinol (FaOH) and falcarindiol (FaDOH), exhibit significant anti-inflammatory properties. FaOH from *Saposhnikovia divaricata* has been shown to reduce LPS-induced TNF-α and IL-6 levels in BV-2 microglia cells and murine serum. Similarly, FaOH and FaDOH from carrots prevented colorectal precancerous lesions in azoxymethane-induced rat models by inhibiting NF-κB activity and downstream inflammatory markers, including TNF-α, IL-6, and COX-2 (Kang et al. 2012; Kobaek-Larsen et al. 2019).

In BV-2 microglia cells, FaOH from *Radix Saposhnikoviae* inhibited NF-κB p65 activation and IκB-α phosphorylation (Zhao et al. 2021). Similarly, FaOH from *Heracleum moellendorffii* roots inhibited LPS-induced NF-κB signaling by preventing IκB-α degradation and nuclear p65 accumulation in RAW264.7 cells (Kim et al. 2019a, b). Additionally, FaDOH suppressed LPS/IFNγ-induced NF-κB, IKK-α, and IKK-β activation in rat primary astrocytes, further supporting its anti-inflammatory and chemopreventive potential (Shiao et al. 2005). These findings highlight the therapeutic promise of PAs in cancer prevention and inflammation management.

2.6 Glucosinolates

Glucosinolates (GSs), anionic secondary metabolites prevalent in plants of the order *Brassicales*, and their hydrolysis products (GSHPs) have garnered significant attention for their diverse biological activities and mechanisms of action. These compounds, available in natural, pure, synthetic, and derivative forms, exhibit promising roles in human and animal health, plant defense, and food preservation. In human health, GSs and GSHPs have been extensively studied for their potential in disease prevention and therapy, particularly as anti-cancer and anti-microbial agents. In plants, they function as biofumigants and biocides, while in food industries, they are explored as natural preservatives. Strategies to enhance GS/GSHP levels in plants often involve biotic and abiotic stressors and targeted phytohormone treatments. The bio-availability and efficacy of these compounds depend significantly on their source concentrations, which are influenced by food preparation, processing, and extraction methods (Maina et al. 2020).

Animal studies further highlight the therapeutic potential of GSs in cardiovascular health. For instance, daily supplementation of dried broccoli sprouts (200 mg/day, approximately 12 μmol glucoraphanin/g dry weight) to spontaneously hypertensive stroke-prone (SHRSP) rats had significantly reduced oxidative stress as evidenced by an increased glutathione (GSH) to oxidized glutathione (GSSG) ratio (10.6–14.0 compared to 4.0–7.2 in control) and lowered blood pressure by 20 mmHg, alongside improved endothelial function. These effects were attributed to sulforaphane, a glucoraphanin metabolite (Wu et al. 2004). Sulforaphane administration (10 μmol/kg body weight) over 4 months has further prevented the hypertension-induced vascular remodeling in SHRSP rats (Senanayake et al. 2012).

Similarly, erucin, a sulforaphane analog in *Eruca sativa* (arugula), demonstrated anti-hypertensive effects. Salt-induced hypertensive rats treated with 10 and 30 mg/kg of *E. sativa* extract has exhibited a significant reduction in mean arterial pressure (40.33 ± 1.15% and 59.43 ± 0.77%, respectively) (Salma et al. 2018). These findings have underscore the potential of GSs and their metabolites in addressing hypertension and related vascular disorders.

2.7 Carotenoids

Carotenoids, a prominent class of phytochemicals, are widely recognized for their diverse biological activities. Naturally occurring in plants, algae, fungi, and microorganisms. They are structurally categorized into oxygenated (xanthophylls) and non-oxygenated (carotenes) forms. They are demonstrated to be a potential therapeutic agents due to their ability to prevent and mitigate the symptoms of various diseases, including cancer, neurodegenerative conditions (like Alzheimer's disease and cerebral ischemia), diabetes associated with obesity and hypertension, and ophthalmic disorders. Carotenoids hold promise as dietary supplements, nutraceuticals, and pharmaceutical compounds (Bhatt and Patel 2020). However, further research and translational studies are necessary to fully harness their therapeutic potential for human health.

Beta-carotene, one of the most extensively studied carotenoids, has been evaluated in various supplementation trials. In a study involving cystic fibrosis patients, supplementation with water-miscible beta-carotene beadlets (0.5 mg/kg body weight for 3 months) has significantly increased LDL lag time and decreased plasma malondialdehyde (MDA) levels, indicating improved oxidative stress markers (Winklhofer-Roob et al. 1995). Also the supplementation of 1 mg/kg body weight of beta-carotene daily (up to 50 mg) for 3 monthshas resulted in a significant reduction in MDA levels, although this effect was not sustained with lower maintenance doses of 10 mg/day (Renner et al. 2001).

In type 2 diabetes patients, beta-carotene supplementation (60 mg/day for 3 weeks) has reduced MDA levels and enhanced LDL stability against copper-induced oxidation, demonstrating potential benefits in oxidative stress management (Levy et al. 2000). These findings have underscore the need for higher doses of beta-carotene to achieve therapeutic effects, a level challenging to attain through a typical diet.

3 Disease-Specific Mechanisms of Phytochemicals

3.1 Cardiovascular Disease Treatment

Cardiovascular diseases (CVDs) encompass a range of disorders affecting the heart and blood vessels, including coronary artery disease, hypertension, and heart failure. These conditions are driven by complex interactions of genetic, environmental, and lifestyle factors. Oxidative stress, inflammation, and endothelial dysfunction are central to their pathophysiology, necessitating innovative approaches to prevention, diagnosis, and treatment for improved global health outcomes. A study by Pagliaro et al. (2015) highlights the bio-active potential of *Brassica oleracea*, a genus that includes broccoli and kale. These vegetables are rich in health-promoting compounds such as vitamin C and glucosinolates, particularly sulforaphane (Akhlaghi and Bandy 2010). In broccoli sprouts, glucoraphanin undergoes hydrolysis to produce sulforaphane, a compound with notable anti-oxidant and detoxifying properties (Houghton et al. 2013). Sulforaphane enhances the activity of detoxifying enzymes and activates the Nrf2 signaling pathway, thereby mitigating oxidative stress and reducing ox-LDL levels, which are associated with atherosclerosis (Zakkar et al. 2009).

In the experimental models, the administration of broccoli sprouts has been shown to elevate glutathione levels, and enhance the activity of antioxidant enzymes. These effects significantly reduce oxidative stress in rats and lower blood pressure in pregnant stroke-prone SHRSP rats and their offspring. Additionally, *Brassica oleracea* contains anthocyanins, which exhibit anti-inflammatory and cardioprotective properties. These compounds improve platelet function, inhibit pro-inflammatory signaling, suppress NF-kB activation, and reduce nitric oxide production. Anthocyanin extracts have further safeguard cardiac tissue from oxidative damage, and apoptosis during ischemia-reperfusion injury in animal studies (Hämäläinen et al. 2007). These findings have underscore the therapeutic potential of *Brassica oleracea* in managing oxidative stress-related disorders.

3.2 Diabetes Mellitus Treatment

Diabetes mellitus (DM) is a multi-factorial metabolic disorder characterized by chronic hyperglycemia due to impaired insulin secretion, resistance, or both. It disrupts glucose homeostasis leading to systemic complications including neuropathy, nephropathy, and cardiovascular diseases. Oxidative stress, inflammation, and epigenetic modifications play pivotal roles in its pathogenesis, necessitating precision medicine and integrative therapeutic strategies to mitigate its global health burden. A study by Aba and Asuzu (2018) highlights the diverse therapeutic potential of terpenoids, presenting an extensive array of compounds with promising anti-diabetic properties. For instance, *Stevia rebaudiana* contains stevioside and rebaudioside, which exert anti-diabetic effects by enhancing GLUT-4 translocation, a crucial mechanism for glucose uptake (Huang et al. 2010). Additionally, trans-dehydrocrotonin (t-DCTN) from *Croton cajucara* and dehydroabietic acid demonstrate both hypoglycemic and hypolipidemic activities, contributing to metabolic regulation (Silva et al. 2005). *Salacia oblonga* offers a unique blend of bicyclic diterpenoids, including salvicin, salvin, and salvifolin, alongside compounds like salacinol and kotalanol. These constituents effectively inhibit α-glucosidase, a key enzyme in carbohydrate digestion, thereby exhibiting significant anti-diabetic potential (Sheffler and Roth 2003). Moreover, bitter melon and the root bark of *Cussonia arborea*, rich in triterpenoids, have shown remarkable efficacy in activating the AMPK pathway, a critical regulator of energy metabolism, further enhancing their anti-diabetic activities (Tan et al. 2008). These findings underscore the versatility of terpenoids and triterpenes as potential antidiabetic agents, functioning through diverse biochemical pathways. Their multi-faceted mechanisms offer promising avenues for developing novel therapeutic strategies to combat diabetes effectively.

3.3 Neuroprotection

Neuroprotection encompasses strategies aimed at preserving neuronal structure and function against degeneration caused by ischemia, oxidative stress, inflammation, or excitotoxicity. It is vital in mitigating neurodegenerative disorders such as Alzheimer's and Parkinson's diseases. Targeted approaches, including anti-oxidant therapies, modulation of signaling pathways, and phytochemical interventions, hold promise for enhancing neuronal survival and functional recovery in neuropathological conditions. A study by Naoi et al. (2019) highlights the critical role of neurotrophic factors (NTFs), such as Brain-Derived Neurotrophic Factor (BDNF) and Glial Cell Line-Derived Neurotrophic Factor (GDNF), in maintaining neuronal health. Deficiencies in these factors are associated with neurodegenerative disorders

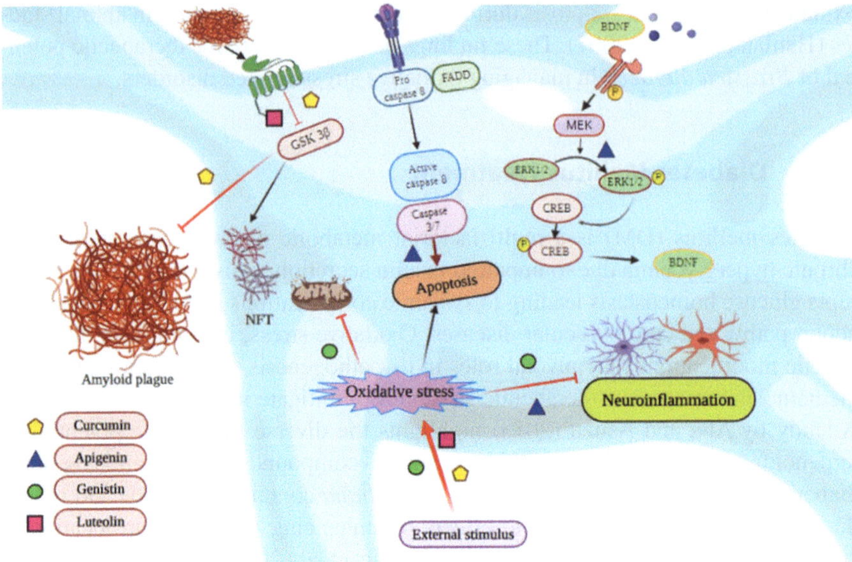

Fig. 2 Several mechanisms associated with Alzheimer's Disease (AD) and their possible prevention by some phytochemicals. (Reused from Khan et al. 2020, under the Creative Commons CC-BY license)

like Alzheimer's and Parkinson's disease. Figure 2 illustrates the proposed mechanisms through which various phytochemicals exert neuroprotective effects (Khan et al. 2020).

Phytochemicals, including flavonoids, curcumin, and resveratrol, have the ability to cross the blood-brain barrier (BBB) and emulate the actions of NTFs by binding to their receptors, triggering downstream signaling pathways, and enhancing NTF expression (Menard et al. 2013). For instance, compounds like 7,8-dihydroxyflavone and fisetin, has exhibited anti-depressant effects by binding to BDNF receptors, while epicatechin inhibits neurotrophin receptors, thus mitigating neurodegeneration. Additionally, flavonoids are known to suppress monoamine oxidase (MAO) activity, further activating neuroprotective pathways (Carradori et al. 2016). Anthocyanins and baicalin contribute to neuroprotection through receptor-mediated mechanisms, reinforcing their potential in combating neuronal damage (Ali Shah et al. 2013). These findings underline the therapeutic promise of phytochemicals in addressing neurodegenerative diseases, showcasing diverse mechanisms of action that target critical neuronal pathways and support neuroprotective functions.

3.4 Anti-carcinogenic Mechanisms

Cancer is a multi-faceted disease characterized by uncontrolled cellular proliferation, evasion of apoptosis, and metastatic potential. It arises from genetic and epigenetic alterations driving tumorigenesis. Advances in molecular oncology have illuminated pathways like PI3K/AKT, Wnt, and MAPK, unveiling targets for precision therapies. Integrative approaches, including immunotherapy and phytochemicals, offer promising avenues for combating malignancies effectively. A study by Al-Ishaq et al. (2020) identified lycopene, abundant in red-colored foods such as tomatoes, as a potent anti-oxidant capable of neutralizing reactive oxygen species, and enhancing detoxification pathways. Its ability to inhibit cytochrome P450 2E1 plays a critical role in cancer prevention by reducing xenobiotic activation (Slattery et al. 2012). Lycopene also exhibits significant anti-inflammatory and anticancer effects, particularly against colorectal cancer (Yoon 2012). The therapeutic efficacy of lycopene is enhanced when delivered via gold nanoparticles, which facilitate the apoptotic pathway by modulating pro-caspase and BAX expression (Lin et al. 2011). Furthermore, lycopene influences cellular growth regulation through the downregulation of HmgCo-A reductase expression, an enzyme critical for cholesterol biosynthesis (Palozza et al. 2010). Its anti-inflammatory action includes the suppression of NF-κB and JNK signaling pathways, resulting in reduced levels of pro-inflammatory cytokines, COX-2, and nitric oxide (NO) production (Cha et al. 2017). Lycopene's protective role extends to mitigating gastric carcinogen-induced damage by enhancing the glutathione (GSH)-dependent detoxification system and guarding cells against oxidative injury. These multi-faceted mechanisms highlight lycopene's potential in preventing cancer and inflammation, underscoring its importance as a dietary therapeutic for promoting health.

3.5 Anti-inflammatory Mechanisms

Inflammatory diseases encompass a spectrum of disorders driven by dysregulated immune responses and chronic inflammation, often mediated by cytokine imbalances and oxidative stress. These conditions, including arthritis, inflammatory bowel disease, and psoriasis, involve complex pathways such as NF-κB and MAPK signaling. Advancements in molecular therapeutics and phytochemical interventions offer promising strategies to modulate inflammation and restore immune homeostasis. A study by Hossen et al. (2020) highlighted the critical role of cytokines in immune regulation, with anti-inflammatory cytokines such as IL-4, IL-10, IL-11, and IL-13 maintaining physiological balance. Impairment in the activity of these cytokines can lead to an overproduction of pro-inflammatory cytokines like TNF-α, IL-1β, IL-6, and IFN-γ, triggering inflammatory responses. Phenolic phytochemicals have demonstrated a significant potential in modulating inflammation by suppressing pro-inflammatory cytokine production, and enhancing anti-inflammatory cytokine activity. For instance, asiatic acid at a dosage of 30 mg/kg reduced TNF-α mRNA expression by 50%, while mangiferin at 20 mg/kg decreased IL-6 mRNA expression by 88% (Guo et al. 2015; Jeong et al. 2014). Oxidative stress, a key contributor to gastrointestinal inflammation and tissue damage, can be mitigated by phenolic compounds like flavonoids, which act by modulating enzyme activity and donating electrons. Crocetin and soyasaponin I were found to reduce myeloperoxidase (MPO) activity, whereas mangiferin and berberine enhanced catalase (CAT) activity, thereby alleviating oxidative stress. Phytochemicals also target critical inflammatory pathways, including NF-κB, MAPK, and STAT, offering a promising approach to mitigating inflammation. These properties underscore their potential as alternatives to conventional immuno-suppressive therapies, especially in preserving intestinal barrier integrity (He et al. 2015).

3.6 Anti-bacterial Effects of Phytochemicals

Bacterial diseases, caused by pathogenic bacteria, pose significant global health challenges, manifesting as infections ranging from mild to life-threatening conditions. These pathogens disrupt host homeostasis through virulence factors, biofilm formation, and antibiotic resistance mechanisms. Advances in molecular biology and alternative therapies, including phytochemicals and bacteriophage technology, offer innovative approaches to combat bacterial infections and mitigate resistance. Phytochemicals have been extensively studied for their anti-bacterial properties, as highlighted by Humann and Lenz (2009). Flavonoids, in particular, exhibit anti-bacterial effects by interacting with bacterial cell walls and increasing membrane permeability (Upadhyay et al. 2014). Anthraquinones derived from *Cassia italica* demonstrate both bacteriostatic and bactericidal activities, whereas tannins inhibit the proliferation of ruminal microbes. Alkaloids further contribute by intercalating with bacterial cell walls and DNA, disrupting their structural integrity. The synergistic potential of phytochemicals is notable. For instance, combining amoxicillin with kaempferide results in greater cell wall damage than when paired with galangin, significantly enhancing membrane permeability and causing cellular deformation (Eumkeb et al. 2012). Similarly, phytochemicals such as thymol, carvacrol, catechins, and eugenol compromise membrane structure, leading to reduced cellular ATP levels. Cinnamaldehyde disrupts ATPase-dependent metabolism and inhibits glucose uptake, further depleting ATP reserves (Negi 2012). Terpenes and alcohols found in tea tree oil impair membrane permeability, damaging the structural integrity of microbial cells. This effect has proven effective against resistant pathogens such as *Escherichia coli, Staphylococcus aureus,* and *Candida albicans* (Özfenerci et al. 2018). Table 1 provides an overview of select phytochemicals and their target microbial pathogens, illustrating their significant anti-microbial potential.

4 Future Directions for Phytomedicines

Phytomedicines, derived from bio-active compounds in plants, have shown remarkable potential in treating various diseases, including infectious, inflammatory, metabolic, and neurodegenerative disorders. However, their widespread clinical adoption is limited by challenges such as low bioavailability, stability, and potential off-target effects. Future research should focus on advanced drug delivery systems, such as

Table 1 List of selected phytochemicals with their plant source, target microbial species, and their mechanism of action

Phytochemical	Plant source	Target microbial species	Mechanism of action	References
Kaempferol	*Alpina calcarata*	*Methicillin-resistant Staphylococcus aureus*	Inhibition of efflux pumps	Randhawa et al. (2016)
Allicin	*Allium sativum*	*Pseudomonas aeruginosa*	Degradation of sulfhydryl-dependent enzymes	Reiter et al. (2017)
Tomatidine	*Solanaceous plants*	*Bacillus and staphylococcus* sp.	Inhibition of ATP synthase	Boulet et al. (2018)
Resveratrol	*Vitis vinifera*	*Campylobacter jejuni*	Inhibition of efflux pumps	Klančnik et al. (2017)
Carvacrol	*Thymus capitatus*	*Enterobacter aerogenes*	Disruption of cell membrane	Althunibat et al. (2016)
Conessine	*Holarrhena antidysenterica*	*Pseudomonas aeruginosa*	Inhibition of efflux pumps	Siriyong et al. (2017)

nanoparticle-based carriers and liposomal formulations, to enhance phytochemical bioavailability and target specificity. For example, encapsulation techniques like poly(lactic-co-glycolic acid) nanoparticles have already demonstrated improved therapeutic outcomes for compounds like curcumin and lycopene, and similar strategies could be extended to other promising phytochemicals. In addition, omics technologies, such as genomics, proteomics, and metabolomics, can facilitate the identification of novel phytochemicals and elucidate their molecular mechanisms of their medicinal properties. High-throughput screening and in silico modeling can further streamline drug discovery processes, reducing time and cost. Moreover, integration with systems biology approaches could reveal synergistic effects when phytochemicals are combined with conventional drugs, enabling combination therapies with reduced side effects.

Ethnopharmacological studies should continue to explore under-utilized medicinal plants, particularly in biodiversity-rich regions, as they hold untapped reservoirs of therapeutic compounds. Rigorous standardization and clinical trials are imperative to validate the efficacy and safety of phytomedicines, bridging the gap between traditional knowledge and modern medicine. In conclusion, phytochemicals offer a promising avenue for developing safe, effective, and sustainable therapeutics. Their multi-faceted biological activities, coupled with advancements in drug delivery and molecular characterization, position phytomedicines as pivotal tools in addressing global health challenges. Continued inter-disciplinary collaboration and innovation will be essential for unlocking their full potential, paving the way for integrative medicine that combines the best of natural and synthetic therapeutics.

References

Aba PE, Asuzu IU (2018) Mechanisms of actions of some bioactive anti-diabetic principles from phytochemicals of medicinal plants: a review. IJNPR 9(2):85–96
Akhlaghi M, Bandy B (2010) Dietary broccoli sprouts protect against myocardial oxidative damage and cell death during ischemia-reperfusion. Plant Foods Hum Nutr 65(3):193–199
Alfurayhi R, Huang L, Brandt K (2023) Pathways affected by falcarinol-type polyacetylenes and implications for their anti-inflammatory function and potential in cancer chemoprevention. Foods 12(6):1192
Ali Shah S, Ullah I, Lee HY, Kim MO (2013) Anthocyanins protect against ethanol-induced neuronal apoptosis via GABAB1 receptors intracellular signaling in prenatal rat hippocampal neurons. Mol Neurobiol 48(1):257–269
Al-Ishaq RK, Overy AJ, Büsselberg D (2020) Phytochemicals and gastrointestinal cancer: cellular mechanisms and effects to change cancer progression. Biomolecules 10(1):105
Althunibat OY, Qaralleh H, Al-Dalin SYA, Abboud M, Khleifat K, Majali IS, Aldal'In HKH, Rayyan WA, Jaafraa A (2016) Effect of thymol and carvacrol, the major components of Thymus capitatus on the growth of *Pseudomonas aeruginosa*. JPAM 10(1):1–9
Bhatt T, Patel K (2020) Carotenoids: potent to prevent diseases review. Nat Prod Bioprospect 10(3):109–117
Bhattarai K, Bastola R, Baral B (2020) Antibiotic drug discovery: challenges and perspectives in the light of emerging antibiotic resistance. Adv Genet 105:229–292
Borges A, Saavedra M, Simoes M (2015) Insights on antimicrobial resistance, biofilms and the use of phytochemicals as new antimicrobial agents. Curr Med Chem 22(21):2590–2614
Boulet ML, Isabelle C, Guay I, Brouillette E, Langlois JP, Jacques PE, Rodrigue S, Brzezinski R, Beauregard PB, Bouarab K, Boyapelly K, Boudreault PL, Marsault E, Malouin F (2018) Tomatidine is a lead antibiotic molecule that targets staphylococcus aureus ATP Synthase subunit C. Antimicrob Agents Chemother 62(6):e02197-17
Caceres AI, Liu B, Jabba SV, Achanta S, Morris JB, Jordt SE (2017) Transient receptor potential cation channel subfamily M member 8 channels mediate the anti-inflammatory effects of eucalyptol. Br J Pharmacol 174(9):867–879
Carradori S, Gidaro MC, Petzer A, Costa G, Guglielmi P, Chimenti P, Alcaro S, Petzer JP (2016) Inhibition of human monoamine oxidase: biological and molecular modeling studies on selected natural flavonoids. J Agric Food Chem 64(47):9004–9011
Cha JH, Kim WK, Ha AW, Kim MH, Chang MJ (2017) Anti-inflammatory effect of lycopene in SW480 human colorectal cancer cells. Nutr Res Pract 11(2):90–96
Cho JY, Chang HJ, Lee SK, Kim HJ, Hwang JK, Chun HS (2007) Amelioration of dextran sulfate sodium-induced colitis in mice by oral administration of beta-caryophyllene, a sesquiterpene. Life Sci 80(10):932–939
Del Prado-Audelo ML, Cortés H, Caballero-Florán IH, González-Torres M, Escutia-Guadarrama L, Bernal-Chávez SA, Giraldo-Gomez DM, Magaña JJ, Leyva-Gómez G (2021) Therapeutic applications of terpenes on inflammatory diseases. Front Pharmacol 12:704197
Eumkeb G, Siriwong S, Phitaktim S, Rojtinnakorn N, Sakdarat S (2012) Synergistic activity and mode of action of flavonoids isolated from smaller galangal and amoxicillin combinations against amoxicillin-resistant *Escherichia coli*. J Appl Microbiol 112(1):55–64
Fadaeinasab M, Basiri A, Kia Y, Karimian H, Ali HM, Murugaiyah V (2015) New indole alkaloids from the bark of *Rauvolfia Reflexa* and their cholinesterase inhibitory activity. Cell Physiol Biochem 37(5):1997–2011
Gasmi A, Mujawdiya PK, Noor S, Lysiuk R, Darmohray R, Piscopo S, Lenchyk L, Antonyak H, Dehtiarova K, Shanaida M, Polishchuk A, Shanaida V, Peana M, Bjørklund G (2022) Polyphenols in metabolic diseases. Molecules 27(19):6280
Pawase PA, Goswami C, Shams R, Pandey VK, Tripathi A, Rustagi S, Darshan G (2024) A conceptual review on classification, extraction, bioactive potential and role of phytochemicals in human health. Future Foods 9:100313

Guo W, Liu W, Jin B, Geng J, Li J, Ding H, Wu X, Xu Q, Sun Y, Gao J (2015) Asiatic acid ameliorates dextran sulfate sodium-induced murine experimental colitis via suppressing mitochondria-mediated NLRP3 inflammasome activation. Int Immunopharmacol 24(2):232–238

Hämäläinen M, Nieminen R, Vuorela P, Heinonen M, Moilanen E (2007) anti-inflammatory effects of flavonoids: Genistein, kaempferol, quercetin, and daidzein inhibit STAT-1 and NF-κB activations, whereas flavone, isorhamnetin, naringenin, and pelargonidin inhibit only NF-κB activation along with their inhibitory effect on iNOS expression and NO production in activated macrophages. Mediat Inflamm 2007(1):45673

He S, Hou X, Xu X, Wan C, Yin P, Liu X, Chen Y, Shu B, Liu F, Xu J (2015) Quantitative proteomic analysis reveals heat stress-induced injury in rat small intestine via activation of the MAPK and NF-κB signaling pathways. Mol BioSyst 11:826–834

Hossen I, Hua W, Ting L, Mehmood A, Jingyi S, Duoxia X, Yanping C, Hongqing W, Zhipeng G, Kaiqi Z, Fang Y, Junsong X (2020) Phytochemicals and inflammatory bowel disease: a review. Crit Rev Food Sci Nutr 60(8):1321–1345

Houghton CA, Fassett RG, Coombes JS (2013) Sulforaphane: translational research from laboratory bench to clinic. Nutr Rev 71(11):709–726

Huang XY, Fu JF, Di DL (2010) Preparative isolation and purification of steviol glycosides from *Stevia rebaudiana Bertoni* using high-speed counter-current chromatography. Sep Purif Technol 71(2):220–224

Humann J, Lenz LL (2009) Bacterial peptidoglycan-degrading enzymes and their impact on host muropeptide detection. J Innate Immun 1(2):88–97

Jeong JJ, Jang SE, Hyam SR, Han MJ, Kim DH (2014) Mangiferin ameliorates colitis by inhibiting IRAK1 phosphorylation in NF-κB and MAPK pathways. Eur J Pharmacol 740:652–661

Juergens UR, Stöber M, Schmidt-Schilling L, Kleuver T, Vetter H (1998) Anti-inflammatory effects of euclyptol (1.8-cineole) in bronchial asthma: inhibition of arachidonic acid metabolism in human blood monocytes *ex vivo*. Eur J Med Res 3(9):407–412

Kang H, Bang TS, Lee JW, Lew JH, Eom SH, Lee K, Choi HY (2012) Protective effect of the methanol extract from *Cryptotaenia japonica* Hassk. against lipopolysaccharide-induced inflammation in vitro and in vivo. BMC Complement Altern Med 12:199

Kennedy-Feitosa E, Okuro RT, Pinho Ribeiro V, Lanzetti M, Barroso MV, Zin WA, Porto LC, Brito-Gitirana L, Valenca SS (2016) Eucalyptol attenuates cigarette smoke-induced acute lung inflammation and oxidative stress in the mouse. Pulm Pharmacol Ther 41:11–18

Khan A, Jahan S, Imtiyaz Z, Alshahrani S, Makeen HA, Alshehri BM, Kumar A, Arafah A, Rehman MU (2020) Neuroprotection: targeting multiple pathways by naturally occurring phytochemicals. Biomedicines 8(8):284

Kim MG, Kim SM, Min JH, Kwon OK, Park MH, Park JW, Ahn HI, Hwang JY, Oh SR, Lee JW, Ahn KS (2019a) Anti-inflammatory effects of linalool on ovalbumin-induced pulmonary inflammation. Int Immunopharmacol 74:105706

Kim HN, Kim JD, Yeo JH, Son HJ, Park SB, Park GH, Eo HJ, Jeong JB (2019b) *Heracleum moellendorffii* roots inhibit the production of pro-inflammatory mediators through the inhibition of NF-κB and MAPK signaling, and activation of ROS/Nrf2/HO-1 signaling in LPS-stimulated RAW264.7 cells. BMC Complement Altern Med 19(1):310

Klančnik A, Pogačar MS, Trošt K, Žnidarič MT, Vodopivec BM, Možina SS (2017) Anti-campylobacter activity of resveratrol and an extract from waste pinot noir grape skins and seeds, and resistance of Camp. jejuni planktonic and biofilm cells, mediated via the CmeABC efflux pump. J Appl Microbiol 122(1):65–77

Kobaek-Larsen M, Baatrup G, Notabi MK, El-Houri RB, Pipó-Ollé E, Arnspang EC, Christensen LP (2019) Dietary polyacetylenic oxylipins falcarinol and falcarindiol prevent inflammation and colorectal neoplastic transformation: a mechanistic and dose-response study in a rat model. Forum Nutr 11(9):2223

Koche D, Shirsat R, Kawale M (2016) An overview of major classes of phytochemicals: their types and role in disease prevention. Hislopia J 9(1/2):1

Kong YR, Tay KC, Su YX, Wong CK, Tan WN, Khaw KY (2021) Potential of naturally derived alkaloids as multi-targeted therapeutic agents for neurodegenerative diseases. Molecules 26(3):728

Kumar A, Nirmal P, Kumar M, Jose A, Tomer V, Oz E, Proestos C, Zeng M, Elobeid T, Sneha V, Oz F (2023) Major phytochemicals: recent advances in health benefits and extraction method. Molecules 28(2):887

Levy Y, Zaltsberg H, Ben-Amotz A, Kanter Y, Aviram M (2000) Dietary supplementation of a natural isomer mixture of beta-carotene inhibits oxidation of LDL derived from patients with diabetes mellitus. Ann Nutr Metab 44(2):54–60

Liew SY, Khaw KY, Murugaiyah V, Looi CY, Wong YL, Mustafa MR, Litaudon M, Awang K (2015) Natural indole butyrylcholinesterase inhibitors from *Nauclea officinalis*. Phytomedicine 22(1):45–48

Lin MC, Wang FY, Kuo YH, Tang FY (2011) Cancer chemopreventive effects of lycopene: suppression of MMP-7 expression and cell invasion in human colon cancer cells. J Agric Food Chem 59(20):11304–11318

Maina S, Misinzo G, Bakari G, Kim HY (2020) Human, animal and plant health benefits of glucosinolates and strategies for enhanced bioactivity: a systematic review. Molecules 25(16):3682

Matsuda A, Kuno A, Kawamoto T, Matsuzaki H, Irimura T, Ikehara Y, Zen Y, Nakanuma Y, Yamamoto M, Ohkohchi N, Shoda J, Hirabayashi J, Narimatsu H (2010) *Wisteria floribunda* agglutinin-positive mucin 1 is a sensitive biliary marker for human cholangiocarcinoma. Hepatology 52(1):174–182

Matsuda A, Kuno A, Matsuzaki H, Kawamoto T, Shikanai T, Nakanuma Y, Yamamoto M, Ohkohchi N, Ikehara Y, Shoda J, Hirabayashi J, Narimatsu H (2013) Glycoproteomics-based cancer marker discovery adopting dual enrichment with *Wisteria floribunda* agglutinin for high specific glyco-diagnosis of cholangiocarcinoma. J Proteome 85:1–11

Menard C, Bastianetto S, Quirion R (2013) Neuroprotective effects of resveratrol and epigallocatechin gallate polyphenols are mediated by the activation of protein kinase C gamma. Front Cell Neurosci 7:281

Mulani MS, Kamble EE, Kumkar SN, Tawre MS, Pardesi KR (2019) Emerging strategies to combat ESKAPE pathogens in the era of antimicrobial resistance: a review. Front Microbiol 10:539

Nabi-Afjadi M, Heydari M, Zalpoor H, Arman I, Sadoughi A, Sahami P, Aghazadeh S (2022) Lectins and lectibodies: potential promising antiviral agents. Cell Mol Biol Lett 27(1):37

Naoi M, Shamoto-Nagai M, Maruyama W (2019) Neuroprotection of multifunctional phytochemicals as novel therapeutic strategy for neurodegenerative disorders: Antiapoptotic and anti-amyloidogenic activities by modulation of cellular signal pathways. Future Neurol 14(1):FNL9

Nathan C, Cars O (2014) Antibiotic resistance--problems, progress, and prospects. N Engl J Med 371(19):1761–1763

Negi PS (2012) Plant extracts for the control of bacterial growth: efficacy, stability and safety issues for food application. Int J Food Microbiol 156(1):7–17

Özfenerci M, Koca U, Çalışkan K (2018) Tea tree oil and its use in aromatherapy. CUPMAP 2:90–102

Pagliaro B, Santolamazza C, Simonelli F, Rubattu S (2015) Phytochemical compounds and protection from cardiovascular diseases: a state of the art. BioMed Res Int 2015:918069

Palozza P, Colangelo M, Simone R, Catalano A, Boninsegna A, Lanza P, Monego G, Ranelletti FO (2010) Lycopene induces cell growth inhibition by altering mevalonate pathway and Ras signaling in cancer cell lines. Carcinogenesis 31(10):1813–1821

Patra S, Pradhan B, Nayak R, Behera C, Das S, Patra SK, Efferth T, Jena M, Bhutia SK (2021) Dietary polyphenols in chemoprevention and synergistic effect in cancer: clinical evidences and molecular mechanisms of action. Phytomedicine 90:153554

Quitério E, Grosso C, Ferraz R, Delerue-Matos C, Soares C (2022) A critical comparison of the advanced extraction techniques applied to obtain health-promoting compounds from seaweeds. Mar Drugs 20(11):677

Rahman MM, Rahaman MS, Islam MR, Rahman F, Mithi FM, Alqahtani T, Almikhlafi MA, Alghamdi SQ, Alruwaili AS, Hossain MS, Ahmed M, Das R, Bin ET, Uddin MS (2022) Role of phenolic compounds in human disease: current knowledge and future prospects. Molecules 27(1):233

Rajput A, Sharma R, Bharti R (2021) Pharmacological activities and toxicities of alkaloids on human health. Mater Today Proc 48:1407–1415

Randhawa HK, Hundal KK, Ahirrao PN, Jachak SM, Nandanwar HS (2016) Efflux pump inhibitory activity of flavonoids isolated from *Alpinia calcarata* against methicillin-resistant *Staphylococcus aureus*. Biologia 71(5):484–493

Reiter J, Levina N, Van Der Linden M, Gruhlke M, Martin C, Slusarenko AJ (2017) Diallylthiosulfinate (Allicin), a volatile antimicrobial from garlic (*Allium sativum*), kills human lung pathogenic bacteria, including MDR strains, as a vapor. Molecules 22(10):1711

Renner S, Rath R, Rust P, Lehr S, Frischer T, Elmadfa I, Eichler I (2001) Effects of beta-carotene supplementation for six months on clinical and laboratory parameters in patients with cystic fibrosis. Thorax 56(1):48–52

Rice LB (2008) Federal funding for the study of antimicrobial resistance in nosocomial pathogens: no ESKAPE. JID 197(8):1079–1081

Salma U, Khan T, Shah AJ (2018) Antihypertensive effect of the methanolic extract from *Eruca sativa* Mill., (*Brassicaceae*) in rats: muscarinic receptor-linked vasorelaxant and cardiotonic effects. J Ethnopharmacol 224:409–420

Senanayake GV, Banigesh A, Wu L, Lee P, Juurlink BH (2012) The dietary phase 2 protein inducer sulforaphane can normalize the kidney epigenome and improve blood pressure in hypertensive rats. Am J Hypertens 25(2):229–235

Shaari KM, Paetz C, Stanslas J, Abas F, Lajis NH (2009) Naturally occurring labdane diterpene and benzofuran from Globba pendula. Nat Prod Commun 4(8):1031–1036

Sheffler DJ, Roth BL (2003) Salvinorin A: the "magic mint" hallucinogen finds a molecular target in the kappa opioid receptor. TIPS 24(3):107–109

Shiao YJ, Lin YL, Sun YH, Chi CW, Chen CF, Wang CN (2005) Falcarindiol impairs the expression of inducible nitric oxide synthase by abrogating the activation of IKK and JAK in rat primary astrocytes. Br J Pharmacol 144(1):42–51

Silva RM, Oliveira FA, Cunha KMA, Maia JL, Maciel MAM, Pinto AC, Nascimento NRF, Santos FA, Rao VSN (2005) Cardiovascular effects of trans-dehydrocrotonin, a diterpene from *Croton cajucara* in rats. Vasc Pharmacol 43(1):11–18

Siriyong T, Srimanote P, Chusri S, Yingyongnarongkul BE, Suaisom C, Tipmanee V, Voravuthikunchai SP (2017) Conessine as a novel inhibitor of multidrug efflux pump systems in Pseudomonas aeruginosa. BMC Complement Altern Med 17(1):405

Slattery ML, Lundgreen A, Welbourn B, Wolff RK, Corcoran C (2012) Oxidative balance and colon and rectal cancer: interaction of lifestyle factors and genes. Mutat Res 734(1–2):30–40

Sur S, Pal D, Roy R, Barua A, Roy A, Saha P, Panda CK (2016) Tea polyphenols EGCG and TF restrict tongue and liver carcinogenesis simultaneously induced by N-nitrosodiethylamine in mice. Toxicol Appl Pharmacol 300:44–46

Tan MJ, Ye JM, Turner N, Hohnen-Behrens C, Ke CQ, Tang CP, Chen T, Weiss HC, Gesing ER, Rowland A, James DE, Ye Y (2008) Anti-diabetic activities of triterpenoids isolated from bitter melon associated with activation of the AMPK pathway. Chem Biol 15(3):263–273

Upadhyay A, Upadhyaya I, Kollanoor-Johny A, Venkitanarayanan K (2014) Combating pathogenic microorganisms using plant-derived antimicrobials: a minireview of the mechanistic basis. Biomed Res Int 2014:761741

Varghese R, Efferth T, Ramamoorthy S (2023) Carotenoids for lung cancer chemoprevention and chemotherapy: promises and controversies. Phytomedicine 116:154850

Venkatalakshmi P, Vadivel V, Brindha P (2016) Role of phytochemicals as immunomodulatory agents: a review. Int J Green Pharm 10(1):1–18

Winklhofer-Roob BM, Puhl H, Khoschsorur G, van't Hof MA, Esterbauer H, Shmerling DH (1995) Enhanced resistance to oxidation of low-density lipoproteins and decreased lipid peroxide formation during beta-carotene supplementation in cystic fibrosis. Free Radic Biol Med 18(5):849–859

Wu L, Ashraf MHN, Facci M, Wang R, Paterson PG, Ferrie A, Juurlink BH (2004) Dietary approach to attenuate oxidative stress, hypertension, and inflammation in the cardiovascular system. Proc Natl Acad Sci USA 101(18):7094–7099

Yoon SW (2012) Systemic inflammatory response as a prognostic factor in patients with cancer. In J Korean Tradit Oncol 17(1):1–7

Zakkar M, Van Der Heiden K, Luong LA, Chaudhury H, Cuhlmann S, Hamdulay SS, Krams R, Edirisinghe I, Rahman I, Carlsen H, Haskard DO, Mason JC, Evans PC (2009) Activation of Nrf2 in endothelial cells protects arteries from exhibiting a proinflammatory state. ATVB 29(11):1851–1857

Zhang Y, Duan W, Owusu L, Wu D, Xin Y (2015) Epigallocatechin-3-gallate induces the apoptosis of hepatocellular carcinoma LM6 cells but not non-cancerous liver cells. Int J Mol Med 35(1):117–124

Zhao Y, Sun X, Zhang T, Liu S, Cai E, Zhu H (2021) Study on the antidepressant effect of panaxynol through the IκB-α/NF-κB signaling pathway to inhibit the excessive activation of BV-2 microglia. Biomed Pharmacother 138:111387

Potential of Plant Extracts Against Viruses: Innovative and Ecological Strategy in Modern Agriculture and Healthcare

Nikolay Manchev Petrov, Mariya Ivanova Stoyanova, and Rajarshi Kumar Gaur

Abstract

Viruses have always been a major threat to the economy and global health due to their ability to mutate, resistance to therapies, and ability to spread rapidly. Several viruses have caused epidemics/pandemics among humans, resulting in millions of deaths. On the other hand, other viruses circulate continuously in the human population, reducing the quality of life and weighing down the health and economic systems of countries through hundreds of thousands of hospitalizations. Currently, there are only a few preferred and proven antiviral drugs at all, which, unfortunately, are only effective against a limited range of viruses. Natural products research can be an effective strategy for the development of new potent antiviral drugs.

Plant pathogenic viruses also present a challenge. They can cause epiphytotics and significant production losses, putting food resources in uncertainty. Antiviral substances in the agricultural sector are completely absent. The search for new antiviral substances from plant metabolites with improved efficacy and reduced side effects is of utmost importance due to the increasing species and strain biodiversity and expanded worldwide distribution. Many natural products have been identified as potential drugs in the recent past or used as models for the

N. M. Petrov (✉)
Virology Laboratory, New Bulgarian University, Sofia, Bulgaria
e-mail: npetrov@nbu.bg

M. I. Stoyanova
Institute of Soil Science, Agrotechnologies and Plant Protection (ISSAPP) "Nikola Pushkarov," Agricultural Academy, Sofia, Bulgaria

R. K. Gaur
Department of Biotechnology, Deen Dayal Upadhyaya Gorakhpur University, Gorakhpur, Uttar Pradesh, India

Uttar Pradesh Council of Agricultural Research, Uttar Pradesh, India

R. K. Gaur et al. (eds.), *Plant Molecular Farming*,
https://doi.org/10.1007/978-981-95-3823-2_4

development of new synthetic drugs. However, less than 15% of natural sources have been studied so far.

Keywords

Plant extracts · Viruses · Antiviral activity · Plant secondary metabolites · Plant farming · Ecological strategy · Agriculture · Healthcare

1 Introduction

Infectious diseases, and in particular viral diseases of humans, animals, and plants, in general, continue to be one of the main causes of high mortality in humans and animals and the greatly deteriorated quality and quantity of plant production in the agricultural sector. The constant emergence of new viral strains and variants that change the etiological characteristics of diseases is a major challenge for global health and requires increased efforts by a combined team of researchers to deal with them, due to their potential for the development of a pandemic and for millions of deaths annually (Baker et al. 2022).

Prevention and therapy against viral infections are not yet effectively developed, despite the rapid development of science and technology, so they still represent a major challenge to health. Natural products with promising antiviral effects against the replication of various viruses are increasingly being sought and developed. Despite the strong development of conventional antiviral drugs, their limitations have recently been caused by problems such as drug resistance, reduced efficacy, increased financial costs, and unwanted side effects (Strasfeld and Chou 2010; Chitalia and Munawar 2020; Fleßa and Marschall 2009; Morris 1994).

The identification and demonstration of bioactive compounds and their mechanism of action against specific sites in viral replication may help in the development of effective antiviral drugs. (Qiu et al. 2016). Plants have, during their evolutionary development, defended themselves against various biotic and abiotic stress factors by synthesizing a wide range of secondary metabolites with intricate chemical structures. These metabolites have been used by different cultures throughout history to seek healing and pain relief against various diseases. This highlights their potential as a source of valuable antiviral compounds and innovative products (Kumar Shakya and Arvind Kumar Shakya 2016; Guerriero et al. 2018; Isah 2019; Ramakrishna and Ravishankar 2011). Thus, in various natural sources of plant origin, different classes of chemical compounds with antiviral potential can be identified, such as alkaloids, terpenoids, inhibitors of genome synthesis, phenols, and others (Lin et al. 2014a, b; Chinthakindi et al. 2018; Song et al. 2005; Elbe and Buckland-Merrett 2017; Haque et al. 2019; Pal and Lal 2024).

On the other hand, natural products are the basis for uncovering new chemical compounds as better therapeutic agents against viral diseases. Unlike in the past, when there was no other alternative, such as plants to combat various diseases, in our times, due to various economic, commercial, and other factors, their potential against infectious diseases has been forgotten. Many viral infections, such as the common cold and some enterovirus ailments with viral etiology, simply disappear on their own without treatment as a result of an adequate and strong immune response and do not require treatment at all costs. In addition, a significant part of viral infections is spread by vectors, and through their insecticidal and other control measures, the spread and development of the viral disease are limited (Jaime et al. 2013; Pal and Lal 2024).

Despite the widespread availability of long-established antiviral drugs for the treatment of different viral diseases, the rapid evolution and change of viral strains and the subsequent emergence of new resistances necessitate the search and testing of new natural sources with combinations of substances, such as plant extracts, with promising sustained antiviral activity.

Numerous scientific studies have demonstrated that natural substances have promising antiviral properties. These compounds, derived from sources such as plants, animals, fungi, marine life, and bacteria, have shown the ability to inhibit various types of viruses. Plant-based antiviral agents, in particular, are appealing due to their broad availability, affordability, and generally lower toxicity compared to synthetic medications. Furthermore, plant extracts typically contain a variety of active compounds, increasing the probability of synergistically enhancing their effectiveness and lowering the chance of viruses developing resistance (Pal et al. 2014a, b; Saha et al. 2017; Jaime et al. 2013; Ntie-Kang and Svozil 2020; Atanasov et al. 2021; Ji et al. 2009; Dias et al. 2012; Pal and Lal 2024).

However, in comparison with the successful obtaining of antibacterial drugs from natural sources, the discovery of effective antiviral drugs has been relatively less successful. As obligate intracellular parasites, viruses depend on the host cell's metabolism for their replication and survival. Therefore, an effective antiviral drug must specifically block and inhibit viral processes without affecting the processes in normal cellular metabolism. For this reason, unlike microorganisms such as bacteria and fungi, broad-spectrum antimicrobial substances in viruses have a high toxicity due to the interference with cellular processes, which is not a desired effect in their development. Plants offer a huge and still unexplored resource of antiviral compounds, due to their evolutionary development of secondary metabolites for protection against various adverse environmental factors. About 1/5 of the tested plants show activities with antiviral potential (Rajbhandari et al. 2001).

For centuries, people have relied on medicinal plants (estimated at 20,000 species) to address diverse health conditions, due to the plants' rich content of various chemical substances (Gurjar et al. 2020). Synthetic drugs are more expensive, less accessible, and, because they contain active mono-substances, often lead to resistance, which necessitates the search for new alternatives of plant origin (Pal et al. 2006; Mohanta et al. 2007). A variety of in vitro and in vivo tests are used to identify and characterize newly discovered bioactive compounds with antiviral potential

(Pal et al. 2014a, b). About 80% of the world's population uses plants from traditional medicine to combat various infectious agents, thanks to their metabolic products with antiviral activities. In this way, pharmaceutical companies also produce novel antimicrobial drugs from natural plant sources (Denaro et al. 2020; Pal and Lal 2024).

The use of medicinal plants against various viral infections dates back several millennia, such as in Chinese medicine, the Egyptian Ebers Papyrus, and India (Mukherjee et al. 2013; Stiefel et al. 2006; Chung et al. 2014; Reid 1996). Such common examples are plants used to treat viral infections in ancient China, such as *Ephedra sinica, Andrographis paniculata,* and *Camellia sinensis* (Bensky et al. 2004; Fashner et al. 2012; Song and Seong 2007; Xu et al. 2008; Tran 2013). In Egypt, plant extracts (such as *Allium sativum* and *Echinacea purpurea* against various respiratory infections) have been prescribed against many diseases, including viral infections (Aboelsoud 2010; Rivlin 2001; Anadón et al. 2016; Hudson and Vimalanathan 2011).

Ayurveda, a holistic medical therapy dating back three millennia in India, describes numerous plants against various viral infections (Mukherjee et al. 2013; Singh et al. 2011; Alzohairy 2016; Manandhar et al. 2018). On the other hand, due to different climatic and geographical factors and the distribution of different plants, in Europe, the elderberry plant (*Sambucus nigra*) has been prescribed against respiratory viral infections (Helman 1978). The rising occurrence of viral infections and the growing resistance to standard treatments have led to a renewed interest in the use of medicinal plants as alternative therapeutic options (Strasfeld and Chou 2010; Sagaya Jansi et al. 2021; Wang et al. 2018).

In recent years, metabolomics has emerged as a vital tool for identifying various classes of antiviral secondary metabolites in plants, including flavonoids, polyphenols, alkaloids, and terpenoids, contributing to the discovery of approximately one-quarter of all antiviral drugs. The interaction between medicinal plants, viral infections, and metabolomics, together with the identified gaps in current scientific knowledge, offers a comprehensive perspective on the untapped potential of plants for developing new antiviral therapeutics and guiding future research efforts (Ghildiyal et al. 2020; Yeshi et al. 2022; Hussein and El-Anssary 2019).

2 Main Groups of Phytochemical Classes of Substances with Antiviral Activity

Several major groups of phytochemical classes with antiviral activities are found in plant species. They are obtained from different plant species and anatomical parts and have demonstrated distinct antiviral effects at specific stages of the viral replication cycle. The major groups of phytochemical classes known for their antiviral activities are shown in Table 1 (Pal and Lal 2024).

Table 1 Main groups of phytochemical classes with pronounced antiviral activity (mainly polyphenols)

Phytochemical	Antiviral activity	References
Phenolic compounds (mostly polyphenols)		
Apigenin	Influenza A, Hepatitis C Virus (HCV), enterovirus 71, FMDV	Ali et al. (2017)
Catechins	HBV, HSV, EBV, Adenovirus, HIV, HCV, influenza virus, DENV, JEV, TBEV, Zika virus (ZIKV), CHIKV, HTLV-1, Rotavirus, Enterovirus EV71, EBOV, PRRSV, VHSV, IHNV, SVCV, GCRV	Xu et al. (2017)
Curcumin	ZIKV, CHIKV, Norovirus, Influenza, DENV type 2, RSV, HBV, HCV, HPV, HIV, Cucumber Mosaic Virus (CMV), EV71	Nabila et al. (2020), Pal and Lal (2024)
Delphinidin	Influenza virus, HSV type 1 and 2, RV, Adenovirus 36, BT2, T4 Simian rotavirus SA-11, HAV, FCV-F9, MNV-1, HCV, WNV, DENV, ZIKV	Sharif et al. (2020)
Epigallocatechin gallate	Influenza, HSV, Hepatitis C virus	Song et al. (2005)
Elderberry	Influenza virus, Common cold	Tiralongo et al. (2016)
Hesperetin	HSV, Dengue, RSV	Qiu et al. (2016)
Myricetin	HIV, RLV, Influenza	Semwal et al. (2016)
Naringenin	DENV, HCV	Kerdudo et al. (2014)
Punicalagin	HSV 1 and 2, Influenza A virus	Sreekanth et al. (2018)
Quercetin	JEV, Influenza A, EBV, MAYV, RV, HCV, HSV, Coronaviruses	Ben-Shabat et al. (2020), Wu et al. (2020a, b)
Kaempferol	Influenza, Hepatitis B virus, HIV	Lin et al. (2014a, b)
Resveratrol	HIV, HSV, Influenza virus	Kerdudo et al. (2014), Pal and Lal (2024)
Routine	HSV, Influenza virus, Coxsackievirus	Elbe and Buckland-Merrett (2017)
Silibinin	Hepatitis C Virus	Ripoli et al. (2016)
Terpenes		
Andrographolide	DENV, CHIKV, HBV, HCV, HSV1, EBV, HIV, HPV16, Influenza	Panraksa et al. (2017), Li et al. (2020)
Artemisinin	Hepatitis B virus, Hepatitis C virus, Dengue virus	Ghosh and Basak (2013)
Betulinic acid	HIV, Herpes simplex virus, influenza virus	Ghosh and Basak (2013)

(continued)

Table 1 (continued)

Phytochemical	Antiviral activity	References
Celastrol	Dengue virus, HIV, Hepatitis C virus	Yu et al. (2017), Narayan et al. (2011) and Tseng et al. (2017)
Glycyrrhizin	Coronaviruses	Rizzato et al. (2017), Wang et al. (2015), Thyagarajan et al. (2002), Marisa et al. (2013), Fiore et al. (2008), Richard (2021) and Schmidt et al. (1997
Camphene Limonene, α-Pinene, α-Terpinene,	Herpes simplex virus, influenza	Astani and Schnitzler (2014), Pal and Lal (2024)
Isoborneol	HSV-1	Armaka et al. (1999),
Carvacrol, Thymol	HSV, Influenza virus, Enteroviruses	Ghosh and Basak (2013)
Coumarins		
Coumestrol	HSV 1 and 2	Argenta et al. (2018)
Alkaloids		
Berberine	Influenza, HSV-1, HCMV, Hepatitis B Virus, Hepatitis C Virus, HIV	Cecil et al. (2011), Wu et al. (2011), Chin et al. (2010), Lin et al. (2014a, b), Mukhopadhyay et al. (2016) and Hung et al. (2019(
Homonojirimycin	Influenza	Zhang et al. (2013a, b)
Camptothecin	Herpesviruses, Poxviruses	Zandi et al. (2011)
Allantoin Atropine, Capsaicin, Trigonelline, Vincamine	HSV 1	Ozcelik et al. (2011)
Homoharringtonine	HSV 1	Amorim et al. (2018), Dong et al. (2018)
Tubingensin A	HSV 1	Tepaske et al. (1989)
Harmine	HSV 2	Wu et al. (2020a, b)
Anisomycin, Castanospermine, Tomatadine Cherylline	DENV	Ka et al. (2021), Quintana et al. (2020), Whitby et al. (2005) and Diosa-Toro et al. (2019)
Schizanthine Z	SARS-CoV-2	Garg and Roy (2020)
Skimmianine, Matrine-type alkaloids	Hepatitis B Virus	Zhang et al. (2017), Zhang et al. (2018)
Cyclopamine	RSV	Risso-Ballester et al. (2021)

(continued)

Table 1 (continued)

Phytochemical	Antiviral activity	References
6-*O*-butanoyl Castanospermine (celgosivir), *n*-butyl deoxynojirimycin	Bovine viral diarrhea virus (BVDV)	Whitby et al. (2004), Ouzounov et al. (2002)
Matrine-type alkaloids, 12-Hydroxyoxysophocarpine, Aranotin, Caffeine	Coxsackieviruses	Pan et al. (2015), Swallow (1978), Becker (1980) and Perez (2003)
Emetine	Newcastle disease virus (NDV), Ebola	Khandelwal et al. (2017), Yang et al. (2018)
Lycorine	Poliovirus	Ieven et al. (1982)
Vinblastine, Vincristine	HCMV, Influenza A virus, HIV	Haque et al. (2019)
Quinine	Influenza virus, Coxsackieviruses	Henss et al. (2018)
Polysaccharides		
β-glucans, carrageenans, fucoidans	Influenza virus, HPV, HSV	Cheung et al. (2015)
Tannins		
Tannins	Influenza virus	Pal and Lal (2024)
Saponins		
Saponins	HSV, Influenza virus	

2.1 Polyphenols

Polyphenols are the most abundant group of compounds found in many plant species and products, which have low toxicity and high antiviral activity. The group includes diverse chemical compounds such as flavonoids, phenolic acids, and tannins. They have a high potential to block viral replication of various viral infections, including orthomyxoviruses, herpesviruses, retroviruses, hepatitis, and coronaviruses. Their mechanism of antiviral activity is the inhibition of specific viral enzymes, control of viral gene expression, and blocking the attachment and entry of the virion into the host cell. Polyphenols also have antioxidant and immunomodulatory properties, through which the immune response against viral infections is enhanced (Cushnie and Lamb 2011).

Phenolic compounds are organic molecules of plant origin that have at least one phenolic group in their structure. They are the most widely distributed plant secondary metabolites and are found in all plant organs of fruits, vegetables, cereals, and legumes and their products. These include flavonoids, tannins, and phenolic acids. The specific antiviral mechanism of phenols varies depending on both the specific type of phenolic compound and the virus involved, mainly inhibiting viral infection, including the disruption of the viral supercapsid and preventing the adsorption of the virus to the host cell (Kumar et al. 2023; Badshah et al. 2021; Vilhelmova-Ilieva et al. 2020; Orfali et al. 2021; Goc et al. 2021; Yamada et al. 2009; Lin et al. 2005). Inhibition of various activities of viral proteins leads to the cessation of the viral replication cycle, which ultimately prevents its multiplication and release from the cell (Wink 2020; Tirado-Kulieva et al. 2022; Montenegro-Landívar et al. 2021).

Notable polyphenolic compounds that disrupt viral replication include the tannin epigallocatechin-3-gallate, a catechin derived from green tea. This compound inhibits the M2 protein of the influenza A virus by promoting IFN-λ expression in human lung epithelial cells. It also inhibits the binding of the HIV-1 supercapsid glycoprotein gp120 to the cellular CD4 receptor on the surface of immune cells. The flavonoids quercetin, baicalein, myricetin, and quercetagetin inhibit the enzymes reverse transcriptase and cellular DNA and RNA polymerases of HIV-1 (Tirado-Kulieva et al. 2022; Steinmann et al. 2013; Wu et al. 2021; Zhu et al. 2020; More 2021; Ono et al. 1990).

Sodium ferulate, a phenolic acid found in plants, inhibits influenza virus replication by activating TLR7 and TLR9 receptors, which leads to the translocation of the transcription factor IRF7 into the cell nucleus and the production of type-1 interferons. Chicoric acid and several dicaffeoylquinic acids interact with the HIV integrase enzyme, which is essential for the integration of viral DNA into the host genome. In addition, gallic acid exhibits anti-HSV-2 activity mainly by preventing the virus from attaching to host cells, thereby restricting its spread. Various polyphenolic compounds have immunomodulatory effects and induce apoptosis in infected cells, limiting the spread of viral infection. This activates the production of various cytokines and chemokines of the host (Zhu et al. 2019; Kratz et al. 2008; Giovinazzo et al. 2020; Dejani et al. 2021; Wang et al. 2022).

2.1.1 Flavonoids

Flavonoids are a class of natural polyphenolic compounds in plants with potential antiviral activities.

Quercetin: Quercetin is a flavonoid found in various fruits, vegetables, and medicinal plants that has antiviral activity against influenza, herpes simplex, and coronaviruses by inhibiting viral replication, modulating the immune response, and reducing inflammation (Wu et al. 2020a, b).

Epigallocatechin gallate is a flavonoid abundant in green tea and has antiviral properties against influenza, herpes, and hepatitis C by inhibiting viral entry and replication, and exerting immunomodulatory effects (Song et al. 2005).

Kaempferol: Kaempferol, a flavonoid present in numerous fruits, vegetables, and medicinal plants, exhibits antiviral effects against influenza, hepatitis B, and HIV by blocking viral replication, limiting viral entry, and exerting immunomodulatory effects (Lin et al. 2014a, b).

Hesperetin: Hesperetin, a flavonoid present in citrus fruits, demonstrates pronounced antiviral activity against herpes simplex, dengue, and respiratory syncytial viruses by inhibiting viral replication, preventing viral attachment, and modulating the immune response (Qiu et al. 2016).

Routine: Rutin, a flavonoid widely present in buckwheat, citrus fruits, and various other plants, exhibits antiviral effects against herpes simplex, influenza, and enteroviruses such as coxsackievirus (Elbe and Buckland-Merrett 2017).

2.2 Alkaloids

Alkaloids are a class of nitrogen-containing organic compounds naturally found in many plant species, known for their diverse biological activities, including antiviral properties. Compounds such as berberine, vinblastine, and reserpine have been investigated for their ability to inhibit the replication of various viruses, including influenza, herpes simplex, and dengue virus. These naturally occurring molecules typically contain at least one nitrogen atom and exhibit pharmacological effects against a range of microbial diseases, showing broad-spectrum activity against both DNA and RNA viruses. Some of the most well-known plant-derived alkaloids include cocaine, morphine, and quinine (Moradi et al. 2018; Abookleesh et al. 2022). The antiviral mechanisms of alkaloids primarily involve the inhibition of:

(1) DNA and RNA replication
(2) RNA translation
(3) Protein synthesis
(4) DNA intercalation
(5) Enzymatic activities
(6) Ribonucleoprotein complex translocation
(7) DNA synthesis
(8) Protein synthesis

Numerous studies have focused on the role of alkaloids in the treatment and prevention of viral infections. In addition to their direct antiviral effects, alkaloids can modulate the human immune response, particularly by enhancing humoral immunity, thereby contributing to resistance against viral pathogens (Abookleesh et al. 2022; Gürü et al. 2020; Peng et al. 2003).

Alkaloids represent a chemically diverse group of naturally occurring compounds widely recognized for their pharmacological properties, including significant antiviral activity.

Berberine: Berberine is an alkaloid primarily found in plants, such as *Berberis* spp., *Coptis* spp., and *Hydrastis canadensis*, which demonstrated antiviral effects against viruses such as HSV, influenza, HIV, and hepatitis B and C. Its mechanism of action involves blocking the attachment of virions to host cells (Lin et al. 2014a, b).

Camptothecin: Camptothecin is an alkaloid extracted from the Chinese tree Camptotheca acuminata. It has been scientifically proven to have anticancer properties and has demonstrated antiviral activity against several DNA viruses, such as herpesviruses and poxviruses, by targeting their viral DNA topoisomerase and blocking the viral replicative cycle (Zandi et al. 2011).

Vinblastine and vincristine: Vinblastine and vincristine are alkaloids obtained from the Madagascar periwinkle (Catharanthus roseus). Like camptothecin, they have anticancer effects and antiviral activity against various viruses, such as human cytomegalovirus (HCMV), influenza A, and HIV. They inhibit viral protein synthesis and prevent virion entry into the cell (Haque et al. 2019).

Quinine: Quinine is an alkaloid isolated from the bark of the Cinchona tree and is used primarily as an antimalarial agent. Its antiviral activity is directed against influenza A, enteroviruses such as coxsackie, and chikungunya virus by inhibiting the synthesis of viral nucleic acids (Henss et al. 2018).

2.3 Polysaccharides

Polysaccharides are complex carbohydrates made up of monosaccharide residues and are found in various plant species, including fungi and algae. Polysaccharides with antiviral activity include β-glucans, carrageenans, and fucoidans, which have the potential to inhibit influenza, human papillomavirus (HPV), and herpes simplex virus. Their main mechanism of action is to inhibit the attachment and entry of virions into host cells, block viral replication and specific viral translation, and stimulate the immune system and have immunomodulatory properties by enhancing the activity of macrophages, natural killer cells, and lymphocytes. Ultimately, this can lead to increased production of cytokines and antibodies, inducing and directing the antiviral immune response (Cheung et al. 2015).

2.4 Terpenoids

Terpenoids are highly volatile and flammable organic compounds and are a major component of all living cells, both in primary and secondary cellular metabolism, located primarily in the leaves and fruits of higher plants, where they sometimes contribute to their colors. They exist in six categories: hemiterpenes, monoterpenes, sesquiterpenes, diterpenes, sesterpenes, triterpenes, and tetraterpenoids. As aromatic compounds, they are responsible for the taste and aroma of plants. Common terpenoids include citral, menthol, camphor, and salvinorin (Abdallah and Quax 2017; Rodriguez-Concepcion et al. 2018; Yazaki et al. 2017; Huang et al. 2021; Ashour et al. 2010; Hassan et al. 2015).

Glycyrrhizin is a terpenoid with antiviral activity found in the roots of the licorice plant (Glycyrrhiza glabra). In China and India, it has been used primarily for the treatment of viral respiratory infections and liver infections caused by hepatitis. It inhibits the adsorption and entry of coronaviruses (Rizzato et al. 2017; Wang et al. 2015; Thyagarajan et al. 2002; Marisa et al. 2013; Fiore et al. 2008; Richard 2021; Schmidt et al. 1997).

Isoborneol, a monoterpene found in various essential oils, has been shown to completely inhibit herpes simplex virus type 1 (HSV-1) replication at a concentration as low as 0.06% (Armaka et al. 1999). Similarly, the monoterpenes β-pinene and limonene, also present in essential oils, exhibit strong anti-HSV-1 activity by reducing viral infectivity by 100%, primarily through direct interaction with free viral particles (Astani and Schnitzler 2014).Celastrol, a pentacyclic triterpenoid, has demonstrated antiviral activity against dengue virus, human immunodeficiency virus (HIV), and hepatitis C virus (HCV) by inhibiting viral replication (Yu et al.

2017; Narayan et al. 2011; Tseng et al. 2017).In addition, a carotenoid pigment extracted from *Natrialba* sp. in Egypt has shown antiviral potential against both HCV and HBV by inhibiting RNA and DNA polymerase activity (Hegazy et al. 2020; Yim et al. 2021).

Thymol and carvacrol, as well as monoterpenoids found in thyme and oregano, exhibit strong antiviral activity against herpes simplex virus, influenza virus, and enteroviruses (Ghosh and Basak 2013).

Artemisinin, a sesquiterpenoid from *Artemisia annua*, has shown potent antiviral effects against hepatitis B virus, hepatitis C virus, and dengue virus (Ghosh and Basak 2013).

Betulinic acid, a triterpenoid found in many plants, has shown antiviral activity against HIV, herpes simplex virus, and influenza virus (Ghosh and Basak 2013).

3 Antiviral Activity of Plant Extracts against Human Viruses

The antiviral activity of plant extracts against human viruses is presented in Table 2 (Pal and Lal 2024).

4 Mechanism of Action of Plant Extracts Against Human Viruses

The life cycle of a virus can be represented as follows:

I. *Attachment*: The virus binds to specific receptors on the surface of the host cell through its glycoprotein ligands.
II. *Entry*: The virus penetrates the host cell either by fusing with the cell membrane or via endocytosis.
III. *Uncoating*: The viral capsid breaks down, releasing the viral genome into the host cell.
IV. *Replication*: The viral genome is replicated using a viral RNA-dependent RNA polymerase or DNA polymerase, generating new copies of viral RNA or DNA.
V. *Translation*: Viral RNA is translated into proteins, including structural components and enzymes essential for replication.
VI. *Assembly*: Newly synthesized viral genomes and proteins are assembled into complete virus particles.
VII. *Cell exit*: Newly assembled virions are released from the host cell, usually by budding through the host cell membrane. To develop new antiviral drugs, researchers are focusing their efforts on specific targets in the viral life cycle. Overall, a complete and comprehensive understanding of the viral replication cycle and its interactions with the host cell is essential for the development of effective antiviral drugs.

Table 2 Antiviral activity of plant extracts against human viruses

Family	Plant species	Antiviral activity
Acanthaceae	*Andrographis paniculata*	Influenza, respiratory infections, HSV
Acoraceae	*Acorus calamus*	HSV, influenza
Adoxaceae	*Sambucus ebulus*	Respiratory infections, influenza
	Sambucus nigra	Influenza, respiratory infections, HSV
Amaryllidaceae	*Allium cepa*	SARS-COV, Newcastle disease virus
	Allium sativum	Influenza, DENV, common cold virus, influenza virus A and B, HIV, HSV-1, HSV-2, Newcastle disease virus, HSV
Anacardiaceae	*Rhus coriaria*	HSV, influenza
Apiaceae	*Anethum graveolens*	HSV, influenza
	Angelica archangelica	Respiratory infections, influenza
	Bupleurum chinense	Respiratory infections, influenza
	Centella asiatica, Foeniculum vulgare	HSV, influenza
Apocynaceae	*Vinca minor*	HSV, influenza
Araliaceae	*Eleutherococcus senticosus*	Respiratory infections, influenza
	Hedera colchica	Influenza, respiratory infections, hepatitis viruses
	Panax ginseng	Respiratory infections, influenza
Arecaceae	*Areca catechu*	HSV, influenza
Asparagaceae	*Anemarrhena asphodeloides*	HSV, influenza
Asphodelaceae	*Aloe vera*	HSV, respiratory infections, HIV
Asteraceae	*Achillea millefolium, Arctium lappa, Artemisia parviflora*	HSV, influenza
	Echinacea purpurea	Respiratory infections, influenza
	Eclipta alba, Matricaria chamomilla, Silybum marianum, Tagetes erecta, Taraxacum mongolicum, Taraxacum officinale	HSV, influenza
	Tussilago farfara	Respiratory infections, influenza
Berberidaceae	*Berberis vulgaris, Epimedium* spp.	HSV, influenza
Bignoniaceae	*Tabebuia avellanedae*	HSV, influenza
Burseraceae	*Boswellia serrata*	HSV, influenza
	Commiphora mukul	HSV, influenza
Caesalpiniaceae	*Hardwickia binata*	HSV, influenza
Capparidaceae	*Capparis spinosa*	HSV-2, HSV-1
Combretaceae	*Terminalia arjuna*	HSV, influenza
	Terminalia bellirica	Respiratory infections, influenza
	Terminalia chebula	HSV, influenza
Crassulaceae	*Rhodiola rosea*	Respiratory infections, influenza
	Rhodiola rosea	HSV, respiratory infections, influenza
Cucurbitaceae	*Cucumis sativus*	HSV, influenza

(continued)

Table 2 (continued)

Family	Plant species	Antiviral activity
Cupressaceae	*Cryptomeria japonica, Cupressus torulosa, Thuja occidentalis*	HSV, influenza
Dipterocarpaceae	*Shorea robusta Triterpenic acid, tannic acid*	HSV, influenza
Ebenaceae	*Diospyros kaki*	Influenza virus H3N2, H5N3, HSV-1, vesicular stomatitis virus, Sendai virus, poliovirus, coxsackievirus, adenovirus, rotavirus, feline calicivirus
Eucommiaceae	*Eucommia ulmoides*	HSV, influenza
Euphorbiaceae	*Euphorbia hirta*	HSV, influenza
Fabaceae	*Astragalus membranaceus, Astragalus propinquus*	Respiratory infections, influenza
	Glycyrrhiza glabra	Respiratory infections, influenza
	Glycyrrhiza glabra	Influenza, respiratory infections, hepatitis viruses
	Glycyrrhiza uralensis	Respiratory infections, influenza
	Trigonella foenum-graecum	HSV, influenza
Gentianaceae	*Swertia chirayita*	HSV, influenza
Geraniaceae	*Pelargonium sidoides*	Respiratory infections, influenza
Ginkgoaceae	*Ginkgo biloba*	Respiratory infections, influenza, HSV
Hypericaceae	*Hypericum perforatum*	HSV, influenza
Juglandaceae	*Juglans regia*	HSV, influenza
	Zataria multiflora	HSV-1
Lamiaceae	*Hyssopus officinalis*	Respiratory infections
	Melissa officinalis	HSV, influenza
	Mentha piperita	HSV, respiratory infections, infections, influenza
	Ocimum sanctum	Respiratory infections, influenza
	Prunella vulgaris	HSV, influenza
	Salvia officinalis	HSV, influenza, respiratory infections
	Satureja hortensis, Scutellaria baicalensis	HSV, influenza
	Vitex negundo	Respiratory infections, influenza
Lauraceae	*Cinnamomum cassia, Cinnamomum verum*	HSV, influenza
Lythraceae	*Punica granatum*	HSV, influenza
Malvaceae	*Althaea officinalis*	Respiratory infections, influenza
	Hibiscus sabdariffa	HSV, influenza
Menispermaceae	*Tinospora cordifolia*	Respiratory infections, influenza
Moraceae	*Artocarpus integrifolia*	(SA-11) and human (HCR3), rotaviruses, HIV
	Morus alba	HSV, influenza
Moringaceae	*Moringa oleifera*	HSV, influenza

(continued)

Table 2 (continued)

Family	Plant species	Antiviral activity
Myrtaceae	*Eucalyptus globulus*	Respiratory infections, influenza
	Syzygium aromaticum, Syzygium cumini	HSV, influenza
Oleaceae	*Ligustrum lucidum*	HSV, influenza
	Olea europaea	HSV, respiratory infections, infections, influenza
Paeoniaceae	*Paeonia lactiflora*	HSV, influenza
Passifloraceae	*Passiflora caerulea, Passiflora edulis, Passiflora incarnata*	HSV, influenza
Phyllanthaceae	*Phyllanthus emblica*	Respiratory infections, influenza
Piperaceae	*Piper longum*	HBV
Plantaginaceae	*Picrorhiza kurroa*	HSV, influenza
Poaceae	*Cynodon dactylon*	HSV, influenza
Polygonaceae	*Polygonum cuspidatum, Rheum palmatum,* *Rheum palmatum,* *Rumex crispus*	HSV, influenza
Ranunculaceae	*Actaea racemosa, Cimicifuga racemosa, Coptis chinensis*	HSV, influenza
Rosaceae	*Rosa damascena, Rubus idaeus*	HSV, influenza
Rubiaceae	*Rubia cordifolia*	HSV, influenza
	Uncaria tomentosa	HSV, influenza
Rutaceae	*Aegle marmelos*	HCVB 1–6
	Citrus aurantium	HSV, influenza
	Zanthoxylum bungeanum	HSV, influenza
Schisandraceae	*Schisandra chinensis*	Respiratory infections, influenza
Solanaceae	*Withania somnifera*	Respiratory infections, influenza
Theaceae	*Camellia sinensis*	Influenza, respiratory infections, HSV
Urticaceae	*Urtica dioica*	Respiratory infections, influenza
Violaceae	*Viola diffusa*	HBV
Vitaceae	*Cissus quadrangularis, Vitis vinifera*	HSV, influenza
Zingiberaceae	*Curcuma aromatica*	HSV, influenza
	Curcuma longa	Influenza, respiratory infections, HSV
	Zingiber officinale	Respiratory infections, influenza
Zygophyllaceae	*Balanites aegyptiaca*	VSV T2
	Tribulus terrestris	HSV, influenza

4.1 Inhibition of Virus Attachment and Entry

Therapeutic regimens for inhibiting the viral replication cycle target various important steps, such as adsorption and fusion of the virion with the host cell, through interaction with surface receptors or coreceptors, which leads to fusion of the supercapsid with the cell membrane and uncoating of the viral genome. Such are mannose-specific lectins of plant origin from the genera *Galanthus* and *Hippeastrum* that block the glycoproteins of the viral capsid and inhibit the entry of the virus into the host. Seaweed extracts, carrageenans, and heparin have demonstrated antiviral effects against the dengue virus. The viral capsid and supercapsid are specific targets for antiviral agents against viruses such as herpes, orthomyxoviruses, paramyxoviruses, rhabdoviruses, coronaviruses, retroviruses, arenaviruses, togaviruses, flaviviruses, and bunyaviruses (Frederico et al. 2017; Zheng et al. 2019).

4.2 Interfering with Viral Replication

A key antiviral mechanism is the inhibition of viral replication, commonly exerted by plant alkaloids. They act at various stages of viral replication, making them promising broad-spectrum antiviral agents, particularly effective in early stages of infection. Compounds like emetine, piperine, and lycorine disrupt viral RNA and DNA synthesis, preventing the virus from replicating within host cells. Emetine blocks viral RNA synthesis by interfering with host ribosomes, effectively halting the replication of viruses such as Zika (Shen et al. 2024). Piperine, found in black pepper, inhibits viral polymerases, reducing replication of HSV and HCV (Kumar et al. 2018). Lycorine, from *Amaryllis* species, interferes with viral RNA and protein synthesis, impeding the production of essential viral components (Chen et al. 2024).

4.3 Viral Genome Modifiers and Protein Processing

Other chemical substances target viral transcription and translation. They can inhibit reverse transcription, integration, replication, transcription, and translation. Viral nucleic acid is a major target for effective and specific antiviral chemotherapy by various medicinal plants. The binding of *Calophyllum lanigerum* to the active site of the reverse transcriptase enzyme is irreversible, thus inhibiting the activity of the enzyme. (Meganck and Baric 2021; Taliansky et al. 2021; Alamgir 2017; Mohan et al. 2020).

4.4 Inhibitors of Virus Assembly and Release

Many antiviral compounds act by disrupting the assembly of newly synthesized viral proteins and preventing their release from host cells. Protease inhibitors, a key class of antiviral agents, function by blocking the cleavage of viral polyproteins, thereby interfering with virus assembly. In contrast, neuraminidase inhibitors like oseltamivir and zanamivir inhibit the release of influenza viruses from infected cells, effectively preventing the spread of the virus between cells. Over 30 plant-derived protease inhibitors with antiretroviral activity have been identified from species such as *Eclipta prostrata, Alpinia galanga, Zingiber zerumbet, Coccinia grandis, Boesenbergia* (Menéndez-Arias and Gago 2013; Zeedan et al. 2020),

Pandurata, *Cassia garretiana*, and *Orostachys japonicus* (Menéndez-Arias and Gago 2013; Lou et al. 2014; Singh et al. 2020).

4.5 Other Targets Such as Immunomodulators

The most commonly used immunomodulators are interferons, interleukins, and colony-stimulating factors. Interferon-induced polypeptides and glycoproteins serve as mediators for inducing the production of enzymes that inhibit viral replication in the cell (Zeedan et al. 2017, Zeedan et al. 2019a, b). Interleukins are involved in the stimulation, growth, differentiation, maturation, and regulation of immune cells, which can help neutralize viruses (Andersen et al. 2020). β-sitosterol from the genus Nigella improves cellular immune responses by enhancing the activity of natural killer (NK) cells and cytotoxic lymphocytes (CTL). Extracts obtained from sitosterolia exhibit the ability to activate lymphocytes (Zeedan et al. 2017, Zeedan et al. 2019a, b; Ali et al. 2021a, b). *Echinacea purpurea* has strong immunostimulatory properties. Proteins extracted from *Allium sativum* (garlic) have mitogenic activity against human lymphocytes, splenocytes, and thymocytes (El-Sherbiny et al. 2021).

5 Antiviral Effects of Herbal Plants against a Number of DNA and RNA Veterinary Viruses

5.1 Herpesviruses

Several members of the human herpesvirus (HHV) family are responsible for a range of human diseases. The majority of HHV infections are caused by HSV-1 and HSV-2. Other herpesviruses that contribute to various diseases include HHV-3 (varicella zoster virus), HHV-4 (Epstein-Barr virus), HHV-5 (cytomegalovirus), HHV-6, HHV-7, and HHV-8. Traditionally, most studies on medicinal plants have focused on their effects against a single herpesvirus, namely HSV. For instance, an aqueous extract from plant material has been used to treat HSV-related infections. The phytochemical podophyllotoxin, isolated from an aqueous extract of *Podophyllum peltatum*, has been found to inhibit HSV-1 (Bedows and Hatfield 1982). In addition to aqueous extracts, organic solvent extracts from various plants have also demonstrated anti-HSV activity, indicating the presence of diverse antiviral substances in medicinal plants. For example, two compounds, geraniin and 1,3,4,6-tetra-O-galloyl-β-D-glucose (1346TOGDG), isolated from the acetone extract of *Phyllanthus urinaria*, have shown inhibitory effects against both HSV-1 and HSV-2 (Yang et al. 2007). HSV infections are particularly problematic for immunocompromised individuals, such as those infected with HIV. Moreover, the emergence of resistant strains of the herpesvirus, which are also resistant to conventional antiviral treatments, presents an ongoing challenge. As a result, there is a continuous search for antiviral agents capable of combating resistant virus strains or preventing the development of viral resistance. One promising example is the aqueous extract of *Carissa edulis*, a medicinal plant native to Kenya, which has demonstrated significant anti-HSV activity in vitro and in vivo against both wild-type and resistant strains of HSV (Tolo et al. 2006).

5.2 Viral Hepatitis

Viral hepatitis, marked by liver inflammation, is caused by five major viruses: hepatitis A, B, C, D, and E. While all these viruses can lead to acute infections, hepatitis B, C, and D are particularly concerning due to their ability to cause chronic infections. In traditional medicine, plants from the *Phyllanthus* genus (*Euphorbiaceae* family) have been widely used to treat viral hepatitis. Clinical studies have been conducted to compare the effects of different *Phyllanthus* species, such as *P. amarus*, *P. niruri*, and *P. urinaria*, in managing hepatitis B infections (Wang et al. 1995). Molecular research has shown that *P. amarus* exhibits anti-HBV activity by inhibiting HBV polymerase and suppressing mRNA transcription, likely through interactions with HBV enhancer I and transcription factors C/EBPα and C/EBPβ (Lee et al. 1996; Ott et al. 1997). Similarly, microarray analysis of ethanol extracts from Phyllanthus nanus revealed antiviral effects against HBV, notably through the upregulation of genes such as annexin 7 (Anx7) (Lam et al. 2006). However, the screening of antiviral compounds remains challenging, largely due to the limited availability of suitable small animal models. However, for human HBV infection, the duck hepatitis B virus (DHBV) model provides an effective system for testing potential antiviral agents (Niu et al. 1990). This model has been instrumental in evaluating the anti-HBV effects of various medicinal plants. Despite prior observations suggesting antiviral activity in *Phyllanthus* species, early studies revealed that *P. amarus* did not significantly inhibit DHBV-DNA replication in vivo (Niu et al. 1990).

Further studies comparing the antiviral effects of *Phyllanthus amarus* and *Phyllanthus maderaspatensis* in the DHBV model concluded that these plants lack the therapeutic potential against HBV previously suggested (Munshi et al. 1993a, b). However, *Phyllanthus nanus*, another species in the genus, demonstrated significant inhibitory activity against DHBV in a primary cell culture model (Lam et al. 2006). These conflicting findings highlight the need for further molecular-level investigations to confirm the preliminary clinical data obtained from the DHBV model.

Beyond assessing the therapeutic potential of *Phyllanthus* species, the DHBV model has been instrumental in detecting other plants with potential antiviral properties suitable for further clinical trials. For instance, a screening of 56 Chinese medicinal herbs identified *Ardisia chinensis* and *Pithecellobium clypearia* as promising anti-DHBV candidates (Leung et al. 2006). Additionally, the model facilitated the discovery of flavones from *Oenanthe javanica Blume DC* (OjF) as potent inhibitors of HBsAg and HBeAg secretion—key factors in HBV pathogenesis—in 2.2.15

cells. OjF also reduced DHBV-DNA levels in an HBV-infected duck model (Wang et al. 2005).

In some cases, medicinal plants are combined to enhance their therapeutic efficacy. This approach has also been explored for viral hepatitis treatment. A liquid fermentation broth of *Ganoderma lucidum* combined with an aqueous extract of *Radix Sophorae flavescentis* demonstrated potent anti-HBV activity in both in vitro and in vivo models. Notably, co-fermentation of these two medicinal plants produced a more potent antiviral effect than simple mixing (Li et al. 2006).

Hepatitis C, another major cause of viral hepatitis, has also been the focus of herbal medicine research. A study evaluating the efficacy and safety of herbal treatments for viral hepatitis analyzed data from the Cochrane Collaboration, MEDLINE, EMBASE, and BIOSIS, along with additional searches in five Chinese and one Japanese journal, incorporating findings from randomized clinical trials (Liu et al. 2003). Among 13 previously completed trials, only four adhered to proper methodologies. Few herbal compounds showed direct anti-HCV activity; however, *Silybin* and *Oxymatrine* demonstrated significant therapeutic effects, including reductions in serum aspartate aminotransferase (AST) and gamma-glutamyl transpeptidase levels, as well as HCV clearance. Additionally, Chinese herbal mixtures such as Bing Gan Tang, Yi Zhu Decoction, and Yi Er Gan Tang, either alone or in combination with antiviral agents, contributed to HCV RNA clearance and normalization of serum alanine aminotransferase (ALT) levels.

This study also summarized various adverse reactions associated with herbal treatments (Liu et al. 2003). Such findings emphasize the necessity of strict pharmacovigilance programs and the involvement of international regulatory agencies to ensure the safety and efficacy of herbal medicines.

Efforts to identify medicinal plants with anti-HCV properties are ongoing, particularly in resource-limited regions where infection rates are high and an effective vaccine is unavailable. Studies have shown that methanolic extracts from *Acacia nilotica* L. Willd ex Delile, *Boswellia carterii*, *Embelia schimperi*, *Quercus infectoria*, and *Trachyspermum ammi*, as well as aqueous extracts from *Piper cubeba*, *Q. Infectoria,* and *Syzygium aromaticum*, exhibit inhibitory effects against HCV (Hussein et al. 2000). Additionally, phytochemicals such as catechin, glycyrrhizin, polyterols, and silymarin have been explored as potential anti-HCV agents (Jassim and Naji 2003; Patrick 1999).

6 Antiviral Activity of Plant Extracts Against Plant Viruses

6.1 Antiviral Activity of Plant Extracts Against Potato Virus Y (PVY)

PVY most commonly infects susceptible pepper cultivars that lack resistance genes such as *pvr22* and *Pvr4* (Ben Khalifa et al. 2009; Petrov et al. 2024). Among viruses in its genus, PVY infects the widest range of hosts—around 500 species across 36 botanical families—with plants from the *Solanaceae*, *Chenopodiaceae*, and

Commelinaceae families being the most frequently affected. In potatoes, typical symptoms include leaf mosaic and wrinkling; mild to severe mottling that often causes leaf petiole deformation and folding; chlorosis and necrosis; vein necrosis; necrotic spots and rings on tubers; and leaf drying and drop. Secondary infections lead to stunted, brittle plants. In peppers grown both in greenhouses and outdoors, symptoms include leaf spots, mosaics, whitening, wrinkling, vein necrosis, defoliation, petiole necrosis, and fruit deformation and necrosis, rendering the fruits unsalable. In some cases, systemic PVY infection causes necrosis in the stem and apical bud of pepper plants, which can be fatal. Depending on the infection's timing and severity, yield losses in pepper crops can reach up to 100% (Arogundade et al. 2019; Petrov et al. 2024).

The antiviral activity of *Hypericum perforatum* was demonstrated by 10% aqueous concentrations of its liquid fraction and 5% of the soft fraction from methanol extracts, both of which significantly reduced PVY levels in virus-inoculated tobacco plants, as measured by DAS-ELISA (Petrov et al. 2015a, 2024).

The antiviral properties of the methanol extract from *Tanacetum vulgare* were evaluated against CMV and PVY. Concentrations above 20% significantly reduced DAS-ELISA values in CMV- and PVY-inoculated tomato plants (cv. Ideal), bringing them below the cut-off threshold. A 5% concentration was sufficient to suppress CMV infection, while 10% effectively inhibited PVY. However, concentrations below 5% were inadequate for controlling either virus. No phytotoxic effects were observed in any of the tested plants, even at the highest concentration of 30% (Petrov et al. 2016, 2021, 2024).

Spraying with a minimum of 5% aqueous dilution of the liquid fraction and 10% dilution of the soft fraction from *Chelidonium majus* methanol extracts significantly reduced PVY levels in virus-inoculated tobacco plants, as measured by DAS-ELISA (Petrov et al. 2015b).

6.2 Antiviral Activity of Plant Extracts against Cucumber Mosaic Virus (CMV)

Cucumber mosaic virus (CMV) is the type representative and the most widely distributed member of its genus globally. It can infect over 1000 plant species across 85 families, including numerous weeds and wild plants that may serve as potential reservoirs for the virus. CMV isolates are classified into two subgroups, I and II, which share approximately 70% nucleotide identity and differ in their geographic distribution (Palukaitis and García-Arenal 2003; Petrov et al. 2024).

References

Abdallah II, Quax WJ (2017) A glimpse into the biosynthesis of terpenoids. KnE Life Sci 3:81
Aboelsoud NH (2010) Herbal medicine in ancient Egypt. J Med Plants Res 4:082–086
Abookleesh FL, Al-Anzi BS, Ullah A (2022) Potential antiviral action of alkaloids. Molecules 27:903

Alamgir ANM (2017) Medicinal, nonmedicinal, biopesticides, color-and dyeyielding plants; secondary metabolites and drug principles; significance of medicinal plants; use of medicinal plants in the systems of traditional and complementary and alternative medicines (CAMs). In: Therapeutic use of medicinal plants and their extracts: volume 1. Springer, Cham, pp 61–104
Ali F, Rahul NF, Jyoti S, Siddique YH (2017) Health functionality of apigenin: a review. Int J Food Prop 20:1197–1238
Ali SA, Singh G, Datusalia AK (2021a) Potential therapeutic applications of phytoconstituents as immunomodulators: pre-clinical and clinical evidences. Phytother Res (Wiley Online Library) 35(7):3702–3731. https://doi.org/10.1002/ptr.7068
Ali SI, Sheikh WM, Rather MA, Venkatesalu V, Muzamil Bashir S, Nabi SU (2021b) Medicinal plants: treasure for antiviral drug discovery. Phytother Res 35(7):3447–3483
Alzohairy MA (2016) Therapeutics role of Azadirachta Indica (Neem) and their active constituents in diseases prevention and treatment. Evid Based Complement Alternat Med 2016:7382506
Amorim IS, Lach G, Gkogkas CG (2018) The role of the eukaryotic translation initiation factor 4E (EIF4E) in neuropsychiatric disorders. Front Genet. https://doi.org/10.3389/fgene.2018.00561
Anadón A, Martínez-Larrañaga MR, Ares I, Martínez MA (2016) Interactions between nutraceuticals/nutrients and therapeutic drugs. In: Nutraceuticals: efficacy, safety and toxicity. Elsevier Inc, Amsterdam, pp 855–874. ISBN 9780128021477
Andersen PI, Ianevski A, Lysvand H, Vitkauskiene A, Oksenych V, Bjørås M et al (2020) Discovery and development of safein-man broad-spectrum antiviral agents. Int J Infect Dis 93:268–276
Argenta DF, Bidone J, Koester LS, Bassani VL, Simões CMO, Teixeira HF (2018) Topical delivery of coumestrol from lipid nanoemulsions thickened with hydroxyethylcellulose for antiherpes treatment. AAPS Pharm Sci Tech 19:192–200
Armaka M, Papanikolaou E, Sivropoulou A, Arsenakis M (1999) Antiviral properties of Isoborneol, a potent inhibitor of herpes simplex virus type 1. Antivir Res 43:79–92
Arogundade O, Balogun OS, Kumar LP (2019) Seed transmissibility of *Cucumber mosaic virus* in *Capsicum* spp. Int J Veg Sci 25(2):146–153
Ashour M, Wink M, Gershenzon J (2010) Biochemistry of terpenoids: monoterpenes, sesquiterpenes and diterpenes. In: Biochemistry of Plant secondary metabolism, vol 40, 2nd edn. Wiley Blackwell, Hoboken, pp 258–303. ISBN 9781444320503
Astani A, Schnitzler P (2014) Antiviral activity of monoterpenes beta-pinene and limonene against herpes simplex virus in vitro. Iran J Microbiol 6:149
Atanasov AG, Zotchev SB, Dirsch VM, Orhan IE, Banach M, Rollinger JM, Barreca D, Weckwerth W, Bauer R, Bayer EA et al (2021) Natural products in drug discovery: advances and opportunities. Nat Rev Drug Discov 20:200–216
Badshah SL, Faisal S, Muhammad A, Poulson BG, Emwas AH, Jaremko M (2021) Antiviral activities of flavonoids. Biomed Pharmacother 140:111596
Baker RE, Mahmud AS, Miller IF, Rajeev M, Rasambainarivo F, Rice BL, Takahashi S, Tatem AJ, Wagner CE, Wang LF et al (2022) Infectious disease in an era of global change. Nat Rev Microbiol 20:193–205
Becker Y (1980) Antiviral agents from natural sources. Pharmacol Ther 10:119–159
Bedows E, Hatfield GM (1982) An investigation of the antiviral activity of Podophyllum peltatum. J Nat Prod 45(6):725–729
Ben Khalifa M, Simon V, Marrakchi M, Fakhfakh H, Moury B (2009) Contribution of host plant resistance and geographic distance to the structure of Potato virus Y (PVY) populations in pepper in Northern Tunisia. Plant Pathol 58:763–772
Ben-Shabat S, Yarmolinsky L, Porat D, Dahan A (2020) Antiviral effect of phytochemicals from medicinal plants: applications and drug delivery strategies. Drug Deliv Transl Res 10:1–14
Bensky D, Clavey S, Stõger E, Lai L (2004) Materia Medica 3rd edition compiled and translated by. Eastland Press Publisher, Seattle
Cecil CE, Davis JM, Cech NB, Laster SM (2011) Inhibition of H1N1 influenza A virus growth and induction of inflammatory mediators by the isoquinoline alkaloid berberine and extracts of goldenseal (*Hydrastis canadensis*). Int Immunopharmacol 11:1706–1714. https://doi.org/10.1016/j.intimp.2011.06.002

Cheung RCF, Ng TB, Wong JH, Chan WY (2015) Marine natural products with anti-herpes activity. Mar Drugs 13(6):4006–4043
Chin LW, Cheng YW, Lin SS, Lai YY, Lin LY, Chou MY, Chou MC, Yang CC (2010) Anti-herpes simplex virus effects of berberine from Coptidis Rhizoma, a major component of a Chinese herbal medicine, Ching-Wei-San. Arch Virol 155:1933–1941
Chinthakindi PK, Khan SI, Tekwani BL (2018) Phytochemical diversity and pharmacological activities of plant species in genus corydalis. J Ethnopharmacol 221:205–226
Chitalia VC, Munawar AH (2020) A painful lesson from the COVID-19 pandemic: the need for broad-spectrum, host-directed antivirals. J Transl Med 18:390
Chung VCH, Ma PHX, Lau CH, Wong SYS, Yeoh EK, Griffiths SM (2014) Views on traditional Chinese medicine amongst Chinese population: a systematic review of qualitative and quantitative studies. Health Expect 17:622–636
Cushnie TPT, Lamb AJ (2011) Antimicrobial activity of flavonoids. IJAA 38(2):99–107
Dejani NN, Elshabrawy HA, da Bezerra Filho CSM, de Sousa DP (2021) Anticoronavirus and immunomodulatory phenolic compounds: opportunities and Pharmacotherapeutic perspectives. Biomolecules 11:1254
Denaro M, Smeriglio A, Barreca D, De Francesco C, Occhiuto C, Milano G, Trombetta D (2020) Phytother Res 34(4):742–768
Dias DA, Urban S, Roessner U (2012) A historical overview of natural products in drug discovery. Metabolites 2:303–336
Diosa-Toro M, Troost B, van de Pol D, Heberle AM, Urcuqui-Inchima S, Thedieck K, Smit JM (2019) Tomatidine, a novel antiviral compound towards dengue virus. Antivir Res 161:90–99. https://doi.org/10.1016/j.antiviral.2018.11.011
Dong HJ, Wang ZH, Meng W, Li CC, Hu YX, Zhou L, Wang XJ (2018) The natural compound homoharringtonine presents broad antiviral activity in vitro and in vivo. Viruses. https://doi.org/10.3390/v10110601
Elbe A, Buckland-Merrett G (2017) Data, disease and diplomacy: GISAID's innovative contribution to global health. Global Chall 1(1):33–46
El-Sherbiny EM, Osman HF, Taha MS (2021) Effectiveness of Echinacea purpurea extract on immune deficiency induced by azathioprine in male albino rats. Biosci J 37:e37029–e37029
Fashner J, Ericson K, Werner S, Joseph S (2012) Treatment of the Common. Cold in children and adults. Am Fam Phys 86:153–159
Fiore C, Eisenhut M, Krausse R, Ragazzi E, Pellati D, Armanini D, Bielenberg J (2008) Antiviral effects of glycyrrhiza species 141 antiviral effects of glycyrrhiza species. Phytother Res 22:141–148
Fleßa S, Marschall P (2009) Socio-economic impact of antiviral intervention. Springer, Berlin/Heidelberg
Frederico ÉHFF, Cardoso ALBD, Moreira-Marconi E, de Sá-Caputo DDC, Guimarães CAS, da Fontoura Dionello C et al (2017) Anti-viral effects of medicinal plants in the management of dengue: a systematic review. Afr J Tradit Complement Altern Med 14(4S):33–40
Garg S, Roy A (2020) In silico analysis of selected alkaloids against main protease (Mpro) of SARS-CoV-2. Chem Biol Interact. https://doi.org/10.1016/j.cbi.2020.109309
Ghildiyal R, Prakash V, Chaudhary VK, Gupta V, Gabrani R (2020) Phytochemicals as antiviral agents: recent updates. In: Plant-derived bioactives: production, properties and therapeutic applications. Springer, Singapore, pp 279–295. ISBN 9789811517617
Ghosh S, Basak P (2013) Emerging concepts of plant-derived antivirals against HIV, denguevirus, and hepatitis-Cvirus. Front Microbiol 4:1–11
Giovinazzo G, Gerardi C, Uberti-Foppa C, Lopalco L (2020) Can natural polyphenols help in reducing cytokine storm in COVID-19 patients? Molecules 25:5888
Goc A, Sumera W, Rath M, Niedzwiecki A (2021) Phenolic compounds disrupt spike-mediated receptor-binding and entry of SARS-CoV-2 Pseudo-Virions. PLoS One 16:e0253489
Guerriero G, Berni R, Muñoz-Sanchez JA, Apone F, Abdel-Salam EM, Qahtan AA, Alatar AA, Cantini C, Cai G, Hausman JF et al (2018) Production of Plant secondary metabolites: examples, Tips and Suggestions for Biotechnologists. Genes 9:309

Gurjar VK, Pal D, Patel AD (2020) Recent advances in chemistry and synthesis of pyrazole derivatives as potential promising antimicrobial agents. In: Pal D (ed) Pyrazole preparation and uses. NOVA Science. ISBN Number: 978-1-53618-250-7

Gürü M, Gürü S, Yılmaz Aydın D (2020) Effect of alkaloids on SARS-CoV-2. In: Naturengs Mtu journal of engineering and natural sciences. Malatya Turgut Ozal University, Malatya

Haque A, Barman N, Ali ME, Saha A, Islam MA (2019) Antiviral activity of Catharanthus roseus against some human pathogenic viruses. Pharmacogn Mag 15(63):507–511

Hassan STS, Masarˇcíková R, Berchová K (2015) Bioactive natural products with anti-herpes simplex virus properties. J Pharm Pharmacol 67:1325–1336

Hegazy GE, Abu-Serie MM, Abo-Elela GM, Ghozlan H, Sabry SA, Soliman NA, Abdel-Fattah YR (2020) In vitro dual (Anticancer and Antiviral) activity of the carotenoids produced by haloalkaliphilic Archaeon Natrialba Sp. M6. Sci Rep 10:5986

Helman CG (1978) Feed a cold, starve a fever—folk models of infection in an English suburban community, and their relation to medical treatment. Cult Med Psychiatry 2:107–137

Henss L, Scholz T, Rittinghausen S, Schädler S (2018) Antiviral effects of quinine sulfate on influenza a virus-infected green monkey kidney cells. Biochem Pharmacol 150:96–108

Huang Y, Xie FJ, Cao X, Li MY (2021) Research Progress in biosynthesis and regulation of plant terpenoids. Biotechnol Biotechnol Equip 35:1800–1809

Hudson J, Vimalanathan S (2011) Echinacea-A source of potent antivirals for respiratory virus infections. Pharmaceuticals 4:1019–1031

Hung TC, Jassey A, Liu CH, Lin CJ, Lin CC, Wong SH, Wang JY, Yen MH, Lin LT (2019) Berberine inhibits hepatitis C virus entry by targeting the viral E2 glycoprotein. Phytomedicine 53:62–69. https://doi.org/10.1016/j.phymed.2018.09.025

Hussein RA, El-Anssary AA (2019) Plants secondary metabolites: the key drivers of the pharmacological actions of medicinal plants. In: Herbal medicine. IntechOpen, London

Hussein G, Miyashiro H, Nakamura N, Hattori M, Kakiuchi N, Shimotohno K (2000) Inhibitory effects of Sudanese medicinal plant extracts on hepatitis C virus (HCV) protease. Phytother Res 14(7):510–516

Ieven M, Vlietinck AJ, Vanden Berghe DA, Totte J, Dommisse R, Esmans E, Alderweireldt F (1982) Plant antiviral agents. III. Isolation of alkaloids from *Clivia miniata* Regel (Amaryllidaceae). J Nat Prod 45:564–573. https://doi.org/10.1021/np50023a009

Isah T (2019) Stress and defense responses in Plant secondary metabolites production. Biol Res 52:39

Jaime MF, Redko F, Muschietti LV, Campos RH, Martino VS, Cavallaro LV (2013) In vitro antiviral activity of plant extracts from Asteraceae medicinal plants. Virol J 10(1):1–10

Jassim SAA, Naji MA (2003) Novel antiviral agents: a medicinal plant perspective. J Appl Microbiol 95:412–427

Ji H-F, Li X-J, Zhang H-Y (2009) Natural products and drug discovery can thousands of years of ancient medical knowledge Lead us to new and powerful drug combinations in the fight against cancer and dementia? EMBO Rep 10:194–200

Ka S, Merindol N, Sow AA, Singh A, Landelouci K, Plourde MB, Pepin G, Masi M, Di Lecce R, Evidente A et al (2021) Amaryllidaceae alkaloid cherylline inhibits the replication of dengue and zika viruses. Antimicrob Agents Chemother

Kerdudo A, Dingas A, Fernandez X, Faure C (2014) Encapsulation of rutin and naringenin in multilamellar vesicles for optimum antioxidant activity. Food Chem 159:12–19

Khandelwal N, Chander Y, Rawat KD, Riyesh T, Nishanth C, Sharma S, Jindal N, Tripathi BN, Barua S, Kumar N (2017) Emetine inhibits replication of RNA and DNA viruses without generating drug-resistant virus variants. Antivir Res 144:196–204. https://doi.org/10.1016/j.antiviral.2017.06.006

Kratz JM, Andrighetti-Fröhner CR, Leal PC, Nunes RJ, Yunes RA, Trybala E, Simões CMO (2008) Evaluation of anti-HSV-2 activity of Gallic acid and Pentyl Gallate. Biol Pharm Bull 31:903–907

Kumar Shakya A, Arvind Kumar Shakya C (2016) Medicinal plants: future source of new drugs. Int J Herb Med 4:59–64

Kumar K, Debnath P, Singh S, Kumar N (2023) An overview of plant phenolics and their involvement in abiotic stress tolerance. Stress 3:570–585

Lam WY, Leung KT, Law PT, Lee SM, Chan HL, Fung KP, Ooi VE, Waye MM (2006) Antiviral effect of Phyllanthus nanus ethanolic extract against hepatitis B virus (HBV) by expression microarray analysis. J Cell Biochem 97(4):795–812

Lee CD, Ott M, Thyagarajan SP, Shafritz DA, Burk RD, Gupta S (1996) Phyllanthus amarus down-regulates hepatitis B virus mRNA transcription and replication. Eur J Clin Investig 26(12):1069–1076

Leung KT, Chiu LC, Lam WS, Li Y, Sun SS, Ooi VE (2006) In vitro antiviral activities of Chinese medicinal herbs against duck hepatitis B virus. Phytother Res 20(10):911–914

Li Y, Yang Y, Fang L, Zhang Z, Jin J, Zhang K (2006) Anti-hepatitis activities in the broth of Ganoderma lucidum supplemented with a Chinese herbal medicine. Am J Chin Med 34(2):341–349

Li DD, Yu P, Xiao W, Wang ZZ, Zhao LG (2020) Berberine: a promising natural isoquinoline alkaloid for the development of hypolipidemic drugs. Curr Top Med Chem 20(28):2634–2647. https://doi.org/10.2174/1568026620666200908165913

Lin CW, Tsai FJ, Tsai CH, Lai CC, Wan L, Ho TY, Hsieh CC, Chao PDL (2005) Anti-SARS coronavirus 3C-like protease effects of Isatis Indigotica root and plant-derived phenolic compounds. Antivir Res 68:36–42

Lin CW, Wu CF, Hsiao NW, Chang CY, Li SW (2014a) Antiviral activity of cepharanthine against severe acute respiratory syndrome coronavirus in vitro. Antivir Ther 19(7):697–703

Lin LT, Hsu WC, Lin CC (2014b) Antiviral natural products and herbal medicines. JTCM 4(1):24–35

Liu J, Manheimer E, Tsutani K, Gluud C (2003) Medicinal herbs for hepatitis C virus infection: a Cochrane hepatobiliary systematic review of randomized trials. Am J Gastroenterol 98(3):538–544

Lou Z, Sun Y, Rao Z (2014) Current progress in antiviral strategies. Trends Pharmacol Sci 35(2):86–102

Manandhar B, Paudel KR, Sharma B, Karki R (2018) Phytochemical profile and pharmacological activity of Aegle Marmelos Linn. J Integr Med 16:153–163

Marisa R, Assessor D, Calapai G, Delbò M (2013) Assessment Report on Glycyrrhiza glabra L. and/or Glycyrrhiza inflata Bat. and/or Glycyrrhiza uralensis Fisch, Radix. Committee on Herbal Medicinal Products (HMPC), Basel

Meganck RM, Baric RS (2021) Developing therapeutic approaches for twentyfirst-century emerging infectious viral diseases. Nat Med 27(3):401–410

Menéndez-Arias L, Gago F (2013) Antiviral agents: structural basis of action and rational design. Struct Phys Viruses 68:599–630

Mohan S, Elhassan Taha MM, Makeen HA, Alhazmi HA, Al Bratty M, Sultana S et al (2020) Bioactive natural antivirals: an updated review of the available plants and isolated molecules. Molecules 25(21):4878

Mohanta TK, Patra JK, Rath SK, Pal D, Thatoi HN (2007) Evaluation of antimicrobial activity and phytochemical screening of oils and nuts of Semicarpus anacardium L.f. Sci Res Essays 2(11):486–490

Montenegro-Landívar MF, Tapia-Quirós P, Vecino X, Reig M, Valderrama C, Granados M, Cortina JL, Saurina J (2021) Polyphenols and their potential role to fight viral diseases: an overview. Sci Total Environ 801:149719

Moradi MT, Karimi A, Lorigooini Z (2018) Alkaloids as the natural anti-influenza virus agents: a systematic review. Toxin Rev 37:11–18

More GK (2021) Metabolomics and cell-based assay analysis of selected South African plants to determine anti-rift valley fever virus activity. PhD thesis, University of South Africa, Johannesburg, South Africa

Morris DJ (1994) Adverse effects and drug interactions of clinical importance with antiviral drugs. Drug Saf 10:281–291

Mukherjee A, Banerjee M, Mandal V, Shukla AC, Mandal SC (2013) Modernization of Ayurveda: a brief. overview of Indian Initiatives. Nat Prod Commun 9:1934578X1400900239
Mukhopadhyay R, Roy S, Venkatadri R, Su YP, Ye W, Barnaeva E, Mathews Griner L, Southall N, Hu X, Wang AQ et al (2016) Efficacy and mechanism of action of low dose emetine against human cytomegalovirus. PLoS Pathog. https://doi.org/10.1371/journal.ppat.1005717
Munshi A, Mehrotra R, Panda SK (1993a) Evaluation of Phyllanthus amarus and Phyllanthus maderaspatensis as agents for postexposure prophylaxis in neonatal duck hepatitis B virus infection. J Med Virol 40(1):53–58
Munshi A, Mehrotra R, Ramesh R, Panda SK (1993b) Evaluation of antihepadnavirus activity of Phyllanthus amarus and Phyllanthus maderaspatensis in duck hepatitis B virus carrier Pekin ducks. J Med Virol 41(4):275–281
Nabila N, Suada NK, Denis D, Yohan B, Adi AC, Veterini AS, Anindya AL, Sasmono RT, Rachmawati H (2020) Antiviral action of curcumin encapsulated in nanoemulsion against four serotypes of dengue virus. Pharm Nanotechnol 8:54–62
Narayan V, Ravindra KC, Chiaro C, Cary D, Aggarwal BB, Henderson AJ, Prabhu KS (2011) Celastrol inhibits tat-mediated human immunodeficiency virus (HIV) transcription and replication. J Mol Biol 410:972–983
Niu JZ, Wang YY, Qiao M, Gowans E, Edwards P, Thyagarajan SP, Gust I, Locarnini S (1990) Effect of Phyllanthus amarus on duck hepatitis B virus replication in vivo. J Med Virol 32(4):212–218
Ntie-Kang F, Svozil D (2020) An enumeration of natural products from microbial, marine and terrestrial sources. Phys Sci Rev 5:20180121
Ono K, Nakane H, Fukushima M, Chermann J-C, Barré-Sinoussi F (1990) Differential inhibitory effects of various flavonoids on the activities of reverse transcriptase and cellular DNA and RNA polymerases. Eur J Biochem 190:469–476
Orfali R, Rateb ME, Hassan HM, Alonazi M, Gomaa MR, Mahrous N, Gaballah M, Kandeil A, Perveen S, Abdelmohsen UR et al (2021) Antibiotics Sinapic acid suppresses SARS-CoV-2 replication by targeting its envelope protein. Antibiotics 10:25
Ott M, Thyagarajan SP, Gupta S (1997) Phyllanthus amarus suppresses hepatitis B virus by interrupting interactions between HBV enhancer I and cellular transcription factors. Eur J Clin Investig 27(11):908–915
Ouzounov S, Mehta A, Dwek RA, Block TM, Jordan R (2002) The combination of interferon α-2b and n-butyl deoxynojirimycin has a greater than additive antiviral effect upon production of infectious bovine viral diarrhea virus (BVDV) in vitro: implications for hepatitis C virus (HCV) therapy. Antivir Res 55:425–435
Ozcelik B, Kartal M, Orhan I (2011) Cytotoxicity, antiviral and antimicrobial activities of alkaloids, flavonoids, and phenolic acids. Pharm Biol 49:396–402. https://doi.org/10.3109/13880209.2010.519390
Pal D, Lal P (2024) Plants showing anti-viral activity with emphasis on secondary metabolites and biological screening. In: Anti-viral metabolites from medicinal plants, Reference series in phytochemistry. Springer Nature Switzerland AG. https://doi.org/10.1007/978.3.031.12199.9
Pal D, Mandal M, Senthilkumar MGP, Padhiari A (2006) Antibacterial activity of methanol extract of Cuscuta reflexa Roxb. Stem and Corchorus olitorius Linn. seed. Fitoterapia 77(7–8):589–591
Pal D, Singh V, Pandey DD, Maurya RK (2014a) Synthesis, characterization and antimicrobial evaluation of some 1, 2, 4-triazole derivatives. Int J Pharm Pharm Sci 6(8):213–216
Pal D, Tripathy R, Pandey DD, Mishra P (2014b) Synthesis, characterization, antimicrobial and pharmacological evaluation of some 2,5-disubstituted sulphonyl amino 1,3,4- oxadiazole and 2-amino-disubstituted 1,3,4-thiadiazole derivatives. J Adv Pharm Technol Res 5:196–201
Palukaitis P, García-Arenal F (2003) Cucumoviruses. Adv Virus Res 62:241–323
Pan QM, Li YH, Hua J, Huang FP, Wang HS, Liang D (2015) Antiviral matrinetype alkaloids from the rhizomes of *Sophora tonkinensis*. J Nat Prod 78:1683–1688. https://doi.org/10.1021/acs.jnatprod.5b00325
Panraksa P, Ramphan S, Khongwichit S, Smith DR (2017) Activity of andrographolide against dengue virus. Antivir Res 139:69–78

Patrick L (1999) Hepatitis C: epidemiology and review of complementary/alternative medicine treatments. Altern Med Rev 4:220–238

Peng J, Hu JF, Kazi AB, Li Z, Avery M, Peraud O, Hill RT, Franzblau SG, Zhang F, Schinazi RF et al (2003) Manadomanzamines A and B: a novel alkaloid ring system with potent activity against mycobacteria and HIV-1. J Am Chem Soc 125:13382–13386

Perez RM (2003) Antiviral activity of compounds isolated from plants. Pharm Biol 41:107–157

Petrov N, Stoyanova M, Valkova M (2015a) The antiviral activity of extract from St. John's wort against potato virus Y. Proc Union Sci Ruse 7(3):229–232. ISSN 1311-1086

Petrov N, Stoyanova M, Valkova M (2015b) Antiviral activity of plant extract from *Chelidonium majus* against potato virus Y. Agric Univ – Plovdiv Sci Works LIX(book 3):104–109

Petrov N, Stoyanova M, Valkova M (2016) Antiviral activity of plant extract from *Tanacetum vulgare* against cucumber mosaic virus and potato virus Y. J BioSci Biotechnol 5(2):189–119. ISSN 1314-6246

Petrov N, Stoyanova M, Gaur R (2021) Chapter 19: Ecological methods to control viral damages in tomatoes. In: Plant virus-host interaction. Molecular approaches and viral evolution, 2nd edn. Elsevier Inc. ISBN 978-0-12-821629-3

Petrov N, Stoyanova M, Gaur R (2024) Chapter 15: Characterization and management of economically important viruses on sweet pepper cultivars in Europe. In: Epper Virome, 1st edn. Elsevier, pp 445–476. ISBN: 978-0-443-15576-5 B978-0-443-15576-5.00010-1

Qiu Y, Wang Z, Owens RJ (2016) Antiviral activity of chrysin derivatives against herpes simplex virus type 1. Antivir Chem Chemother 25(3):81–88

Quintana VM, Selisko B, Brunetti JE, Eydoux C, Guillemot JC, Canard B, Damonte EB, Julander JG, Castilla V (2020) Antiviral activity of the natural alkaloid anisomycin against dengue and zika viruses. Antivir Res 176:104749. https://doi.org/10.1016/j.antiviral.2020.104749

Rajbhandari M, Wegner U, Jülich M, Schoepke T, Mentel R (2001) Screening of Nepalese medicinal plants for antiviral activity. J Ethnopharmacol 74(3):251–255

Ramakrishna A, Ravishankar GA (2011) Influence of abiotic stress signals on secondary metabolites in plants. Plant Signal Behav 6:1720–1731

Reid DP (1996) The Shambhala guide to traditional Chinese medicine, 1st edn. Shambhala, Boulder

Richard SA (2021) Exploring the pivotal immunomodulatory and anti-inflammatory potentials of Glycyrrhizic and Glycyrrhetinic acids. Mediat Inflamm 2021:6699560

Ripoli M, Angelico R, Sacco P, Ceglie A, Mangia A (2016) Phytoliposome-based Silibinin delivery system as a promising strategy to prevent hepatitis C virus infection. J Biomed Nanotechnol 12:770–780

Risso-Ballester J, Galloux M, Cao J, Le Goffic R, Hontonnou F, Jobart-Malfait A, Desquesnes A, Sake SM, Haid S, Du M et al (2021) A condensatehardening drug blocks RSV replication in vivo. Nature 595(7868):596–599. https://doi.org/10.1038/s41586-021-03703-z

Rivlin RS (2001) Recent advances on the nutritional effects associated with the use of garlic as a supplement historical perspective on the use of garlic 1,2. J Nutr 131:951S–954S

Rizzato G, Scalabrin E, Radaelli M, Capodaglio G, Piccolo O (2017) A new exploration of licorice metabolome. Food Chem 221:959–968

Rodriguez-Concepcion M, Avalos J, Bonet ML, Boronat A, Gomez-Gomez L, Hornero-Mendez D, Limon MC, Meléndez Martínez AJ, Olmedilla-Alonso B, Palou A et al (2018) A global perspective on carotenoids: metabolism, biotechnology, and benefits for nutrition and health. Prog Lipid Res 70:62–93

Sagaya Jansi R, Khusro A, Agastian P, Alfarhan A, Al-Dhabi NA, Arasu MV, Rajagopal R, Barcelo D, Al-Tamimi A (2021) Emerging paradigms of viral diseases and paramount role of natural resources as antiviral agents. Sci Total Environ 759:143539

Saha S, Pal D, Kumar S (2017) Antifungal and antibacterial activities of phenyl and ortho hydroxyl phenyl linked imidazolyl triazolo hydroxamic acid derivatives. Inventi Rapid: Med Chem 1:1–8

Schmidt R, Enzinger C, Ropele S, Schmidt H, Fazekas F, Cinatl J, Morgenstern B, Bauer G, Chandra P, Rabenau H et al (1997) Glycyrrhizin, an active component of Liquorice roots, and replication of SARS-associated coronavirus. Lancet 71:2045–2046

Semwal DK, Semwal RB, Combrinck S, Viljoen A (2016) Myricetin: a dietary molecule with diverse biological activities. Nutrients 8:90
Sharif N, Khoshnoudi-Nia S, Jafari SM (2020) Nano/microencapsulation of anthocyanins; a systematic review and meta-analysis. Food Res Int 132:109077
Singh D, Singh B, Goel RK (2011) Traditional uses, phytochemistry and pharmacology of Ficus Religiosa: a review. J Ethnopharmacol 134:565–583
Singh P, Gupta E, Mishra N, Mishra P (2020) Shikimic acid as intermediary model for the production of drugs effective against influenza virus. In: Phytochemicals as lead compounds for new drug discovery. Elsevier, pp 245–256. https://doi.org/10.1016/B978-0-12-817890-4.00016-0
Song JM, Seong BL (2007) Tea Catechins as a potential alternative anti-infectious agent. Expert Rev Anti-Infect Ther 5:497–506
Song JM, Lee KH, Seong BL (2005) Antiviral effect of catechins in green tea on influenza virus. Antivir Res 68(2):66–74
Sreekanth T, Nagajyothi P, Muthuraman P, Enkhtaivan G, Vattikuti S, Tettey C, Kim DH, Shim J, Yoo K (2018) Ultra-sonicationassisted silver nanoparticles using Panax ginseng root extract and their anti-cancer and antiviral activities. J Photochem Photobiol B: Biol 188:6–11
Steinmann J, Buer J, Pietschmann T, Steinmann E (2013) Anti-infective properties of Epigallocatechin-3-Gallate (EGCG), a component of green tea. Br J Pharmacol 168:1059–1073
Stiefel M, Shaner A, Schaefer SD (2006) The Edwin Smith papyrus: the birth of analytical thinking in medicine and otolaryngology. Laryngoscope 116:182–188
Strasfeld L, Chou S (2010) Antiviral drug resistance: mechanisms and clinical implications. Infect Dis Clin N Am 24:413–437
Swallow DL (1978) Antiviral agents. Prog Drug Res 22:267–326
Taliansky M, Samarskaya V, Zavriev SK, Fesenko I, Kalinina NO, Love AJ (2021) RNA-based technologies for engineering plant virus resistance. Plants 10(1):82
Tepaske MR, Gloer JB, Wicklow DT, Dowd PF (1989) Tubingensin A: an antiviral carbazole alkaloid from the sclerotia of *Aspergillus tubingensis*. J Org Chem 54:4743–4746. https://doi.org/10.1021/jo00281a010
Thyagarajan SP, Jayaram S, Gopalakrishnan V, Hari R, And J, Sripathi MS (2002) Conference proceedings herbal medicines for liver diseases in India. J Gastroenterol Hepatol 17:370–376
Tirado-Kulieva VA, Hernández-Martínez E, Choque-Rivera TJ (2022) Phenolic compounds versus SARS-CoV-2: an update on the Main findings against COVID-19. Heliyon 8:e10702
Tiralongo E, Wee SS, Lea RA (2016) Elderberry supplementation reduces cold duration and symptoms in air-Travellers: a randomized, double-blind placebo-controlled clinical trial. Nutrients 8:182
Tolo FM, Rukunga GM, Muli FW, Njagi EN, Njue W, Kumon K, Mungai GM, Muthaura CN, Muli JM, Keter LK, Oishi E, KofiTsekpo MW (2006) Anti-viral activity of the extracts of a Kenyan medicinal plant Carissa edulis against herpes simplex virus. J Ethnopharmacol 104(1/2):92–99
Tran J (2013) Green tea: A potential alternative anti-infectious agent Catechins and viral infections. Adv Anthropol 3:198–202
Tseng CK, Hsu SP, Lin CK, Wu YH, Lee JC, Young KC (2017) Celastrol inhibits hepatitis C virus replication by upregulating Heme Oxygenase-1 via the JNK MAPK/Nrf2 pathway in human Hepatoma cells. Antivir Res 146:191–200
Vilhelmova-Ilieva N, Galabov AS, Mileva M (2020) Tannins as antiviral agents. In: Tannins—structural properties, biological properties and current knowledge. IntechOpen, London
Wang M, Cheng H, Li Y, Meng L, Zhao G, Mai K (1995) Herbs of the genus Phyllanthus in the treatment of chronic hepatitis B: observations with three preparations from different geographic sites. J Lab Clin Med 126(4):350–352
Wang WN, Yang XB, Liu HZ, Huang ZM, Wu GX (2005) Effect of Oenanthe javanica flavone on human and duck hepatitis B virus infection. Acta Pharmacol Sin 26(5):587–592
Wang L, Yang R, Yuan B, Liu Y, Liu C (2015) The antiviral and antimicrobial activities of licorice, a widely-used Chinese herb. Acta Pharm Sin B 5:310–315

Wang Y, Dong T, Qi G, Qu L, Liang W, Qi B, Zhang Z, Shang L, Gao H, Du X et al (2018) Prevalence of common respiratory viral infections and identification of adenovirus in hospitalized adults in Harbin, China 2014 to 2017. Front Microbiol 9:2919
Wang B, Ding Y, Zhao P, Li W, Li M, Zhu J, Ye S (2022) Systems pharmacology-based drug discovery and active mechanism of natural products for coronavirus pneumonia (COVID-19): an example using flavonoids. Comput Biol Med 143:105241
Whitby K, Taylor D, Patel D, Ahmed P, Tyms AS (2004) Action of celgosivir (6 O-butanoyl castanospermine) against the pestivirus BVDV: implications for the treatment of hepatitis C. Antivir Chem Chemother 15:141–151. https://doi.org/10.1177/095632020401500304
Whitby K, Pierson TC, Geiss B, Lane K, Engle M, Zhou Y, Doms RW, Diamond MS (2005) Castanospermine, a potent inhibitor of dengue virus infection in vitro and in vivo. J Virol 79:8698–8706
Wink M (2020) Potential of DNA intercalating alkaloids and other Plant secondary metabolites against SARS-CoV-2 causing COVID-19. Diversity 12:175
Wu Y, Li JQ, Kim YJ, Wu J, Wang Q, Hao Y (2011) In vivo and in vitro antiviral effects of berberine on influenza virus. Chin J Integr Med 17:444–452
Wu W, Li R, Li X, He J (2020a) Antiviral effect of quercetin on viruses. J Immunol Res:1–9
Wu ZN, Chen NH, Tang Q, Chen S, Zhan ZC, Zhang YB, Wang GC, Li YL, Ye WC (2020b) β-Carboline alkaloids from the seeds of *Peganum harmala* and their anti-HSV-2 virus activities. Org Lett 22:7310–7314. https://doi.org/10.1021/acs.orglett.0c02650
Wu Y, Crich D, Pegan SD, Lou L, Hansen MC, Booth C, Desrochers E, Mullininx LN, Starling EB, Chang KY et al (2021) Polyphenols as potential inhibitors of SARS-CoV-2 Rna dependent Rna polymerase (Rdrp). Molecules 26:7438
Xu J, Wang J, Deng F, Hu Z, Wang H (2008) Green tea extract and its major component epigallocatechin Gallate inhibits hepatitis B virus in vitro. Antivir Res 78:242–249
Xu J, Xu Z, Zheng W (2017) A review of the antiviral role of green tea Catechins. Molecules 22:1337
Yamada K, Ogawa H, Hara A, Yoshida Y, Yonezawa Y, Karibe K, Nghia VB, Yoshimura H, Yamamoto Y, Yamada M et al (2009) Mechanism of the antiviral effect of Hydroxytyrosol on influenza virus appears to involve morphological change of the virus. Antivir Res 83:35–44
Yang CM, Cheng HY, Lin TC, Chiang LC, Lin CC (2007) The in vitro activity of geraniin and 1,3,4,6-tetra-O-galloyl-beta-d-glucose isolated from Phyllanthus urinaria against herpes simplex virus type 1 and type 2 infection. J Ethnopharmacol 110(3):555–558
Yang S, Xu M, Lee EM, Gorshkov K, Shiryaev SA, He S, Sun W, Cheng Y-S, Hu X, Tharappel AM et al (2018) Emetine inhibits zika and Ebola virus infections through two molecular mechanisms: inhibiting viral replication and decreasing viral entry. Cell Discov 4:31
Yazaki K, Arimura GI, Ohnishi T (2017) "Hidden" Terpenoids in plants: their biosynthesis, localization and ecological roles. Plant Cell Physiol 58:1615–1621
Yeshi K, Crayn D, Ritmejeryte E, Wangchuk P (2022) Plant secondary metabolites produced in response to abiotic stresses has ˈpotential application in pharmaceutical product development. Molecules 27:313
Yim SK, Kim I, Warren B, Kim J, Jung K, Ku B (2021) Antiviral activity of two marine carotenoids against SARS-CoV-2 virus entry in Silico and in vitro. Int J Mol Sci 22:6481
Yu JS, Tseng CK, Lin CK, Hsu YC, Wu YH, Hsieh CL, Lee JC (2017) Celastrol inhibits dengue virus replication via up-regulating type I interferon and downstream interferon-stimulated responses. Antivir Res 137:49–57
Zandi K, Ramedani E, Mohammadi K, Tajbakhsh S, Deilami I, Rastian Z (2011) Evaluation of antiviral activities of curcumin derivatives against HSV-1 in Vero cell line. Nat Prod Commun 6(6):867–870
Zeedan GSG, Abd El K, Razik AEH, Abdel-Shafy S, Koreny Farag T (2019a) The effects of green tea and Propolis extracts on pro-inflammatory cytokines TNFα, IFN-γ, IL2, and immunoglobulin production in experimentally infected rabbits with bovine Herpesvirus-1. World Vet J 9(4):329–339

Zeedan GSG, Mahmoud AH, Abdalhamed AM, Abd El K AEH (2019b) Detection of lumpy skin disease virus in cattle using real-time polymerase chain reaction and serological diagnostic assays in different governorates in Egypt in 2017. Vet World 12(7):1093

Zeedan GS, Mahmoud AH, Abdalhamed AM, Ghazy AA, Abd El-Razik KA (2020) Rapid detection and differentiation between sheep pox and goat pox viruses by real-time qPCR and conventional PCR in sheep and goat in Egypt. World Vet J 10(1):80–87

Zhang GB, Tian LQ, Li YM, Liao YF, Li J, Bing FH (2013a) Protective effect of homonojirimycin from *Commelina communis* (dayflower) on influenza virus infection in mice. Phytomedicine 20:964–968. https://doi.org/10.1016/j.phymed.2013.04.009

Zhang GB, Zhang B, Zhang XX, Bing FH (2013b) Homonojirimycin, an alkaloid from dayflower inhibits the growth of influenza A virus in vitro. Acta Virol 57:85–86

Zhang YB, Zhang XL, Chen NH, Wu ZN, Ye WC, Li YL, Wang GC (2017) Four matrine-based alkaloids with antiviral activities against HBV from the seeds of *Sophora alopecuroides*. Org Lett 19:424–427. https://doi.org/10.1021/acs.orglett.6b03685

Zhang YB, Luo D, Yang L, Cheng W, He LJ, Kuang GK, Li MM, Li YL, Wang GC (2018) Matrine-type alkaloids from the roots of *Sophora flavescens* and their antiviral activities against the hepatitis B virus. J Nat Prod 81:2259–2265. https://doi.org/10.1021/acs.jnatprod.8b00576

Zheng S-C, Xu J-Y, Liu H-P (2019) Cellular entry of white spot syndrome virus and antiviral immunity mediated by cellular receptors in crustaceans. Fish Shellfish Immunol 93:580–588. https://doi.org/10.5772/intechopen.11173215

Zhu Y, Shao Y, Qu X, Guo J, Yang J, Zhou Z, Wang S (2019) Sodium Ferulate protects against influenza virus infection by activation of the TLR7/9-MyD88-IRF7 signaling pathway and inhibition of the NF-KB signaling pathway. Biochem Biophys Res Commun 512:793–798

Zhu J, Ou L, Zhou Y, Yang Z, Bie M (2020) (-)-Epigallocatechin-3-Gallate induces interferon-Λ2 expression to anti-influenza A virus in human bronchial epithelial cells (BEAS-2B) through P38 MAPK signaling pathway. J Thorac Dis 12:989–997

Biopharmaceutical Production from Cell Engineering and Molecular Pharming

Anita Mishra, Ayushi Malik, Avinash Marwal, and Rajarshi Kumar Gaur

Abstract

Biopharmaceuticals are produced in living cells using biotechnological processes and molecular biology methods, whereas synthetic drugs are products of chemical processes. Eukaryotic cells are becoming useful platforms for recombinant protein production. Production of biopharmaceuticals from *E. coli* or mammalian cell lines is achieved by the introduction of a gene or cDNA coding for the protein of interest into a well-characterized strain of producer cells. With molecular tools of genomics and proteomics, mammalian cells and plants can be used for the production of life-saving therapeutics. Production of biopharmaceuticals with plants provides benefits of cost-effectiveness, mass production at rapid rate, posttranslational modifications, and a product that is free from harmful pathogen contamination. Biopharmaceuticals show high specificity and target only specific molecules, which reduces side effects associated with conventional small-molecule drugs. The application of biopharmaceuticals has facilitated the treatment of patients who poorly respond to traditional synthetic drugs. The production of biopharmaceuticals in plants—known as molecular farming—offers benefits such as scalability, speed, and improved safety compared with insect and mammalian cell expression systems.

A. Mishra (✉)
Department of Science (Biotechnology), Biyani Girls College, Jaipur, Rajasthan, India

A. Malik · A. Marwal
Nanoparticle Synthesis and Bioinformatics Laboratory, Department of Biotechnology, Vigyan Bhawan, Block – B, New Campus, Mohanlal Sukhadia University, Udaipur, Rajasthan, India

R. K. Gaur
Department of Biotechnology, Deen Dayal Upadhyaya Gorakhpur University, Gorakhpur, Uttar Pradesh, India

Uttar Pradesh Council of Agricultural Research, Uttar Pradesh, India

R. K. Gaur et al. (eds.), *Plant Molecular Farming*,
https://doi.org/10.1007/978-981-95-3823-2_5

Keywords

Biopharmaceuticals · Recombinant · Molecular farming · Cost-effectiveness · Specificity

1 Introduction of Biopharmaceutical Production

Biopharmaceuticals are proteins of pharmaceutical importance, such as monoclonal antibodies and vaccines, produced by recombinant expression in living cells with help of molecular biology methods. Molecular farming is a term used for molecular biology methods used for the production of biopharmaceuticals in mammalian cell lines, microbial cells, and plants. Biopharmaceuticals were first introduced in the 1980s. Today, biopharmaceuticals are produced for a broad range of medicines. Biopharmaceuticals offer many advantages as compared to conventional medicines. Biopharmaceuticals, as compared to conventionally produced drugs, do not show any side effects. Biopharmaceuticals are highly specific for the target and do not show any side effects. In many cases, patients are more responsive to biopharmaceuticals as compared to conventional medicines (Kesik-Brodacka 2018). Biopharmaceuticals differ from synthetically produced drugs in different aspects. Biopharmaceuticals are more complex as compared to synthetic drugs due to the polymeric chain present in their structure. The purity of a conventional pharmaceutical drug can be verified easily. It is not easy to verify the purity of biopharmaceuticals. Certain degree of variation may occur in the product from one batch to the next, depending on the manufacturing process used. These variations must be monitored to ensure the specificity of the product. Synthetic pharmaceuticals are not associated with any immunogenicity. Biopharmaceuticals have been found to be immunogenic, and even a small difference in the structure of active constituents of a drug or the presence of impurities may produce immunogenicity (Crommelin et al. 2005). In place of the term "generic," the term "biosimilars" is used for biopharmaceuticals because they may be different in glycosylation pattern and electrical potential of their ingredients when compared with their original reference, which may affect the quality and safety of the product. "Biobetters" represent a group of biopharmaceuticals in which proteins are deliberately altered to make them equivalent to their innovative references so as to make pharmaceutical proteins more target-specific and improve their efficacy. The alterations introduced in biopharmaceutical proteins include variations in the amino acid sequence as well as the glycosylation pattern in the protein (Gottlieb 2008).

1.1 Overview and Importance of Biopharmaceuticals

Biopharmaceuticals are important because they are effective, have fewer side effects, and can be used as medication in a wide range of diseases. Biopharmaceuticals is the term used for medical drugs produced using biotechnology. Biopharmaceutical products typically include vaccines, monoclonal antibodies, therapeutic proteins, and products related to gene therapies. Unlike traditional pharmaceuticals, which are usually small molecules synthesized chemically, biopharmaceuticals are large, complex molecules expressed and purified from living organisms (Zhu 2012). Various steps involved in the production of biopharmaceutical proteins involve identifying the gene responsible for expressing the pharmaceutical protein. Transgenic technology is used to express the gene in a suitable host. After gene expression, the biopharmaceutical protein is purified and analyzed (Fig. 1). Bacteria, yeast, or mammalian cells can be used for the production of biopharmaceuticals (Butler and Meneses-Acosta 2012).

Monoclonal antibodies, which are used to target specific antigens in diseases like cancer and autoimmune disorders, are one of the biopharmaceutical products. Recombinant proteins include hormones like insulin and enzymes used in therapy. Vaccines provide immunity against infectious diseases. Gene therapy involves modifying or replacing genes to treat or prevent disease. Biopharmaceuticals allow for more precise targeting of diseases at the molecular level, often resulting in better efficacy and fewer side effects compared to traditional drugs (Kesik-Brodacka 2018). They enable the development of treatments for conditions that were previously difficult or impossible to treat, such as certain cancers, genetic disorders, and chronic illnesses (Kumar et al. 2020; Ranjith et al. 2022). Vaccines and monoclonal antibodies have significantly improved public health by preventing and treating infectious diseases, thereby reducing morbidity and mortality rates. The biopharmaceutical industry is a significant driver of economic growth, contributing to job creation and technological advancements in healthcare. Ongoing research in biopharmaceuticals fosters innovation, leading to new therapeutic modalities and improvements in existing treatments. Regulatory frameworks for biopharmaceuticals are evolving, helping ensure safety and efficacy while encouraging innovation and timely access to new therapies (Lee et al. 2019). Thus, biopharmaceuticals represent a crucial frontier in modern medicine, offering hope for patients with challenging health conditions. Their continued development is essential for advancing healthcare, improving quality of life worldwide, and addressing the complex health challenges of the future.

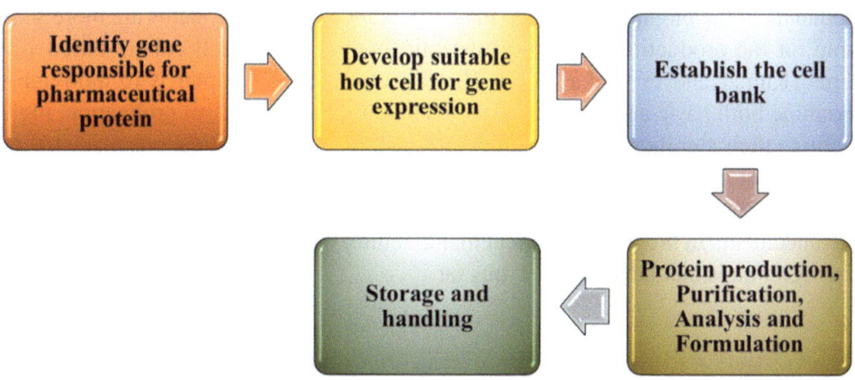

Fig. 1 Overview of the biopharmaceutical production process

1.2 Cell Engineering for Biopharmaceutical Production

Cell engineering involves modifying cells to enhance their ability to produce biopharmaceuticals. This process leverages techniques in genetic engineering, synthetic biology, and cell culture optimization to create more efficient, robust, and high-yielding cell lines. Cell engineering methods that are used in biopharmaceutical production are genetic modification, cell line development, metabolic engineering, synthetic biology, and cell culture optimization. Genetic modification, transgenesis, and use of gene-editing tool CRISPR/Cas9 enable the expression of foreign genes in host cells used for the synthesis of pharmaceutical proteins. CRISPR/Cas 9, allowing for targeted modifications that can enhance production or reduce unwanted byproducts. Different cell lines are developed for the production of biopharmaceuticals. CHO cells (Chinese Hamster Ovary) are widely used due to their ability to perform posttranslational modifications like human cells (Butler and Meneses-Acosta 2012). HEK293 cells are human embryonic kidney cells that are also popular for producing complex proteins (Tan et al. 2021). Metabolic engineering includes modifying metabolic pathways within the cells to optimize the yield of desired products while minimizing waste and byproduct formation. Cell culture optimization is done to enhance growth conditions (e.g., media formulation, bioreactor design) to maximize cell density and productivity. Fed-batch and perfusion culture systems are used to maintain optimal nutrient levels and remove waste products.

Cell engineering increases product yield and makes the process more cost-effective. Cell engineering technology improves the efficacy and reduces the immunogenicity of biopharmaceutical products, as it can alter the glycosylation pattern and posttranslational modifications (Kuriakose et al. 2016; Van Landuyt et al. 2019). Optimized cell lines and culture systems facilitate easier scale-up from laboratory to industrial production, ensuring a consistent supply of biopharmaceuticals. Cell engineering is a vital component of modern biopharmaceutical production, enabling the development of innovative therapies with improved efficacy and safety. By harnessing advanced techniques in genetic modification, metabolic engineering, and synthetic biology, researchers can create robust cell lines that meet the growing demands of the biopharmaceutical industry. Plant cell-based systems are more scalable for the production of biopharmaceuticals than animal cell culture-based systems (Kumar and Priyanka 2024). The objective of this chapter is to thoroughly examine both basic and advanced practices in biopharmaceutical production, highlighting their significance. This paper also addresses the modifications made to cell lines, which necessitate the optimal design of bioreactors for the effective multiplication of subsequent products. Additionally, it underscores the technological advancements in biopharmaceutical manufacture, emphasizing the sustainability of the techniques that have evolved thus far.

2 Genetic Modification Techniques for Cell Lines

Genetic modification techniques for cell lines used in biopharmaceutical production are essential for enhancing the yield, efficiency, and quality of biologics. Genetic engineering is an alteration in the genome of an animal cell or plant cell (Das et al. 2011; Sahu et al. 2014). Gene-editing tools such as CRISPR-Cas9 are used to improve the host cell line expression system, cell stability, humanize the glycosylation pathway of proteins produced, and enhance productivity. It is an ideal tool that can also be used to knock out undesirable genes and insert beneficial traits, which can enhance protein production or improve cell line stability (Khan et al. 2020). The transfection process can also be used to introduce nucleic acid sequences and produce desirable cell lines for enhanced productivity. Recombinant DNA technology is used to produce pharmaceuticals, which include recombinant forms of natural proteins such as human growth hormones, insulin, and cytokines. Derivatives of natural proteins with pharmaceutical applications are also produced using recombinant DNA technology. Protein mutants containing an amino acid sequence different from that of the wild protein with therapeutic applications can be produced (Lakra et al. 2019). Virus-like particles are protein derivatives that are not pathogenic but resemble viruses and bacteriophages. Cancer vaccines and immunotoxins provide immunity and treatment against cancer. Protein derivative IgG fusion proteins consist of two domains. The transport domain delivers the drug to the brain by binding to a receptor on the blood-brain barrier, and the therapeutic domain provides treatment (Liu and Timko 2022).

Hybridoma technology is used to produce monoclonal antibodies with high specificity and purity in large quantities for diagnostic and therapeutic applications (Zhu 2012; Mitra and Tomar 2021). Chimeric nucleases that include zinc-finger nucleases (ZFNs) and transcription activator-like effector nucleases (TALENs) can be used to make site-specific cuts in DNA for genome editing and biopharmaceutical production (Sivanandhan et al. 2016). Metabolic engineering (ME) can enhance product yield in biopharmaceuticals by modifying the metabolic pathways of cell lines. Overexpression or deletion of the genes involved in metabolic pathways can increase biopharmaceutical product yield in cell lines. Different approaches used for metabolic engineering are homologous recombination, use of nucleases, and short hairpin RNA (shRNA) to improve the efficiency and specificity of biopharmaceutical products (Nielsen 2012; Beuger et al. 2009; Tummala et al. 2013).

Recently, transgenic technology is used to produce biopharmaceutical proteins like antibodies and vaccines in plants. Expression of biopharmaceutical protein in plant cells is achieved by both stable and transient gene transfection. Stable transfer of a foreign gene is achieved by integration of the gene into the nuclear or chloroplast genome. Transgenic technology has been used to produce plant-based edible vaccines. Agrobacterium-mediated gene transfer or microprojectile bombardment is used for transferring the genes responsible for the expression of biopharmaceutical proteins into the nucleus. Transient gene transfer allows expression of recombinant protein in a very short time, so that plants can be used as bioreactors for the expression of pharmaceutical proteins. These techniques are often used in combination to create cell lines that can efficiently produce high-quality biopharmaceuticals, including monoclonal antibodies, vaccines, and other therapeutic proteins. Advantages and challenges are associated with every method, and the choice of method to be used during production depends on the specific requirements of the production process. Metabolic engineering has become an important tool for altering the metabolic pathway in cell lines. Genetic engineering tools can be used for deleting or overexpressing genes involved in metabolic pathways and the formation of biopharmaceutical products. Genetic engineering is used to create expression systems with more precise, faster, and enhanced product quality. Glycosylation is considered to be the most important aspect during biopharmaceutical production. Proper protein carboxylation, phosphorylation, and sulfation also need to be considered to produce novel products (Amann et al. 2019).

3 Bioreactor Design and Operation

Biopharmaceutical products are required in large quantities for the production of therapeutics, industrial enzymes, vaccines, and antibiotics, as their demand is increasing day by day in the healthcare industry to treat various diseases. To fulfill the demand for these products, a scale-up of biopharmaceutical production is necessary. Fermentation technology is used for large-scale production and scale-up of biopharmaceuticals. Different types of bioreactors with different host systems that are used for large-scale production include batch, fed-batch, and continuous modes of bioreactors. The most commonly used host systems for expression of biopharmaceuticals are mammalian cells, insect cells, yeast, and bacterial cells. Many valuable products, including hepatitis vaccine, growth hormones, insulin, and interferons, have been produced using recombinant microbial cultures in bioreactors. Effective bioreactor design and operation play a very important role in the successful production of biopharmaceuticals. Different parameters need to be considered before designing a bioreactor for process development, production host to be used, challenges associated with scale up of biopharmaceuticals. Attention to detail in design, strict adherence to operational protocols, and continuous monitoring are essential to optimize yields and ensure product quality (Mitra and Murthy 2022).

3.1 Types of Bioreactors and Purification Techniques

Several different types of bioreactors have been developed for large-scale production of biopharmaceuticals. The design of bioreactors for industrial-scale production of biopharmaceuticals depends on the product and type of biocatalyst used. Bioreactor with continuous flow of media is required to maintain the growth of cells in culture. Different bioreactors are used based on the requirements of the process used for biopharmaceutical production. Bubble column reactors, shaken devices, stirred-tank reactors, single-use micro-bioreactors, fixed-bed bioreactors, and airlift bioreactors are different bioreactors used for biopharmaceutical production.

Operating conditions for bioreactor are optimized according to the requirements of the process at the cellular level (Mitra and Murthy 2022). The yield of biopharmaceuticals can be increased by optimization of the bioreactor, which is done by controlling parameters critical to cell growth, such as pH, temperature, dissolved oxygen, rate of nutrient uptake, agitation, and pressure (Murthy et al. 2012). An integrative and smart control system is used to make the process more efficient and economic. Single-use micro-bioreactors are used to improve productivity during processes that need complex operating conditions. Single-use micro-bioreactors increase the speed of the process and make the process more flexible (Pazmiño et al. 2021). The configuration of the bioreactor depends on the biocatalyst used. Single-use bioreactors are preferred to produce high-value biopharmaceuticals. Bioreactors used for the production of biopharmaceutical products related to regenerative medicine should be supplied with a continuous flow of media, and culturing needs to be done under highly controlled conditions to maintain cells in a continuously growing state, as products in this case are regenerated cells and tissues themselves. These bioreactors range in size from 50 to 2000 L. Airlift bioreactors and fixed-bed bioreactors with advanced sensors are also used for biopharmaceutical production. In biopharmaceutical production, bioreactors and purification techniques play crucial roles in the development and manufacturing of therapeutic proteins, vaccines, and other biologics. The choice of bioreactor and purification technique depends on the specific biopharmaceutical product being produced, the desired scale, and the characteristics of the target biomolecule (Fang et al. 2022; Mitra and Murthy 2022). Advances in biotechnology continue to improve the efficiency and effectiveness of these processes. Purification technologies used in the biopharmaceutical industry are based on the characteristics and properties of the biopharmaceutical product to be purified, as well as the scale of production, to maximize yield and purity of the product. Affinity chromatography, size-exclusion chromatography, ion exchange chromatography, and filtration are the most commonly used methods in the biopharmaceutical industry. Affinity chromatography takes advantage of specific interactions between the target protein and a ligand immobilized on a chromatographic matrix (Łącki and Riske 2020). It is commonly used for purifying monoclonal antibodies (mAbs) and other recombinant proteins. Ion exchange chromatography separates proteins based on their net charge. Proteins are bound to charged resin and are eluted with a gradient of salt or pH changes that disrupt the protein-resin interaction. Exclusion or gel filtration chromatography can separate molecules based on size. This technique is used to remove aggregated impurities. Different molecular weight proteins are separated because smaller molecules enter the pores of the stationary phase and are delayed, while larger molecules are excluded and eluted faster. Different types of filters are used for purification in the biopharmaceutical industry. Different filtration systems are used according to process and product requirements. Single-use filtration systems are discarded after use and reduce the risk of contamination. Inline filtration is used to filter the product continuously during manufacturing so that the product can be analyzed during production, and quality can be maintained. Batch filtration is used for the sterilization of products produced by batch culture, where highly sterilized conditions need to be maintained (Ottinger et al. 2022).

4 Technological Advances in Biopharmaceutical Production

The ultimate aim of biopharma development is to improve the quality of life and to extend longevity. The quest for new drugs is never-ending, as is the need to understand disease causes beyond the symptoms. The rapid emergence of new technologies is revolutionizing the biopharma industry (Abdullah et al. 2008).

4.1 CRISPR-Cas9 Technology

After extensive development, CRISPR/Cas has transitioned from a technology primarily focused on cleaving DNA strands to a diverse array of methodologies for single-base gene editing (Cong et al. 2013), transcriptional regulation (Qi et al. 2013), and RNA strand cleavage (East-Seletsky et al. 2016). As a result, these systems hold promise for the treatment of numerous human afflictions, including cancer (Li et al. 2021), chronic diseases (Luo et al. 2021; Wang et al. 2022; Schoger et al. 2020), and genetic disorders attributable to single gene mutations (Gootenberg et al. 2017; Wu et al. 2019) (Fig. 2).

Within cellular environments, CRISPR/Cas9 and analogous systems exhibit exceptional capabilities in gene editing. Nonetheless, the formulation of a safe, stable, and efficient strategy for delivering gene-editing tools to affected cells in vivo constitutes a significant challenge for their clinical application. CRISPR is typically administered in the form of plasmids, mRNA, or RNP; however, all three modalities are subject to rapid immunological clearance from both the circulatory and digestive systems (Verma et al. 2024).

Adeno-associated virus serves as one of the most widely utilized vectors for gene therapy delivery, and even when excluding its loading capacity, the integration of AAV into the human genome may result in adverse health effects and poses challenges for public acceptance (Gillmore et al. 2021). Nonviral vectors with enhanced loading capacities and favorable safety profiles have emerged as promising alternatives, particularly lipid nanoparticles (LNPs), which have been evaluated in clinical trials (Das et al. 2015a, b; Tailor et al. 2021, 2022). Given that vectors such as LNPs and gold nanoparticles (AuNPs) are not of biological origin, they frequently provoke an immune response in vivo. These vectors are also vulnerable to incorrect uptake by digestive organs, although researchers have implemented modifications or incorporated peptides to mitigate immune clearance. Exosomes derived from organisms or biofilms utilized as nonviral vector coatings have proven to be particularly effective in evading immune response, and the proteins and peptides enriched within the biofilm may facilitate the delivery of gene therapies to target cells.

The occurrence of disease results in alterations to the microenvironment surrounding the affected cells, distinguishing it from normal tissue and creating conditions favorable for the development of delivery vectors. Additionally, the onset of disease is often accompanied by the overexpression of inflammatory factors and

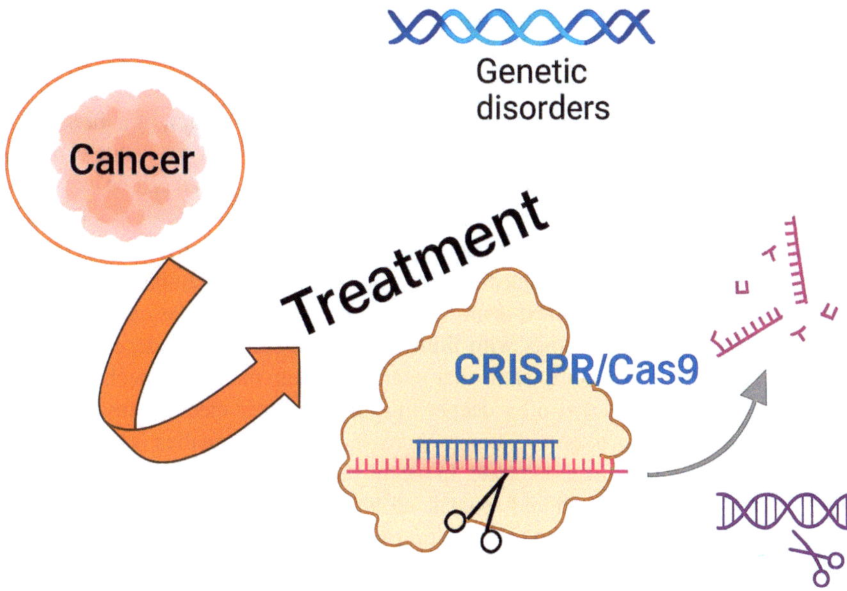

Fig. 2 CRISPR/Cas9 administered for the treatment of multiple chronic diseases

other disease-associated membrane proteins. The incorporation of peptides that selectively recognize these membrane proteins enables nanomaterials to specifically target diseased cells and facilitates the delivery of CRISPR/Cas to execute gene-editing functions. Furthermore, employing nanomaterials that respond to disease-specific conditions may enhance the accumulation of these materials at the site of pathology (Marwal and Gaur 2019, 2021). Consequently, a comprehensive investigation into the mechanisms underlying disease onset, various biological pathways involved, changes in protein expression, and the specific microenvironment surrounding diseased cells could significantly inform the development of vectors for targeted gene delivery (Chen et al. 2020).

Moreover, the efficiency and safety of CRISPR/Cas9 are critical considerations for clinical applications. Potential off-target effects associated with CRISPR/Cas9 could lead to severe consequences, necessitating the identification and rectification of such issues prior to clinical implementation. Researchers have thoroughly explored the cleavage mechanisms of Cas9 and have developed several variants with reduced likelihood of off-target effects while maintaining efficacy. Adjusting the sgRNA has proven to be an effective strategy for minimizing off-target effects. In conclusion, in-depth studies of disease mechanisms, the development of more efficient and specific delivery vectors, and enhancements to Cas9 variants for broader and safer applications are essential. With these foundational advancements, CRISPR/Cas9 technology is well-positioned for comprehensive clinical application to assist in the treatment of human diseases (Wienert et al. 2019).

4.2 High-Throughput Screening Methods

Numerous technologies are available for high-throughput screening (HTS). In the process of selecting one, the most critical criterion is the formulation of a therapeutic hypothesis and the design of an appropriate screening method to evaluate it. For instance, if the objective is to identify an inhibitor of an enzyme that can be expressed and quantified, an enzymatic screening approach is logically suited for this purpose. In contrast, if the aim is to inhibit the pathway containing that enzyme or if the enzyme cannot be isolated in its biologically relevant form, a pathway-based screening approach within a cellular context is more fitting (Wei et al. 2007; Yasgar et al. 2016).

Moreover, every assay comes with limitations, necessitating the development of counter assays to exclude compounds that interact through undesired mechanisms, as well as secondary assays that offer deeper insights into their mechanisms of action. HTS assays should strive to provide comprehensive insights, given their ongoing utility throughout the lead optimization phase. In recent years, HTS campaigns have increasingly relied on target-directed and specialized libraries, in addition to assay formats that minimize artifacts and yield richer information. For example, in fluorescence-based screenings, the acquisition of multiple parameters can provide additional insights into the authenticity of a hit, thereby reducing costs associated with the validation of numerous compounds. Additionally, high-content screening is becoming more prevalent with the incorporation of primary cells and complex three-dimensional cultures. Beyond the enhancement of existing technologies like molecular markers (Marwal and Gaur 2020), there remains potential for the introduction of novel detection strategies in HTS, including time-resolved anisotropy analysis (TRA) (Smith and Ghiggino 2015), multiphoton excitation techniques (Bullen 2008; Lin et al. 2019), intrinsic protein fluorescence (Ray et al. 2009), localized surface plasmon resonance (SPR) (Cao et al. 2011), and electrochemical screenings (Richter 2008; Golla and Seethala 2004).

Advancements in analysis software may also provide significant benefits to various HTS methodologies. For example, new analytical approaches to assess AS-MS screening data could reduce the number of experiments necessary for hit prioritization. In cell-based screening, linking phenotypic responses from small-molecule screenings to genomic perturbations through machine-learning models has the potential to accelerate target deconvolution. Determining the binding site of a compound identified via biochemical assays or identifying the molecular target following a phenotypic screening is highly desirable. For instance, with a ligand believed to influence cell migration, one might collect cells that migrated to a different chamber in a Boyden chamber after treatment and compare the relative abundance of sgRNA to those that did not migrate, thus gaining profound insights into the effects of the small molecule on cell migration. Generally, HTS represents the initiating phase of a drug discovery endeavor, which will involve substantial chemical optimization prior to considering a drug candidate for clinical trials. By commencing with carefully designed screening assays and subsequently employing complementary methods to validate hits and elucidate molecular targets, the drug developer is well-positioned to advance toward achieving their objectives (Srivastava et al. 2024; Blay et al. 2020).

5 Sustainability and Environmental Impact

It has been posited that the concept of sustainability encompasses the facilitation of conditions that allow life to thrive on this planet over an extended time frame, specifically for multiple future generations (Ehrenfeld 2008). A crucial aspect of sustainability thinking is to promote an "integrational perspective" (Lozano 2008) or an "integrative approach" (Hedlund-de Witt 2013, 2014; Byrne 2016); this signifies an acknowledgment of the influence and repercussions associated with various interconnected elements within the operation of a larger, complex system. Within the prevailing market-based socio-economic paradigm that influences our world, microeconomic benefit-risk assessments are typically the primary considerations in decision-making, with the allocation of benefits and risks being a subject of contention. Consequently, the societal and environmental ramifications of human activities are often overlooked in decision-making processes or are treated as secondary to the financial objectives of investors. The impacts of traditional economic practices have been sharply illuminated by recent reports detailing the speed and implications of climate change (IPCC et al. 2018) and the alarming decline in biodiversity (WWF 2018), as well as emerging threats to global freshwater availability (Rodell et al. 2018). In response to the growing recognition of such existential risks, a significant number of businesses are now rigorously evaluating their production systems to mitigate environmental harm associated with their operations. In this context, one must consider how sustainability principles can be effectively integrated into the biopharmaceutical manufacturing sector. By their intrinsic nature, biopharmaceuticals aim to enhance quality of life and life expectancy, particularly for individuals suffering from severe illnesses, including vulnerable populations such as children. Nonetheless, numerous existing business and scientific practices within the pharmaceutical industry, including the biopharmaceutical sector, have contributed to environmental degradation and have undermined public trust in their pursuit of profitability. This includes various practices that may be deliberate or the result of unforeseen outcomes. For example, the increase in pharmaceutical consumption and subsequent excretion by patients into wastewater has resulted in unintended consequences for the aquatic environment, disrupting biological processes and harming certain aquatic species (Fabbri 2015). Additionally, pharmaceutical effluents have been linked to the development and proliferation of antimicrobial-resistant strains of microorganisms. A salient obstacle to addressing these concerns is the additional financial burden associated with implementing environmentally benign manufacturing processes, which may affect access to generic versions of established drugs, particularly antibiotics (Milmo 2017). The issue of pharmaceutical pricing poses a universal challenge that impacts access across both high-income and low-income countries (Hurst 2017). Allegations of price-gouging are prevalent, with the case of Turing Pharmaceuticals' dramatic 5000% increase in the price of pyrimethamine attracting global attention due to the apparent exploitative practices involved (Saltiel and Finnefrock 2018).

Although there are multiple factors influencing pharmaceutical pricing, including market size, recovery of development expenses, and supply chain issues, the overlap of various pricing strategies with more notorious instances has resulted in diminished public confidence in pharmaceutical companies charging fair prices for their products. Additionally, the issue of transparency related to treatments offered by pharmaceutical firms is contentious, with some researchers suggesting that details of clinical trials are often published only when outcomes are favorable, claiming that published clinical trial data may be selectively presented to bolster approval arguments, and expressing the concern that patient safety is at risk due to insufficient public access to regulatory hearings, investigations, and approvals (Löfstedt and Way 2016). From an environmental standpoint, the increase in the number and volume of biopharmaceuticals needing global manufacturing and transport heightens the cradle-to-grave impact due to rising emissions affecting land, air, and water, along with a higher consumption of natural resources. These issues combine to undermine the overall sustainability of the pharmaceutical sector, driven by environmental harm and weak socio-economic relationships, the former causing damage to the natural ecosystem essential for human survival and well-being, and the latter eroding societal trust and limiting safe, equitable access to treatments globally. Aside from unethical practices, the routine "business-as-usual" mindset in biopharmaceutical innovation and production also contributes to heightened fragility in the sector. The pace of biopharmaceutical approvals has remained steady since the launch of Humulin in 1982, with FDA approvals in 2017 nearing a record high (DeFrancesco 2018). While each approval marks a significant achievement in disease treatment, there are related consequences for the environment and society. As mentioned earlier, the financial viability of biopharmaceutical companies depends on robust product pipelines and consistently obtaining regulatory approval for new products (Thakor et al. 2017). This dependency has led to soaring development expenses that outpace inflation (DiMasi et al. 2016) and consequently increased prices to recover these costs, necessitating greater investment from governments and affecting access to treatments. In summary, to foster a more sustainable model within the sector, the integration and application of sustainability principles (economically, socially, environmentally, and ethically) in biopharmaceutical manufacturing are essential to tackle the socio-economic and ethical-environmental challenges posed by current manufacturing, innovation, and economic paradigms to ensure accessible treatments for those affected by diseases worldwide.

6 Future Trends in Biopharmaceutical Production

The biopharma sector has reached technological maturity as a result of decades of interdisciplinary scientific collaboration and invention. The foundational knowledge established by early biological product manufacturers and scientific discoveries from industry and academic researchers in areas like chemical engineering, biology, chemistry, bioinformatics (Prajapat et al. 2011a, b), and statistics (Jain et al. 2021) enabled modern processes capable of reliably producing complex, high-quality biopharmaceutical products. They used innovative technical solutions to make up for the technological shortcomings of earlier procedures. Manufacturers effectively incorporated these technologies into processes on a regular basis to produce treatments at the sizes needed to satisfy growing market demand, despite the fact that the unit operations utilized were diverse, intricate, and difficult to scale-up. A shift to product-focused strategies has been brought about by the advancement of processing technology and improved contact with regulators, which encourages ongoing studies to better understand the connections between product quality and processing conditions. Since mAb manufacturing methods eventually converge to a single set of unit activities, it is likely that mAb processing has reached its peak and that a significant breakthrough in process technology would be needed to upend the current state of affairs. The rise in the use of these treatments in clinical settings is proof that the insights gathered from conventional mAb processing have aided in the creation of innovative therapeutic approaches. Innovation may be stimulated to boost production capacity, guarantee dependable supply chains, and lower the cost of medications by continuing to use industrial expertise and partnerships between academia, business, and the government (Szkodny and Lee 2022). In the future, the sector will keep pushing the creation and use of vital products, providing patients with immunizations and medications that can save and change their lives for many decades. Cell engineering contributes to the establishment of highly efficient cell factories capable of generating substantial quantities of high-quality biopharmaceuticals. Through the optimization of metabolic pathways and the enhancement of protein expression levels, it is possible to substantially reduce production costs and increase yield. Furthermore, the capacity to engineer cells with specific attributes, such as increased stress resistance and enhanced protein secretion, further bolsters the efficiency and scalability of biopharmaceutical production (Swapnil et al. 2021; Marwal et al. 2014; Mahmood et al. 2017).

7 Conclusion

The integration of cell engineering and molecular pharming signifies a significant paradigm shift in the production of biopharmaceuticals. By leveraging the capabilities of genetic engineering and advanced cell culture methodologies, researchers can design cells to produce intricate therapeutic proteins with exceptional efficiency and accuracy. This novel approach not only offers considerable advantages over traditional biopharmaceutical production methods but also facilitates the development of innovative treatments for a wide range of diseases. In contrast, molecular pharming capitalizes on the inherent capabilities of plants to generate therapeutic proteins. By introducing foreign genes into plant genomes, plants can be transformed into living bioreactors capable of producing valuable biopharmaceuticals at a significantly lower cost than traditional approaches. This methodology provides numerous advantages, including reduced production costs, scalability, and the potential for oral delivery of therapeutics. As the field of biopharmaceutical production advances, it is anticipated that additional groundbreaking developments will emerge. By synthesizing the strengths of cell engineering and molecular pharming, novel therapies targeting previously challenging diseases can be developed. The future of biopharmaceutical production appears promising, with immense potential to enhance human health outcomes.

Acknowledgments The authors are thankful to Mohanlal Sukhadia University, Udaipur, for providing the necessary facilities during the course of study.

Declaration of Competing Interest The authors declare that there are no conflicts of interest. The authors agree to be accountable for all aspects of the work to ensure that the questions related to the accuracy or integrity of any part of the work are appropriately investigated and resolved.

Funding This research did not receive any specific grant from funding agencies in the public, commercial, or not-for-profit sectors.

References

Abdullah MA, ur Rahmah A, Sinskey AJ, Rha C (2008) Cell engineering and molecular pharming for biopharmaceuticals. Open Med Chem J 2:49

Amann T, Schmieder V, Faustrup Kildegaard H, Borth N, Andersen MR (2019) Genetic engineering approaches to improve posttranslational modification of biopharmaceuticals in different production platforms. Biotechnol Bioeng 116(10):2778–2796

Beuger V, Künkele KP, Koll H, Gärtner A, Bähner M, Burtscher H, Klein C (2009) Short-hairpin-RNA-mediated silencing of fucosyltransferase 8 in Chinese-hamster ovary cells for the production of antibodies with enhanced antibody immune effector function. Biotechnol Appl Biochem 53(Pt 1):31–37

Blay V, Tolani B, Ho SP, Arkin MR (2020) High-throughput screening: today's biochemical and cell-based approaches. Drug Discov Today 25(10):1807–1821

Bullen A (2008) Microscopic imaging techniques for drug discovery. Nat Rev Drug Discov 7:54–67

Butler M, Meneses-Acosta A (2012) Recent advances in technology supporting biopharmaceutical production from mammalian cells. Appl Microbiol Biotechnol 96:885–894

Byrne EP (2016) Sustainability as contingent balance between opposing though interdependent tendencies; a process approach to progress and evolution. In: Byrne E, Mullally G, Sage C (eds) Transdisciplinary perspectives on transitions to sustainability. Routledge, pp 41–62

Cao J, Galbraith EK, Sun T, Grattan KTV (2011) Comparison of surface plasmon resonance and localized surface plasmon resonance-based optical fibre sensors. J Phys Conf Ser 307(1):012050. IOP Publishing

Chen F, Alphonse M, Liu Q (2020) Strategies for nonviral nanoparticle-based delivery of CRISPR/Cas9 therapeutics. Wiley Interdiscip Rev Nanomed Nanobiotechnol 12:e1609

Cong L, Ran FA, Cox D, Lin S, Barretto R, Habib N, Hsu PD, Wu X, Jiang W, Marraffini LA, Zhang F (2013) Multiplex genome engineering using CRISPR/Cas systems. Science 339(6121):819–823

Crommelin D, Bermejo T, Bissig M, Damiaans J, Krämer I, Rambourg P (2005) Pharmaceutical evaluation of biosimilars: important differences from generic low-molecular-weight pharmaceuticals. Eur J Hosp Pharm Sci 11:11–17

Das S, Marwal A, Choudhary DK, Gupta VK, Gaur RK (2011) Mechanism of RNA interference (RNAi): current concept. Int Proc Chem Biol Environ Eng 9:240–245

Das AK, Marwal A, Verma R (2015a) Preparation and characterization of nano-bio hybrid-based magneto liposome. Int J Pharm Sci Res 6(1):367

Das AK, Marwal A, Pareek V (2015b) Nanoparticles-protein hybrid based magnetic liposome. Int J Chem Mol Nucl Mater Metall Eng 9(2):230–233

DeFrancesco L (2018) Drug pipeline: 4Q17. Nat Biotechnol 36:130

DiMasi JA, Grabowski HG, Hansen RW (2016) Innovation in the pharmaceutical industry: new estimates of R & D costs. J Health Econ 47:20–33

East-Seletsky A, O'Connell MR, Knight SC, Burstein D, Cate JH, Tjian R, Doudna JA (2016) Two distinct RNase activities of CRISPR-C2c2 enable guide-RNA processing and RNA detection. Nature 538(7624):270–273

Ehrenfeld J (2008) Sustainability by design: a subversive strategy for transforming our consumer culture. Yale University Press
Fabbri E (2015) Pharmaceuticals in the environment: expected and unexpected effects on aquatic fauna. Ann N Y Acad Sci 1340(1):20–28
Fang Z, Lyu J, Li J, Li C, Zhang Y, Guo Y, Wang Y, Zhang Y, Chen K (2022) Application of bioreactor technology for cell culture-based viral vaccine production: present status and future prospects. Front Bioeng Biotechnol 10:921755
Gillmore JD, Gane E, Taubel J, Kao J, Fontana M, Maitland ML, Seitzer J, O'Connell D, Walsh KR, Wood K, Phillips J (2021) CRISPR-Cas9 in vivo gene editing for transthyretin amyloidosis. N Engl J Med 385(6):493–502
Golla R, Seethala R (2004) A sensitive, robust high-throughput electrochemiluminescence assay for rat insulin. J Biomol Screen 9(1):62–70
Gootenberg JS, Abudayyeh OO, Lee JW, Essletzbichler P, Dy AJ, Joung J, Verdine V, Donghia N, Daringer NM, Freije CA, Myhrvold C (2017) Nucleic acid detection with CRISPR-Cas13a/C2c2. Science 356(6336):438–442
Gottlieb S (2008) Biosimilars: policy, clinical, and regulatory considerations. Am J Health Syst Pharm 65(14 Suppl 6):S2–S8
Hedlund-de Witt A (2013) Worldviews and the transformation to sustainable societies: an exploration of the cultural and psychological dimensions of our global environmental challenges. Doctoral dissertation, VU University Amsterdam
Hedlund-de Witt A (2014) Rethinking sustainable development: considering how different worldviews envision "development" and "quality of life". Sustainability 6(11):8310–8328
Hurst DJ (2017) Restoring a reputation: invoking the UNESCO universal declaration on bioethics and human rights to bear on pharmaceutical pricing. Med Health Care Philos 20(1):105–117
Intergovernmental Panel on Climate Change (IPCC) (2018) Global warming of 1.5°C [In press]
Jain K, Desai N, Sharma K, Marwal A (2021) Development and screening of byproduct for its secondary metabolites, antioxidant and anti-diabetic potential from anthracnose-infected fruits of pomegranate: a sustainable approach. 3 Biotech 11:1–11
Kesik-Brodacka M (2018) Progress in biopharmaceutical development. Biotechnol Appl Biochem 65(3):306–322
Khan AH, Tye GJ, Noordin R (2020) CRISPR-Cas9 genome editing tool for the production of industrial biopharmaceuticals. Mol Biotechnol 62(4):401–411
Kumar SA, Priyanka D (2024) Biopharmaceutical production in plant-based systems: current status and future prospects. Int J Biol Environ Invest 10(1):157–167
Kumar R, Bhanti P, Marwal A, Gaur RK (2020) Gene expression-based supervised classification models for discriminating early-and late-stage prostate cancer. Proc Natl Acad Sci India Sect B: Biol Sci 90:541–565
Kuriakose A, Chirmule N, Nair P (2016) Immunogenicity of biotherapeutics: causes and association with posttranslational modifications. J Immunol Res 2016:1298473
Łącki KM, Riske FJ (2020) Affinity chromatography: an enabling Technology for Large-Scale Bioprocessing. Biotechnol J 15(1):e1800397
Lakra N, Mahmood S, Marwal A, Sudheep NM, Anwar K (2019) Bioengineered plants can be an alternative source of omega-3 fatty acids for human health. In: Plant and human Health, volume 2: phytochemistry and molecular aspects, pp 361–382
Lee M, Ly H, Möller CC, Ringel MS (2019) Innovation in regulatory science is meeting evolution of clinical evidence generation. Clin Pharmacol Ther 105(4):886–898
Li X, Pan Y, Chen C, Gao Y, Liu X, Yang K, Luan X, Zhou D, Zeng F, Han X, Song Y (2021) Hypoxia-responsive gene editing to reduce tumor thermal tolerance for mild-photothermal therapy. Angew Chem Int Ed Engl 60(42):21200–21204
Lin H, Fan T, Sui J, Wang G, Chen J, Zhuo S, Zhang H (2019) Recent advances in multiphoton microscopy combined with nanomaterials in the field of disease evolution and clinical applications to liver cancer. Nanoscale 11(42):19619–19635
Liu H, Timko MP (2022) Improving protein quantity and quality—the next level of plant molecular farming. Int J Mol Sci 23(3):1326

Löfstedt R, Way D (2016) Transparency and trust in the European pharmaceutical sector: outcomes from an experimental study. J Risk Res 19(7–8):1082–1103
Lozano R (2008) Envisioning sustainability three-dimensionally. J Clean Prod 16(18):1838–1846
Luo N, Li J, Chen Y, Xu Y, Wei Y, Lu J, Dong R (2021) Hepatic stellate cell reprogramming via exosome-mediated CRISPR/dCas9-VP64 delivery. Drug Deliv 28(1):10–18
Mahmood S, Lakra N, Marwal A, Sudheep NÁ, Anwar K (2017) Crop genetic engineering: an approach to improve fungal resistance in plant system. In: Plant-microbe interactions in agro-ecological perspectives: volume 2: microbial interactions and agro-ecological impacts, pp 581–591
Marwal A, Gaur RK (2019) Nanophytovirology: an emerging field for disease management. In: Plant diseases-current threats and management trends. https://doi.org/10.5772/intechopen.86653
Marwal A, Gaur RK (2020) Molecular markers: tool for genetic analysis. In: Animal biotechnology. Academic, pp 353–372
Marwal A, Gaur RK (2021) Plant viruses as an engineered nanovehicle (PVENVs). In: Plant virus-host interaction: molecular approaches and viral evolution. Academic, pp 525–536
Marwal A, Sahu A, Gaur RK (2014) New insights in the functional genomics of plants responding to abiotic stress. In: Molecular approaches in plant abiotic stress. CRC Press/Science Publishers/Taylor & Francis Group, Boca Raton, pp 158–180
Milmo S (2017) Pharmaceutical environmental pollution and antimicrobial resistance. Pharm Technol Eur 29:6–8
Mitra S, Murthy GS (2022) Bioreactor control systems in the biopharmaceutical industry: a critical perspective. Syst Microbiol Biomanuf 2(1):91–112
Mitra S, Tomar PC (2021) Hybridoma technology: advancements, clinical significance, and future aspects. J Genet Eng Biotechnol 19(1):159
Murthy GS, Johnston DB, Rausch KD, Tumbleson ME, Singh V (2012) Design and evaluation of an optimal controller for simultaneous saccharification and fermentation process. Appl Biochem Biotechnol 166(1):87–111
Nielsen J (2012) Production of biopharmaceutical proteins by yeast: advances through metabolic engineering. Bioengineered 4(4):207–211
Ottinger M, Wenk I, Carvalho Pereira J, John G, Junne S (2022) Single-use technology in the biopharmaceutical industry and sustainability: a contradiction? Chem Ing Tech 94(12):1883–1891
Pazmiño M, Terán-Maza VA, Calderón J, Gaona K, Mihai R, Pais-Chanfrau JM, Trujillo Toledo L (2021) Single-use systems bioreactors in the biopharmaceutical industry and its use in SARS-CoV-2 vaccine candidates' production: a review. Prensa Med Argent 107:204–218
Prajapat R, Marwal A, Sahu A, Gaur RK (2011a) Phylogenetics and in silico docking studies between coat protein of Mimosa yellow vein virus and whey α-lactalbumin. Am J Biochem Mol Biol 1(3):265–274
Prajapat R, Gaur RK, Marwal A (2011b) Homology modeling and docking studies between AC1 rep protein of Begomovirus and Whey á-lactalbumin. Asian J Biol Sci 4(4):352–361
Qi LS, Larson MH, Gilbert LA, Doudna JA, Weissman JS, Arkin AP, Lim WA (2013) Repurposing CRISPR as an RNA-guided platform for sequence-specific control of gene expression. Cell 152(5):1173–1183
Ranjith S, Sathyan A, Duraisamy S, Peter AS, Marwal A, Jain K, Chidambaram P, Kumarasamy A (2022) Exploring a computational method for evaluating the Epinecidin-1 and its variants binding efficacy with breast Cancer receptor (HER-2). Int J Pept Res Ther 28(4):118
Ray K, Szmacinski H, Lakowicz JR (2009) Enhanced fluorescence of proteins and label-free bioassays using aluminum nanostructures. Anal Chem 81(15):6049–6054
Richter MM (2008) Chapter 7: Electrochemiluminescence. In: Ligler FS, Taitt CR (eds) Optical biosensors. Today and tomorrow. Elsevier Science, pp 317–384
Rodell M, Famiglietti JS, Wiese DN, Reager JT, Beaudoing HK, Landerer FW, Lo MH (2018) Emerging trends in global freshwater availability. Nature 557:651–659

Sahu AK, Marwal A, Nehra C, Choudhary DK, Sharma P, Gaur RK (2014) RNAi mediated gene silencing against betasatellite associated with Croton yellow vein mosaic begomovirus. Mol Biol Rep 41:7631–7638

Saltiel M, Finnefrock M (2018) Hyperinflation—definitions and causes. J Pharm Pract 31:370–373

Schoger E, Carroll KJ, Iyer LM, McAnally JR, Tan W, Liu N, Noack C, Shomroni O, Salinas G, Groß J, Herzog N (2020) CRISPR-mediated activation of endogenous gene expression in the postnatal heart. Circ Res 126(1):6–24

Sivanandhan G, Selvaraj N, Lim YP, Ganapathi A (2016) Targeted genome editing using site-specific nucleases, ZFNs, TALENs, and the CRISPR/Cas9 system Takashi Yamamoto (ed.). Ann Bot 118(2):vii–viii

Smith TA, Ghiggino KP (2015) A review of the analysis of complex time-resolved fluorescence anisotropy data. Methods Appl Fluoresc 3(2):022001

Srivastava A, Pandey V, Singh N, Marwal A, Shahid MS, Gaur RK (2024) In silico identification of papaya genome-encoded microRNAs to target begomovirus genes in papaya leaf curl disease. Front Microbiol 15:1340275

Swapnil P, Meena M, Singh SK, Dhuldhaj UP, Marwal A (2021) Vital roles of carotenoids in plants and humans to deteriorate stress with its structure, biosynthesis, metabolic engineering and functional aspects. Curr Plant Biol 26:100203

Szkodny AC, Lee KH (2022) Biopharmaceutical manufacturing: historical perspectives and future directions. Ann Rev Chem Biomol Eng 13(1):141–165

Tailor S, Marwal A, Meena M (2021) Application of nanotechnology in management of various plant diseases. In: Innovative approaches in diagnosis and management of crop diseases. Apple Academic Press, pp 1–17

Tailor S, Jain K, Marwal A, Meena M, Anbarasu K, Gaur RK (2022) Outlooks of nanotechnology in organic farming management. Def Life Sci J 7(1):52–60

Tan E, Chin CSH, Lim ZFS, Ng SK (2021) HEK293 cell line as a platform to produce recombinant proteins and viral vectors. Front Bioeng Biotechnol 9:796991

Thakor RT, Anaya N, Zhang Y, Vilanilam C, Siah KW, Wong CH, Lo AW (2017) Just how good an investment is the biopharmaceutical sector? Nat Biotechnol 35(12):1149–1157

Tummala S, Titus M, Wilson L, Wang C, Ciatto C, Thill G, Foster D, Li C, Szabo Z, Guttman A, Bettencourt B (2013) Evaluation of exogenous siRNA addition as a metabolic engineering tool for modifying biopharmaceuticals. Biotechnol Prog 29(2):415–424

Van Landuyt L, Lonigro C, Meuris L, Callewaert N (2019) Customized protein glycosylation to improve biopharmaceutical function and targeting. Curr Opin Biotechnol 60:17–28

Verma RK, Mishra M, Marwal A, Akhtar A, Manchev N, Petrov RK (2024) CRISPR/Cas9 protection system for Chilli leaf curl virus. Acta Microbiol Bulg 40(1):112–115

Wang D, Chen L, Li C, Long Q, Yang Q, Huang A, Tang H (2022) CRISPR/Cas9 delivery by NIR-responsive biomimetic nanoparticles for targeted HBV therapy. J Nanobiotechnol 20(1):1–16

Wei M, Wynn R, Hollis G, Liao B, Margulis A, Reid BG, Klabe R, Liu PC, Becker-Pasha M, Rupar M, Burn TC (2007) High-throughput determination of mode of inhibition in lead identification and optimization. SLAS Discovery 12(2):220–228

Wienert B, Wyman SK, Richardson CD, Yeh CD, Akcakaya P, Porritt MJ, Morlock M, Vu JT, Kazane KR, Watry HL, Judge LM (2019) Unbiased detection of CRISPR off-targets in vivo using DISCOVER-Seq. Science 364(6437):286–289

Wu Y, Zeng J, Roscoe BP, Liu P, Yao Q, Lazzarotto CR, Clement K, Cole MA, Luk K, Baricordi C, Shen AH (2019) Highly efficient therapeutic gene editing of human hematopoietic stem cells. Nat Med 25(5):776–783

WWF (2018) In: Grooten M, Almond REA (eds) Living planet report 2018: aiming higher. Author, Gland

Yasgar A, Jadhav A, Simeonov A, Coussens NP (2016) AlphaScreen-based assays: ultra-high-throughput screening for small-molecule inhibitors of challenging enzymes and protein-protein interactions. In: Janzen W (ed) High throughput screening, vol 1439. Humana Press, pp 77–92

Zhu J (2012) Mammalian cell protein expression for biopharmaceutical production. Biotechnol Adv 30(5):1158–1170

Biopharmaceuticals of Plant Origin with Health Importance and Potential Benefits for the Development of Modern Agriculture

Stefka A. Ivanova, Danka P. Obreshkova, Boris Seryozhev Mladenov, Nikolay Manchev Petrov, and Mariya Ivanova Stoyanova

Abstract

The study of products of plant origin in all fields of modern medicine, agriculture, and plant agriculture is increasingly relevant due to the great application possibilities. For example, some substances of natural origin successfully affect a number of progressive polypathogenic neurodegenerative diseases in which the combination of various mechanisms and risk factors is found to induce anatomical, cellular, and molecular changes leading to disruption of higher cortical functions, deficits in memory, learning, behavioral and functional disorders, and total breakdown of intellect and mental activity. Bulgarian pharmacologist D. Paskov studied the alkaloid galantamine, which is most widely distributed in representatives of the family Amaryllidaceae: Caucasian snowdrop (*Galanthus woronowii*

S. A. Ivanova
Bulgarian Pharmaceutical Science Society, Sofia, Bulgaria

D. P. Obreshkova
Clinic research of Drugs in Medicine, University Hospital "N. I. Pirogov", Sofia, Bulgaria
e-mail: phddanka@abv.com

B. S. Mladenov
University Multiprofile Hospital for Active Treatment and Emergency Medicine "N.I.Pirogov", Sofia, Bulgaria

N. M. Petrov (✉)
Virology Laboratory, New Bulgarian University, Sofia, Bulgaria
e-mail: npetrov@nbu.bg

M. I. Stoyanova
Institute of Soil Science, Agrotechnologies and Plant Protection (ISSAPP) "Nikola Pushkarov," Agricultural Academy, Sofia, Bulgaria

R. K. Gaur et al. (eds.), *Plant Molecular Farming*,
https://doi.org/10.1007/978-981-95-3823-2_6

Losinsk), snowdrop (*Galanthus nivalis* L., *Galanthus elwesii*) Hook. Fil. Marsh snowdrop (*Leucojum aestivum* L.).

Keywords

Galantamine · Omega-3 unsaturated fatty acids · Amino acids · Oxidative stress · Cellular changes · Molecular changes · Amyloidogenesis · Endogenous antioxidant systems · Dicarboxylic acids · Phenolcarboxylic acids

Lipid peroxidation, DNA covalent bond breakage, and oxidation-induced cellular damage demonstrate acceleration in the aging process and progression of multiple diseases (including Alzheimer's disease and Parkinson's disease) (Lobo et al. 2010).

Parkinson's disease (PD) is known as the second most important neurodegenerative disorder. It has been reported that 1% of people older than 65 years and 2.6% of the population over 85 years suffer from Parkinson's disease. The importance of age as a risk factor has been proven by the tendention of the enchance of cases with the disease with the increasing of age. It has been reported that patients with Parkinson's disease make up two-thirds of all people suffering from movement disorders. PD has become increasingly more common with advances in age, reaching the proportions of 2.6% of the population over 85 years old.

The antioxidant enzymes, such as glutathione and superoxide dismutase, play an important role in the protection of oxidation of DNA, proteins, and lipids in dopamine neurons. But the development of more reactive oxygen species in brain areas during the pathogenesis of the disease cannot be overcome by the decreased function of these antioxidant enzymes. The results of the increased oxidative stress are mitochondrial dysfunction and damage to the substantia nigra. Some of the more detrimental reactive oxygen species are superoxides (O_2) in cellular mitochondria and H_2O_2, O_2, and hydroxyl (OH) radicals (Lobo et al. 2010).

Oxidative damage is the earliest event in Alzheimer's disease and causes synaptic dysfunction, dendritic spine loss, affects cell membranes, and leads to cell death. Oxidative stress plays a role in the progression of Alzheimer's disease in the frontal cortex (Ansari and Scheff 2010).

The search for possibilities to enlarge the therapeutic options is of great importance. The new trends must be connected with the synthesis of substances or use of compounds from plant origin that can be more effective and less toxic than the current drugs, especially in decreasing neurodegenerative processes and for the protection of the brain from oxidative stress.

The role of antioxidants is important against oxidative stress (Pisoschi and Pop 2015). The new trends can be a combined therapy of compounds with different targets, such that some of them are neuroprotectors and others are used to reduce oxidative stress.

Another important perspective can be multitarget compounds, which exhibit simultaneously anti-neurogenerative and antioxidant properties.

It has been described that perspective properties for obtaining therapeutic beneficial effect exert Galantamine and Omega-unsaturated acids.

Galantamine exerts antioxidant and neuroprotective effects (Tsvetkova et al. 2013). Pharmacological studies of galantamine prove: nervous system: (1) Reversibly inhibits acetylcholinesterase, which leads to the accumulation of acetylcholine and to indirect cholinomimetic effects, (2) Allosterically potentiates α7-subtype nicotinic acetylcholine receptors, thereby enhancing dopamine and γ-aminobutyric acid neurotransmission, (3) Improves learning and memory processes (Tsvetkova et al. 2016).

Unsaturated higher fatty acids can modulate physiological processes and can influence some immunological parameters (La Guardia et al. 2005). Omega-6 unsaturated fatty acids are known as vitamin F and are found in vegetable oils (Simopoulos 2008).

1 Oxidative Stress

Amino acids are the same for all organisms, and there is at least one specific codon in the DNA for each of them. By further chemical modifications, some amino acids included in a polypeptide chain are converted into derived amino acids: L-cystine (by oxidation of two cysteine residues), hydroxyproline and hydroxylysine in collagen, and desmosine in elastin, α-carboxyglutamate in prothrombin, i.e., they are the result of posttranslational modification. Many of the processes occurring in living organisms are the result of the participation of metal ions in electron exchange reactions. A number of metals with reversibly variable degrees of oxidation are associated with oxygen transport (Fe^{2+}/Fe^{3+}- hemoglobin, Cu^{+}/Cu^{2+}- hemocyanin). In catalytic enzymatic processes, electron transport depends on biogenic ions such as Cu^{2+}, Co^{2+}, Co^{3+}, Mo^{4+}, and Ni^{2+}. Electron exchange reactions in physiological processes in living organisms are associated with the formation of reactive particles (RP). Curcumin induces the antioxidant defense system glutathione transferase, lowers cholesterol levels, and binds ferric ions into complexes (Ak and Gülçin 2008).

Oxidative stress is involved in tumor development because free radicals cause direct DNA damage and are mutagenic and can activate processes such as proliferation, invasiveness, and metastasis. For example, *Helicobacter pylori* infection, which increases the formation of reactive oxygen and nitrogen free radicals in the stomach, is a cause of gastric cancer development. In hematologic cancers, such as leukemia, free radicals disrupt the function of immune cells by increasing immune evasion of leukemic cells. Oxidative stress is a risk factor for the development of atherosclerosis, heart failure, myocardial infarction, diabetes, macular degeneration, arthritis, and lupus erythematosus (Lobo et al. 2010).

Reactive oxygen-containing free radicals are formed in the cells of living organisms as a result of normal cellular metabolism and are essential for the maintenance of cellular homeostasis. Under physiological conditions, low-to-moderate concentrations of reactive oxygen species are important for protein phosphorylation and cell differentiation and are involved in immune response and inflammation processes, while high levels are released locally by macrophages and phagocytes to protect the organism. Reactive particles play an important role in cellular redox signaling. For the maintenance of cellular homeostasis, a balance between the formation of reactive oxygen species and their consumption is needed (Lobo et al. 2010).

Reactive oxygen-containing free radicals (ROS), such as superoxide radical anion and hydrogen peroxide, are formed in biological systems by partial reduction of molecular oxygen. Hydrogen peroxide is produced by multiple enzymes, including several oxidases. The superoxide anion is produced by one-electron reduction of molecular oxygen or arises by one-electron reduction of peroxide, which is facilitated by the transition metals copper and iron (Lobo et al. 2010).

2 Antioxidants

In Alzheimer's disease (AD), antioxidants are used to reduce oxidative stress: curcumin, ginkgo biloba, melatonin, coenzyme Q10 and coenzyme Q10 derivatives, herbal products containing some dicarboxylic and phenolcarboxylic acids and other components (essential oils) with antioxidant and anti-inflammatory effects, vitamin E, and selegiline. Curcumin induces the antioxidant defense system glutathione transferase, lowers cholesterol levels, and binds ferric ions in complexes. Idebenone exerts neuroprotective effects. The activity of peptides has also been investigated for their antioxidant activity. The antihistamine latrepirdine (dimebon) is an antagonist of N-methyl-D-aspartate glutamate receptors and protects mitochondria from oxidative stress by blocking calcium channels. The selective estrogen receptor modulator raloxifene (Evista) protects neurons from degeneration via an antioxidant effect. Galantamine, silymarin, and tribestan have also proven antioxidant activity (Obreshkova 2012).

2.1 Direct Antioxidants

Radical scavengers, metal chelating agents, and antioxidant enzymes have a direct antioxidant effect.

Antioxidant radical scavengers.

The mechanism of action of antioxidant radical scavengers, such as phenolic compounds, is by hydrogen atom donation or electron exchange with the free radical (-R) to Free radical (R)-H (Hydrogen) (RH) or (:R), where the reactivity of the reactive species is reduced.

2.2 Chelating Antioxidants

In chelation, the participation of ions in electron exchange reactions is prevented. Chelating agents, such as albumin, transferrins, ceruloplasmin, glutamic acid, histidine, and polysaccharides, bind metal ions of variable valence with coordination bonds to form stable complexes.

2.3 Antioxidant Enzyme Systems in the Body

The organism possesses the ability to protect itself with the help of enzymes, which attach to highly reactive agents and neutralize and damage free radicals. It has been reported that during evolution, organisms have developed some mechanisms to protect themselves against the harmful action of ROS and Reactive Nitrogen Species (RNS) in such a way that the total content of prooxidants is under control. Antioxidant activity is calculated by determining the decrease in absorption of specific compounds at different concentrations.

3 Relationship Between Oxidative Stress and AD

Extensive damage caused by oxidative stress and high production of ROS in the brain is associated with the progression of AD. Oxidative stress is thought to contribute to the acceleration of AD pathogenesis in three ways: macromolecular peroxidation, high metal ion redox potential, and mitochondrial dysfunction. All of these mechanisms affect cellular homeostasis, ROS generation, and increased expression of Aβ and tau formation (Petersen et al. 2007).

Metal ions such as copper, zinc, and iron are involved in the biological functions of metalloproteins and in neuronal processes. The body's metal homeostasis is severely impaired in neurodegenerative diseases (including AD). It has been described that Cu^{2+} and Zn^{2+} levels in the brain of Alzheimer's disease patients are three times higher (Kozlowski et al. 2012). These cations are able to bind to the hydrophilic N-terminal residues of Aβ peptides, generating high levels of ROS. Moreover, binding of metal ions to Aβ monomers allows for easier aggregation into neurotoxic oligomers and subsequently into plaques.

Similar to the aforementioned mechanisms, metal ions can bind to tau proteins. Zn^{2+} has been demonstrated to promote phosphorylation of tau proteins, leading to induced fibrillization. Fe^{3+} cations also lead to the initiation of aggregation and oligomer formation of cytochrome C oxidase levels in (Bader et al. 2011). A paper published by Mutisya et al. reported drastic decreases in the hypothalamus of AD sufferers.

Aβ oligomers can accumulate in the inner mitochondrial membrane, leading to the failure of key electron transport and a subsequent increase in ROS concentrations. Rationalizing on the given data, as well as the lack of an effective therapeutic regimen against AD, has directed many author teams to obtain new compounds possessing antioxidant properties. The body possesses the ability to protect itself with the help of enzymes that attach to highly reactive agents, damage, and neutralize free radicals.

Antioxidant molecules are divided into two groups: enzymatic and nonenzymatic compounds. Singlet oxygen attacks linoleic acid to produce racemic mixtures of 13-hydroxy-9Z, 11E-octadecadienoic acid, 9-hydroxy-10E,12Z-octadecadienoic acid, 10-hydroxy-8E, 12Z-octadecadienoic acid, and 12-hydroxy-9Z-13-E-octadec adienoic acid. Similar attacks on arachidonic acid produce a much larger set of products, including various isoprostanes, hydroperoxy- and hydroxy-eicosatetraenoates, and 4-hydroxyalkenals. Although many of these products are used as markers of oxidative stress, linoleic acid-derived products are more prevalent than arachidonic acid products (Barcelo-Coblijn and Murphy 2009).

4 Acetylcholinesterase Inhibitors

Classical therapy is compensatory with the reversible acetylcholinesterase inhibitors galantamine, Donepezil, Rivastigmine, which, by inhibiting the breakdown of intrasynaptic acetylcholine, increase the possibility of conducting a signal to the postsynaptic cholinergic neuron. Rivastigmine with acetylcholinesterase forms a carbamoylated instead of an acylated complex, which is hydrolyzed much more slowly. The tertiary amine Tacrine is not administered due to hepatotoxicity. Clinical trials have shown that acetylcholinesterase inhibitors modestly affect cognitive dysfunction and behavioral disturbances in patients with early and intermediate disease within 1 month of starting therapy (Tsvetkova and Obreshkova 2019).

Indirect antioxidant effects are produced by systems that stimulate the formation, action, or regeneration of direct antioxidants. Indirect antioxidants do not react directly with reactive particles. An indirect antioxidant is glucose-6-phosphate dehydrogenase, which regenerates reduced form of Nicotinamide adenine dinucleotide phosphate (NADP) by restoring the glutathione-reduced. Indirect antioxidant effects are produced by systems that stimulate the formation, action, or regeneration of direct antioxidants. Indirect antioxidants do not react directly with reactive particles. An indirect antioxidant is glucose-6-phosphate dehydrogenase, which regenerates NADPH by restoring the glutathione-reduced form.

4.1 Galantamine

4.1.1 Distribution of Galantamine in Nature

In nature, the alkaloid galantamine is most widely distributed in representatives of the family Amaryllidaceae: Caucasian snowdrop (*Galanthus woronowii Losinsk*), snowdrop (*Galanthus nivalis L., Galanthus elwesii Hook,* and marsh snowdrop (*Leucojum aestivum L.*). Galantamine was first isolated by Proskurnina and Yakovleva in 1952 from the tubers and aerial parts of *Galanthus woronowii Losinsk*, and in 1956 by Paskov and Ivanova-Bubeva, it was obtained from the tubers of *Galanthus nivalis L.* In Bulgaria, the genus Galanthus is represented by three species: *Gracilis Čelak*, *Galanthus elwesii Hook.* Fil., and *Galanthus graecus Orph. ex Boiss*. Later, galantamine was found in *Lycoris aurea L., Lycoris radiata L., Lycoris guangxiensis Y. Xu et G. J. Fan, Narcissus pseudonarcissus L., Narcissus tazetta L., Narcissus jonquilla Pipit, Pancratium maritimum L., Ungernia victoris Vved. ex Artjush, Ungernia severtzovii (Regel) B. Fedtsch, Ungernia trisphera Bunge, Hippeastrum equestre Linn. Fil. (Amaryllis equestris Linn.Fil.),* and *Cyrtanthus elatus (Jacq.) Traub. (Vallota speciosa L. f., T. Durand et Schinz).* The following alkaloids have been isolated from *Leucojum aestivum L.*: Galantamine, Apogalantamine, N-Desmethylgalantamine, Dihydrogalantamine, Epigalantamine, Epinorgalantamine, Aestivine, Homolycorine, Lycorine, Lycorenine, Narwedine, and Nivalidine.

4.1.2 Pharmacological Effects and Administration of Galantamine

The main pharmacological effects of galantamine are on the following: (I) nervous system: (1) reversibly inhibits acetylcholinesterase, which leads to accumulation of acetylcholine and to indirect cholinomimetic effects; (2) allosterically potentiates α7-subtype nicotinic acetylcholine receptors, by which it enhances dopamine and γ-aminobutyric acid neurotransmission (Schilström et al. 2007); (3) improves learning and memory processes: alone and in combination with Pymadin; (4) through anticholinesterase action, it facilitates the conduction of nerve impulses at the level of neuromuscular synapses, has a potent antagonistic action against nondepolarizing (competitive) myorelaxants (antagonizes Tubocurarine-induced muscle paralysis), and potentiates the action of depolarizing myorelaxants; (5) inhibits amyloid aggregation and cytotoxicity (Matharu et al. 2009); (II) cardiovascular system: induces weak bradycardia followed by weak tachycardia; (III) digestive tract: increases smooth muscle tone and contractility and enhances secretion of salivary and gastric glands; (IV) respiratory system: (1) excites the respiratory center via the central pathway and via the sino-carotid and aortic reflexogenic zone, accelerating and deepening respiration (Ago et al. 2011).

The ability to cross the blood–brain barrier and the different routes of administration (oral, intravenous (i.v.), intramuscular (i.m.), subcutaneous (s.c.), intraocular, electrophoresis) determine the following areas of application of galantamine: (1) Alzheimer's disease (Tsvetkova et al. 2019), Alzheimer's with cerebrovascular syndrome, vascular dementia, Huntington's chorea, (2) neurology: poliomyelitis, cerebral palsies in children, pyramidal and extra-pyramidal paresis, trigeminal paresis, residual paresis after cerebral hemorrhage, myasthenia, multiple sclerosis, myopathies: progressive muscular dystrophy, congenital myotonia, dystrophic myotonia, neuromuscular atrophy, myotonic muscular atrophy, spinal muscular atrophy (Werdnig-Hoffman disease), neuritis, polyneuropathy, radiculoneuropathy, plexitis, and enuresis nocturna; (3) surgery and anesthesiology: to antagonize the effect of competing neuromuscular blockers; (4) treatment of postoperative gastrointestinal and bladder pareses; (5) ophthalmology: glaucoma, myasthenic ptosis of the eyelids, optic nerve atrophy; (6) physiotherapy: angioneurosis, acrocyanosis, neurodermatitis, obliterative endarteritis; (7) radiology: for functional diagnostics—cholecystography and cholangiography; (8) Toxicology: In poisoning with cholinolytics, morphine, and its analogues (Ago et al. 2011).

Galantamine has antioxidant and neuroprotective effects. The antioxidant activity was determined by the hydroxyl group at the sixth position (galantamine and galantamine hydrobromide). It was absent when oxygen was substituted with a carbonyl group (Narwedine, Narwedine hydrobromide). The hydrogen bromide salt of galantamine has enhanced antioxidant activity compared to the base. Galantamine can bind to -O^{2-} (superoxide), -O_2^{2-} (peroxide), -OH (hydroxyl), and RO- (alkoxy) (Tsvetkova et al. 2013).

According to studies by Geerts, 1.5–5.0 mg/kg is the optimal dose for activation of α7-subtype nicotinic acetylcholine receptors, by which galantamine protects neurons from the influence of superoxide reactive oxygen species (Schilström et al. 2007).

The neuroprotective effect of galantamine toward different cytotoxic agents (β-amyloid, glutamate, and hydrogen peroxide) is mediated by allosteric activation of α7-subtype nicotinic acetylcholine receptors (Schilström et al. 2007), leading to (1) enhanced expression of the protective protein Bcl-2 and (2) activation of phosphoinositol-3-kinase through induction of phosphorylation of L-Serine-L-Threonine protein kinase.

Hydrogen peroxide is formed during mitochondrial dysfunction and autooxidation of dopamine, with the resulting -OH reactive oxygen species being a neurotoxic factor. In a model of hydrogen peroxide-induced oxidative stress in the human neuroblastoma cell line SK-N-SH, galantamine exerted a protective effect through the following mechanisms: (1) reduction of superoxide anion and nitric oxide formation; (2) protection of the mitochondrial membrane by restoring membrane potential; (3) suppression of P2X7-receptor activation; (4) increase in acetylcholine levels by dose-dependent inhibition of acetylcholinesterase; (5) potentiation of α7-subtype nicotinic acetylcholine receptors (Schilström et al. 2007). Nakao demonstrated that galantamine at concentrations of 0.1–10 μM dose-dependently decreased hydrogen peroxide-induced apoptosis of rat brain endothelial cells. This effect was attributed to the activation by galantamine of the antioxidant enzyme oxygenase 1, which protects neurons by improving vascular microcirculation (Tsvetkova et al. 2016).

Galantamine exerts neuroprotective effects by stimulating α7-subtype nicotinic acetylcholine receptors (Schilström et al. 2007), activating cytoplasmic Janus protein tyrosine kinase, resulting in (1) reduction of superoxide anion synthesis in astrocytes by inhibition of nicotinadenine dinucleotide phosphate oxidase and (2) suppression of nuclear factor Nk-B3 (p65, REL A), which prevents the expression of genes for proinflammatory factors: cytokines and cyclooxygenase.

In an in vitro model of oxidative stress in rat hippocampal cells, resulting from oxygen and glucose deprivation, 5 mM galantamine showed neuroprotective activity through a 50% reduction in the release of lactate dehydrogenase, which is a marker of apoptosis. Lorrio's studies showed that 10 mg/kg galantamine administered 3 h after cerebral ischemia dose-dependently reduced the loss of pyramidal neurons under the action of superoxide anion, decreased DNA fragmentation, improved memory, and restored superoxide dismutase levels. This fact is of great importance because in humans, 3 h is the therapeutic window for effective therapy in cerebral ischemia. In a model of cerebral ischemia, drug administration alone and the combination of 10 mg/kg galantamine/10 mg/kg Memantine dose-dependently protected pyramidal neurons from apoptosis by lowering caspase-3 activity. Iliev's studies found that administration of galantamine 25 min after induced ischemia resulted in improved learning ability in rats. Li demonstrated that by stimulating α7-subtype nicotinic acetylcholine receptors, galantamine protects the PC12 cell line from ethanol-induced oxidation (Schilström et al. 2007).

5 Unsaturated Fatty Acids

5.1 Factors that Affect the Metabolism of Unsaturated Fatty Acids

Oleic acid belongs to the Omega-9 unsaturated fatty acids and is a precursor to its fatty acid series but is not an essential fatty acid, as the body can synthesize it from simple precursors. Linoleic acid (LA), α-linolenic acid (ALA), and oleic acid are metabolized by the same enzymes: Δ6 and Δ5 desaturases. The enzymes metabolize the fatty acids with the following sequence of preference: Omega-3 unsaturated fatty acids > Omega-6 unsaturated fatty acids > Omega-9 unsaturated fatty acids, and under normal physiological conditions, the metabolites of Omega-9 unsaturated fatty acids are formed only in trivial amounts in cells. This is of important biological significance, as the presence of significant amounts of eicosatrienoic acid (EtE) (an Omega-9 unsaturated fatty acid) indicates a deficiency of Omega-3 unsaturated fatty acids and Omega-6 unsaturated fatty acids and is used as an indicator of essential fatty acids. The activity of Δ6 and Δ5 desaturases is slow in humans (Δ5 > Δ6); therefore, the conversion of linoleic acid and α-linolenic acid to their respective metabolites: γ-linolenic acid (GLA) and eicosapentaenoic acid (EPA), may be inappropriate under certain conditions. To avoid blocking the action of Δ6 and Δ5 desaturases in some diseases and in aging individuals, it may be necessary to supplement the metabolites: γ-linolenic acid and Dihomo-γ-linolenic acid (DGLA) (to avoid Δ6 desaturase) and arachidonic (AA), docosahexaenoic (DHA), and eicosapentaenoic (EPA) acids (to avoid Δ5 desaturase) (Lee and Park 2014). Arachidonic acid can be supplemented with a proper diet, and DHA and EPA can be obtained from marine fish (Fotuhi et al. 2009). Many of the unsaturated fatty acids are lost in food preparation, and direct oral supplementation of DHA and EPA may be necessary.

Omega-3 unsaturated fatty acids, DHA and EPA, which were first described in Greenland in the diet of Eskimos who consumed large amounts of seafood, have beneficial health effects (La Guardia et al. 2005).

Omega-3 unsaturated fatty acids are essential for normal growth and development of the body and play an important role in the protection and therapy of coronary artery disease, hypertension, arthritis, autoimmune diseases, breast tumors, inflammatory processes, bowel cancer, rheumatoid arthritis, and psoriasis (Ruxton et al. 2005). The most important and nutritionally relevant Omega-3 unsaturated (essential) fatty acids are DHA and EPA, which are required for efficient brain and body functioning at the cellular level.

Low DHA levels are associated with depression, schizophrenia, memory loss, increased risk of developing Alzheimer's disease, preterm birth, low birth weight, and hyperactivity in children.

α-Linolenic acid is the main building block of DHA and EPA, essential for the functioning of the immune, cardiovascular, and nervous systems. Only plants can synthesize Omega-3 unsaturated fatty acids and Omega-6 unsaturated fatty acids. For the complete conversion of ALA into DHA and EPA, the satisfactory presence of certain nutrients is required: Vitamins B3, B6, and C, zinc, and magnesium. When one or more of these nutrients are deficient, the human body is unable to provide optimal amounts of both DHA and EPA, even with sufficient intake of ALA.

In the diet of the average inhabitant of civilized countries, the ratio between Omega-3 unsaturated fatty acids and Omega-6 unsaturated fatty acids ranges from 1:20 to 1:40 and should be at least 1:2, and according to new data, even close to 1:1. The American "Western diet" contains 14–25 times more Omega-6 unsaturated fatty acids than Omega-3 unsaturated fatty acids. There was a relative balance between Omega-6 unsaturated fatty acids and Omega-3 unsaturated fatty acids. The Mediterranean type of diet has a healthier balance between Omega-3 unsaturated fatty acids and Omega-6 (Ruxton et al. 2005).

5.2 Actions of Unsaturated Fatty Acids

5.2.1 Action of Omega-3 Unsaturated Fatty Acids and Omega-6 Unsaturated Fatty Acids as a Secondary Messenger (Mediator)

Essential fatty acids and their long-chain metabolites act as secondary messengers. Several hormones and growth factors activate phospholipase A2 (PLA2), which triggers the release of dihomo-γ-linolenic, arachidonic, docosahexaenoic, and eicosapentaenoic acids from the lipid region of the cell membrane. These fatty acids are used to form various eicosanoids and contribute to their activity. Inhibition of PLA2 interferes with the action of various growth factors, cytokines, and proteins. The tumoricidal action of tumor necrosis factor α (TNF-α) is mediated by its ability to activate PLA2. Inhibitors of PLA2 completely stop the tumoricidal action of TNF-α. Studies have shown that TNF-α-resistant tumor cells can be sensitized to TNF-α by the addition of various unsaturated fatty acids, especially γ-linolenic acid. Dihomo-gamma-linolenic acid inhibits tumor necrosis factor-alpha production by human leukocytes. This indicates that these acids are essential for some effects of various growth factors and cytokines. Unsaturated fatty acids increase the activity of the secondary mediator protein kinase C and can activate macrophages and polymorphonuclear leukocytes (Dooper et al. 2003).

5.2.2 Antimicrobial Action

Linolenic acid rapidly kills cultures of *Staphylococcus aureus*. Hydrolyzed linseed oil, which contains linoleic and linolenic acids, inactivates methicillin-resistant *Staphylococcus aureus* cultures. Linolenic acid stimulates the adhesion of

Lactobacillus casei to mucosal surfaces and thus increases their growth. Lactobacilli inhibit the growth of pathogenic bacteria: *Clostridioides difficile, Escherichia coli, Helicobacter pylori, Pseudomonas aeruginosa, Salmonella typhimurium, and Shigella flexneri.* Unsaturated fatty acids inactivate enveloped viruses and exert antifungal activity. The anti-inflammatory, antibacterial, antiviral, and antifungal actions of unsaturated fatty acids may explain their beneficial properties (Sun et al. 2003). This antimicrobial action has been confirmed in the case of basil oil, and we have also demonstrated its chemical composition.

5.2.3 Other Actions of Omega-3 Unsaturated Fatty Acids and Omega-6 Unsaturated Fatty Acids

Unsaturated fatty acids inhibit angiotensin-converting enzyme (ACE) activity in leukocytes. They may function as endogenous regulators of ACE activity, regulating angiotensin II formation and enhancing endothelial nitric oxide production. Omega-6 unsaturated fatty acids also exhibit beneficial effects in cardiovascular disease and diabetes (Ruxton et al. 2005).

5.2.4 Omega-3 Unsaturated Fatty Acids in Alzheimer's Disease

The brain is protected by a network of blood vessels called the blood–brain barrier. It helps filter out potentially dangerous substances and chemicals that can damage the fine balance of chemical neurotransmitters and electrical impulses that are responsible for the body's processes of learning, remembering, and coordinating movements. Scientists from the Karolinska Institute in Sweden have published the results of their study in the Journal of General Medicine. They have shown the importance of Omega-3 unsaturated fatty acids crossing the blood–brain barrier, their use in building the membranes of nerve cells that ensure the proper flow of oxygen and nutrient saturation for optimal brain health, and in preventing dementia and Alzheimer's disease (Cole et al. 2005). Omega-3 unsaturated fatty acids lower the risk of chronic inflammation and thus help fight chronic diseases (Ruxton et al. 2005).

It has been shown that people with Alzheimer's disease have very low levels of the Omega-3 unsaturated fatty acid DHA. Trials with 33 volunteers showed that intake of Omega-3 unsaturated fatty acids through diet or supplementation for 6 months resulted in a marked improvement in DHA levels. Even more interesting was the discovery that after DHA intake, there was a significant decrease in plasma concentrations of proinflammatory cytokines (Cole et al. 2005).

Taking supplements containing Omega-3 unsaturated fatty acids prevents cellular inflammation in the brain, which would have a positive result in Alzheimer's disease. Omega-3 unsaturated fatty acids may have neuroprotective properties against dementia, which is becoming a major global public health problem. In rat models, Omega-3 unsaturated fatty acids have been found to reduce cardiovascular risk factors (triglyceridemia) and improve cerebral blood circulation (Cole et al. 2005).

Docosahexaenoic acid (DHA) improves the age-related impairment (Tsukada et al. 2000). It has been reported that supplementation with long-chain ω-3 polyunsaturated fatty acids exerts a beneficial effect on behavior and cognition in children with attention deficit/hyperactivity disorder (Widenhorn-Müller et al. 2014). Fish consumption leads to an increase in long-chain Omega-3 fatty acids and a decrease in the risk of cognitive decline or Alzheimer's disease (Fotuhi et al. 2014). Omega-3 fatty acids play an important role in the treatment of depressive disorders (Grosso et al. 2014).

Consumption of Omega-3 unsaturated fatty acids is associated with a decreased incidence of white matter abnormalities. A meta-analysis of cohort studies showed a reduced risk of stroke with substantial fish consumption. Risk factors for cerebrovascular disease may lead to increased cognitive impairment, and the neuroprotective role of Omega-3 unsaturated fatty acids may be mediated by modification of these risk factors (Cole et al. 2005).

Polyunsaturated fatty acids play an important role in preventing inflammatory diseases. Omega-3 unsaturated fatty acids reduce inflammation by inhibiting leukocyte migration (James et al. 2000). EPA and DHA also reduce inflammation (Li et al. 2005). Omega-3 unsaturated fatty acids may directly limit pathology in AD (Cole et al. 2005). It has been described as the anti-inflammatory effect of docosahexaenoic acid on cytokine-induced adhesion molecule expression in human retinal vascular endothelial cells (Chen et al. 2005).

Omega-3 unsaturated fatty acids exhibit beneficial effects in cardiovascular disease, protect against premature mortality after heart attack (Simopoulos 2008), diabetes, retinopathies and glomerular dysfunction, prostatitis and ulcerative colitis, lower cholesterol and triglyceride levels, and reduce the risk of stroke and asthma, improve bone mineral density, and reduce the risk for osteoporotic fractures by decreasing the expression in osteoblasts of IL-1, IL-6, and TNF-λ, which stimulate osteoclasts and bone resorption. During intrauterine fetal development, Docosahexaenoic acid (DHA) and arachidonic fatty acids are incorporated into the cell membranes of the retina and brain.

6 Phytocompounds and Amino Acids with Antioxidant Activity

Given the involvement of amyloid-induced oxidative stress in AD pathology, one of the most promising approaches for preventive intervention is antioxidant therapy, which inhibits the deleterious effects of reactive oxygen species by inducing endogenous antioxidant enzymes. In this regard, phytocompounds with antioxidant, anti-amyloidogenic, anti-inflammatory, and antiapoptotic properties are promising as a new alternative therapy. Antioxidants, both drugs and dietary supplements, improve thinking, memory, and mental abilities. Dietary supplements with antioxidant activity include plant extracts containing polyphenols, phenolcarboxylic acids, Omega-unsaturated fatty acids, amino acids, and vitamins. Polyphenols are a class of natural antioxidants found in fruits, vegetables, and grains. They are classified into non-flavonoids such as phenolic acids, stilbenes, and lignans and flavonoids such as flavones, flavanols, isoflavones, and anthocyanins. In the last decade, polyphenols have been proposed in the prevention and treatment of cognitive diseases due to their antioxidant, anti-inflammatory, and anti-amyloidogenic functions (Perry et al. 1999). Curcumin, a non-flavonoid compound from *Curcuma longa L.*, has been used for centuries in traditional Indian and Chinese medicine and has neuroprotective effects in Alzheimer's disease due to antioxidant action, anti-inflammatory effect, and binding of redox-active metals: iron and copper (Ak and Gülçin 2008). There are some facts about the antioxidant and radical-scavenging properties of curcumin (Obreshkova 2012).

The above pharmacological actions of *Ocimum basilicum L.*, *Clinopodium vulgare L.*, and *Echium vulgare L.* make them interesting for further studies for other pharmacological effects.

Resveratrol belongs to a non-flavonoid class of polyphenolic compounds, stilbenes, and is prevalent in grapes, cranberries, and wine. The antioxidant properties of Resveratrol have been linked to its ability to stimulate the activity of the endogenous antioxidant enzymes superoxide dismutase and catalase and to reduce the activity of the prooxidant enzyme malondialdehyde in the brain in mouse models of Alzheimer's disease, thereby acting neuroprotectively and reducing cognitive impairment (Petersen et al. 2009).

The flavanol epigallocatechin-3-gallate is the most important catechin in *Camellia sinensis L.* (Perry et al. 1999).

Phytocannabinoids from *Cannabis sativa L.* are terpene-phenolic neuroprotective compounds. Cannabidiol inhibits the hyperphosphorylation of proteins and possesses antioxidant and antiapoptotic effects.

The influence of plant extracts on the aging process and Alzheimer's disease has been studied, with results showing that higher intake of antioxidants improves cognitive function and lowers the risk of developing the disease. The mechanisms of this effect are related to enhancing neural communication and lowering oxidative stress (Ansari et al. 2010; Petersen et al. 2007).

In a mouse model of Alzheimer's disease, it has been shown that decreasing L-arginine levels leads to amyloid deposition, hyperphosphorylation of τ-protein, behavioral changes, and neuronal loss. The cognitive function of 16 patients with a mean age of 79 years was assessed using the Hasegawa Dementia Scale (HDS-R) before initiation of oral L-arginine (1.6 g/day), after 3 months of therapy, and 3 months after stopping treatment.

7 Conclusion

Synthetic galantamine (Reminyl) was first registered for the treatment of Alzheimer's disease in Sweden in 2000 and is approved for this indication in the European Union, the United States, Canada, and Japan.

The study of products of this kind is a contribution to the compartmental therapy of Alzheimer's disease by reversible acetylcholinesterase inhibitors. The present studies further develop a segment of classical therapy in some neurodegenerative diseases and outline guidelines for prevention and adjunctive therapy by the application of natural sources that, through their antioxidant activity, resist oxidative modification in diseases. This is a contribution to the efforts to combat these socially important diseases.

Protective antioxidant effects in Alzheimer's disease are exerted by extracts of different plants and by Omega-3 fatty acids.

References

Ago Y, Koda K, Takuma K, Matsuda T (2011) Pharmacological aspects of the acetylcholinesterase inhibitor Galantamine. J Pharmacol Sci 116(1):6–17

Ak T, Gülçin I (2008) Antioxidant and radical scavenging properties of curcumin. J Chem Biol Interact 174(1):27–37

Ansari MA, Scheff SW (2010) Oxidative stress in the progression of Alzheimer disease in the frontal cortex. J Neuropathol Exp Neurol 69(2):155–167

Bader B, Nübling G, Mehle A, Nobile S, Kretzschmar H, Giese A (2011) Single particle analysis of tau oligomer formation induced by metal ions and organic solvents. Biochem Biophys Res Commun 411(1):190–196

Barcelo-Coblijn G, Murphy EJ (2009) Alpha-linolenic acid and its conversion to longer chain n-3 fatty acids: benefits for human health and a role in maintaining tissue n-3 fatty acid levels. Prog Lipid Res 48(6):355–374

Chen W, Esselman WJ, Jump DB, Busik JV (2005) Anti-inflammatory effect of docosahexa-enoic acid on cytokine-induced adhesion molecule expression in human retinal vascular endothelial cells. Invest Ophthalmol Vis Sci 46(11):4342–4347

Cole GM, Lim GP, Yang F, Teter B, Begum A, Ma Q, Harris-White ME, Frautschy SA (2005) Prevention of Alzheimer's disease: Omega-3 fatty acid and phenolic antioxidant interventions. Neurobiol Aging 26(1):133–136

Dooper MMBV, van Riel B, Graus YMF, M'Rabet L (2003) Dihomo-gamma-linolenic acid inhibits tumour necrosis factor-alpha productionby human leucocytes independently of cyclooxygenase activity. Ther Immunol 110(3):348–357

Fotuhi M, Mohassel P, Yaffe K (2009) Fish consumption, long-chain omega-3 fatty acids and risk of cognitive decline or Alzheimer disease: a complex association. Nat Clin Pract Neurol 5(3):140–152

Grosso G, Pajak A, Marventano S, Castellano S, Galvano F, Bucolo C, Drago F, Caraci F (2014) Role of omega-3 fatty acids in the treatment of depressive disorders: a comprehensive meta analysis of randomized clinical trials. PLoS One 9(5):e96905

James MJ, Proudman SM, Neumann MA, Gibson RA (2000) Dietary polyunsaturated fatty acids and inflammatory mediator production. Am Soc Clin Nutrit 71(1):343s–348s

Kozlowski H, Luczkowski M, Remelli M, Valensin D (2012) Copper, zinc and iron in neurodegenerative diseases (Alzheimer's, Parkinson's and prion diseases). Coordinat Chem Rev 256(19–20):2129–2141

La Guardia M, Giammanco S, Di Majo D, Tabacchi G, Tripoli E, Giammanco M (2005) Omega 3 fatty acids: biological activity and effects on human health. Panminerva Med 47(4):245–257

Lee H, Park WJ (2014) Unsaturated fatty acids, desaturases, and human health. J Med Food 17(2):189–197

Li H, Ruan XZ, Powis SH, Fernando R, Mon WY, Wheeler DC, Moorhead JF, Varghese Z (2005) EPA and DHA reduce LPS-induced inflammation responses in HK-2 cells: evidence for a PPAR-gamma-dependent mechanism. Kidney Int 67(3):8670–8874

Lobo V, Patil A, Phatak A, Chandra N (2010) Free radicals, antioxidants and functional foods: impact on human health. Pharmacogn Rev 4(8):118–126

Matharu B, Gibson G, Parsons R, Huckerby TN, Moore SA, Cooper LJ, Millichamp R, Allsop D, Austen B (2009) Galantamine inhibits-amyloid aggregation and cytotoxicity. J Neurol Sci 280(1–2):49–58

Obreshkova D (2012) Analytical study and quality control of bulgarian drugs with antioxidant activity. Pharmacia LVII(1-4):107114

Perry EK, Pickering AT, Wang WW, Houghton PJ, Perry NS (1999) Medicinal plants and Alzheimer's disease: from ethnobotany to phytotherapy. J Pharm Pharmacol 51(5):527–534

Petersen RB, Nunomura A, Lee HG, Casadesus G, Perry G, Smith MA, Zhu X (2007) Signal transduction cascades associated with oxidative stress in Alzheimer's disease. J Alzheimer's Dis 11(2):143–152

Pisoschi AM, Pop A (2015) The role of antioxidants in the chemistry of oxidative stress: a review. Eur J Med Chem 97:55–74

Ruxton C, Calder PC, Reed SC, Simpson MJA (2005) The impact of long-chain n-3 polyunsaturated fatty acids on human health. Nutri Res Rev 18(1):113–129

Schilström B, Ivanov VB, Wiker C, Svensson TH (2007) Galantamine enhances dopaminergic neurotransmission in vivo via allosteric potentiation of nicotinic acetylcholine receptors. Neuropsychopharmacol 32(1):43–53

Simopoulos AP (2008) The omega-6/omega-3 fatty acid ratio, genetic variation, and cardiovascular disease. Asia Pac J Clin Nutr 17(Suppl. 1):131–134

Sun CQ, O'Connor CJ, Robertson A (2003) Antibacterial actions of fattyacids and monoglycerides against Helicobacter pylori. FEMS Immunol Med Microbiol 36:9–17

Tsukada H, Kakiuchi T, Fukumoto D, Nishiyama S, Koga K (2000) Docosahexaenoic acid (DHA) improves the age-related impairment ofthecoupling mechanism between neuronal activation andfunctional cerebral blood flow response: a PET studyin conscious monkeys. Brain Res 862(1–2):180–186

Tsvetkova D, Obreshkova D (2019) Modern approaches and strategies for prevention and therapeutic influence of Alzheimer's disease. Int J Pharm Res Allied Sci 8(1):1–16

Tsvetkova D, Obreshkova D, Zheleva-Dimitrova D, Saso L (2013) Antioxidant activity of Galanthamine and some of it's derivatives. Curr Med Chem 20(36):4595–4608

Tsvetkova D, Obreshkova D, Ivanova St, Hadjieva B (2016) Benefits of acethylcholin-esterase inhibitor Galantamine in treatment of Alzheimer disease and instrumental methods for its analysis in medicinal plants. J Med Pharm Allied Sci 5(7):99–116

Widenhorn-Müller K, Schwanda S, Scholz E, Spitzer M, Bode H (2014) Effect of supplementation with long-chain ω-3 polyunsaturated fatty acids on behavior and cognition in children with attention deficit/hyperactivity disorder (ADHD): a randomized placebo-controlled intervention trial. Prostaglandins Leukot Essent 91(1–2):49–60

Orchids: A Source of Biopharmaceuticals and Their Conservation Strategies

Madhu Kamle, Rituraj Borah, Mansi Singh, and Pradeep Kumar

Abstract

India is one of the world's largest pool of orchid genetic resources. The Orchidaceae is the largest family of flowering plants. Wild orchids are highly valued for their sheer beauty and variety. They are ecologically important since they offer a variety of medicinal applications, serve as climatic indicators, and signal a healthy forest ecosystem. Because of their beneficial properties, orchids are overexploited for trade and are classified as endangered. Unfortunately, the country's genetic endowment is eroding at a rapid pace due to risks like habitat degradation and climate change. However, many orchids are at risk due to unsustainable harvests for horticulture, food, or medicinal purposes.These risks have grown too large to tackle on a species-by-species basis within such a vast group as Orchidaceae. To conserve orchids effectively, we must adopt broader strategies that support targeted efforts for the most threatened species. To conserve orchid genetic resources from extinction, a variety of conservation strategies should be implemented with solid scientific support. This chapter highlights the importance of orchids, focusing on their bioactive compounds, historical and pharmaceutical uses, ecological roles, and conservation strategies. Conservation strategies not only safeguard various orchid species, but they also provide the necessary amount of phytochemical extracts for future applications and keep these species from becoming endangered.

M. Kamle
Department of Biochemistry, University of Lucknow, Lucknow, India

R. Borah
Applied Microbiology Laboratory, Department of Forestry, North Eastern Regional Institute of Science and Technology, Nirjuli, Arunachal Pradesh, India

M. Singh · P. Kumar (✉)
Applied Microbiology Laboratory, Department of Botany, University of Lucknow, Lucknow, UP, India

R. K. Gaur et al. (eds.), *Plant Molecular Farming*,
https://doi.org/10.1007/978-981-95-3823-2_7

Keywords

Orchids · Medicinal uses · Northeast India · Pharmaceutical compound · Human health

1 Introduction

Orchids, which belong to the Orchidaceae family, are the most beautiful flowers among the flowering plants. These economically important plants are distributed from the tropical to the temperate zones. There are about 800 genera with about 25,000 species (Chung et al. 2009). As the literal meaning of orchid is "testicles" in Greek, Theophrastus first coined the term orchid as the anatomy of the plant resembles testicles. Orchids are both epiphytic and terrestrial. Orchids are not only ornamental plants but also serve as a noble source of medicine, flavoring agents, bioactive compounds, etc. Orchids are mainly cultivated for their beautiful flowers, but they are not well explored for their medicinal properties (Kalaiarasan and John 2012; Bahera et al. 2013).

The distribution of orchids is cosmopolitan and is known to occur in a varied range of climatic conditions, from the alpine regions to humid climates, except the ice-covered Antarctica and the dry, sandy deserts of Africa and Australia. Orchids are abundantly distributed in the tropical regions of Southeast Asia, encompassing countries such as Bhutan, China, India, Japan, Laos, Malaysia, Myanmar, Nepal, and the Philippines, and extend to tropical areas of South and Central America, Australia, and South Africa.

The scientific account of Indian orchids was first documented by Von Rheede (1678–1703), the then Dutch Governor of Malabar, in his monumental work "*Hortus Malabaricus.*" In his book "*Flora Indica, vol. III,*" William Roxburgh (1832), the "Father of Indian Botany," provided a treatment of 57 species of orchids. Sir J. D. Hooker (1888, 1890) made the most significant contribution to Indian orchids; he described about 1600 orchid species in his book "*Flora of British India (vols. 5 and 6).*"

In India, the family Orchidaceae is widely distributed from alpine to coastal regions, but its maximum diversity occurs in the Eastern Himalayan and Peninsular regions, respectively (https://nrcorchids.nic.in/images/Orchid-Preparations_Technical-Bulletin1.pdf).

2 Traditional Uses of Orchids in Medicine

Jīvantī and *Rāsnā*, the very common drugs of Ayurveda, are orchids, which have been used in Ayurvedic drug preparation for centuries. Orchids are widely used in many Chinese and South African medicines. A total of 365 plants were listed in *the Chinese Materia Medica* (*Shennon bencaojing* [ca. 100 AD], or the *Divine Husbandman's Classic of the Materia Medica*), of which many were orchids.

However, after Li Shizhen (1518–19) published *Bencao gangmu*, the number of plants used in Chinese medicines increased. *Dendrobium nobile, Gastrodia elata,* and *Bletilla striata* are the orchids that are extensively used in Chinese medicine. They were mentioned in both the *Shennon bencaojing* and Li's *Bencao gangmu. Bletilla striata,* also known as Pe-chi, is used to treat tuberculosis, coughing up of blood from the respiratory tract, and gastric and duodenal ulcers. Some Flickingeria species and all Dendrobium species in China are described by the term Shih Hu. Shih Hu is medicinally used for alleviating gastric problems, fever, inflammation, pain, and epilepsy. *Gastrodia elata,* commonly called Tian-Ma, is used as a herbal medicine to treat headache, dizziness, blackout, numbness of limbs, paralysis of one side of the body, epilepsy, cramps, spasms, migraine, rheumatism, vertigo, sharp pain along the course of one or more nerves, infantile convulsion, and pain in the lower back (Hew and Yong 2006).

In traditional Chinese medicine, five major categories of orchid-derived treatments have been identified, namely Shi-Hu, Tian-Ma, Bai-Ji, Jin-Xian-Lin, and Shan-Ci-Gu. The Shi-Hu group encompasses five species of the genus Dendrobium, while Tian-Ma refers specifically to preparations derived from the tubers of Gastrodia elata. Bai-Ji is the most effective substance from *Bletilla striata.* Two species of the genus *Anoectochilus,* viz. *Anoectochilus formosanus* and *A. koshunensis* constitute the Jin-Xian-Lian, and *Cremastra appendiculata*, also called the Chinese tulip, is known as Shan-Ci-Gu in China (Bulpitt et al. 2007).

3 Phytochemicals in Orchids

Phytochemicals that are extracted from orchids can generally be classified as alkaloids, anthocyanins, flavonoids, carotenoids, and sterols. Among these, alkaloids and flavonoids contribute most to their biological properties. E. de Wildeman started the investigation of orchid alkaloids in 1892 in some European species. He identified 104 species in 78 genera for medicinal purposes.

Dendrobine was the first alkaloid extracted from *Dendrobium nobile Lindl.*, which is used in the preparation of the Chinese drug "Chin-Dhi-Hu" (Chen and Chen, 1935). In the Science Laboratories of the University of Reading in the UK, Williams and Harborne surveyed 142 species from 75 genera for their leaf flavonoids. The most common constituents were flavones C-glycoside and flavonols (Williams 1979). In phytochemical content, *Dendrobium* is the leading genus among orchids. In total, 32 alkaloids, 15 bibenzyls, 6 coumarins, 4 fluorenones, 22 phenanthrenes, and 7 sesquiterpenoids, which include the 100 compounds reviewed by Zhang et al. from 42 *Dendrobium* species (Zhang et al. 2003). Various secondary metabolites produced by the *Dendrobium* species, such as alkaloids, bibenzyls, fluorenones, phenanthrenes, and sesquiterpenes, are responsible for their medicinal properties.

Further studies carried out on *Dendrobium* orchids in the last three decades show that many other phenanthrene compounds are isolated, such as amoenumin, crepaditin, cumalatin, defuscin, denfigenin, dendrophenol, gigantol, moscatin, dihydrophenanthrene, ephemeranthoquinone, moscatalin, rotundatin, shihunidine, and shihunine (Li et al. 2008). 2,6-dimethoxy-1,4-benzoquinone isolated from *Cymbidium* sp. is reported to have an allergic reaction (Hausen et al. 1984). Calanthoside isolated from *Calanthe liukiuensis* and *Calanthe discolour* shows an activating effect on skin blood flow (Yoshikawa et al. 1998). Habenariol, which inhibits lipid peroxidation of low-density lipoprotein of humans, is isolated from Habenaria repens (Johnson et al. 1999). Scoparone and scopoletin isolated from *Dendrobium densiflorum* show in vitro antiplatelet aggregation activity (Fan et al. 2001). Gatrol isolated from *Gastrodia elata* rhizomes has relaxant effects on smooth muscle preparations of guinea pig ileum (Hayashia et al. 2002).

9,19-cyclolanosta-24,24-dimethyl-25-en-3-yl-trans-p-hydoxyxinnamate and nidemin are extracted from *Scaphyglottis livida* and *Nidema boothi* (Estrada et al. 2002). Cirrhopetalanthrin isolated from Cremastra appendiculata shows cytotoxicity against human breast cancer (MCF-7), human lung adenocarcinoma (A546), human colon cancer (HCT-8), human hepatoma (Bel7402), human stomach cancer (BGC-823), and human ovarian cancer (A2780) cell lines (Xia et al. 2005). Kinsenoside isolated from Anoectochilus formosanus shows significant antihepatotoxic activity (Wu et al. 2007). Dose-dependent antinociceptive and anti-inflammatory effects shown by 5-lansota-24,24-dimethyl-9(11),25-dien-3-ol and Cyclobalanone are isolated from *Scaphyglottis livida* (Deciga-campos et al. 2007). Cymbidine A isolated from Cymbidium goeringii shows hypotensive and diuretic properties (Chemical Pharmaceutic Bulletin 55: 780–783) (https://nrcorchids.nic.in/images/Orchid-Preparations_Technical-Bulletin1.pdf),

Majumder and Sen (1987) first isolated and described Moscatilin, a bibenzyl compound from *Dendrobium moscatum and* later, the compound was purified from the stems of *Dendrobium loddigesii.* Shriram et al. reported the presence of an active molecule identified as phenanthrene derivative 9,10-dihydro-2,5-dimethoxyp henanthrene-1,7-diol in the tubers of *Eulophia nuda* (Shrirama et al. 2010). The leaf and stem extracts of *Dendrobium panduratum* subsp. *vilosum* were treated with acetone, benzene, petroleum ether, ethanol, and distilled water. The phytochemical analysis revealed that alkaloids, tannins, steroids, triterpenoids, phenols, and flavonoids in different extracts are present (Johnson and Janikiraman 2013). Similarly, phytochemical analysis of *Vanda tessellata* polar (Methanol) and nonpolar (Chloroform) root extract shows the presence of phenols, alkaloids, tannins, flavonoids, anthocyanins, terpenoids, steroids, and cryogenic glycosides. Methanol extract showed more phytochemicals than the chloroform extract. The cryogenic glycosides are present in the nonpolar extract, while absent in the polar extract (Chaitanya et al. 2013). Additionally, Kiran et al. reported the presence of phytoconstituents like tannins, glycosides, and saponins in *Sarcanthus pauciflorus.* They also reported that the total phenol content was found to be 258.65 mg Gallic Acid Equivalents (GAE)/g of dry weight, which was estimated by the Folin–Ciocalteu reagent (FCR) method (Kiran et al. 2013).

In another phytochemical investigation of the plant *Habenaria longicorniculata,* carbohydrates, alkaloids, glycosides, saponins, flavonoids, phenols, vitamin E, and vitamin C have been identified (Jagtap et al. 2014). Preliminary phytochemical screening of *Geodorum densiflorum* shows the presence of alkaloids, steroids, carbohydrates, flavonoids, tannins, and saponins in different extracts. The test showed that ethanol has a higher extractive value than water and chloroform (Keerthiga and Anand 2014). Different extracts (petroleum ether, chloroform, ethyl acetate, ethanol) of leaves and bulbs of *Nervilla plicata* showed the presence of 12 different secondary metabolites, viz. Alkaloid, flavonoid, phenol, tannin, glycoside, saponin, resin, steroids, terpenoids, cardiac glycoside, triterpenoids, and reducing sugar (Haridas et al. 2014). Borkar et al. also tested different extracts of *Geodorum densiflorum,* like petroleum, acetone, benzene, chloroform, ethanol, and distilled water, for phytochemical analysis. The tests confirmed the presence of alkaloids, glycosides, tannins, proteins, and phytosterols in the different extracts (Borkar and Masirkar 2015). The presence of saponins, tannins, steroids, flavonoids, and glycosides in *Acampe papillosa*; saponins, tannins, terpenoids, steroids, and flavonoids in *Cymbidium aloifolium,* while *Rhynchostylis retusa* has glycosides, flavonoids, phlobatannins, saponins, tannins, terpenoids, and steroids was also reported (Marjoka et al. 2016). Similarly, phytochemical screening of *Luisia zeylanica* Lindl showed the presence of phytochemicals, mainly alkaloids and other secondary metabolites, viz. flavonoid, steroid, saponin, phlobatannin, terpenoid, tannin, glycoside, anthraquinone, quinine, and coumarin, were screened out in this study. Among the three studied parts, leaf extract was found superior to others in terms of ten other studied secondary metabolites (Sohag et al. 2017). Subha et al. documented that various extracts of *C. aloifolium* showed the presence of 11 phytochemicals, viz. flavonoids, coumarins, saponins, phenols, quinones, alkaloids, carbohydrates, tannin, cardiac glycosides, oxalate gum, and mucilage. The leaf exhibited the maximum number of phytochemicals (8/11)—flavonoids, coumarins, quinones, oxalate, alkaloids, carbohydrates, tannins, and cardiac glycosides (72.7%), followed by capsule cover and root (5/11)—45.4%; the seed exhibited the minimum (3/11)—27.27% of phytochemicals (Shubha and Chowdappa 2016).

From the roots of *Cyrtopodium paniculatum,* four new stillbenoids were extracted viz. cyrtopodinone, cyrtopodinol, coeludols A and B together with 21 known stillbenoids identified as batatasin III, confusaridin, gigantol, ephemeranthoquinone, erianthridin, nudol, denthyrsinin, cyrtopodin, 3,4,6-trimethoxy-9,10-dihydrophena nthrene-2,7-diol, lusidol A and B, coelonin, confusarin, 9′,10′-dihydro-4,5′-dimethoxy-[1,3′-biphenantherene]-2,2′,7,7′-tetrol, lusianthridin, blestriarenes A, B, and C, 2.4-dimethoxyphenantherene-3,7-diol, gymconopin and monbarbatain C. (Auberon et al. 2017). In another work of Schuster et al., it was revealed that the most active compounds from the ethanolic extract of *Eulophia macrobulbon* could be identified as 4-methoxy-9,10-dihydro-2,7-phenanthrenediol, 4-methoxy-2,7-phenanthrenediol, 1,5-dimethoxy-2,7-phenantherenediol, 1,5,7-trimethoxy-2,6 -phenanthrenediol, and 1-(4-hydroxybenzyl)-4,8-dimethoxy-2,7-phenanthrenediol (Schuster et al. 2017).

4 Pharmaceutical Importance of Orchids

Orchid species are highly valuable to pharmaceutical industries in terms of pharmaceutically important values due to the presence of a diverse range of bioactive compounds that possess therapeutic properties. In Fig. 1, the pharmacological importance of orchids is mentioned, which consists of a wide range of properties containing compound extraction, such as antimicrobial property, anti-inflammatory property, antioxidant-rich compound extraction, and anticancer property-carrying compounds.

(a) *Antimicrobial Properties of Orchids*

Microorganisms are seen to have become resistant to the synthetic chemicals used as antimicrobial agents; therefore, there is always a search for compounds that can serve the same purpose, and plants are always the best option. It is seen that among the medicinal plants available, orchids produce a number of bioactive compounds that work against a diverse group of organisms. There are many experimental data that prove the orchid-based compounds show remarkable inhibition against different microbes.

Antibacterial

Fitzgerald et al. (2004) described bacteriostatic inhibition against the *E. coli, Lactobacillus plantarum,* and *Lactobacillus innocua.* Vanillin worked in a

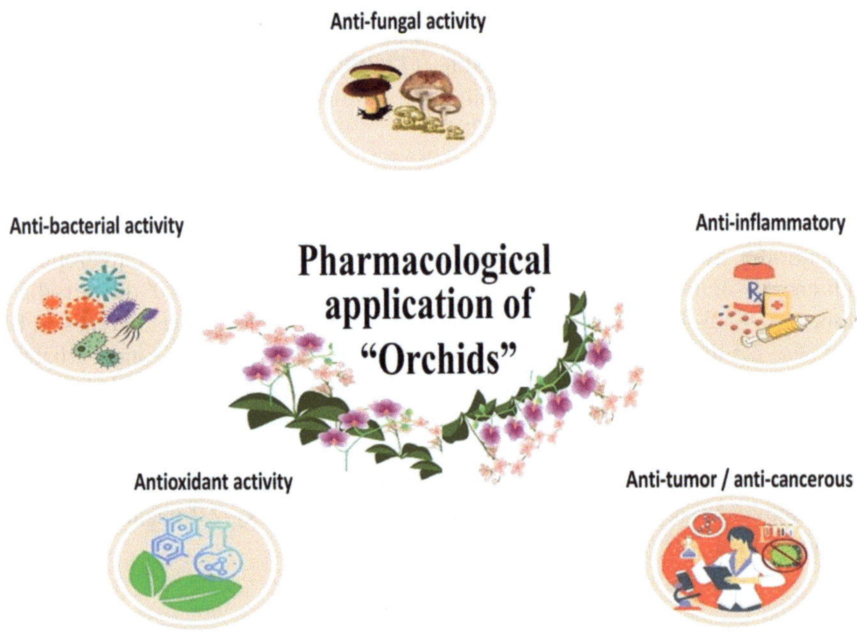

Fig. 1 Pharmacologically important applications of orchids

dose-dependent manner against the microbes, and the effect was not bactericidal. Similarly, aqueous extract of orchid *Dendrobium nobile* flower and stem showed antibacterial activity against *Bacillus subtilis, E. coli*, *Proteus*, *Staphylococcus aureus,* and *Salmonella typhi* (Devi et al. 2009). Additionally, antibacterial activities of *Vanda tessellata* methanolic root extract were tested against *Bacillus subtilis, Klebsiella pneumonia, E. coli, Pseudomonas putida,* and *Staphylococcus aureus,* showing broad inhibition as compared to chloroform extract (Chaitanya et al. 2013). Methanolic extract of *Sarcanthus pauciflorus* tested against *Bacillus subtilis, Escherichia coli, Pseudomonas aeruginosa,* and *Staphylococcus aureus* by the agar well diffusion method showed a remarkable inhibition zone (Kiran et al. 2013). Behera et al. also tested the antimicrobial activity of *Acampe praemorsa* and *Vanda tessellata* against 13 pathogenic bacteria, viz. *Vibrio cholerae* (Inaba), Vibrio cholerae (Ogawa), *Vibrio cholerae* (0139), *Staphylococcus aureus*, *Shigella sonnei*, *Shigella dysentriae*, *Shigella boydii*, Enterotoxigenic Escherichia coli (ETEC), Enteropathogenic *Escherichia coli* (EPEC), *Escherichia coli* O157:H7, and *Bacillus subtilis.* It was also noted that all the pathogens were moderately susceptible to the plant extracts. *Bacillus subtilis* was found to be the most susceptible, while *Shigella sonnei* was the most resistant (Behera et al. 2013).

Similarly, *Aerides odorata, Acampe papilosa, Acampe ochracea* against kanamycin-resistant *E. coli*, ampicillin-resistant *E. coli*, and antibiotic-susceptible *E. coli* showed inhibition zone suggesting good antimicrobial properties (Paul et al. 2013) In another work different parts (leaf, stem, root, and pseudo-bulb) of *Dendrobium crumenatum* were tested against *Enterobacter aerogenes* (lab strain), *Escherichia coli* (ATCC 25922), *Klebsiella pneumoniae*, *Pseudomonas aeruginosa* (ATCC 27853), *Saccharomyces cerevisiae* (lab strain), *Shigella flexeneri* (ATCC 12022), *Staphylococcus aureus* (ATCC 25933), and *Streptococcus pneumoniae* (lan strain) and was reported that the plant extracts showed better antibiotic activity when compared to amoxicillin, chloramphenicol and kanamycin as standards (Sandrasagaran et al. 2014). Chloroform extract of *Vanda tessellata* showed better inhibition against *Bacillus subtilis, Escherichia coli, Klebsiella pneumonia, Staphylococcus aureus, and Vibrio cholerae* compared to methanol, ethanol, and hexane extract (Bhattacharjee et al. 2014). It was also reported that methanolic and ethanolic extracts of *Geodorum densiflorum* showed better inhibition against *Bacillus cereus, Bacillus subtilis, Escherichia coli, Pseudomonas aeruginosa, Salmonella typhi, and Staphylococcus aureus* compared to water, chloroform, ethanol, and petroleum ether (Borkar and Masirkar 2015). The varying degrees of inhibition of methanolic extracts of *Eria mysorensis* Lindl., *Bulbophyllum fischeri* Seidenf., *B. neilgherrense* Wight, *Acampe praemorsa* (Roxb.) Blatter & McCann, *Dendrobium herbaceum* Lindl., *Oberonia brunoniana* Wight, *Luisia macrantha* Blatt. McCann., *Pholidota imbriacata* Lindl., *Coelogyne nervosa* A. Rich., and *Vanda roxburghii* R.Br. against *Escherichia coli* MTCC-1610, *Pseudomonas aeruginosa* MTCC-7296, and *Staphylococcus aureus* MTCC-96 was also recorded (Ranjitha et al. 2016). In the work of Hoque et al. *Pholidota pallida, Aerides odoratum* and *Acampe papillosa* were tested against *Shigella dysenteriae*, *Salmonella typhi*, *Staphylococcus aureus* and *Bacillus subtilis* and INABA-ET (*Vibrio*). Highest inhibition was shown by

Aerides odorata and *Pholidota pallida* against *Salmonella typhi* (Hoque et al. 2016). In another work, Subha et al. recorded that when different parts of *Cymbidium aloifolium* (leaf, capsule cover, seed and root) were tested for antibacterial property, the methanol and acetone extracts of capsule cover showed remarkable inhibition against *Escherichia coli* (MTCC 443), *Pseudomonas aeruginosa* (MTCC 4996) and *Staphylococcus aureus* (MTCC 3160) (Shubha and Chowdappa 2016).

Antifungal

Bhattacharrjee et al. reported that when tested against *Penicillium* sp., *Rhizopus* sp. *Aspergillus niger,* the chloroform extract of *Vanda tessellata* showed significant antimicrobial activity (Bhattacharjee et al. 2014). Similarly, aqueous extracts of *Acampe papillosa, Aerides odoratum, and Pholidota pallida* were tested against *Macrophomina phaseolina* (Maubl.) Ashby, *Fusarium equiseti* (Corda), *Curvularia lunata* (Wakker), *Colletotrichum corchori* Ikata (Yoshida), *Botryodiplodia theobromae* Pat., and *Alternaria alternata* (Fr.) Kedissler, and it was recorded that *Aerides odoratum* showed better inhibition against *Colletotrichum corchori* among others (Hoque et al. 2016). The methanolic extract of *Sarcanthus pauciflorus* has also shown antifungal activity against Cryptococcus neoformans and Candida albicans by the agar well diffusion method (Kiran et al. 2013). Additionally, methanolic (polar) and Chloroform (nonpolar) extracts of *Vanda tessellata* showed antifungal activity against *Aspergillus flavus, Aspergillus niger, Fusarium, Colletotrichum, Rhizopus,* and *Mucor* (Chaitanya et al. 2013).

(b) *Antitumor/Anticancer*

Ho and Chen (2003) reported that *Dendrobium loddigessii* has a natural antiplatelet agent, Moscatilin, which shows a potential cytotoxic effect against cancer cell lines derived from the placenta, stomach, and lung. Devi et al. (2009) recorded the cytotoxic effect of aqueous extracts of *Dendrobium nobile* flowers against Dalton's Lymphoma Ascites (DLA) cell-induced mice. Similarly, Shrirama et al. (2010) reported that the antiproliferative activity of the phenanthrene derivative 9,10-dihydro-2,5-dimethoxyphenanthrene-1,7-diol extracted from *Eulophia nuda* tubers showed good cytotoxicity against human breast cancer cell lines MCF-7 and MDA-MB-231 at a 1000 μg/ml concentration. Prasad and Koch (2014) also reported that at a concentration of 350 μg/ml, the ethanolic extract of *Dendrobium formosum* induces a cytotoxic effect against Dalton's lymphoma. Four active stilbenoids, viz. blestriarenes A, lusidol A, blestriarenes B, and blestriarenes C, extracted from *Cyrtopodium paniculatum,* showed moderate cytotoxicity on the human glioblastoma U-87MG cell line (Auberon et al. 2017). Prasad and Koch (2016) recorded the cytotoxic effects of ethanolic extracts of *Dendrobium crepidatum* and *Dendrobium chrysanthum* in T-cell lymphoma and, in a dose- and time-dependent manner, increased the generation of intracellular reactive oxygen species.

Similarly, it was revealed that the presence of different bioactive compounds, like 4-methoxy-9,10-dihydro-2,7-phenanthrenediol, 4-methoxy-2,7-phenanthrenediol, 1,5-dimethoxy-2,7-phenanthrenediol, 1-(4-hydroxybenzyl)-4, 8-dimethoxy-2,

7-phenanthrenediol, 1,5,7-trimethoxy-2,6-phenanthrenediol, in *Eulophia macrobulbon* showed cytotoxic effects against three different cancer lines, CaCo-2, MCF-7, and HeLa (Schuster et al. 2017). Wei et al. (2018) recorded that DOPA-1, a polysaccharide extracted from *Dendrobium officinale*, in a dose-dependent manner, inhibits the growth of HepG-2 cells. Reactive oxygen species (ROS) levels increase and mitochondrial membrane potential decreases when HepG-2 cells are treated with DOPA-1.

Additionally, it was reported that Gigatol, a bibenzyl compound from many medicinal orchids, decreases the levels of cytosolic β-catenin and phosphorylated LRP6 in HEK293 cells. In a dose-dependent manner, Gigatol reduces phosphorylated LRP6, total LRP6, and cytosolic β-catenin in MDA-MB-231 and MDA-MB-468 cells of breast cancer, which decreases the expression of Wnt expression genes Axin2 and Survivin. They also demonstrated that the viability and migration of breast cancer cells are suppressed by Gigatol (Yu et al. 2018).

(c) *Anti-inflammatory*

Chinswamy et al. tested the enzyme-based cyclooxygenase assay (COX-1 and COX-2) of seven orchid species—*Tridactyle tridentate* (Harv.) Schltr., *Polystachya pubescens* (Lindl.) Rchb.f., *Eulophia petersii* (Rchb.f.) Rchb.f., *Eulophia hereroensis* Schltr., *Cyrtorchis arcuata* (Lindl.) Schltr., *Bulbophyllum scaberulum* (Rolfe) Bolus, and *Ansellia africana* Lindl.— and found that *E. hereroensis* showed significant inhibition, while *B. scaberulum* and *A. africana* showed poor or no inhibition at all (Chinsamy et al. 2014).

Schuster et al. also evaluated the anti-inflammatory activity of the ethanolic extract of *Eulophia macrobulbon* roots (EME) in a lipopolysaccharide (LPS)-stimulated macrophage model by pro- and anti-inflammatory cytokine secretion and inducible nitric oxide synthase (iNOS) expression, as determined by Enzyme-Linked Immunosorbent Assay (ELISA) and Western blot. They found that EME and its compounds reduce proinflammatory cytokines interleukin 6 (IL-6) and tumor necrosis factor alpha (TNF-α) production, iNOS expression and subsequently increase the anti-inflammatory cytokine interleukin 10 (IL-10) production in LPS-stimulated macrophages (Schuster et al. 2017).

(d) *Antioxidant*

Chimsook reported that *Dendrobium signatum* extract, extracted by sonication–maceration, showed better free radical scavenging activity compared to ascorbic acid used as a chemical standard (Chimsook 2016). Different bioactive compounds extracted from the ethanolic extract of *Eulophia macrobulbon* (EME) showed strong radical scavenging activity when the final concentration of EME was diluted to 100 μg/ml. 1 ml of 2,2-Diphenyl-1-Picrylhydrazyl (DPPH) (0.3 mM) was added to 2.5 ml of the sample (Schuster et al. 2017). Sohag et al. reported that the methanolic root extract of *Luisia zeylanica* Lindl. showed remarkable free radical scavenging activity of DPPH, as compared to n-hexane, butanol-1, and dichloromethane (DCM) extract of root, leaf, and stem, as well as ascorbic acid (Sohag et al. 2017).

Advanced Conservation Strategies for Orchids

In order to ensure the survival of rare orchid species, conservation of orchids is entailed by preserving the natural habitat of the species, pollinator protection, prevention of ill practices, reintroduction of endangered species in their natural habitats, and more. All these preservation events are performed in two major ways to conserve orchids: in situ and ex situ.

Various actions are taken by the government to protect endangered species of orchids from extinction and prevent threatened species from becoming extinct. The Convention on International Trade in Endangered Species of Wild Fauna and Flora (CITES) program is one of them, under which wild plants and animals are managed by the IUCN (De and Pathak 2018).

(a) *In Situ Conservation*

Effective biodiversity conservation requires knowing what exists and what is at risk. Systematic and taxonomic studies play an important role in circumscribing species and in identifying high priorities for conservation (Fay 2018). The goal of in situ conservation is to preserve all living things in their native habitat, with a focus on wild plants, animals, and endangered species. The first nation with a national plan for protecting biodiversity in its native environment is Turkey (Kaya and Raynal 2001). National parks, sanctuaries, and biosphere reserves are a few examples of in situ preservation. A biosphere reserve is a major protected area set aside for the conservation of plant and animal resources, whereby specific activities are strictly forbidden, including planting, farming, grazing, harvesting timber, foraging, and poaching. The concept of biosphere reserves was first promoted in 1971 by the United Nations Educational, Scientific, and Cultural Organization (UNESCO) through its "Man and Biosphere" campaign (De and Pathak 2019).

Pollinators play an important role in orchid protection because orchids need particular pollinators in order to get fertilized. Their natural habitat requires pollinators to ensure orchid multiplication. The orchids must adjust to a completely new environment where they are transplanted; they may or may not survive as they do in their native habitat (Ram et al. 2011). India has an elaborate Protective Area Network (PAN) comprising a total of 1014 protected areas, including 106 national parks, 573 wildlife sanctuaries, 115 conservation reserves, and 220 community reserves, covering approximately 5.32% of the total geographical area of the country (National Wildlife Database Cell 2023). Under the Wildlife Protection Act, 1972 (amended in 1992), a few State Governments, including those of Arunachal Pradesh, Sikkim, Karnataka, and West Bengal, have designated the orchid-rich habitats as "Orchid Sanctuaries." Any effort to eradicate them from their natural habitat should be treated as stealing or trafficking (Medhi et al. 2012).

Yeung (2017) reviews the development of orchid seeds and protocorms, discussing the survival of seeds and plantlets in their native environments as well as under symbiotic mycorrhizal relationships. With time, it comes to knowledge that mycorrhizal fungi play a critical role in the survival of orchid populations that are self-sustaining; nevertheless, further study is needed to gain a complete understanding of these relationships, particularly with regard to epiphytic orchid species. The processes involved are not clear, even with temperate terrestrial species. Plants are capable of performing photosynthesis, but at the same time, their dependency on mycorrhizal associations is also found using techniques like measuring carbon, nitrogen, and hydrogen isotope enrichment (Gebauer et al. 2016).

Individual species approaches to conservation will remain the best for species that have been designated at risk of extinction in the near future, but there is a need to supplement these with other approaches in order to conserve groups of species that face similar threats, or are closely related, or coexist sympatrically. This is true for larger families like Orchidaceae, as not much is known about population genetics, etc., for many species. Methods addressing the conservation of populations (rather than individual species) may be suitable in groups that are undergoing relatively rapid evolution due to hybridization and/or polyploidization (Ennos et al. 2012).

(b) *Ex Situ Conservation*

The strategy to conserve species (plants or animals) distant from their natural habitat is known as ex situ conservation. This type of conservation make germplasm easily accessible for use in research and other purposes; genetic material must be moved from its original site. It is regarded as the most useful, economical, and extensively applied type of conservation strategy (De and Pathak 2019).

Cryopreservation, in vitro propagation, gene banks, botanical gardens, and seed storage are among the most widely used ex situ conservation methods (Swarts 2018). India's Botanical Survey operates three National Orchidaria and Experimental Gardens in Yercaud, Howrah, and Shillong. The State Forest Research Institute of Arunachal Pradesh preserves orchid species through ex situ conservation at their Orchid Research Centers in Tipi, Itanagar, Dirang, Jenging, and Sessa (De and Pathak 2019).

A single orchid capsule can yield millions of seeds. Nevertheless, in order to germinate in the natural environment, they need a symbiotic relationship with mycorrhiza because they lack the metabolic apparatus and functional endosperm. As a result, the germination rate is minimal (between 0.01 and 0.2%). Asymbiotic techniques have been used to grow many orchids, with germination rates as high as 90%. The world's largest seed bank is the Millennium Seed Bank, located in Sussex, and it is overseen by the Royal Botanic Gardens, Kew. To retain their high seed germination viability, the orchid seeds (*Dactylorhiza, Dendrobium, Eulophia, and Paphiopedilum*) are stored at −70 °C. These seeds have demonstrated outstanding potential under preservation for a long time at low temperatures (De and Pathak 2019).

A botanical garden is a documented collection of living plants for the purpose of scientific research, and it plays a vital role in conservation. The New York Botanical

Garden and the Royal Botanic Garden (Kew) both preserve collections of live orchids and carry out continuing research projects (Swarts 2018). Currently, in India, there are 13 botanical gardens that conserve a wide variety of orchid species. The Lloyd Botanical Garden in Darjeeling, West Bengal, is one of them, harboring 43 different varieties of orchid species (De and Pathak 2018).

The in vitro technique enables the production of healthy and disease-free plants on a large scale. Different orchid species have been micropropagated using a variety of explants, such as seeds, shoots, roots, leaf apexes, nodal segments, rhizomes, and pseudobulbs (Lal and Singh 2021). The in vitro technique is also used to create secondary metabolites, eradicate viruses, and for mass propagation (Oseni 2018). It works quite well to preserve biodiversity and the gene pool. In vitro culture is the only practical way to protect and reintroduce threatened genetic material into the environment (Bhowmik and Rahman 2017).

Plant organs and reproductive components can be preserved by a process called cryopreservation, which involves keeping them in liquid nitrogen at a temperature of −196 °C. High moisture content seeds work especially well for seed preservation when using cryopreservatives. The physiological and metabolic processes are fully halted in these circumstances, allowing the plant material to be preserved indefinitely. During subculturing, there is less likelihood of contamination and genetic instability. Cryoprotectants (Dimethyl Sulfoxide or DMSO) and Plant Vitrification Solutions (PVS, e.g., glycerol) can be used to successfully cryopreserve orchid tissues and explants in liquid nitrogen cylinders (De and Pathak 2019).

5 Futuristic Prospects

The orchids are seen to be used in many ethnomedicines and have many antimicrobial, anti-inflammatory, anticancer, and antioxidant properties. Due to anthropogenic activities, the orchids are becoming endangered, and those that are already endangered are nearing extinction. From many experiments, it is evident that orchids are the source of many bioactive compounds, which have the potential to make many commercial drugs available. Therefore, further steps could be taken to test these compounds on experimental models such as guinea pigs so that the medicines could be made globally available. Scientific validation is also required for the medicines from orchid species used by the local people. The medicines from orchids can replace synthetic drugs, which can reduce the side effects. It is seen that orchids are used locally as a cure for many ailments. Therefore, steps could be taken to carry on further research on the orchids and a process to save them and extract their useful compounds for the betterment of human health.

References

Auberon F, Olatunji OJ, Raminoson D, Christian D, Muller CD, Soengas B, Bonté F, Lobstein A (2017) Isolation of novel stilbenoids from the roots of *Cyrtopodium paniculatum* (Orchidaceae). Fitoterapia 116(2017):99–105

Behera D, Rath CC, Tayung K, Mohapatra UB (2013) Ethnomedicinal uses and antibacterial activity of two orchid species collected from Similipal Biosphere Reserve Odisha, India. Int J Agricul Technol 9(5):1269–1283

Bhattacharjee B, Islam T, Rahman Z, Islam SMS (2014) Antimicrobial activity and phytochemical screening of whole plant extracts of vanda tessellata (roxb.) hook.ex.g.don. World J Pharm Pharmaceut Scie 4(1):72–83

Bhowmik TK, Rahman MM (2017) Effect of different basal media and PGRs on in vitro seed germination and seedling development of medicinally important orchid Cymbidium aloifolium (L.) Sw. J Pharmacogn Phytochem 6:167–172

Borkar SU, Masirkar DR (2015) Studies on phytochemical investigations and antimicrobial activity of an endangered orchid Geodorum densiflorum (Lam) Schltr. Int J Res Biosci Agric Technol 1:2347–517X

Bulpitt CJ, Li Y, Bulpitt PF, Wang J (2007) The use of orchids in Chinese medicine. J R Soc Med 100:558–563

Chaitanya PJ, Lalagoud M, Chandrashekar R, Lakshmi Bhavani RN, Kudle KR (2013) Phytochemical and antimicrobial screening of the polar and non-polar solvent Root extracts of Vanda tessellate (Roxb.)Hook.ex G.Don. Int J Pharmacogn Phytochem Res 5(4):315–320

Chen K. K., Chen A. L. (1935). The Alkaloid of Chin-Shih-Hu. J. Biol. Chem. 111, 653–658. 10.1016/s0021-9258(18)75010-2

Chimsook T (2016) Phytochemical screening, total phenolic content, antioxidant activities and cytotoxicity of Dendrobium signatum leaves. MATEC Web Conf 62:03005

Chinsamy M, Finnie JF, Staden JV (2014) Anti-inflammatory, antioxidant, anti-cholinesterase activity and mutagenicity of South African medicinal orchids. S Afr J Bot 91:88–98

Das AJ, Kumar R, Athar M, Singh RD, Kumar M, Khan MA, Jai P (2013) Ethno medicinal study of threatened plants of Sonitpur district in Assam, Northeast India. Int Res J Pharm 4(1):146

De LC, Pathak P (2018) Conservation, management, and utilization of orchid genetic resources. J Orchid Soc India 32:81–91

De L, Pathak P (2019) Conservation, management and utilization of orchid genetic resources. J Orchid Soc India 32:81

Deciga-Campos M, Palacios-Espinosa JF, Reyes-Ramırez A, Matac R (2007) Antinociceptive and anti-inflammatory effects of compounds isolated from Scaphyglottis livida and Maxillaria densa. J Ethnopharmacol 114:161–168

Devi PU, Selvi S, Devipriya D, Murugan Sand Suja S (2009) Antitumor and antimicrobial activities and inhibition of in-vitro lipid peroxidation by Dendrobium nobile. Afr J Biotechnol 8(10):2289–2293

Estrada S, Acevedo L, Rodriguez M, Toscano R, Mata R (2002) New Triterpenoids from the Orchids Scaphyglottis Livida and Nidema Boothii. Nat Prod Lett 16(2):81–86

Fan C, Wang W, Wang Y, Qin G, Zhao W (2001) Chemical constituents from Dendrobium densiflorum. Phytochemistry 57:1255–1258

Fay MF (2018) Orchid conservation: how can we meet the challenges in the twenty-first century? Bot Stud 59(1):16. https://doi.org/10.1186/s40529-018-0232-z

Fitzgerald DJ, Stratford M, Gasson MJ, Ueckert J, Bos A, Narbad A (2004) Mode of antimicrobial action of vanillin against Escherichia coli, Lactobacillus plantarum and Listeria innocua. J Appl Microbiol 97:104–113

Gebauer G, Preiss K, Gebauer AC (2016) Partial mycoheterotrophy is more widespread among orchids than previously assumed. New Phytol 211(1):11–15. https://doi.org/10.1111/nph.13865

Haridas R, Thangapandian V, Thomas B (2014) Phytochemical evaluation of a medicinal orchid: Nervilia plicata (andrews) schltr. Int J Pharm Dev Technol 4(4):204–209

Hausen BM, Shoji A, Jarchow O (1984) Orchid Allergy. Arch Dermatol 120:1206–1208
Hayashia J, Sekinea T, Deguchia S, Lin Q, Horiea S, Tsuchiyaa S, Yanoa S, Watanabea K, Ikegamia F (2002) Phenolic compounds from Gastrodia rhizome and relaxant effects of related compounds on isolated smooth muscle preparation. Phytochemistry 59:513–519
Hew CS, Yong JWH (2006) Orchids in Chinese Medicine. https://www.researchgate.net/publication/236982474
Ho CK, Chen CC (2003) Moscatilin from the Orchid Dendrobrium loddigesii is a potential anticancer agent. Cancer Investig 21(5):729–736
Hoque MM, Khaleda L, Al-Forkan M (2016) Evaluation of pharmaceutical properties on microbial activities of some important medicinal orchids of Bangladesh. J Pharmacogn Phytochem 5(2):265–269
Jagtap SS, Satpute RA, Rahatgaonkar AM, Lanjewar KR (2014) Phytochemical screening, antioxidant, antimicrobial and quantitative multi-elemental analysis of Habenaria longicorniculata J. Graham. J Acad Indus Res 3(3):108
Johnson M, Janikiraman N (2013) Phytochemical and TLC studies on stem and leaves of the orchid Dendrobioum panduratum subsp. villosum Goplalan & A. N. Henry. Indian J Nat Prod Resour 4(3):250–254
Johnson MK, Alexander KE, Lindquist N, Loo G (1999) A phenolic antioxidant from the freshwater orchid, Habenaria repens. Compar Biochem Physiol Part C 122:211–214
Kalaiarasan A, John SA (2012) In vitro screening for anti-inflammatory activity of Bulbophyllum kaitenseRechib. Pseudobulb extract by HRBC method, Eastern Paninsular flora of South India. Int J Sci Res Publ 2(7):1–7
Kaya Z, Raynal D (2001) Biodiversity and conservation of Turkish forests. Biol Conserv 97:131–141. https://doi.org/10.1016/S0006-3207(00)00069-0
Keerthiga M, Anand SP (2014) Physicochemical, preliminary phytochemical analysis and antibacterial activity against clinical pathogens of medicinally important orchid Geodorum densiflorum (lam) schltr. Int J Pharm Pharm Sci 6(8):2014
Kiran R, Kekuda TRP, Kumar HGP, Hosetti BB, Krishnaswamy K (2013) Biological activities of Sarcanthus pauciflorus. J Appl Pharma Sci 3(07):105–110
Lal N, Singh M (2021) Prospects of plant tissue culture in orchid propagation: a review. Indian J Biol 7:103–110. https://doi.org/10.21088/ijb.2394.1391.7220.15
Li Y, Wang CL, Guo SX, Yang JS, Xiao PG (2008) Two new compounds from Dendrobium candidum. Chem Pharm Bull 56(10):1477–1479
Majumder PL, Sen RC (1987) Moscatilin, a bibenzyl derivative from the orchid Dendrobium moscatum. Phytochemistry 26(7):2121–2124
Marjoka A, Alam O, Huda MK (2016) Phytochemical screening of three medicinally important epiphytic orchids of Bangladesh. Jahangirnagar Univ J Biol Sci 5(1):95–99
Medhi RP, Chakraborti M, Rampal (2012) Orchid biodiversity in India: conservation and utilization. Indian J Genet Plant Breed 72:148–156
Mohanty JP, Pal P, Barma AD (2015) An overview on orchids. UJPSR 1(1):45–45
National Wildlife Database Cell, Wildlife Institute of India (2023). https://wii.gov.in/nwdc_aboutus
Oseni O (2018) A review on plant tissue culture, a technique for propagation and conservation of endangered plant species. Int J Curr Microbiol App Sci 7. https://doi.org/10.20546/ijcmas.2018.707.438
Paul P, Chowdhury A, Nath D, Bhattacharjee MK (2013) Antimicrobial efficacy of orchid extracts as potential inhibitors of antibiotic resistant strains of Escherichia coli. Asian J Pharm Clin Res 6(3):108
Prasad R, Koch B (2014) Antitumor activity of Ethanolic extract of Dendrobium formosum in T-cell lymphoma: an in vitro and in vivo study. Biomed Res Int 2014:Article ID753451
Prasad R, Koch B (2016) In vitro anticancer activities of Ethanolic extracts of Dendrobium crepidatum and Dendrobium chrysanthum against T-cell lymphoma. J Cytol Histol 7:4
Ram RB, Lata R, Meena ML (2011) Conservation of floral biodiversity of Himalayan Mountain regions with special reference to Orchids. Asian Agri-History 15:231–241

Ranjitha MC, Akarsh S, Kekuda PTR, Darshini SM, Vidya P (2016) Antibacterial activity of some plants of Karnataka, India. J Pharmacogn Phytochem 5(4):95–99

Sandrasagaran UM, Subramaniam S, Murugaiyah V (2014) New perspective of Dendrobium crumenatum orchid for antimicrobial activity against selected pathogenic bacteria. Pak J Bot 46(2):719–724

Schuster R, Zeindl L, Holzer W, Khumpirapang W, Okonogi S, Viernstein H, Mueller AM (2017) Eulophia macrobulbon—an orchid with significant anti-inflammatory and antioxidant effect and anticancerogenic potential exerted by its root extract. Phytomedicine 24(2017):157–165

Shrirama V, Kumarb V, Kishor PBK, Suryawanshid SB, Upadhyay AK, Bhat MK (2010) Cytotoxic activity of 9,10-dihydro-2,5-dimethoxyphenanthrene-1,7-diol from Eulophia nuda against human cancer cells. J Ethnopharmacol 128(2010):251–253

Shubha J, Chowdappa S (2016) Phytochemical analysis and antibacterial activity of Cymbidium aloifolium L. a medicinal orchid from Western Ghats of Karnataka, India. Int J Adv Sci Res Publ 2:19. ISSN-2454–9878

Sohag SI, Hoque MM, Huda MK (2017) Phytochemical screening and antioxidant activity of rare medicinal orchid Luisia zeylanica Lindl. J Pharmacogn Phytochem 6(4):688–692

Swarts N (2018) Conservation methods for terrestrial orchids. J Ross Publishing, Plantation, Florida

Tsering J, Tam N, Tag H, Gogoi BJ, Apang O (2017) Medicinal orchids of Arunachal Pradesh. Bullet Arunachal Forest Res 32(1&2):1–16

Xia WB, Xue Z, Li S, Wang SJ, Yang YC, He DX, Ran GL, Kong LZ, Shi JG (2005) Chemicalconstituents from tuber of Cremastra appendiculata [in Chinese]. Zhongguo Zhong Yao Za Zhi, 30(23):1827–1830

Wei Y, Wang L, Wang D, Wang D, Wen C, Han B, Ouyang Z (2018) Characterization and anti-tumor activity of a polysaccharide isolated from Dendrobium officinale grown in the Huoshan County. Chin Med 13:47

Williams CA (1979) The leaf flavonoids of the Orchidaceae. Phytochemistry 18:803–813

Wu B, He S, Pan YJ (2006) New Dihydrodibenzoxepins from Bulbophyllum kwangtungense. Planta Med 72:1244–1247

Wu JB, Lin WL, Hsieh CC, Ho HY, Tsay HS, Lin WC (2007) The Hepatoprotective activity of Kinsenoside from Anoectochilus formosanus. Phytother Res 21:58–61

Yeung EC (2017) A perspective on orchid seed and protocorm development. Bot Stud 58(1):33. https://doi.org/10.1186/s40529-017-0188-4

Yoshikawa M, Murakami T, Kishi A, Sakurama T, Matsuda H, Nomura M, Matsuda H, Kubo M (1998) Novel Indole S,O-Bisdesmoside, Calanthoside, the precursor glycoside of Tryptanthrin, Indirubin and Isatin, with increasing skin blood flow promoting effects, from two Calanthe species (Orchidaceae). Chem Pharm Bull 46(5):886–888

Yu S, Wang Z, Su Z, Song J, Zhou L, Sun Q, Liu S, Li S, Li Y, Wang M, Zhang GQ, Zhang X, Liu ZJ, Lu D (2018) Gigantol inhibits Wnt/β-catenin signalling and exhibits anticancer activity in breast cancer cells. BMC Complement Altern Med 18:59

Zhang HJ, Zhou JJ, Li XS (2003) Study advance of gastrodia elata b1. Amino Acids and Biotic Resources.25(1):17–20

Chloroplast Biogenesis in Plants Responding to Abiotic Stresses

Indrajeet Sahani, Rajarshi Kumar Gaur, Dinesh Yadav, and Ramwant Gupta

Abstract

The apparatus of the electron transport chain for photosynthesis process is made up of various protein complexes, such as photosystem II (PSII), Cytb6f, Plastocyanin (PC), Photosystem-I (PSI), and photoprotection-related proteins. These protein complexes are prone to damage due to different abiotic stresses and are continuously resynthesized to replace the damaged ones. Oxygenic photosynthesis uses sunlight, CO2, and water to produce food and oxygen for most life forms. In photosynthetic plants, chloroplasts are the primary reactive oxygen species (ROS)-producing sites, involving PSII and PSI, which hamper photosynthesis and cause photoinhibition. The plant may adjust its thylakoid membrane's lipid content and structural configuration to more effectively dissipate excess energy (non-photochemical quenching), improve membrane fluidity, and optimize the photosynthetic process, thus preventing potential damage from abiotic stresses. Chloroplasts shift their orientation by migrating perpendicular to incident light when exposed to high light intensity. Chloroplast biogenesis and development require coordinated expression of genes encoded by the plastid and nuclear systems and are necessary

I. Sahani · R. Gupta (✉)
Department of Botany, Deen Dayal Upadhyaya Gorakhpur University, Gorakhpur, Uttar Pradesh, India
e-mail: ramwant.bot@ddugu.ac.in

R. K. Gaur
Department of Biotechnology, Deen Dayal Upadhyaya Gorakhpur University, Gorakhpur, Uttar Pradesh, India

Uttar Pradesh Council of Agricultural Research, Uttar Pradesh, India

D. Yadav
Department of Biotechnology, Deen Dayal Upadhyaya Gorakhpur University, Gorakhpur, Uttar Pradesh, India

R. K. Gaur et al. (eds.), *Plant Molecular Farming*,
https://doi.org/10.1007/978-981-95-3823-2_8

for the growth of chloroplasts in higher plants. A new D1 protein is generated and placed inside PSII again after the damaged D1 protein in PSII is eliminated and broken down by specific proteases. These findings suggest that there is a specific pathway involved in chloroplast biogenesis after stress-induced damage. This insight could inform the development of more stable protein complexes of the photosynthetic electron transport chain through genetic engineering, which might increase the number of important constituents of core photosynthetic machinery, improving photosynthetic rates and helping the scientific community to better understand the mechanisms to boost photosynthetic efficiency and accelerate the stress tolerance techniques for crop improvement.

Keywords

Chloroplast · Photo systems · D1 protein · Biogenesis · Abiotic stresses

1 Introduction

The chloroplast is a semiautonomous cell organelle, like mitochondria, and chloroplasts are derived from blue-green algae. It is a double-membrane-bound structure including the inner and outer envelopes (Bobik and Burch-Smith 2015). The chloroplast network, which consists of membranous structures known as thylakoids. The thylakoid is a structure where the light reaction of photosynthesis takes place. The carbon fixation reactions primarily occur in the stroma of the chloroplast (Mechela et al. 2019). Photosynthesis is a very complex process involving many protein complexes. Chloroplasts are the site of many important processes, viz., the metabolism of carbohydrates, lipids, tetrapyrroles, amino acids, and isoprenoid compounds (Heldt and Piechulla 2021). Chloroplasts are also involved in the production of specialized metabolites, the regulation of photosynthesis, the biogenesis and maintenance of their own structures, and interactions with other cellular components (Song et al. 2021). Chloroplasts have their own DNA, which is extranuclear; therefore, they are regarded as semiautonomous cellular organelles like mitochondria because they synthesize some of their own proteins and lipids essential for the development of the chloroplast membrane (Singh et al. 2022).

Chloroplasts have a complex membrane network made up of grana, which are stacked thylakoid membranes, and stromal lamellae, which are unstacked structures and which connect granum lamellae (Mazur et al. 2021). Thylakoid structure is typically seen through electron microscopy, influences the effectiveness of photosynthesis, and is controlled by a number of physiological and developmental variables (Daum and Kühlbrandt 2011).

Oxygenic photosynthesis uses sunlight, CO2, and water to produce food and oxygen for the majority of life forms. Chloroplasts are the cell organelles where photosynthesis takes place (Blankenship 2021). The apparatus for photosynthetic processes, i.e., chloroplasts, is made up of several proteins that are encoded by the nucleus and chloroplasts. A significant coordination between the expression levels of the two genomes is required for chloroplast function (Dobrogojski et al. 2020). Nevertheless, a large number of proteins (photosystem II (PSII), PSI, Cyt b6f, PC, and photoprotection-related protein) have been identified (Pinnola and Bassi 2018) and have been implicated in the control of the thylakoid and its lumen in response to several abiotic stresses. Photosystem II (protein complex) is also known as water-plastoquinone oxidoreductase. At least 20 proteins, photosynthetic pigments, quinones, and metal atoms make up the PSII core complex. The light reactions of photosynthesis start from here. It is found in the thylakoid membranes of plants. Within the photosystem, chlorophyll captures photosynthetically active radiation (PAR), which is utilized to excite electrons that are then carried through a system of electron carriers, including coenzymes and cofactors, to complete the reaction that reduces plastoquinone to plastoquinol (Shen 2015).

2 Photosystem II

Photosystems are the main constituents of the light reaction of photosynthesis and produce assimilatory power, viz., ATP and NADPH (Singh and Thakur 2018). From photosystem II (PSII), the light reactions of photosynthesis start. Its primary role is light harvesting, which leads to the hydrolysis of water molecules (photolysis) to yield oxygen, which participates further in the electron transport chain (Zhang and Reisner 2020). On the other hand, PSII is highly vulnerable to multiple abiotic stressors such as excessive light, extreme temperature, drought, and salinity (Muhammad et al. 2021). The integrity and activity of photosystems, e.g., PSI and PSII, are essential components of the photosynthetic apparatus of plants necessary for efficiently converting light energy into chemical energy. In the PSII reaction center, the D1 protein is an essential and highly labile component. Two homologous proteins, D1 and D2, form a pseudo-symmetric heterodimer in the center of PSII (Chauhan et al. 2023).

3 Abiotic Stresses

Stress can be defined as any kind of unfavorable conditions or elements that disturb a plant's metabolism, development, and growth. In basic terms, stress is a change in physiological state brought about by many living and nonliving elements, which disrupts the equilibrium. Many abiotic stresses, including drought, salt, heat stress, cold temperatures, heavy metal toxicity, flooding, and extreme soil pH, are often experienced by plants (Rathod and Verma 2023).

4 Drought Stress

One of the drastic abiotic factors that significantly lowers plant yields worldwide is drought, which is defined as a period of very little or no rainfall (Yadav et al. 2020). During the period of drought, plants exhibit a wide range of physiological, morphological, and biochemical alterations, including changed gene expression and decreased photosynthetic yield (Wahab et al. 2022), which eventually results in poor grain yield and decreased growth. The generation of reactive oxygen species

(ROS) is triggered by drought conditions, which disrupts the dynamic balance between ROS production and ROS scavenging (Nawaz et al. 2020). Plants have evolved a number of adaptive strategies that provide them resistance from the negative effects of drought stress, such as tolerance, avoidance, and escape (Bandurska 2022). Due to drought stress, a majority of farmers face high yield loss for their crops, and it is a great threat to global food security (Leng and Hall 2019).

5 High Temperature Stress

There are three types of temperature stress, viz., high temperature stress, chilling temperature stress, and freezing temperature stress (Devireddy et al. 2021). Plants subjected to temperature stress exhibit declining germination rates, stunted growth, and reduced photosynthetic activity, which often leads to death (Waraich et al. 2012). Understanding the mechanisms through which temperature stress induces these adverse effects is crucial for understanding the strategies that enable plants to withstand unfavorable temperature conditions(Hassan et al. 2021). In response to temperature stress, plants make changes in their membrane lipid composition, regulate stress-related genes, enhance the accumulation of metabolites, and adopt different detoxification pathways (Bhattacharya 2022). Every living thing that exists has been developed to survive in a specific temperature range.

Even a minute increase or decrease in temperature can lead to serious effects on a plant's physiological and biochemical processes (Nievola et al. 2017). An increase in temperature beyond the limit of tolerance causes cells to lose control over chemical reactions, leading to denaturation of proteins and physical changes that are harmful to the organism (Precht 2013).

It has been discovered that exogenous applications of protectants, such as osmoprotectants like proline (pro), phytohormones like abscisic acid (ABA), and gibberellic acids (GA), signaling molecules like nitric oxide (NO), trace elements viz. selenium, silicon, and nutrients viz. nitrogen, phosphorus, potassium, calcium, etc., have been found to be productive in reducing the damage caused by high temperature stress (Mishra et al. 2022).

6 Salt Stress

Plant development is stunted and growth is low when soil salt is high because it alters physiological and biochemical processes, particularly during photosynthesis (Hailu and Mehari 2021). A 50% loss in agricultural productivity is due to salt stress. Additionally, osmotic stress brought on by salinity inhibits metabolic processes and has an ionic influence on the shape of subcellular organelles, which lowers photosynthesis (Hameed et al. 2021).

It is observed that the chlorophyll and carotenoid content decreased by 82% and 55%, respectively, in various plants, such as tomato, during the period of both drought and high temperature stress (Zhou et al. 2020). These two stresses together also affect carpel and pollen development, grain formation, and ovule functions

(Sinha et al. 2021). In addition, an excess of ROS is produced, which causes lipid peroxidation and protein denaturation, decreased nutrient uptake, membrane structural damage, and stunted plant development and productivity (Yadav et al. 2022). Reduced nutrient uptake, decreased root and shoot length, decreased grain and kernel size, decreased activity of antioxidant enzymes such as catalase and ascorbate peroxidase, and markedly elevated levels of ROS and malondialdehyde were all seen in maize plants under heat and drought stress.

7 Chloroplast Adaptation in Response to Abiotic Stress

Plants have evolved a number of adjustable strategies that provide them resistance to the negative effects of these abiotic stresses (Zhang et al. 2022). Numerous metabolic processes, including photosynthetic electron transport, occur in the chloroplast; as an organelle, its metabolic balance is readily upset by external stressors (Muhammad et al. 2021). When there is a disturbance in chloroplast homeostasis, it is communicated to the nucleus through retrograde signals to adjust all cellular activities, which can be coordinated according to the supply of sugar and other compounds to the chloroplast (Unal et al. 2020). Plants produce ROS during periods of drought, which causes oxidative damage to cellular components. In photosynthetic plants, chloroplasts are the primary ROS-producing sites involving PSII and PSI, which hamper photosynthesis and cause photoinhibition (Chauhan et al. 2023). These photosystems (PSI and PSII) are prone to oxidative damage during abiotic stress like drought, elevated temperatures, high light intensity, and salinity, due to oxidative stress caused by an imbalance in reactive oxygen species (ROS) (Didaran et al. 2024). Reactive oxygen species imbalance may lead to deficiencies in the proportions of the photosystems (Foyer 2018). These properties at the level of photosystem II (PS II) make plants reliable and flexible, as plants have evolved various strategies to repair and replace their photosynthetic apparatus (Robbins and Kelly 2024). PS II activity is mostly inhibited by the formation of reactive oxygen species (ROS), which are by-products of metabolism that become toxic at high concentrations (Sachdev et al. 2021). Redox balance of the cells is impacted by hours of intense radiation exposure in tropical and subtropical crops, which leads to the production of excessive ROS (Goswami et al. 2022).

8 Chloroplast Movement

Higher plants, algae, mosses, and other organisms have been shown to exhibit chloroplast mobility when exposed to light fluctuations (Bag 2021). To maximize light absorption in low-light conditions, chloroplasts position themselves parallel to the incident light by gathering parallel to the leaf's plane (Karabourniotis et al. 2021). The chloroplasts shift their orientation by migrating perpendicular to the incidence when exposed to high light intensity, thus protecting themselves from photooxidation and the leaf from excessive light stress (Chauhan et al. 2023). Such a chloroplast movement can reduce the leaf's absorption of light by roughly 15%. The blue light receptor known as phototropin has been discovered to have a major role in controlling the movements of chloroplasts during photorelocation in plants. The chloroplast surface is home to the cytoplasmic protein *CHLOROPLAST UNUSUAL POSITIONING 1(CHUP1)*, which also has an important role in chloroplast movement. It interacts with actin to assist the chloroplast movement (Kong and Wada 2016). Plants have a very efficient antioxidant system. This antioxidant machinery serves to reduce the oxidative stress-induced damage listed above and builds up the ROS defense mechanism. Enzymatic and nonenzymatic antioxidants are the two arms of the antioxidant machinery. Antioxidant enzymes are essential for plants' ability to withstand stress and are a key ROS scavenger. Enzymatic antioxidants, viz. glutathione peroxidase (GPX), ascorbate peroxidase (APX), catalase (CAT), and superoxide dismutase (SOD), eliminate ROS by a sequence of redox reactions. This strategy effectively reduces the likelihood of further damage and stabilizes the photosystems (Rajput et al. 2021).

9 Non-photochemical Quenching

Although photosynthesis depends on light, an organism may suffer serious damage or even die if its luminescence levels are too high (Blankenship 2021). To reduce such possible harm, plants and microalgae have evolved photoprotective strategies known as non-photochemical quenching (NPQ) (Pinnola 2019). It is energy-dependent quenching (qE), which releases excess light energy as heat (Lu et al. 2022). In NPQ, extra absorbed energy is thermally dissipated in the form of heat by de-exciting singlet-excited chlorophyll in photosystem II (PSII) (Kalariya et al. 2019).

10 Production and Scavenging of ROS During Abiotic Stress

Both photosystems contribute to the production of ROS in chloroplasts: Photosystem I (PSI) sinks electrons to molecular oxygen, whereas photosystem II traps an excess photon. Excess energy in PSII causes P680 chlorophyll (Chl) molecules to develop in the reaction center and form the triplet state of light-excited chlorophyll in the PSII light-harvesting antenna complex (LHC). The absorbed energy can be transferred to ground-state oxygen (O_2) by these Chl molecules due to their longer lifespan, which makes it easier to produce highly reactive singlet oxygen (Barber et al. 2023). The sequential synthesis of $O^{\bullet -}$ and H_2O_2 in PSI depends on ineffective photochemical and non-photochemical quenching (NPQ), which releases electrons derived from PSII to O_2 through an electron transport component linked with PSI. Superoxide dismutases (SODs) catalyze the reduction of O^{2-} to O_2 and H_2O_2, while stromal and thylakoid membrane-associated ascorbate peroxidases (APXs) use the ascorbate–glutathione cycle to detoxify H_2O_2 (Barber et al. 2023). To optimize photosynthetic efficiency, the electron transport chain's oxidation and reduction must be balanced by such an electron sink, commonly called the water–water cycle (Sunil et al. 2023).

11 Chloroplast Biogenesis

A coordinated expression of genes encoded by the plastid and nuclear systems is necessary for the formation of chloroplasts in higher plants. In *Arabidopsis*, it is found that the expression of chloroplast genes and the formation of chloroplasts depend on RNA PROCESSING8 (RP8), which forms new chloroplast proteins (Zhu et al. 2024). When the RP8 gene is altered to lose its function, the mature rpoA transcript, which codes for the alpha subunit of the plastid-encoded RNA polymerase (PEP) complex, accumulates less. PEP-dependent gene transcription, such as that of psaA, psbA, psbB, petB, and rbcL, is very necessary for chloroplast biogenesis. The rp8 mutant either lacks thylakoids or has a very reduced thylakoid phenotype and pale-green cotyledons, with very few leaves in them. Therefore, it was suggested that RP8 plays a crucial part in the transcription of rpoA (Kong et al. 2023). Chloroplast biogenesis includes the D1 protein repair cycle, thylakoid membrane remodeling, and cytochrome b_6f regeneration.

12 Thylakoid Membrane Remodeling

Plants may adjust the lipid content and structural configuration of their thylakoid membranes to more effectively dissipate excess energy, improve membrane fluidity, and optimize the photosynthetic process, thus preventing potential damage by abiotic stresses (Kowalewska et al. 2023). Abiotic stresses experienced by plants, such as extreme heat and drought increase the risk of harm to the photosystem, activating the thylakoid membrane remodeling process, which leads to modification of the lipid composition, which in turn enhances the fluidity of the membrane (Ďúranová et al. 2023). An improvement in systems for energy dissipation occurs, which minimizes the accumulation of unnecessary energy, as a result of which optimization of the photosynthetic process takes place and finally leads to protection and stability of the photosystem (Simkin et al. 2023). Generally, during thylakoid membrane remodeling processes, protein complexes are reorganized due to restructuring in membrane lipids (Muller et al. 2023). These changes include an increase in stacking of grana to capture maximum energy, PSII protein complex repair and replacement, and adjustments in lipid composition to enhance membrane thylakoid fluidity (Nami et al. 2021).

13 D1 Protein Repair Cycle (PSII Repair Cycle)

A new D1 protein is generated and placed again inside PSII after the damaged D1 protein in PSII is eliminated and broken down by particular proteases (Inagaki et al. 2021). One of the most important strategies for restoring PSII functionality under stress is this repair cycle. Photoinhibition has a significant impact on the ATP-dependent D1 protein synthesis. During environmental conditions, the photosynthetic process is maintained by replacing the damaged D1 protein with a newly synthesized D1 protein (Gupta et al. 2023). But when PSII damage rates are higher than repair rates, damaged D1 protein builds up and prevents normal photosynthesis, termed photoinhibition (Li et al. 2023). Damage in the D1 protein occurs due to several abiotic stresses, particularly high temperature stress. Specific proteases remove damaged D1 proteins, resulting in their degradation. Subsequently, the synthesis of new D1 proteins occurs through the transcription and translation of the psbA gene. Finally, the incorporation of the newly synthesized D1 protein into photosystem II (PSII) restores its functionality.

14 Cytochrome b_6f (cytb_6f) Regeneration

Cytochrome b6f (cytb6f) is a key component of the light-dependent processes of oxygenic photosynthesis. It connects the two photosystems, namely photosystem II (PSII) and photosystem I (PSI), and it oxidizes electron carriers like plastoquinol (PQ) and reduces plastocyanin (Pc) (Roach and Krieger-Liszkay 2014). Through an electron splitting mechanism known as the Q-cycle, electron transmission is linked to the formation of a transmembrane proton gradient for ATP synthesis (Crofts 2021). Cytb6f is a critical component of the redox-sensing hub that controls light harvesting, electron transfer, and photosynthetic gene expression. It is also necessary for cyclic electron transfer (CET) and catalyzes the rate-limiting step in linear electron transfer (LET) (Malone et al. 2021). Like the D1 protein, it is also prone to damage by abiotic stresses. The thylakoid contains the 25 kDa protein DEIP1, which interacts with PetA and PetB during Cytb6f's assembly (Yang et al. 2023).

15 Conclusion

In this chapter, we have explained the adjustments made in protein structure, overall chloroplast adaptation, and biogenesis in response to various abiotic stresses. Therefore, comparative information on plant behavior and organelles, as well as stress cross talk networks, will be beneficial in understanding the deep mechanism of chloroplast biogenesis. The fundamental practices, viz., agronomic management, application of growth regulators, and nutrient and water management, are not enough to reduce the yield loss due to the decreased photosynthetic efficiency during the period of immense stress. Our collected information is used to explain the mechanism of damage, adaptation, and biogenesis of chloroplasts under various stressful conditions. This will help the scientific community to better understand the mechanisms to boost photosynthetic efficiency and accelerate the stress tolerance techniques for crop improvement.

16 Future Prospective

Traditional breeding and transgenic methods were successful in improving the photosynthetic rate during periods of abiotic stress in crops, viz., rice, wheat, maize, and soybeans. However, most of these promising lines failed to produce high yields during periods of immense stress. It indicates that there is still a wide scope for novel methods to mitigate the effects of these abiotic stresses on chloroplasts. One such prospect could be the development of more stable protein complexes of the photosynthetic electron transport chain using genetic engineering. Approximately 100 genes are known to be essential for the basic machinery of photosynthesis and its biochemical reactions. These genes produce complexes and enzymes that are necessary for the light and dark reactions that occur during photosynthesis. The specific functions and effects of these enzymes and complexes on the photosynthetic efficiency of different plants have not yet been thoroughly identified. Crop photosynthesis rates may be increased by using transgenic techniques to increase the amount of essential components in the core photosynthetic mechanism. It is also observed that C4 plant chloroplasts are less prone to damage from these abiotic stresses than those of C3 plants, and C4 plants are photosynthetically more efficient than C3 plants. Therefore, well-focused research on developing the C4 pathway of photosynthesis in C3 plants could enhance photosynthetic yield in C3 plants. One example of such research is the *"C4 rice project";* this project takes a period of approximately 15 years for the development of C4 rice by plant breeders (Fischer 2020).

Finally, it is proposed that future studies should concentrate on improving the chloroplast function more efficiently under a variety of environmental conditions. Expanding our knowledge of the genetic foundation of photosynthetic efficiency will lead to a significant increase in grain yield. Improvements in photosynthetic efficiency led to significant increases in crop yields across diverse environmental conditions and periods of extreme stress.

References

Bag P (2021) Light harvesting in fluctuating environments: evolution and function of antenna proteins across photosynthetic lineage. Plants 10(6):1184

Bandurska H (2022) Drought stress responses: coping strategy and resistance. Plants 11(7):922

Barber TM, Hanson P, & Weickert MO (2023) Metabolic-associated fatty liver disease and the gut microbiota. Endocrinology and Metabolism Clinics, 52(3):485–496

Bhattacharya A (2022) Lipid metabolism in plants under low-temperature stress: a review. In: Physiological processes in plants under low temperature stress. Springer, Singapore, pp 409–516

Blankenship RE (2021) Molecular mechanisms of photosynthesis. John Wiley & Sons, Hoboken, New Jersey, USA

Bobik K, Burch-Smith TM (2015) Chloroplast signaling within, between and beyond cells. Front Plant Sci 6:781

Chauhan J, Singh P, Choyal P, Mishra UN, Saha D, Kumar R et al (2023) Plant photosynthesis under abiotic stresses: damages, adaptive, and signaling mechanisms. Plant Stress 10:100296

Crofts AR (2021) The modified Q-cycle: a look back at its development and forward to a functional model. Biochim Biophys Acta Bioenerget 1862(8):148417

Daum B, Kühlbrandt W (2011) Electron tomography of plant thylakoid membranes. J Exp Bot 62(7):2393–2402

Devireddy AR, Tschaplinski TJ, Tuskan GA, Muchero W, Chen JG (2021) Role of reactive oxygen species and hormones in plant responses to temperature changes. Int J Mol Sci 22(16):8843

Didaran F, Kordrostami M, Ghasemi-Soloklui AA, Pashkovskiy P, Kreslavski V, Kuznetsov V, Allakhverdiev SI (2024) The mechanisms of photoinhibition and repair in plants under high light conditions and interplay with abiotic stressors. J Photochem Photobiol B Biol 259:113004

Dobrogojski J, Adamiec M, Luciński R (2020) The chloroplast genome: a review. Acta Physiol Plant 42(6):98

Ďúranová H, Šimora V, Ďurišová Ľ, Olexiková L, Kovár M, & Požgajová M (2023) Modifications in ultrastructural characteristics and redox status of plants under environmental stress: A review. Plants, 12(8):1666

Fischer RA (2020) Advances in the potential yield of grain crops. In: Population, agriculture, and biodiversity:: problem s and prospects. University of Missouri Press, Columbia, pp 149–180

Foyer CH (2018) Reactive oxygen species, oxidative signaling and the regulation of photosynthesis. Environ Exp Bot 154:134–142

Goswami AK, Maurya NK, Goswami S, Bardhan K, Singh SK, Prakash J et al (2022) Physio-biochemical and molecular stress regulators and their crosstalk for low-temperature stress responses in fruit crops: a review. Front Plant Sci 13:1022167

Gupta B, Shrestha J (2023) Abiotic stress adaptation and tolerance mechanisms in crop plants. Frontiers in Plant Science, 14:1278895

Hailu B, Mehari H (2021) Impacts of soil salinity/sodicity on soil-water relations and plant growth in dry land areas: a review. J Nat Sci Res 12:1–10

Hameed A, Ahmed MZ, Hussain T, Aziz I, Ahmad N, Gul B, Nielsen BL (2021) Effects of salinity stress on chloroplast structure and function. Cells 10(8):2023

Hassan MU, Chattha MU, Khan I, Chattha MB, Barbanti L, Aamer M et al (2021) Heat stress in cultivated plants: nature, impact, mechanisms, and mitigation strategies—a review. Plant Biosyst 155(2):211–234

Heldt HW, Piechulla B (2021) Plant biochemistry. Academic Press. Elsevier, Amsterdam, Netherlands

Inagaki Y, Akune R, Matsuda A, Kuroki S, Takisawa K, & Fukushima T (2021) Effect of chloroplast movement on laser speckle method. Transactions of the ASABE, 64(6):1747–1754

Kalariya KA, Goswami N, Mehta D, Singh AL, Saran PL (2019) Chlorophyll fluorescence: a physiological mechanism and a physical tool in plant eco-physiological studies. Adv Plant Physiol 18:203

Karabourniotis G, Liakopoulos G, Bresta P, Nikolopoulos D (2021) The optical properties of leaf structural elements and their contribution to photosynthetic performance and photoprotection. Plants 10(07):1455

Kong SG, Wada M (2016) Molecular basis of chloroplast photorelocation movement. J Plant Res 129:159–166

Kong Y, & Zheng Y (2023) Magic blue light: A versatile mediator of plant elongation. Plants, 13(1):115

Kowalewska A, & Majewska-Smolarek K (2023) Eugenol-based polymeric materials—antibacterial activity and applications. Antibiotics, 12(11):1570

Leng G, Hall J (2019) Crop yield sensitivity of global major agricultural countries to droughts and the projected changes in the future. Sci Total Environ 654:811–821

Li R, He Y, Chen J, Zheng S, & Zhuang C (2023) Research progress in improving photosynthetic efficiency. International Journal of Molecular Sciences, 24(11):9
Lu D, Zhang Y, Zhang A, Lu C (2022) Non-photochemical quenching: from light perception to photoprotective gene expression. Int J Mol Sci 23(2):687
Malone LA, Proctor MS, Hitchcock A, Hunter CN, Johnson MP (2021) Cytochrome b6f–Orchestrator of photosynthetic electron transfer. Biochim Biophys Acta Bioenerg 1862(5):148380
Mazur R, Mostowska A, Kowalewska Ł (2021) How to measure grana–ultrastructural features of thylakoid membranes of plant chloroplasts. Front Plant Sci 12:756009
Mechela A, Schwenkert S, Soll J (2019) A brief history of thylakoid biogenesis. Royal Soc Open Biol 9(1):180237
Mishra UN, Saha D, Chauhan J, Kumar V, Jatav HS, Lal D et al (2022) Emerging roles of osmoprotectants in response to multiple abiotic stress tolerance in plants. In: Omics analysis of plants under abiotic stress. Apple Academic Press. Apple Academic Press Goldenrod Cir NE, Florida, USA, pp 179–206
Muhammad I, Shalmani A, Ali M, Yang QH, Ahmad H, Li FB (2021) Mechanisms regulating the dynamics of photosynthesis under abiotic stresses. Front Plant Sci 11:615942
Muller JD, Rotenberg E, Tatarinov F, Oz I, & Yakir D (2023) Detailed in situ leaf energy budget permits the assessment of leaf aerodynamic resistance as a key to enhance non-evaporative cooling under drought. Plant, cell & environment, 46(10):3128–3143
Nami F, Tian L, Huber M, Croce R, & Pandit A (2021) Lipid and protein dynamics of stacked and cationdepletion induced unstacked thylakoid membranes. BBA advances, 1:100015
Nawaz M, Anjum SA, Ashraf U, Azeem F, Wang Z (2020) Antioxidant defense system and reactive oxygen species (ROS) interplay in plants under drought condition. In: Handbook of climate change management: research, leadership, transformation. Springer, Singapore, pp 1–25
Nievola CC, Carvalho CP, Carvalho V, Rodrigues E (2017) Rapid responses of plants to temperature changes. Temperature 4(4):371–405
Pinnola A (2019) The rise and fall of Light-Harvesting Complex Stress-Related proteins as photoprotection agents during evolution. J Exp Bot 70(20):5527–5535
Pinnola A, Bassi R (2018) Molecular mechanisms involved in plant photoprotection. Biochem Soc Trans 46(2):467–482
Precht H (2013) Temperature and life. Springer Science & Business Media. Springer, Singapore
Rajput VD, Harish S, Verma KK, Sharma L, Quiroz-Figueroa FR et al (2021) Recent developments in enzymatic antioxidant defence mechanism in plants with special reference to abiotic stress. Biology 10(4):267
Rathod A, Verma NS (2023) Impact of abiotic stress on agronomical crops. In: Frontiers of agronomy. Elite Publishing House, Rohini, p 27
Roach T, Krieger-Liszkay A (2014) Regulation of photosynthetic electron transport and photoinhibition. Curr Prot Peptide Sci 15(4):351–362
Robbins EH, Kelly S (2024) Widespread adaptive evolution in angiosperm photosystems provides insight into the evolution of photosystem II repair. Plant Cell 37:koae281
Sachdev S, Ansari SA, Ansari MI, Fujita M, Hasanuzzaman M (2021) Abiotic stress and reactive oxygen species: generation, signaling, and defense mechanisms. Antioxidants 10(2):277
Shen JR (2015) The structure of photosystem II and the mechanism of water oxidation in photosynthesis. Annu Rev Plant Biol 66(1):23–48
Simkin AJ, Alqurashi M, Lopez-Calcagno PE, Headland LR, & Raines CA (2023) Glyceraldehyde-3-phosphate dehydrogenase subunits A and B are essential to maintain photosynthetic efficiency. Plant Physiology, 192(4):2989–3000
Singh J, Thakur JK (2018) Photosynthesis and abiotic stress in plants. In: Biotic and abiotic stress tolerance in plants. Springer, Singapore, pp 27–46
Singh S, Naik J, Pandey A (2022) Genetics of plant organelles: plastid and mitochondrial genomes. In: Plant genomics for sustainable agriculture. Springer Nature Singapore, Singapore, pp 313–330

Sinha R, Fritschi FB, Zandalinas SI, Mittler R (2021) The impact of stress combination on reproductive processes in crops. Plant Sci 311:111007
Song Y, Feng L, Alyafei MAM, Jaleel A, Ren M (2021) Function of chloroplasts in plant stress responses. Int J Mol Sci 22(24):13464
Sunil L, Sadashivaiah Chandrakanth R, & Dorairaj D (2023) Proteomics response of medicinal plants to salt stress. In Stress-responsive factors and molecular farming in medicinal plants (pp. 227–241). Singapore: Springer Nature Singapore
Unal D, García-Caparrós P, Kumar V, Dietz KJ (2020) Chloroplast-associated molecular patterns as concept for fine-tuned operational retrograde signalling. Philos Trans R Soc B 375(1801):20190443
Wahab A, Abdi G, Saleem MH, Ali B, Ullah S, Shah W et al (2022) Plants' physio-biochemical and phyto-hormonal responses to alleviate the adverse effects of drought stress: a comprehensive review. Plants 11(13):1620
Waraich EA, Ahmad R, Halim A, Aziz T (2012) Alleviation of temperature stress by nutrient management in crop plants: a review. J Soil Sci Plant Nutr 12(2):221–244
Yadav S, Modi P, Dave A, Vijapura A, Patel D, Patel M (2020) Effect of abiotic stress on crops. Sustain Crop Prod 3(17):5–16
Yadav N, Monika Kumar A, Kumar N, Mamta H et al (2022) Impacts on plant growth and development under stress. In: Plant stress mitigators: action and application. Springer Nature Singapore, Singapore, pp 61–100
Yang X, Che Y, García VJ, Shen J, Zheng Y, Su Z et al (2023) Cyclophilin 37 maintains electron transport via the cytochrome b6/f complex under high light in Arabidopsis. Plant Physiol 192(4):2803–2821
Zhang JZ, Reisner E (2020) Advancing photosystem II photoelectrochemistry for semi-artificial photosynthesis. Nat Rev Chem 4(1):6–21
Zhang H, Zhu J, Gong Z, Zhu JK (2022) Abiotic stress responses in plants. Nat Rev Genet 23(2):104–119
Zhou R, Yu X, Wen J, Jensen NB, Dos Santos TM, Wu Z et al (2020) Interactive effects of elevated CO 2 concentration and combined heat and drought stress on tomato photosynthesis. BMC Plant Biol 20:1–12
Zhu L, Li X, Yang Z, Hao C, Li H, and Qin X (2024) The yellow-cotyledon gene (ATYCO) is a crucial factor for thylakoid formation and photosynthesis regulation in Arabidopsis. Plant Science, 348, p.112208

Plant Secondary Metabolites: Impact on Human Health and Performance in Connection with Nutrigenomics

Khushboo Jain, Ayushi Malik, Navnidhi Sharma, Mayank Suthar, Sunita Verma, and Avinash Marwal

Abstract

Nutrigenomics delves into the intricate interplay between genes and nutrients, utilizing molecular tools to explore and comprehend the diverse responses that arise when dietary supplements are administered to individuals or populations. The Human Genome Project has revealed that a mere 0.1% of gene sequences differ, contributing to variations in weight, height, hair color, and nutritional needs. Single Nucleotide polymorphisms (SNPs) are pivotal in this genetic diversity and are capable of altering encoded proteins. From a nutrigenomics point of view, nutrients are potent signals or sensors that direct specific cells to undergo metabolic changes, determining whether an individual is healthy or diseased. Gene-nutrient interactions play a crucial role in either enhancing or impeding the ability of transcription factors to interact with elements that govern RNA (Ribonucleic acid) polymerase binding. Such interactions can instigate direct actions, activating nuclear receptors and initiating gene transcription, or regulate molecules associated with a spectrum of diseases, from inflammation to cancer. By interpreting nutrigenomics data as a final step toward restoring sound health and preventing disease, data mining and other bioinformatics techniques

K. Jain
Department of Biotechnology, Faculty of Science, MVSC, Janardhan Rai Nagar Rajasthan Vidyapeeth (Deemed to be University), Udaipur, Rajasthan, India

A. Malik · M. Suthar · A. Marwal (✉)
Nanoparticle Synthesis and Bioinformatics Laboratory, Department of Biotechnology, Vigyan Bhawan, Block – B, New Campus, Mohanlal Sukhadia University, Udaipur, Rajasthan, India

N. Sharma
School of Life Sciences, Coventry University, Coventry, UK

S. Verma
Manas Enclave Residential Society, Indirapuri Street, Bhuwana, Udaipur, Rajasthan, India

R. K. Gaur et al. (eds.), *Plant Molecular Farming*,
https://doi.org/10.1007/978-981-95-3823-2_9

contribute to evidence-based intervention options. This review is an attempt to examine the science behind the evolution of nutrigenomics and its effect on human nutritional status, body composition, and ultimately, their work performance, shedding light on the exploration of genomic tools for complex dynamics of gene-nutrient interactions in this context.

Keywords

Plant secondary metabolites · Plant molecular farming · Nutrients · Diet · Health performance · Nutrigenomics · Gene-nutrient interaction

1 Introduction

Plants have served as sources of food, feed, and medicine for millennia. However, the rapidly increasing global population has placed significant pressure on arable land to meet nutritional demands. Despite the availability of over 20,000 edible plant species, only 15 contribute to 90% of the world's food requirements, with three crops—rice, wheat, and maize—providing 60% of global energy intake (Ferin and Yan 2019). This imbalance, coupled with inadequate nutrition, has fueled a surge in chronic diseases, further escalating the demand for medicinal interventions. Excessive consumption of refined carbohydrates, combined with sedentary lifestyles, contributes to glucotoxicity and a rise in noncommunicable diseases (NCDs) such as diabetes, cancer, and chronic coronary thrombosis (Jukanti et al. 2020; Kumar et al. 2019; Jain et al. 2021; Debras et al. 2020). The global nutritional landscape is further complicated by the triple burden of hidden hunger, malnutrition, and the rising prevalence of NCDs. Among the 7.6 billion people on Earth, nearly 2 billion experience malnutrition, while an additional 1.9 billion suffer from obesity due to carbohydrate-rich, nutrient-poor processed foods (Sreenivasulu and Fernie et al. 2022). Nutrition plays a pivotal role in maintaining the balance between health and disease. Bioactive compounds are obtained both from plants and fungi (Singh et al. 2024). Foods not only provide essential nutrients such as carbohydrates, lipids, proteins, and vitamins, but also contain bioactive compounds that contribute to disease prevention and overall health. Plants produce a wide range of metabolites that influence human health. Primary metabolites, including carbohydrates, proteins, and vitamins, are essential for the plant's biological functions and human sustenance (Lakra et al. 2019; Swapnil et al. 2021).

In contrast, secondary metabolites, synthesized in response to environmental stresses, do not participate in routine plant processes (Marwal et al. 2014a; Maheshwari et al. 2023). These metabolites are produced under specific conditions where changes in the plant's redox state signal the need for adaptation. Prolonged or extreme stress can alter the redox state further, leading to apoptosis and senescence in plants (Simsek and Whitney 2024; Mishra et al. 2024). More than 50,000 secondary metabolites have been identified in plants, categorized as phenolics, flavonoids, terpenes, steroids, alkaloids, lignans, stilbenes, and essential micronutrients such as vitamins, minerals, and trace elements. These compounds exhibit diverse biological activities, including antiaging, anticancer, anti-infective, anti-inflammatory, and antiviral properties, thereby playing a significant role in human health (Teoh 2015; Kumar et al. 2017). While plant-based diets are rich in these beneficial metabolites, ensuring dietary balance is critical for obtaining adequate amounts of essential nutrients. Even those following omnivorous diets may face nutrient deficiencies if their diet lacks diversity (Fig. 1).

The completion of the Human Genome Project (HGP) introduced new perspectives on the role of nutrients in influencing gene expression and overall health (Fig. 2). This led to the emergence of nutrigenomics, which aims to explore how gene expression in response to metabolic processes influences individual health. It investigates the interactions between genotype and environmental factors, including nutrition, and their effects on metabolic responses. Additionally, it seeks to determine whether understanding these interactions can enable the development of personalized dietary recommendations. The study of nutrigenomics uses molecular tools to examine the impact of dietary components on gene expression. Bioactive compounds in diets can influence gene activity, either enhancing or suppressing specific gene expressions. For instance, vitamins A and D, and fatty acids have been shown to activate nuclear receptors and induce gene transcription, while compounds like resveratrol (found in wine) and soy genistein can modulate molecular signaling pathways, such as nuclear factor kappa B (Fialho et al. 2008; Mahan and Scott-Stump 2005).

Although much research has focused on the effects of plant-derived compounds on chronic diseases and their biological impact, the relationship between specific plant secondary metabolites and human growth, development, and athletic performance remains underexplored. This chapter aims to provide a comprehensive

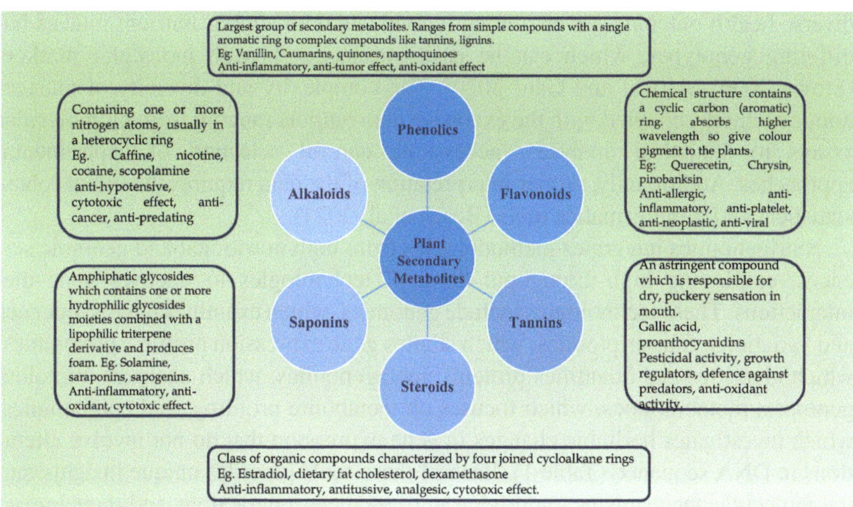

Fig. 1 Classification of plant secondary metabolites

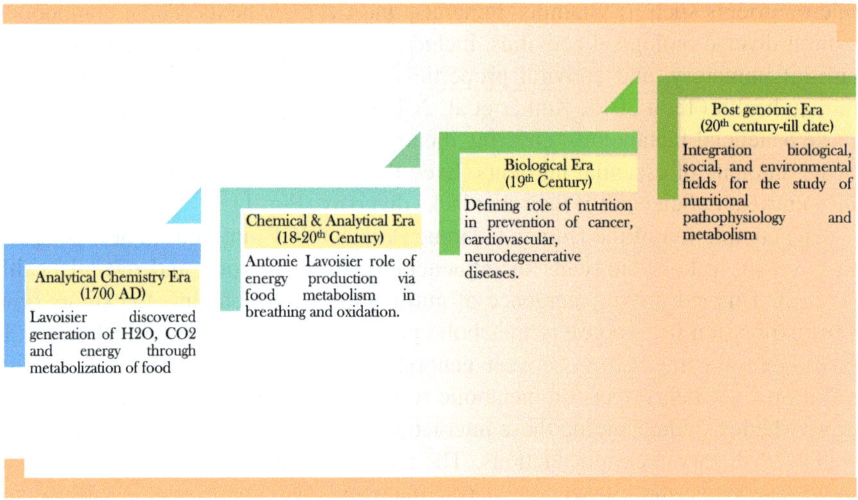

Fig. 2 Timeline for the development of nutrigenomics

overview of the mechanisms by which plant secondary metabolites influence human health, with an emphasis on tools and techniques used in nutrigenomics research. The chapter also highlights their potential role in enhancing athletic performance and cognitive functions, addressing a critical gap in the existing literature.

2 Tools and Techniques in Nutrigenomics Research

Genetic variations significantly influence nutrient metabolism and are linked to diverse health outcomes, even among individuals with similar nutrient intakes but differing genotypes, which can be easily accessed through molecular markers (Stover 2006; Marwal and Gaur 2020). The complexity and diversity of nutrigenomic methods, coupled with the extensive data outputs ranging from biomolecular assays to statistical analyses, necessitate careful selection of experimental approaches. Additionally, proper interpretation of the data requires the use of robust statistical and bioinformatics tools (Bösch et al. 2024).

Nutrigenomics integrates methodologies from both nutritional and genomic sciences, employing high-throughput "omics" technologies to uncover gene–diet interactions. These technologies include genomics, which examines DNA sequences and variations; transcriptomics, which studies gene expression profiles; proteomics, which identifies and quantifies proteins; metagenomics, which analyzes microbial genomes; metabolomics, which focuses on metabolite profiling; and epigenomics, which investigates heritable changes in gene expression that do not involve alterations in DNA sequence (Table 1). Each of these tools provides unique insights into the molecular mechanisms underlying nutrient–gene interactions and their impact on health. To effectively leverage these advanced technologies, researchers must

Table 1 Details of tools and techniques used for nutrigenomics

Genomics: Genomic methods detect genetic variations or changes in chromosomal structure and architecture as a result of alterations in the site interaction matrix within a chromosome			
S. No	*Technique used*	*Advantages*	*Disadvantages*
1	Polymerase chain reaction (PCR)	Simple, cheap, and widely used method in which minute amounts of DNA are amplified and then quantified	Specific primers are needed prior to sequencing; errors in primer design produce misleading results; there is a limited sample number, and a single specific region of interest per experiment
2	SNP array	Provide information on an extended number of SNPs	Expensive and laborious
3	NGS (Next-Generation Sequencing) platforms hotspot/shotgun sequencing platforms	High-throughput methods to sequence small or large DNA segments or the entire genome	High price for equipment and consumables; large amount of data for analysis; issues related to storage and safety of data, and issues related to comparison with a "reference genome"
4	Karyotyping fluorescent in situ hybridization (FISH)	Qualitative methods that provide information on chromosomal structure at different resolutions (chromosomes in karyotype; specific regions in FISH)	Karyotype can show only large abnormalities in chromosome structure, while FISH provides information solely related to the region of interest
5	Microarrays platforms	Comparative technique. Information on many genes/all the genes	High price for equipment and consumables
6	Hi-C platforms	Detection of chromosomal interactions between functional elements, such as promoters and enhancers	Expensive method with relatively few applications
7	Chromatin immunoprecipitation on chip microarrays	Advanced technology allowing genome-wide analysis of protein of interest	Limited resolution compared with sequencing; there may be low quality/affinity of available antibodies for a protein of interest; relatively costly experiments

(continued)

Table 1 (continued)

Transcriptomics: Methods assess (either quantitatively or qualitatively) the amount of specific mRNA (gene expression) or the expression of other noncoding RNA species (microRNA [miRNA], other short noncoding RNA, long noncoding RNA, etc.)			
1	RNA sequencing	Deep sequencing technologies provide transcriptome profiling (high-throughput) by sequencing complementary DNA; provide precise gene expression	Expensive; isoforms of the transcriptome are generally indistinguishable from each other; large RNA molecules need fragmentation before sequencing
2	Microarray platforms	Relatively straightforward and available in most laboratories	There may be a limited range of detection owing to saturation of signals, lack of normalization methods; comparing across experiments is difficult
Proteomics: Platforms measure the number of specific proteins or assess the protein structure (e.g., protein sequencing) or the interaction between specific proteins and other structures (e.g., DNA protein interactions at binding sites).			
1	2D gel electrophoresis; fluorescence 2D difference gel electrophoresis	Semiquantitative, can assess posttranslational modifications	Poor separation of acidic, basic, hydrophobic, and low-abundant proteins; fluorescent gel electrophoresis requires special equipment for visualization, and fluorophores are expensive
2	Isotope-coded affinity tag	Sensitive and reproducible, detects peptides with low expression levels, quantitation is straightforward	Some proteins are not detected
3	Multi-dimensional protein identification technology	High separation potential for large protein complex identification	Not quantitative; difficulties in identifying isoforms
4	Protein array	High-throughput and specific	Limited protein levels are more difficult to detect (sensitivity); antibodies against different proteins have different affinities; difficult to compare levels between different proteins

(continued)

Table 1 (continued)

5	Mass spectrometry	High sensitivity (for most proteins) and specificity; high-throughput, qualitative and quantitative, shows posttranslational modifications	No single method to identify all proteins; not sensitive enough to identify minor or weak spots; matrix-assisted laser desorption/ionization and electrospray ionization do not favor identification of hydrophobic peptides and basic peptides; dedicated equipment (high costs)
Metabolomics: Approaches investigate the level of known metabolites or attempt to identify unknown metabolites, biomarkers, or targets for treatment			
1	Gas/liquid chromatography (GC/LC) coupled with mass spectrometer (MS)	Semiquantitative high-throughput methods that provide targeted or untargeted analysis of metabolome. Can be designed as quantitative with appropriate standards and internal controls	Samples require fractionation before analysis; data interpretation requires skillful bioinformatics; integration of metabolomic profiles with molecular data from other high-throughput technologies is crucial and difficult; require dedicated and costly equipment, a dedicated operator, sometimes expensive standards
Epigenomics: Approaches investigate mechanisms that alter gene expression, other than changes in the DNA sequence			
1	Bisulfite pyrosequencing; methylation-specific PCR; methylation-specific real-time PCR	Are cheap and straightforward techniques. When designed appropriately, they can provide single-site resolution (pyrosequencing and several PCR versions)	Qualitative/ semiquantitative methods; low-throughput
2	High-performance liquid chromatography (HPLH), GC/MS; immunostaining with anti-5mC antibodies; digestion with methylation-sensitive enzymes	Versatile and relatively low costs (except for HPLH-GC/MS)	HPLC-GC/MS platforms: Require dedicated and costly equipment, a dedicated operator, and sometimes expensive standards
3	Microarrays NGS (Next-Generation Sequencing)platforms	Accurate, high-throughput	Expensive and laborious; data interpretation requires skillful bioinformatics

(continued)

Table 1 (continued)

4	Western blot, immunohistochemistry, immunocytochemistry, enzyme-linked immunosorbent assay; chromatin immunoprecipitation (ChIP-on-chip)	Reliable, semiquantitative methods, and fast output	Sometimes there are issues related to specificity and affinity of antigen binding sites; depending on assay, expensive experiments

ensure alignment between experimental design and the chosen methodologies. Proper data interpretation requires the application of specialized statistical and bioinformatics techniques tailored to the specific research question (Prajapat et al. 2011, 2012). This integrated approach enables a comprehensive understanding of the role of nutrients and plant secondary metabolites in modulating gene expression and contributing to human health and performance.

3 Interactions Between Plant Secondary Metabolites and Nutrigenomics

Nutrigenomics explores the complex interplay between bioactive food components and the human genome, with particular attention to plant secondary metabolites (PSMs). Compounds such as polyphenols and carotenoids exhibit the ability to regulate microRNAs (Srivastava et al. 2024a), activate transcription factors, epigenetic modifications, and broader effects on gene expression, collectively contributing to oxidative stress reduction, inflammation control, apoptosis regulation, and metabolic health improvement (Sales et al. 2014; Marwal et al. 2014b). Understanding these pathways provides an opportunity to design personalized dietary strategies aimed at disease prevention and health optimization (Mierziak et al. 2021) (Fig. 3).

Numerous studies highlight the cardiometabolic benefits of polyphenols. For instance, cocoa polyphenols have been shown to enhance endothelial function in patients with nonalcoholic steatohepatitis (Loffredo et al. 2017). Similarly, in hypertensive individuals with impaired glucose tolerance, cocoa polyphenols improved insulin sensitivity, enhanced endothelial function, and reduced both systolic and diastolic blood pressure. Grape seed extract has also demonstrated antihypertensive effects, particularly in individuals with prehypertension (Park et al. 2016). Furthermore, grape polyphenols were found to mitigate fructose-induced oxidative stress and insulin resistance in overweight or obese individuals with a family history of type 2 diabetes (Hokayem et al. 2013). Extra virgin olive oil polyphenols have also been associated with improved cardiometabolic health outcomes (Pinto et al. 2019; Storniolo et al. 2017).

The variability in individual responses to dietary metabolites has been linked to several factors, including gender, age, ethnicity, disease status, metabolic state, gut

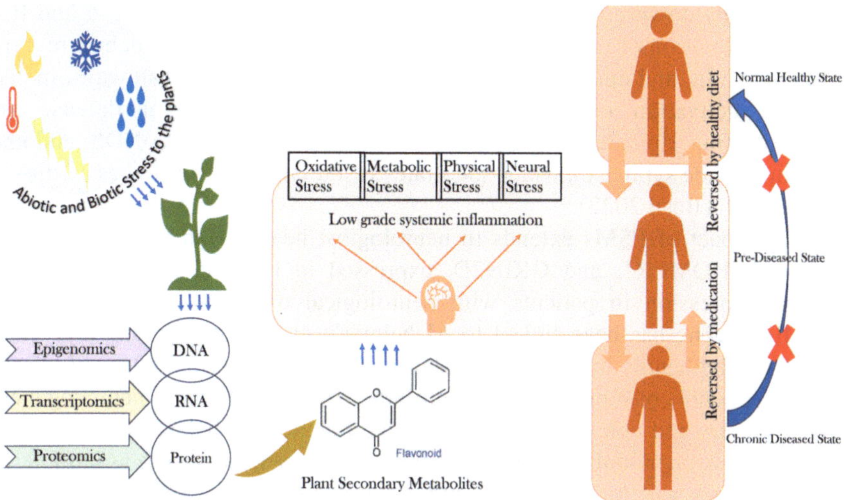

Fig. 3 Mechanism of action of nutrients at the genomic level for health improvement

microbiota composition, and gene polymorphisms (Gibney et al. 2019; Milenkovic et al. 2017). For example, polyphenols can regulate enzymes involved in the arachidonic acid cascade, influencing the biosynthesis of eicosanoids and controlling inflammatory processes associated with cardiovascular diseases. Additionally, polyphenols interact with specific receptors, such as type 2 taste receptors. These receptors, traditionally known for their role in bitter taste detection, have also been identified in extra-oral tissues, including the human intestine, where they mediate nutrient–gut interactions. These interactions have implications for gastrointestinal motility, appetite regulation, and glycemic control (Tarragon and Moreno 2020). This growing body of evidence underscores the potential of PSMs to modulate health outcomes through complex genetic and metabolic pathways, highlighting the value of nutrigenomics in advancing personalized nutrition strategies.

3.1 Gene Expression Modulation

As described earlier, plant secondary metabolites (PSMs) play a pivotal role in modulating cellular processes such as oxidative stress response, inflammation, apoptosis, and cell cycle regulation (Marwal et al. 2019, 2020). These effects are mediated through the regulation of gene expression via diverse signaling pathways (Meena et al. 2022). For example, flavonoids such as quercetin and resveratrol have been shown to activate the Nrf2 pathway, leading to the upregulation of antioxidant genes, including NQO1 (NAD(P)H dehydrogenase quinone 1) and HO-1 (heme oxygenase-1) (Jaiswal 2004). These enzymes are crucial in protecting cells against oxidative damage caused by reactive oxygen species (ROS). Curcumin, another bioactive compound, has demonstrated anti-inflammatory properties by downregulating pro-inflammatory cytokines like TNF-α and IL-6 through inhibition of the NF-κB pathway. Similarly, flavonoids such as resveratrol, quercetin, and epicatechins modulate multiple signaling pathways, including the activation of Nrf2, thereby enhancing antioxidant defense and cytoprotecting. Epicatechins have also been found to upregulate eNOS (endothelial nitric oxide synthase), which facilitates vasodilation and improves cardiovascular health (Basu 2022).

The impact of PSMs extends to neurological health. Genes such as GLUL, SLC1A2, HOMER1, and GRIN2D, expressed in immune cells, have shown altered expression in patients with neurological disorders. For instance, the HOMER1 gene has been linked to Alzheimer's disease, suggesting a potential role for polyphenols in modulating neurological conditions. Evidence indicates that polyphenol consumption may contribute to the prevention of neurodegenerative disorders and enhancement of cognitive functions (Ruskovska et al. 2021). Gene expression modulation by nutrients and PSMs begins as early as intrauterine life. Nutrients and bioactive compounds during this stage can influence gene expression or alter nucleotide sequences, potentially modifying the organism's response to toxic and infectious agents. Furthermore, inherited genetic variations may affect nutrient metabolism, influencing individual dietary requirements and susceptibility to noncommunicable diseases (NCDs). Understanding these genome-food interactions provides a foundation for personalized nutrition strategies aimed at promoting health and mitigating NCD risks (Felisbino et al. 2021). By elucidating the molecular mechanisms underlying the gene-modulating effects of PSMs, nutrigenomics holds promise for advancing personalized dietary interventions that optimize health and reduce disease susceptibility.

3.2 Epigenetic Modifications

Plant secondary metabolites (PSMs) significantly influence cellular functions not only by directly regulating gene expression but also through epigenetic mechanisms. These mechanisms include DNA methylation, histone modifications, and the regulation of noncoding RNAs, which collectively affect chromatin structure and gene activity without altering the DNA sequence itself. Such epigenetic modifications provide long-term regulation of essential cellular processes, including differentiation, cell survival, and stress response (Bind et al. 2022). One prominent example is the inhibition of histone deacetylases (HDACs) by flavonoids like epigallocatechin-3-gallate (EGCG) found in green tea, which leads to the activation of tumor-suppressor genes via histone acetylation.

Similarly, polyphenols can engage with the epigenome in diverse ways to influence gene expression. Compounds like resveratrol and curcumin can modulate histone acetylation and DNA methylation, thereby altering gene expression patterns related to cancer and inflammation. Glucosinolates such as sulforaphane inhibit DNA methyltransferases (DNMTs), preventing the silencing of tumor-suppressor genes. Additionally, polyphenols like quercetin and genistein have demonstrated the capacity to reactivate silenced tumor-suppressor genes by regulating DNA methylation, highlighting their potential long-term health benefits (Bode and Dong 2015; Pop et al. 2019; Ranjith et al. 2022). A study conducted by Felisbino et al. showed bioinformatics analyses of polyphenol-modulated mRNAs. In this process, TNF, NFKBIA, PPARG, and CCR7 genes predominantly participate as central hubs within functional clusters due to their numerous interactions and regulatory significance. These are related to inflammation, lipid metabolism, and endothelial function. Flavonoids such as luteolin and genistein modulate histone acetylation and hypermethylation, influencing gene activity. Similarly, resveratrol and folic acid can enact comparable epigenetic changes, demonstrating the extensive capacity of dietary phytochemicals to influence the epigenome.

Dietary components, either alone or in combination with environmental factors, can trigger epigenetic alterations with profound effects on cellular processes. These changes impact health and disease pathways, including hormonal regulation, cell signaling, carcinogen metabolism, apoptosis, cell cycle control, energy homeostasis, and angiogenesis. Epigenetic modifications during embryonic development can also influence the health of offspring, underscoring the transgenerational effects of diet. Consequently, understanding individual health status and aligning nutritional needs with genetic and epigenetic profiles is crucial for optimizing long-term health outcomes (Felisbino et al. 2021).

3.3 Additional Mechanisms

Plant secondary metabolites (PSMs) extend their influence on gene expression through additional mechanisms that are integral to their therapeutic potential, particularly in modulating oxidative stress, inflammation, and metabolic disorders.

Transcription Factor Activation

PSMs can modulate cellular processes by activating or inhibiting transcription factors. For instance, flavonoids such as quercetin activate the Nrf2 pathway, enhancing antioxidant defenses and protecting cells from oxidative damage. Similarly, curcumin and sulforaphane inhibit the pro-inflammatory NF-κB pathway, thereby reducing inflammation and contributing to disease prevention (Bode and Dong 2015). These interactions demonstrate the potential of PSMs to regulate oxidative stress and inflammation, two critical factors in chronic disease pathogenesis.

MicroRNA Regulation

Another significant mechanism of PSM action involves the regulation of miRNAs, small RNA (Ribonucleic acid) molecules that modulate gene expression posttranscriptionally. For example, quercetin and resveratrol enhance the expression of tumor-suppressor miRNAs, such as miR-34a, while suppressing oncogenic miRNAs like miR-21, thereby exerting anticancer effects. EGCG from green tea and curcumin regulate miRNAs linked to metabolic and inflammatory pathways, offering therapeutic promise for metabolic disorders (Cui et al. 2017; Tili et al. 2010).

Inflammation and Insulin Resistance

In the context of obesity and metabolic disorders, hypertrophied adipocytes secrete free fatty acids and pro-inflammatory cytokines, activating macrophages into a pro-inflammatory M1 phenotype. These macrophages release cytokines such as TNF-α, IL-1, and IL-6, which impair insulin signaling and exacerbate oxidative stress. This process involves the activation of kinases like IKK and JNK, which phosphorylate IRS proteins, disrupt the PI3K-Akt pathway, and promote insulin resistance. Polyphenols counter these effects by inhibiting NF-κB signaling, reducing kinase phosphorylation, and preventing NF-κB translocation to the nucleus (Nandipati et al. 2017; Khan et al. 2020). Monomeric and oligomeric flavanols from grape seeds have been shown to suppress NF-κB activity, likely through interactions with cell signaling proteins (Milenkovic et al. 2014).

Cell Adhesion and Immune Cell Transmigration

PSMs also modulate pathways governing cell adhesion and immune cell transmigration. Bioinformatic analyses have revealed that miRNA targets influenced by PSMs are involved in regulating actin cytoskeleton dynamics, cell adhesion, and vascular permeability. Dysregulated endothelial function, triggered by proatherogenic stimuli like diabetes and oxidative stress, facilitates immune cell migration into the arterial intima, initiating vascular inflammation (Incalza et al. 2018). Polyphenols, by regulating miRNA expression, mitigate these effects and reduce the risk of vascular dysfunction. Flavanol metabolites, for instance, decrease immune cell adhesion and transmigration in endothelial cells, demonstrating their protective role in cardiovascular health (Milenkovic et al. 2022).

Neuronal and Nonneuronal Pathways

PSMs also influence pathways associated with neuronal function, such as axon guidance and synaptic signaling. Certain genes expressed in neuronal cells, such as PPP2R3A, are also found in nonneuronal cells like blood cells, where they can regulate inflammatory responses. This gene, for instance, promotes NF-κB activation via small GTPase proteins, enhancing the expression of adhesion molecules and inflammatory mediators. These multifaceted mechanisms highlight the potential of PSMs to regulate critical molecular pathways. Their influence on transcription factors and miRNAs underpins their ability to prevent or mitigate a range of health disorders, from inflammation and metabolic syndromes to vascular dysfunction and neurodegenerative diseases. Understanding these interactions can pave the way for developing targeted nutritional and therapeutic strategies (Ruskovska et al. 2021).

4 Role of Nutrigenomics in Athletic Performance and Cognitive Function

The connection between diet and athletic performance is well-established, with plant secondary metabolites (PSMs) playing a significant role in improving endurance, energy metabolism, and recovery.

4.1 Athletic Performance

Antioxidant and Anti-inflammatory Effects

Intense physical activity generates reactive oxygen species (ROS), which can lead to oxidative stress, muscle damage, and inflammation. PSMs, such as polyphenols and flavonoids, are known for their antioxidant properties and can neutralize ROS, mitigating exercise-induced oxidative damage. For example, quercetin has been shown to alleviate muscle fatigue and inflammation, thereby potentially enhancing endurance performance (De Oliveira et al. 2018). Curcumin, a potent anti-inflammatory compound derived from turmeric, blocks pro-inflammatory pathways like NF-kB and COX-2, thereby reducing muscle soreness and enhancing recovery after intense physical exertion (Drobnic et al. 2014). Berries, rich in anthocyanins, also provide significant health benefits due to their antioxidant and anti-inflammatory effects. Anthocyanins help combat chronic inflammation and improve immune function, reducing the occurrence of common illnesses that could otherwise impair athletic performance (Naselli et al. 2024).

Enhanced Energy Metabolism

PSMs also influence energy metabolism by impacting mitochondrial function, a crucial component for energy production during exercise. Resveratrol, found in grapes and red wine, activates SIRT1, an enzyme that promotes mitochondrial biogenesis and enhances energy generation in muscle cells, thereby improving aerobic capacity and endurance (Dolinsky et al. 2012). Additionally, catechins in green tea have been shown to enhance fat oxidation during exercise, preserving glycogen stores and prolonging endurance performance (Thielecke and Boschmann 2009). This modulation of energy metabolism is crucial for athletes aiming to optimize their physical performance over extended periods.

Modulation of Nitric Oxide (NO) Production

Nitric oxide (NO), produced by endothelial nitric oxide synthase (eNOS) in blood vessels, plays a vital role in promoting vasodilation, increasing blood flow, and enhancing oxygen delivery to muscles during exercise. Dietary nitrates, such as those found in beetroot and pomegranate, have been shown to increase NO production, which improves blood flow and enhances exercise performance. Beetroot juice, for example, has been found to reduce oxygen consumption during physical activity, thereby improving endurance (Jones 2014). Pomegranate juice also contributes to better postexercise recovery by increasing NO availability, supporting muscle oxygenation, and reducing muscle damage (Trombold et al. 2011). Moreover, PPARs (peroxisome proliferator-activated receptors), activated by long-chain fatty acids and certain PSMs regulate cellular metabolism and lipid oxidation, further improving cardiovascular health and athletic performance (Kaur et al. 2018).

4.2 Cognitive Function

Neuroprotection and Neuroinflammation

In addition to their benefits for athletic performance, PSMs are also crucial for maintaining cognitive health. Chronic neuroinflammation and oxidative stress are key contributors to neurodegenerative diseases like Alzheimer's and Parkinson's. PSMs, including curcumin, carotenoids, and polyphenols, have demonstrated therapeutic potential by targeting pathways involved in neuroinflammation and oxidative stress. Pantothenic acid, a B vitamin involved in neurotransmitter synthesis, highlights the importance of micronutrients in maintaining neuronal function and cognitive performance. Deficiencies in such nutrients can lead to symptoms like fatigue, memory impairment, and neuropathy, underlining the impact of nutrition on brain health (Virmani et al. 2013). Flavonoids, such as anthocyanins in berries, exhibit neuroprotective effects by neutralizing free radicals and reducing oxidative stress in the brain, which is essential for maintaining cognitive function. Studies have shown that diets rich in anthocyanins improve memory and cognitive performance in animal models (Apostolova and Victor 2015). Other compounds, such as epicatechin from cocoa, also support cognitive function by increasing acetylcholine levels, a neurotransmitter crucial for memory and learning (Bisson et al. 2008).

Reducing Neuroinflammation

Neuroinflammation plays a significant role in cognitive decline and neurodegenerative diseases. Curcumin and resveratrol are well-known for their ability to reduce neuroinflammation by lowering pro-inflammatory cytokines and decreasing microglial activation, which protects neurons and supports cognitive health (Belviranli et al. 2013; Braidy et al. 2016). Additionally, flavonoids have been shown to enhance the expression of brain-derived neurotrophic factor (BDNF), which is essential for neurogenesis, synaptic plasticity, and memory formation (Grassi et al. 2010). By modulating inflammatory pathways and promoting neuroplasticity, PSMs offer potential for improving cognitive function and preventing age-related cognitive decline. The incorporation of plant secondary metabolites into an athlete's diet offers a promising avenue for enhancing performance, improving recovery, and supporting cognitive health. By influencing key biological processes such as energy metabolism, oxidative stress, inflammation, and neuroprotection, PSMs provide multifaceted benefits that can optimize both physical and mental performance. The ongoing research in nutrigenomics highlights the potential for personalized nutrition to maximize the health benefits of these bioactive compounds based on an individual's genetic makeup.

5 Conclusion and Future Prospective

In recent years, there has been a marked increase in public awareness regarding health and diet. This shift underscores the importance of relying on nutrient-rich foods rather than conventional medicines for disease prevention and overall physical fitness enhancement. Plants offer an optimal source of essential nutrients due to their rich content of phytochemicals and metabolites. Advances in genetic improvement techniques have further enabled the augmentation of these beneficial compounds. Concurrently, significant progress in plant metabolomics techniques, such as LC-MS, GC-MS, and NMR (Nuclear magnetic resonance), has facilitated the measurement and identification of a broad spectrum of nontargeted metabolites. However, the naturally low accumulation of these metabolites in plants necessitates innovative approaches, such as plant metabolic engineering and molecular breeding, to enhance their production. Understanding the genetic mechanisms underlying the biosynthesis of functional food metabolites is crucial. Biotechnological strategies, including cisgenics, genome editing with TALENs, and the CRISPR-Cas9 system, have emerged as promising tools for producing plants enriched with health-promoting compounds and metabolites (Verma et al. 2024). Bioactive compounds such as resveratrol, quercetin, genistein, catechins, curcumin, and anthocyanins have demonstrated their efficacy in modulating genes involved in oxidative stress, inflammation, apoptosis, and molecular metabolism, thereby contributing to the prevention and treatment of various chronic diseases. Despite this progress, existing studies often focus narrowly on changes in gene expression levels, leaving a significant gap in understanding the precise modes of action of these bioactives.

Moreover, while many genes are directly associated with single metabolic pathways or disease prevention, there is a pressing need to explore their roles in broader metabolic processes like inflammation, apoptosis, and oxidative stress. These processes may indirectly influence the prevention and treatment of other diseases. Genetic variants also exhibit differential responses to bioactive compounds, emphasizing the need for more comprehensive research. Additionally, as many genes are implicated in multiple chronic diseases, further studies in this area could yield valuable insights. Integrating bioinformatics analyses has proven instrumental in adding value to existing data (Srivastava et al. 2024b), offering a deeper understanding of the relationship between nutritional metabolites and their target genes under nutrigenetic studies.

The integration of developments in nutrition science with advancements in communication and marketing has facilitated the emergence of personalized nutritional counseling based on nutrigenomics. This approach marks a significant departure from standardized diets, offering tailored dietary recommendations informed by an individual's genetic profile. Personalized counseling not only encourages healthier lifestyle choices but also supports more accurate disease diagnoses, delays the progression of chronic illnesses, and enhances treatment outcomes. Despite its availability for clinical applications, nutrigenomics is still underutilized in healthcare settings. Continued research is necessary to strengthen the connections between genetic profiles, dietary habits, environmental factors, and disease symptoms. The integration of emerging technologies into health sciences has paved the way for innovative business models centered on personalized nutritional counseling based on DNA analysis. This field's expansion relies on the collaborative efforts of scientists and healthcare professionals who drive advancements in nutrigenomics and develop tools that promote improved quality of life and dietary practices. As the field evolves, it holds immense potential to transform healthcare by fostering a more individualized approach to nutrition and disease management, ultimately contributing to a healthier global population.

Acknowledgments The authors are thankful to Mohanlal Sukhadia University, Udaipur, for providing the necessary facilities during the course of study.

Funding This research did not receive any specific grant from funding agencies in the public, commercial, or not-for-profit sectors.

Declaration of Competing Interest The authors declare that there are no conflicts of interest. They agree to be accountable for all aspects of the work and ensure that questions related to the accuracy or integrity of any part of the work are appropriately investigated and resolved.

References

Apostolova N, Victor VM (2015) Molecular strategies for targeting antioxidants to mitochondria: therapeutic implications. Antioxid Redox Signal 22(8):686–729

Basu A (2022) The power of tea polyphenols: benefits for health and wellness. Codeage. Retrieved from https://www.codeage.com/blogs/education/the-power-of-tea-polyphenols-benefits-for-health-and-wellness

Belviranl M, Okudan N, Atalık KEN, Öz M (2013) Curcumin improves spatial memory and decreases oxidative damage in aged female rats. Biogerontology 14(2):187–196

Bind S, Bind S, Sharma AK, Chaturvedi P (2022) Epigenetic modification: a key tool for secondary metabolite production in microorganisms. Front Microbiol 13:Article 784109

Bisson JF, Nejdi A, Rozan P, Hidalgo S, Lalonde R, Messaoudi M (2008) Effects of long-term administration of a cocoa polyphenolic extract (Acticoa powder) on cognitive performances in aged rats. Br J Nutr 100(1):94–101

Bode AM, Dong Z (2015) Chemopreventive effects of licorice and its components. Curr Pharmacol Rep 1(1):60–71

Bösch ES, Spörri J, Scherr J (2024) Genetic variants affecting iron metabolism in healthy adults: a systematic review to support personalized nutrition strategies. Nutrients 16(22):3793

Braidy N, Jugder BE, Poljak A, Jayasena T, Mansour H, Mohammad Nabavi S, Grant R (2016) Resveratrol as a potential therapeutic candidate for the treatment and management of Alzheimer's disease. Curr Top Med Chem 16(17):1951–1960

Cui J, Zhou B, Ross SA, Zempleni J (2017) Nutrition, microRNAs, and human health. Adv Nutr 8(1):105–112

De Oliveira ACL, Tabrez S, Shakil S, Khan MI, Asghar MN, Matias BD, de Carvalho Melo-Cavalcante AA (2018) Mutagenic, antioxidant, and wound healing properties of *Aloe vera*. J Ethnopharmacol 227:191–197

Debras C, Chazelas E, Srour B, Kesse-Guyot E, Julia C, Zelek L, Hercberg S (2020) Total and added sugar intakes, sugar types, and cancer risk: results from the prospective NutriNet-Santé cohort. Am J Clin Nutr 112(6):1267–1279

Dolinsky VW, Jones KE, Sidhu RS, Haykowsky M, Czubryt MP, Gordon T, Dyck JR (2012) Improvements in skeletal muscle strength and cardiac function induced by resveratrol during exercise training contribute to enhanced exercise performance in rats. J Physiol 590(11):2783–2799

Drobnic F, Riera J, Appendino G, Togni S, Franceschi F, Valle X, Tur J (2014) Reduction of delayed onset muscle soreness by a novel curcumin delivery system (Meriva®): a randomized, placebo-controlled trial. J Int Soc Sports Nutr 11(1):1–10

Felisbino K, Granzotti JG, Bello-Santos L, Guiloski IC (2021) Nutrigenomics in regulating the expression of genes related to type 2 diabetes mellitus. Front Physiol 12:699220

Fernie AR, Yan J (2019) De novo domestication: an alternative route toward new crops for the future. Mol Plant 12(5):615–631

Fialho BE, Moreno FS, Ong TPP (2008) Nutrition in the post-genomics: fundamentals and applications of omics tools. J Nutr 21(6):757–766

Gibney ER, Milenkovic D, Combet E, Ruskovska T, Greyling A, Gonzalez-Sarrias A, Rodriguez-Mateos A (2019) Factors influencing the cardiometabolic response to (poly)phenols and phytosterols: a review of the cost action positive activities. Eur J Nutr 58(1):37–47

Grassi D, Desideri G, Ferri C (2010) Blood pressure and cardiovascular risk: what about cocoa and chocolate? Arch Biochem Biophys 501(1):112–115

Hokayem M, Blond E, Vidal H, Lambert K, Meugnier E, Feillet-Coudray C, Morand C (2013) Grape polyphenols prevent fructose-induced oxidative stress and insulin resistance in first-degree relatives of type 2 diabetic patients. Diabetes Care 36(6):1454–1461

Incalza MA, D'Oria R, Natalicchio A, Perrini S, Laviola L, Giorgino F (2018) Oxidative stress and reactive oxygen species in endothelial dysfunction associated with cardiovascular and metabolic diseases. Vasc Pharmacol 100:1–19

Jain K, Desai N, Sharma K, Marwal A (2021) Development and screening of byproduct for its secondary metabolites, antioxidant and anti-diabetic potential from anthracnose-infected fruits of pomegranate: a sustainable approach. 3 Biotech 11:1–11

Jaiswal AK (2004) Nrf2 signaling in coordinated activation of antioxidant gene expression. Free Radic Biol Med 36(10):1199–1207

Jones AM (2014) Dietary nitrate supplementation and exercise performance. Sports Med 44(Suppl 1):35–45

Jukanti AK, Pautong PA, Liu Q, Sreenivasulu N (2020) Low glycemic index rice—a desired trait in starchy staples. Trends Food Sci Technol 106:132–149

Kaur KK, Allahbadia G, Singh M (2018) Impact of nutrigenomics on various metabolic disorders in relation to lifestyle alteration. Austin J Nutr Food Sci 6(1):1100

Khan H, Ullah H, Castilho P, Gomila AS, D'Onofrio G, Filosa R, Wang F, Nabavi SM, Daglia M, Silva AS (2020) Targeting NF-kappaB signaling pathway in cancer by dietary polyphenols. Crit Rev Food Sci Nutr 60(16):2790–2800

Kumar R, Marwal A, Gaur RK (2017) Identification of potential reference miRNA for qRT-PCR studies in cancers using miRNA-seq data. Int J Curr Microbiol App Sci 6(12):1327–1333

Kumar R, Bhanti P, Marwal A, Gaur RK (2019) Gene expression-based supervised classification models for discriminating early-and late-stage prostate cancer. Proc Natl Acad Sci India Sect B Biol Sci 90:541–565

Lakra N, Mahmood S, Marwal A, Sudheep NM, Anwar K (2019) Bioengineered plants can be an alternative source of omega-3 fatty acids for human health. In: Plant and human health, volume 2: Phytochemistry and molecular aspects, Springer, Cham. pp 361–382

Loffredo L, Baratta F, Ludovica P, Battaglia S, Carnevale R, Nocella C, Angelico F (2017) Effects of dark chocolate on endothelial function in patients with non-alcoholic steatohepatitis. Nutr Metabol Cardiovasc Dis 28(2):143–149

Mahan LK, Scott-Stump S (2005) Food, nutrition, & diet therapy, 6th edn. Saunders USA.

Maheshwari Y, Tailor S, Marwal A, Mishra A (2023) Plant-microbe interactions to reduce salinity stress in plants for the improvement of the agricultural system. In: Plant-microbe interaction-recent advances in molecular and biochemical approaches. Academic Press, Elsevier. pp 297–309

Marwal A, Gaur RK (2020) Molecular markers: tool for genetic analysis. In: Animal biotechnology. Academic Press, Elsevier. pp 353–372
Marwal A, Sahu A, Gaur RK (2014a) New insights in the functional genomics of plants responding to abiotic stress. Molecular approaches in plant abiotic stress. CRC Press, Science Publishers, Taylor & Francis Group, Boca Raton, pp 158–180
Marwal A, Verma SS, Trivedi S, Prajapat R (2014b) Reactive oxygen species: its effects on various diseases. J Adv Biotechnol 3(1):132–134
Marwal A, Kumar R, Verma RK, Mishra M, Gaur RK, Khurana SP (2019) Genomics and molecular mechanisms of plant's response to abiotic and biotic stresses. In: Plant biotechnology: Progress in genomic era, pp 131–146. Springer Singapore.
Marwal A, Srivastava AK, Gaur RK (2020) Improved plant tolerance to biotic stress for agronomic management. Agri 9(2):84–100
Meena M, Yadav G, Sonigra P, Nagda A, Mehta T, Swapnil P, Marwal A (2022) Role of elicitors to initiate the induction of systemic resistance in plants to biotic stress. Plant Stress 5:100103
Mierziak J, Kostyn K, Boba A, Czemplik M, Kulma A, Wojtasik W (2021) Influence of the bioactive diet components on the gene expression regulation. Nutrients 13(11):3673
Milenkovic D, Vanden Berghe W, Boby C, Leroux C, Declerck K, Szarcvel Szic K, Heyninck K, Laukens K, Bizet M, Defrance M (2014) Dietary flavanols modulate the transcription of genes associated with cardiovascular pathology without changes in their DNA methylation state. PLoS One 9(5):e95527
Milenkovic D, Morand C, Cassidy A, Konic-Ristic A, Tomas-Barberan F, Ordovas JM, Rodriguez-Mateos A (2017) Interindividual variability in biomarkers of cardiometabolic health after consumption of major plant-food bioactive compounds and the determinants involved. Adv Nutr 8(4):558–570
Milenkovic D, Rodriguez-Mateos A, Lucosz M, Istas G, Declerck K, Sansone R, Deenen R, Köhrer K, Corral-Jara KF, Altschmied J, Haendeler J (2022) Flavanol consumption in healthy men preserves integrity of immunological-endothelial barrier cell functions: Nutri (epi) genomic analysis. Mol Nutr Food Res 66(21):2100991
Mishra A, Marwal A, Tailor S, Jain K, Malik A, Suthar M, Meena M (2024) Overview of cell signaling response under plant stress. In: Molecular dynamics of plant stress and its management. Springer Nature Singapore, Singapore, pp 121–158
Nandipati KC, Subramanian S, Agrawal DK (2017) Protein kinases: mechanisms and downstream targets in inflammation-mediated obesity and insulin resistance. Mol Cell Biochem 426(1–2):27–45
Naselli F, Cardinale P, Vasto S, Proia P, Baldassano S, Caradonna F (2024) Challenges and future perspectives of sustainable supplements, functional foods, and nutrigenomics in athletic performance. Human Movem 25(1):1–15
Park E, Edirisinghe I, Choy YY, Waterhouse A, Burton-Freeman B (2016) Effects of grape seed extract beverage on blood pressure and metabolic indices in individuals with pre-hypertension: a randomised, double-blinded, two-arm, parallel, placebo-controlled trial. Br J Nutr 115(2):226–238
Pinto X, Fanlo-Maresma M, Corbella E, Corbella X, Mitjavila MT, Moreno JJ, Corella D (2019) A Mediterranean diet rich in extra-virgin olive oil is associated with a reduced prevalence of nonalcoholic fatty liver disease in older individuals at high cardiovascular risk. J Nutr 149(11):1920–1929
Pop S, Enciu AM, Tarcomnicu I, Gille E, Tanase C (2019) Phytochemicals in cancer prevention: modulating epigenetic alterations of DNA methylation. Phytochem Rev 18(4):1005–1024
Prajapat R, Marwal A, Sahu A, Gaur RK (2011) Phylogenetics and in silico docking studies between coat protein of Mimosa yellow vein virus and whey α-lactalbumin. Am J Biochem Mol Biol 1(3):265–274
Prajapat R, Marwal A, Sahu AK, Gaur RK (2012) Molecular in silico structure and recombination analysis of betasatellite in *Calotropis procera* associated with begomovirus. Arch Phytopathol Plant Protect 45(16):1980–1990

Ranjith S, Sathyan A, Duraisamy S, Peter AS, Marwal A, Jain K, Chidambaram P, Kumarasamy A (2022) Exploring a computational method for evaluating the Epinecidin-1 and its variants binding efficacy with breast cancer receptor (HER-2). Int J Pept Res Ther 28(4):118

Ruskovska T, Budić-Leto I, Corral-Jara KF, Ajdžanović V, Arola-Arnal A, Bravo FI, Deligiannidou GE, Havlik J, Janeva M, Kistanova E (2021) Systematic bioinformatic analyses of nutrigenomic modifications by polyphenols associated with cardiometabolic health in humans—evidence from targeted nutrigenomic studies. Nutrients 13(7):2326

Sales NMR, Pelegrini PB, Goersch MC (2014) Nutrigenomics: definitions and advances of this new science. J Nutr Metabol 2014:Article 202759

Simsek M, Whitney K (2024) Examination of primary and secondary metabolites associated with a plant-based diet and their impact on human health. Foods 13(7):1020

Singh NA, Marwal A, Goyal J, Rai N (2024) Bioactive molecules derived from Extremophilic fungi and their agro-biotechnological application. In: Extremophiles for sustainable agriculture and soil health improvement. Springer Nature Switzerland, Cham, pp 137–162

Sreenivasulu N, Fernie AR (2022) Diversity: current and prospective secondary metabolites for nutrition and medicine. Curr Opin Biotechnol 74:164–170. Available online 20th November 2021

Srivastava A, Pandey V, Singh N, Marwal A, Shahid MS, Gaur RK (2024a) In silico identification of papaya genome-encoded microRNAs to target begomovirus genes in papaya leaf curl disease. Front Microbiol 15:1340275

Srivastava A, Pandey V, Marwal A, Ali A, Gaur RK (2024b) In-silico prediction and validation of *Carica papaya* protein domains interaction with the Papaya leaf curl virus and associated betasatellite encoded protein. Discov Appl Sci 6(6):287

Storniolo CE, Casillas R, Bullo M, Castaner O, Ros E, Saez GT, Fito M (2017) A Mediterranean diet supplemented with extra virgin olive oil or nuts improves endothelial markers involved in blood pressure control in hypertensive women. Eur J Nutr 56(1):89–97

Stover PJ (2006) Influence of human genetic variation on nutritional requirements. Am J Clin Nutr 83(2):436S–442S

Swapnil P, Meena M, Singh SK, Dhuldhaj UP, Marwal A (2021) Vital roles of carotenoids in plants and humans to deteriorate stress with its structure, biosynthesis, metabolic engineering and functional aspects. Curr Plant Biol 26:100203

Tarragon E, Moreno JJ (2020) Polyphenols and taste 2 receptors: physiological, pathophysiological, and pharmacological implications. Biochem Pharmacol 178:114086

Teoh ES (2015) Secondary Metabolites of Plants. In: Medicinal Orchids of Asia. Springer, Cham. pp 59–73. https://doi.org/10.1007/978-3-319-24274-3_5

Thielecke F, Boschmann M (2009) The potential role of green tea catechins in the prevention of the metabolic syndrome: a review. Phytochemistry 70(1):11–24

Tili E, Michaille JJ, Alder H, Volinia S, Delmas D, Latruffe N, Croce CM (2010) Resveratrol modulates the levels of microRNAs targeting genes encoding tumor-suppressors and effectors of TGFβ signaling pathway in SW480 cells. Biochem Pharmacol 80(12):2057–2065

Trombold JR, Barnes JN, Critchley L, Coyle EF (2011) The effect of pomegranate juice supplementation on strength and soreness after eccentric exercise. J Strength Cond Res 25(7):1782–1788

Verma RK, Mishra M, Marwal A, Akhtar A, Manchev N, Petrov RK (2024) CRISPR/Cas9 protection system for Chilli Leaf Curl Virus. Acta Microbiol Bulg 40(1):112–115

Virmani A, Pinto L, Binienda Z, Ali S (2013) Food, nutrigenomics, and neurodegeneration—neuroprotection by what you eat! Mol Neurobiol 48(2):353–362

Harnessing Potential Benefits of Plant Molecular Farming for the Challenging Production of Next-Generation Vaccines

Ayushi Malik, Khushboo Jain, Mayank Suthar, Anita Mishra, Sunita Verma, Rajarshi Kumar Gaur, and Avinash Marwal

Abstract

Throughout the annals of history, botanical entities have played an integral role in combating a plethora of diseases. While synthetic treatments have been efficacious in addressing the escalating prevalence of acute and chronic afflictions, the attendant adverse effects pose significant risks, particularly among the elderly. In stark contrast, natural plant-based remedies have evinced efficacy in disease management with negligible or no adverse effects. The advent of plant molecular farming has presented a robust solution to confront these global health challenges. Nevertheless, this approach is not devoid of significant challenges. This chapter endeavours to underscore the potential advantages of cultivating next-generation vaccines through plant molecular farming, providing salient insights into the attendant challenges and benefits. Addressing these impediments and

A. Malik · M. Suthar · A. Marwal (✉)
Nanoparticle Synthesis and Bioinformatics Laboratory, Department of Biotechnology, Vigyan Bhawan, Block – B, New Campus, Mohanlal Sukhadia University, Udaipur, Rajasthan, India

K. Jain
Department of Biotechnology, Faculty of Science, MVSC, Janardhan Rai Nagar Rajasthan Vidhyapeeth (Deemed to be University), Udaipur, Rajasthan, India

A. Mishra
Department of Science (Biotechnology), Biyani Girls College, Jaipur, Rajasthan, India

S. Verma
Manas Enclave Residential Society, Indirapuri Street, Bhuwana, Udaipur, Rajasthan, India

R. K. Gaur
Department of Biotechnology, Deen Dayal Upadhyaya Gorakhpur University, Gorakhpur, Uttar Pradesh, India

Uttar Pradesh Council of Agricultural Research, Uttar Pradesh, India

R. K. Gaur et al. (eds.), *Plant Molecular Farming*,
https://doi.org/10.1007/978-981-95-3823-2_10

leveraging the advantages of plant-based molecular farming holds the promise of propelling the research community into uncharted territories.

Keywords

Disease management · Treatment · Natural remedies · Next-generation vaccine · Synthetic treatments

1 Introduction

Plant molecular farming is a revolutionary biotechnology technique that uses plants as effective platforms to produce a wide range of molecules with biological and economic significance. These consist of human blood products, growth regulators, monoclonal antibodies, biopharmaceuticals, enzymes, polymers, vaccines, hormones, and antimicrobial peptides (AMPs). The idea, which has changed a lot since the initial plant transformation trials, depends on the capacity to use plants as biofactories for valuable biomolecules (Shanmugaraj et al. 2020). The origins of this technology can be found in early genetic engineering achievements, such as the cloning of important proteins like insulin, somatotropin, and human somatostatin (Satheeshkumar 2020). In 1974, the idea of transforming plant cells was initially put forth, and by the middle of the 1980s, De Block et al. reported the first effective *Agrobacterium*-mediated plant transformation. Originally, the goal of plant genetic engineering was to improve agricultural features, including optimizing amino acid composition, modifying lipid content, and increasing yield (Xing et al. 2006; Swapnil et al. 2021). It was used concurrently to create disease resistance, herbicide resistance, and biological insecticides (Srivastava et al. 2024; Nehra et al. 2014). These initiatives eventually grew to incorporate complex manufacturing lines and advanced plant transformation systems, making it possible to use plants for external protein or metabolite production.

Various techniques, including *Agrobacterium*-mediated transformation, direct gene delivery methods, and pollen-mediated procedures, were developed because of early research aimed at increasing transformation efficiency and expanding the variety of host plants. In recent decades, transient expression systems have emerged as a prominent alternative to stable transformation methods. These systems, often based on plant viruses and delivered via *Agrobacterium* infiltration, have proven advantageous by reducing the time and complexity associated with plant regeneration and somaclonal variation (Gao et al. 2007). Over 30 years of advancements have resulted in a robust and versatile molecular farming technology, which retains its initial promise as a cost-effective and scalable platform. Higher plants provide significant advantages even when other production methods, such as bacteria, fungi, and transgenic animals, are used for protein synthesis. The absence of human disease-causing agents (virions and prions) that could infect the finished product, the availability of established agricultural infrastructure, and the significantly lower production costs when compared to fermentation or bioreactors are some of the benefits. Furthermore, plants naturally produce eukaryotic proteins with precise folding, glycosylation, and biological activity. They also offer stable storage conditions for proteins, which lowers the danger of degradation (Horn et al. 2004). The development of edible vaccines is one area receiving heightened interest in plant molecular farming. Immunizations have been essential in preventing and eliminating infectious diseases like rinderpest and smallpox, as well as drastically lowering illnesses like polio, measles, and diphtheria. For instance, in the United States, 13 childhood vaccines have prevented an estimated 20 million diseases and 40,000 deaths, with economic savings reaching $69 billion. Globally, vaccines deployed in low- and middle-income countries have achieved even greater benefits, with reductions in illness-related costs and broader economic impacts amounting to $1.53 trillion (Chung et al. 2022).

Charles Arntzen and Hugh Mason developed the idea of edible vaccinations, which have several benefits over conventional needle-based vaccination techniques. Since mucosal tissues are often the primary points of entry for many infections, these vaccines elicit strong immune responses in these tissues. The cost-effectiveness, ability to produce proteins with appropriate folding and glycosylation, and lack of contamination issues associated with mammalian systems are all advantages of using plant-based systems for vaccine production. Biologics manufacturing in plants is thought to be 32% less expensive than in conventional systems, mostly due to decreased site and batch costs. The stability and usefulness of biologics derived from plants have been significantly improved by developments in glycoengineering. For instance, Elelyso, a medicinal enzyme made in plants, works better than its mammalian counterpart because of its specific glycosylation profile (Margolin et al. 2020). Recent advances in molecular farming have made it possible to produce novel platforms for DNA and mRNA vaccines as well as complicated vaccine candidates, such as envelope glycoproteins from human immunodeficiency virus (HIV)-1, HPV, West Nile virus, and dengue virus. Using modified coat proteins from viruses such as the tobacco mosaic virus (TMV), the latter can be encapsulated in plant-derived systems, offering a novel and scalable method of vaccine production (Ponndorf et al. 2021). Therefore, plant molecular farming, supported by decades of research and technology developments, has enormous potential to address global health issues and transform the biologics production process. This chapter explores the development, methods, and uses of this innovative technology, highlighting how it has the potential to revolutionize the pharmaceutical and vaccine manufacturing industries. This chapter also delves into the success stories of commercialized vaccines.

2 Basic Principles of Plant Molecular Farming in Vaccine Production

Plant molecular farming (PMF) has been advocated since the 1990s as a rapid, cost-effective, and particularly safe alternative to the cultivation of bacteria or animal cells to produce biopharmaceutical proteins. A variety of plant species have been investigated to generate a broad range of protein-based drug candidates. The inherent safety of these products is frequently underscored as a significant advantage of PMF, given that plant viruses do not replicate in human hosts and vice versa.

Nevertheless, a more nuanced examination of this principle is warranted when considering other pathogens, as toxic compounds can present risks even in the absence of replication. It is also imperative to assess the risks associated with the host system, such as the presence of toxic secondary metabolites (Jain et al. 2021), as well as the production methodology, for example, transient expression through bacterial infiltration, which substantially elevates the endotoxin load (Buyel 2023).

2.1 Traditional Vaccine Production Methods

In the early 1980s, Fritz Kreuzaler was inspired by a presentation regarding plant viruses at the Max-Planck Institute in Cologne to formulate the concept of antibody-mediated disease resistance through the expression of pathogen-specific antibodies in plants (as given in Fig. 1). His contemporaries regarded the assembly of functional full-length antibodies from two heavy chains and two light chains, without the aid of mammalian chaperones, as improbable. Despite this scepticism, Kreuzaler and his team successfully confirmed the presence of an assembled full-size antibody following the injection of complementary DNAs (cDNAs) encoding the heavy and light chains of a murine immunoglobulin M (IgM) into the nuclei of *Acetabularia mediterranea*, an algal species characterized by large cells that serve as a model organism in cell biology. To facilitate the accumulation of the recombinant proteins within the cytosol, the sequences encoding the N-terminal signal peptide, which is essential for co-translational translocation into the endoplasmic reticulum, were excised from the antibody genes. The existence of the assembled antibody was validated using an anti-idiotope antibody that specifically binds to the

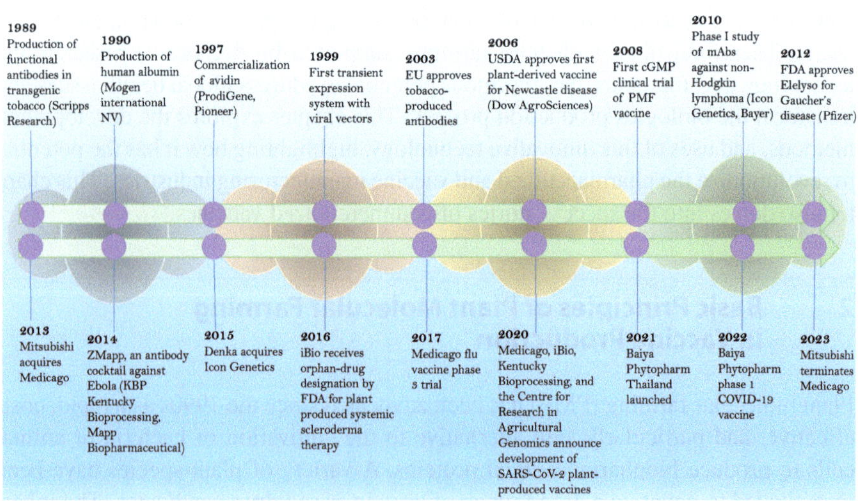

Fig. 1 Historical evidence of plant-based vaccine production

assembled immunoglobulin, distinguishing it from the individual heavy and light chains.

This pioneering research was initially presented as a doctoral thesis (Stieger 1987) and subsequently published in a journal article (Stieger et al. 1991). The initial demonstration established that plant cells possess the capability to produce complex, multimeric mammalian proteins. Subsequently, a complete murine IgG1 antibody was synthesized in transgenic tobacco plants, and recombinant human serum albumin was expressed in transgenic potato and tobacco plants, in addition to tobacco cell cultures derived from these transgenic lines. In these investigations, the recombinant proteins incorporated N-terminal signal peptides, which facilitated their routing to the secretory pathway, thereby enhancing protein folding and stability, and subsequently promoting higher accumulation levels. These pioneering studies laid the groundwork for the concept of molecular farming, which involves utilizing plants as platforms to produce valuable recombinant proteins. A plethora of subsequent studies emerged, predominantly concentrating on pharmaceutical proteins. This approach was based on the premise that the production of high-value proteins in plants would prove more cost-effective than the utilization of microbial or mammalian cell systems, which require expensive fermentation infrastructure and clean-room facilities to meet good manufacturing practices (GMP). The cultivation of plants in greenhouses or open fields is indeed significantly less costly and more scalable than fermentation systems. Nonetheless, it was soon recognized that open-field cultivation presents challenges in controlling product quality and safety, while downstream processing—the extraction and purification of target proteins from the plant matrix—constitutes a notable cost driver necessitating process optimization, as discussed further. Various proteins have been successfully produced in plants, encompassing pharmaceutical proteins such as antibodies, vaccines, hormones, and enzymes, as well as proteins designated for diagnostic, research, and cosmetic applications. A distinguishing feature of plant-based protein expression is the extensive variety of plant species and systems utilized for production. While traditional biopharmaceutical manufacturing relies on a limited array of well-established platforms, such as the bacterium *Escherichia coli*, the yeast *Saccharomyces cerevisiae*, Chinese hamster ovary (CHO) cells, murine NS0, Sp2/0 cells, and human embryonic kidney cells, numerous plant species have been proposed as potential production platforms, including tobacco, potato, tomato, alfalfa, safflower, carrot, lettuce, strawberry, moss, duckweed, maize, wheat, and rice. Furthermore, the systems employed range from whole plants to organ cultures, such as hairy roots, and cell suspension cultures. Various expression strategies can be effectively employed in protein production, including: (i) Stable transgenic or transplastomic lines that afford high scalability, alongside transient expression in non-transgenic plants for rapid manufacturing; (ii) Constitutive, tissue-specific, and inducible expression methods; and (iii) Diverse protein targeting approaches to regulate product accumulation, address post-translational modifications, and facilitate downstream processing (Schillberg et al. 2013).

The comprehensive review of molecular farming platforms and their corresponding product categories has been conducted. The extensive assortment of plant species, systems, expression methodologies, and protein targeting strategies suggests that a suitable platform is likely available for nearly any conceivable protein product. However, the systematic evaluation of various platforms to identify the most suitable option on a case-by-case basis is a labour-intensive and time-consuming endeavour, and the lack of a standardized platform has impeded efforts to optimize productivity and downstream processing. Consequently, the molecular farming community has begun to consolidate around a limited number of platforms, with tobacco (*Nicotiana tabacum*) and its close relative *N. benthamiana* emerging as the predominant species utilized (Spiegel et al. 2018).

2.1.1 Molecular Farming Leading Historically with Demonstrative Success

In the realm of pharmaceutical proteins, additional investigations were often conducted using animal models. For instance, the Hepatitis C virus E1E2 heterodimer, produced transiently in lettuce, elicited immune responses in mice, while domain III of the West Nile virus envelope protein, produced in *N. benthamiana,* offered protection against lethal infections in mice. The transient co-expression of dengue virus structural proteins and truncated non-structural RNA-dependent RNA polymerase in *N. benthamiana* resulted in the formation of virus-like particles (VLPs), which prompted a robust antibody response in mice. These findings and similar studies have clearly demonstrated that plants can serve as effective systems to produce vaccines and other proteins intended for biopharmaceutical applications. Plants are highly suitable for producing animal vaccines, since crude plant extracts can be administered to animals either through feeding or injection, thereby circumventing the need for protein purification and significantly reducing production costs. This cost reduction is critical for the manufacture of animal vaccines, as the market generally accommodates products priced at just a few cents per dose. For instance, research demonstrated that piglets were effectively protected against enterotoxigenic *E. coli* (ETEC) when their diets included Arabidopsis thaliana seeds synthesizing antibodies against ETEC. Furthermore, crude extracts derived from *N. benthamiana* leaves, which transiently expressed the influenza virus H5 trimer, successfully inhibited the transmission of avian influenza (H5N1) when injected into chickens (Ponndorf et al. 2021; Lai et al. 2018).

Oral vaccines have also been the subject of exploration in human clinical trials. Plants that are generally regarded as safe (GRAS), such as tomatoes, lettuce, and bananas, have been selected for this purpose, with bananas offering an additional advantage due to their robust skin, which protects the vaccine-containing fruit from damage and dehydration. Human subjects consuming raw transgenic potato tubers expressing the Norwalk virus capsid protein in the form of virus-like particles showed a significant immune response to the antigen. However, it is important to note that oral vaccines contained within plant tissues may be compromised in the stomach before they reach the gut-associated lymphoid tissue, either necessitating high accumulation levels to ensure effectiveness or encapsulation within storage compartments for added protection. Moreover, achieving batch-to-batch consistency is a challenge when using fruits or tubers. This challenge can potentially be mitigated by producing oral vaccines in plant cell suspension cultures, which facilitate the preparation of homogeneous mixtures with uniform vaccine concentrations. Additional clinical trials have been undertaken to assess plant-derived antibodies or replacement human proteins, including insulin for diabetes, gastric lipase for cystic fibrosis, and glucocerebrosidase for Gaucher's disease (Spiegel et al. 2018; Kurup and Thomas 2020; Ar'evalo-Villalobos et al. 2020a).

The production for pharmaceutical proteins intended for clinical trials must comply with Good Manufacturing Practice (GMP), which is more easily achievable with cell suspension cultures and hairy roots in bioreactors, allowing production to occur entirely within a clean-room environment, akin to the systems used for microbial and mammalian cell cultures. This level of control is not attainable with whole plants cultivated in soil, rockwool blocks, or hydroponic systems. Consequently, regulatory authorities permit upstream manufacturing to transpire in a non-GMP context, provided that clarified extracts of plant tissue are aseptically transitioned to a GMP facility for subsequent processing. Nonetheless, it is imperative that the upstream cultivation stage is carefully regulated to ensure consistent plant growth and to avoid exposure to pests, rodent excretions, and pesticides, as variability in growth conditions and contamination may adversely affect both product yield and quality. The establishment of these guidelines paved the way to produce an HIV-specific monoclonal antibody in transgenic tobacco plants, which was subsequently evaluated in a first-in-human Phase I clinical trial involving a single vaginal administration of the antibody in healthy female participants. This study confirmed the safety and tolerability of the plant-derived pharmaceutical protein (Sack et al. 2015; Ma et al. 2015).

2.2 Selection of Plant Expression Systems

Plant transient expression systems are being increasingly preferred in biotechnological applications due to their rapid output, scalability, cost-effectiveness, ease of purification, and the availability of established protocols for specific plants (Marwal et al. 2013a). The efficacy and reliability of these transient systems in the production of complex and valuable proteins have been successfully demonstrated for human health benefits and for the production of disease-free plants (Marwal et al. 2013b). *Nicotiana benthamiana* is commonly utilized as the primary host for both research and commercial purposes, enabling the production of proteins within days following the introduction of the DNA construct (Marwal et al. 2013c). This characteristic has made it a preferred choice for companies such as Medicago Inc. (https://www.medicago.com), Forte Protein Inc. (www.forteprotein.com), Kentucky BioProcessing (https://kentuckybioprocessing.com), and Icon Genetics (https://www.icongenetics.com), which are actively engaged in the commercial manufacturing of hormones, antibodies, enzymes, therapeutics, and vaccines (Nosaki et al. 2021; Ward et al. 2021).

The optimization of yield is a significant focus in the production of high-value proteins, as the costs associated with downstream processing can rise considerably with proteins sourced from more diluted solutions. The development of transgenic plants usually entails a prolonged timeline, often requiring approximately 2- to 3-years to establish a homozygous line for the host plant. In contrast, transient expression methodologies facilitate the rapid generation of substantial quantities of recombinant proteins. For instance, Medicago Inc. has demonstrated the capability to produce purified influenza vaccine products within 3 weeks of receiving the genetic sequence. Recently, Medicago Inc. illustrated the effectiveness of transient systems in the production of vaccines targeting both SARS-CoV-2 and influenza, showcasing consistent immunogenicity in human clinical trials (Venkataraman 2022).

During the COVID-19 pandemic, various research teams developed the receptor-binding domain (RBD) as a vaccine antigen, utilizing diverse host systems. The RBD of the SARS-CoV-2 spike protein has been shown to elicit humoral immunity in animal models and non-human primates, specifically employing *Nicotiana benthamiana* as the production host. Transient expression systems generally yield recombinant proteins at concentrations quantified in grams per kilogram of fresh biomass, with potential for further enhancement through innovative approaches. A diverse array of recombinant proteins, including vaccines, growth factors, therapeutics, hormones, and antibodies, has been successfully produced using these transient systems. While transient expression systems provide numerous advantages over competing methodologies, there remains a pressing need for innovations in areas such as novel vectors, expression strategies, clarification processes, and especially downstream processing to facilitate large-scale production at reduced costs, thereby making alternative proteins more affordable and accessible (Ruocco and Strasser 2022).

2.2.1 Transgenic Plants for Edible Vaccines

Transgenic plants are being investigated as a possible platform for creating edible vaccines. Engineered seeds or plant tissues can be designed to produce recombinant vaccines that can be taken orally, thereby reducing the need for painful injections and eliminating cold-chain storage requirements, which facilitates quick and effective global distribution during vaccination campaigns. The oral administration route has the potential to transform current vaccine delivery methods. Producing vaccine antigens in plant cells holds great promise for reducing the expensive and complicated fermentation processes, streamlining transportation and distribution, removing the necessity for sterile injections, and extending shelf-life for years at room temperature. Various plant species have successfully expressed recombinant vaccines for oral delivery, including potato, rice, banana, tomato, lettuce, tobacco, alfalfa, wheat, spinach, and carrot (Stöger et al. 2005; Shi et al. 2023).

Vaccines produced in plants are bioencapsulated within the cell wall and organelles, allowing vaccine proteins to resist degradation and traverse the digestive system so that they can be detected by the immune surveillance system. The human intestine, with its substantial mucosal surface area, serves as an ideal entry point for plant-derived vaccines administered orally. The gut-associated lymphoid tissue (GALT), which is the largest immune tissue in the human body (over 70%), acts as a significant reservoir of regulatory T-cells (Tregs) and has an extensive surface area of about 300 m^2. A unique benefit of plant-made vaccine antigens is their bio-encapsulation in plant cell walls; this feature allows them to endure the enzymatic and acidic conditions found in the gut lumen. Because human digestive enzymes cannot break down the glycosidic bonds of carbohydrates found in the cell wall, plant-derived vaccine antigens can successfully navigate the digestive system. The cellulosome, which exists in anaerobic cellulolytic bacteria as an extracellular enzyme complex, possesses binding, structural, and catalytic domains that enable direct interaction with the plant cell wall and the cleavage of glycosidic bonds. When the vaccine arrives at the gut mucosal layer, resident bacteria assist in breaking down the plant cell wall, which helps in releasing the vaccine antigen. *Bacteroides fragilis* facilitates movement through the mucous layer by degrading mucin glycoproteins. Crossing the intestinal epithelium and presenting vaccine antigens to the immune system is essential and is achieved through various tags (receptor-binding proteins) linked to the vaccine, such as CTB (Vighi et al. 2008; Koropatkin et al. 2012; Khan and Daniell 2021).

2.2.2 Plant Suspension Culture Systems

In vitro cultured plant cells, commonly referred to as plant cell suspension culture, have exhibited considerable potential as bio-production platforms for alternative proteins. Suspension cultures can comprise transgenic plants, hairy roots, protoplasts, or cell cultures, thereby integrating the advantages of whole-plant cultivation systems with the benefits seen in microbial and mammalian cell cultures. This combination renders them promising and versatile options to produce therapeutic proteins. Critical considerations for the implementation of plant cell suspension culture include scalability, safety, protein folding, post-translational modifications, production costs, stability, and flexibility. Unlike other eukaryotic systems, such as Chinese hamster ovary (CHO) or human embryonic kidney (HEK) cells, plant cultures can thrive in simple, cost-effective growth media and are readily scalable for rapid biomanufacturing. These attributes collectively contribute to reduced overall production costs. High-value recombinant proteins can be secreted from plant cells through the incorporation of signal peptides into target proteins, facilitating downstream processing. Signal peptides sourced from various origins can direct recombinant proteins into the endoplasmic reticulum (ER) lumen in a manner dependent on the signal recognition particle (SRP), which is further facilitated by SRP receptors. Upon arrival in the ER, signal peptides are cleaved from the precursor proteins, which are subsequently encapsulated within small vesicles. These vesicles bud off towards the Golgi apparatus, where they ultimately release proteins into extracellular compartments. The targeting of recombinant proteins to the ER and Golgi apparatus is critical for ensuring the proper folding necessary for their functionality. Secretory proteins then undergo additional modifications in the ER and Golgi apparatus, including N-glycosylation. Over recent decades, numerous recombinant proteins have been expressed and validated for their functionality. Notably, taliglucerase alfa, the first plant-derived recombinant protein approved by the FDA, was expressed in carrot suspension culture (Breitling and Aebi 2013; Tekoah et al. 2015).

Moreover, plant cell suspension culture has been utilized as a host for vaccine production. Recent reports indicate the successful production of receptor-binding domain (RBD) and spike proteins in tobacco cells (BY-2) and *Medicago truncatula* A17 cells (Rebelo et al. 2022). Considering the advantages of plant suspension culture, various vaccine antigens have been successfully expressed and validated for their functionality through both in vivo and in vitro analyses (Muthamilselvan et al. 2016). Technologies have been developed utilizing cells from diverse sources, including carrot, soybean, tobacco, rice, Medicago, and potato, which have resulted in vaccines targeting a wide array of diseases. Nevertheless, significant potential exists for future research and development (R&D) aimed at optimizing the culture environment to enhance both production levels and protein recovery efficiency. The primary advantages of bioreactors include regulated growth conditions that ensure complete containment, consistency in protein quality, yield, and homogeneity, as well as rapid production timelines ranging from gene introduction to protein within approximately 4 to 5 weeks (Fischer et al. 2009). However, bioreactors encounter constraints like those facing mammalian cell cultures, such as relatively lower concentrations (Xu et al. 2011), instability of secreted protein products, and elevated capital costs in comparison to greenhouse or open-field cultivation. From a downstream processing perspective, significant advantages associated with secreted protein products include more straightforward and cost-efficient purification processes, although the efficiency of secretion may be limited by factors such as protein hydrophobicity, size, and/or charge (Fischer et al. 2009). For non-secreted recombinant proteins, plant cell homogenates face challenges similar to those faced by leafy plants. Consequently, these may not be as economically attractive compared to alternative protein expression systems, as the potential savings from intracellular protein production may be counterbalanced when compared with the advantages associated with secreted proteins.

2.2.3 Expression of Vaccines and Related Recombinant Proteins from Plants

The biotechnological expression of therapeutic proteins necessitates effective expression and purification strategies to ensure that recombinant proteins are produced in their native conformations. The stable or transient expression of recombinant therapeutic or vaccine proteins in plants serves as a promising approach for realizing substantial scalability and reducing production costs (Yusibov and Rabindran 2008). The initial proof-of-concept involved the expression of Hepatitis B surface antigen (HBsAg) in plants, after which numerous subunit vaccines and virus-like particles have been successfully developed (Mbewana et al. 2015). Recently, this methodology has been adopted by Medicago to produce the COVID-19 vaccine and seasonal influenza vaccines, both of which have now successfully completed clinical trials. Transient expression systems that utilize plants to synthesize recombinant pharmaceutical proteins are recognized as advantageous relative to other expression systems. This transient expression method allows for rapid scalability compared to fermenter-derived platforms. Furthermore, transient expression vectors derived from whole or deconstructed plant viruses enable high-level expression of recombinant biopharmaceuticals. Nonetheless, two primary challenges must be addressed to establish plants as effective expression vehicles for recombinant therapeutics: insufficient expression levels and the development of purification methods that are both scalable and efficient (Holtz et al. 2015).

2.2.4 Seed-Based Expression Systems in Plants

Seeds have been effectively employed in the expression of various recombinant proteins, including vaccine antigens, therapeutic antibodies, hormones, industrial enzymes, and proteins for cell culture. The United States Department of Agriculture (USDA) has authorized the field release of transgenic plant seeds that express proteins such as lactoferrin, serum albumin, and human lysozyme in rice, along with apolipoprotein in safflower and both brazzein and Hepatitis B surface antigen in corn (http://www.isb.vt.edu/searchrelease-data.aspx). Seed crops present several advantages over mammalian cell cultures, transgenic animals, and microbial fermentations as host systems due to the extensive knowledge base available regarding their cultivation, harvesting, processing, and storage (Kusnadi et al. 1997; Nikolov and Hammes 2002).

Shi et al. (2023) report the stable expression of amino acid sequences encoded by the TM-1 gene from *Mycoplasma gallisepticum* (MG), which is considered a candidate vaccine antigen for Chronic Respiratory Disease (CRD) in chickens. In this study, wheat seed tissues were utilized as production hosts. The recombinant protein, with a molecular weight of 41.8 kDa, was expressed consistently in endosperm tissues, achieving an expression level of 1.03 mg/g dry weight. Following oral administration to chickens, this plant-derived edible vaccine was effective in eliciting antibody responses without any significant weight loss. Two doses of the TM-1 vaccine candidate elicited an immune response and provided protection against MG challenge at levels comparable to those achieved with commercially produced inactivated vaccines for CRD. This investigation underscores that plant-made edible vaccines are safe, scalable, stable at ambient temperature, and cost-effective.

Shahid et al. (2020) describe a study in which the protective antigens, hemagglutinin-neuraminidase (HN), and fusion (F) proteins of Newcastle Disease Virus (NDV) were expressed in transgenic maize utilizing a constitutive seed-specific Zein promoter and the 35S promoter, respectively. The study observed an approximately 2–7.1-fold increase in F gene mRNA expression in the leaves, while the HN gene mRNA demonstrated an 8–28-fold higher expression in the seeds. The concentration of the F protein was measured at 1.66 μg/ml, constituting 0.5% of the total soluble protein in leaves, whereas the HN protein reached 2.4 μg/ml, making up 0.8% of the total protein in seeds. When chicks were administered the seeds and leaves of transgenic maize, an immune response was successfully induced against both NDV antigens.

Stable transgenic rice seeds expressing the F protein of Newcastle disease (ND) virus were developed through agro-infiltration (Ma et al. 2020). When this vaccine was administered to pathogen-free chickens, it resulted in a significant generation of neutralizing antibody responses against both heterologous and homologous ND virus strains. Two doses of 4.5 μg antigen each provided complete protection to chickens from lethal doses of NDV. Furthermore, chickens immunized with the F protein exhibited greater mean weight gain within 15 days post-challenge compared to those vaccinated with the conventional whole virus vaccine, thus enhancing economic advantages. This study emphasizes the potential of plant-based vaccines as effective platforms for the eradication of NDV. In the context of nutraceuticals, industrial enzymes, and oral vaccines, seed crops substantially mitigate the requirements for processing and purification. Despite lower biomass yields per unit area in comparison to leaf crops, the cultivation of seed crops in open fields offers considerable benefits in terms of protein stability and economic efficiency, thus outweighing the disadvantages associated with reduced biomass yield (Hood and Howard 2009; Nikolov and Hammes 2002).

2.2.5 Leaf-Based and Fruit-Based Expression in Plants

Leafy crops like tobacco and alfalfa offer significant benefits due to their high biomass yields, the potential for multiple growth cycles each year, and the existing agricultural infrastructure comparable to their seed-derived counterparts; additionally, they have a lower risk of pollen dispersion because they can be grown without flowering. Tobacco has become a primary leaf-based platform for the commercial production of recombinant proteins since the development of tobacco that produces monoclonal antibodies. Moreover, significant progress has been made in utilizing leaves for the temporary expression of vaccines, monoclonal antibodies, and other therapeutic agents via tobacco (*Nicotiana benthamiana*) as the expression system. This transient expression method avoids regulatory issues linked to transgenic plants and has become the favoured technique for low-volume protein production. Utilizing leafy tissues for transient expression provides a notable advantage for tobacco cultivated in greenhouses, particularly when there is an urgent need for rapid production of pandemic vaccines. The straightforward process of extracting proteins from leaves is sometimes seen as a benefit when compared to seed extraction, which may require additional steps such as soaking and grinding (Fischer et al. 2009; Dubey et al. 2018; Conley et al. 2011).

Lettuce and tomato are favoured vegetable crops that can be eaten raw. Varieties of lettuce (*Lactuca sativa*) have been employed to create edible vaccines for diseases such as swine fever, stage I hepatitis, and *E. coli*. Likewise, carrots (*Daucus carota*) have been utilized to develop edible vaccines against *Helicobacter pylori* and *E. coli* (Zhang et al. 2010). Bananas (*Musa acuminata*) have become the preferred plant for generating vaccines against Hepatitis B. Bananas, which can be consumed raw and are extensively grown in developing regions, have the significant advantage of producing adequate levels of antigenic protein in their fruits (Altindis et al. 2014). Nonetheless, leafy tissues have significant drawbacks, including high water content, storage instability of harvested leaf biomass, leading to issues with the stability of recombinant proteins, and challenges in separating downstream and upstream processing. Thus, while leaf expression presents certain benefits, it is countered by the potential instability of expressed proteins in metabolically active tissues (and their extracts). Additionally, leaf extracts contain phenolics, pigments derived from chlorophyll, and inherent proteolytic activities that can disrupt downstream processing (Barros et al. 2011), resulting in complex, multi-step purification procedures that negatively affect the yield and quality of the purified proteins. Furthermore, despite the low production costs associated with leafy and seed crops grown in open fields, the manufacturing of plant-based pharmaceuticals has increasingly shifted towards cultivation in controlled environments like bioreactors and greenhouses (Spok and Karner 2008).

3 Plant-Based Vaccine Production Strategies

Plants have been utilized as cost-effective alternatives to produce biopharmaceuticals, minimizing the risk of contamination by pathogens and taking advantage of their substantial industrial capabilities. The application of plants for producing recombinant proteins began in 1990, when human serum albumin was synthesized in tobacco. Numerous plant-derived vaccines have been developed against pathogens affecting both humans and animals. When vaccines are generated in plants, it is crucial to assess the potential for these vaccines to cause allergic reactions (Takeyama et al. 2015). In this context, every plant-produced vaccine, whether purified or in its raw form, should be assessed for safety as one of the initial steps in vaccine development. For example, Ward et al. showed that the plant-produced virus-like particles (VLPs)-influenza vaccine did not provoke allergic or hypersensitivity responses in subjects during Phase I/II clinical trials (Ward et al. 2014).

Subunit vaccines often exhibit low immunogenicity, necessitating the use of adjuvants, increased dosages, or additional booster vaccinations. Virus-like particles (VLPs) represent a type of subunit vaccine that demonstrates higher immunogenicity compared to the 3-dimensional arrangement of individual antigenic proteins. VLPs are composed of repeated structural viral proteins capable of assembling into sizes ranging from 22 to 150 nm, with 40 nm identified as the optimal size for recognition by dendritic cells (Fifis et al. 2004). VLPs are viewed as a safer alternative to vaccines made from killed or inactivated pathogens, as they do not contain genetic material and can possess up to three distinct protein layers derived from a single structural antigen (Zeltins 2013). In this regard, VLPs can be categorized as enveloped or non-enveloped. Non-enveloped VLPs do not contain any elements from the host cell and are most commonly employed in clinical trials. Conversely, enveloped VLPs are more intricate, incorporating membrane components from the producer cell with antigens displayed on the outer surface. Following the pioneering work of Valenzuela et al. (1982) in producing the first VLP with the HBsAg antigen, numerous VLPs have been generated using various production systems. The production of the initial antigen in plants (HBsAg) has given rise to a diverse range of vaccine antigen prototypes featuring VLP structures produced in plants (Scotti and Rybicki 2013). As previously mentioned, the latest research on VLPs in plants has primarily involved *Nicotiana benthamiana* due to its remarkable ability to create complex structures with high yields, although other plants such as *Nicotiana tabacum* and *Arabidopsis thaliana* have also been employed (Greco et al. 2007), along with potato, tomato, and lettuce.

Edible vaccines produced from plants are particularly appealing when oral delivery is the method of inoculation. A newer model of VLPs emphasizes the development of chimeric VLPs, which offer the benefit of multivalence, meaning they can consist of antigens from different pathogens and may even include functional RNA segments. One of the advantages of the chimeric VLP strategy is its multivalence, as some epitopes or antigens might act as adjuvants for others. Nonetheless, certain challenges have been identified regarding the proper folding of the protein, necessitating the incorporation of strategies within the design framework to address these issues. Another variant of chimeric VLPs is created by integrating immunogenic peptides into the VLP of plant viruses rather than those from mammalian viruses. To guarantee that the modified capsid protein (CP) of the plant virus constructs the VLP, the incorporated peptide should be positioned within the surface structure, typically achieved by cloning the antigenic peptide at either the N- or C-terminal of the CP protein (Rutkowska et al. 2019; Balke and Zeltins 2019; Lei et al. 2020). A few of the plant genetic engineering transformation methods for subunit vaccine production are as follows:

3.1 Stable Nuclear Transformation

The process of nuclear transformation in plant cells was established approximately four decades ago, primarily through the transfection method utilizing *Agrobacterium tumefaciens* (Bevan et al. 1983). This technique involves the cloning of DNA that contains the transcriptional unit (promoter–gene (antigen)–terminator) for protein expression into a binary plasmid, commonly utilizing vectors such as pBI121 and pCambia (Bevan 1984). Following this, a plasmid encapsulating the transcriptional unit of the antigen is introduced into *A. tumefaciens*, typically through electroporation. The *Agrobacterium* carrying these plasmids is then co-cultured with fragments derived from plant leaves or stems, which leads to the regeneration of a complete transgenic plant via either organogenesis or embryogenesis processes (Dandekar and Fisk 2004). Ultimately, the detection of transgenes and proteins in fully regenerated plants serves to confirm the successful genetic transformation and expression of the antigen.

Initially, the primary model plant employed for vaccine production was *Nicotiana tabacum* (Ganapathi et al. 2004). Presently, a range of additional plant species has been considered for vaccine manufacturing, with certain species such as lettuce and carrots offering advantages for the formulation of oral vaccines (van Eerde et al. 2019). Within the framework of *Agrobacterium*-mediated nuclear transformation, plasmids contain a segment situated between the right and left T-borders, which undergoes genetic recombination with the nuclear genome upon delivery (Tzfira and Citovsky 2006). This segment comprises a gene responsible for conferring antibiotic resistance, thereby facilitating the selection of transformed cells. Subsequently, these transformed cells are cultured in a growth medium supplemented with regulators and hormones to achieve complete plant regeneration (Gallois and Marinho 1995). Gene expression and antigen production within transgenic plants are regulated by promoters, which may be either constitutive or inducible by specific substances or physical conditions, such as alcohol, NaCl, temperature, or light. It is noteworthy that viral genetic elements can be included in the genetic construct to amplify the number of gene copies (Ar'evalo-Villalobos et al. 2020b) and to enhance protein yields.

3.1.1 Transient Nuclear Transformation

The transient nuclear expression method is employed when it is necessary to maintain transformed plant cells for a limited duration, typically ranging from 2 to 10 days. In this context, the integration of the transgene into the nuclear genome is not induced through antibiotic selection. Nevertheless, numerous copies of the gene are introduced into the nucleus, enabling the production of messenger RNA that facilitates high recombinant-protein yields (Yusibov and Rabindran 2008). Various strategies can be utilized to achieve transient transformation, with the primary objective being the "infection" of a significant number of plant cells using either *Agrobacterium tumefaciens* or viral vectors that carry the gene encoding the antigen (Yusibov et al. 2013; Nehra et al. 2019; Verma et al. 2024). Both transient and stable transformation methodologies can also be applied to plant cell cultures or microalgae, thereby promoting more uniform antigen production and optimizing manufacturing practices, particularly in the containment of genetically modified cells (Sahu et al. 2014a; Massa et al. 2018). One of the most widely recognized viral vectors utilized for transient expression was developed from components of the RNA of the *Tobacco mosaic* virus (TMV). Currently, an extensive array of viral vector types is accessible, allowing for the adaptation of virtually any virus to serve as an expression vector. Recently, the *Bamboo mosaic* virus, which belongs to the *Potexvirus* genus, has been utilized as a viral vector to produce *Japanese Encephalitis* Virus antigens in plants (Chen et al. 2017). Viral vectors derived from plant *geminiviruses* have been significantly employed to produce biopharmaceutical proteins (Chen et al. 2011; Rybicki and Martin 2011). The mechanism of rolling circle replication enables the delivery of replicons (copies of transgenes) and can be engineered to generate multiple proteins within plant cells. Importantly, the distinct capability of each virus to evade plant immune systems and replicate within plant cells is critical for the success of the viral vector in achieving high levels of antigen production.

3.1.2 Chloroplast Transformation

Chloroplast transformation signifies the inaugural instance of genetic modification in a green cell, achieved approximately three decades ago. This process was initially pursued to produce cells that exhibit resistance to antibiotics (Fischer et al. 1996). The first successful chloroplast transformation was accomplished through biolistic techniques, wherein microparticles composed of gold or tungsten, measuring approximately 10 micrometres, are coated with genetic material that includes essential gene expression units and then propelled into target cells for transformation. Upon entering the cell, the DNA molecules can integrate into the chloroplast genome through DNA recombination mechanisms. Additionally, other plastid organelles, such as chromoplasts found in fruits and plastids located within tubers, have also been transformed. Currently, the production of antigens within chloroplasts presents advantages, including a superior yield of recombinant proteins compared to stable nuclear transformation, which is influenced by the copy number of chloroplast genomes within a given plant (Bock 2014). Consequently, chloroplasts have the potential to be utilized to produce proteins intended for subunit vaccines. However, a limitation of chloroplasts is their lack of post-translational modification machinery, which poses a challenge when post-translationally modified proteins are essential for vaccine production.

4 Advantages of Plant Molecular Farming in Vaccine Production

Plant molecular farming is a novel approach for vaccine production. Genetically engineered plants are employed as bioreactors for the production of recombinant proteins or antigens, which can be used as vaccines (Fig. 2). Plant molecular pharming thus reduces the cost of production on a large scale by using microbial cell culture or mammalian cell culture in complex bioreactors. Plant molecular pharming also reduces the risk of contamination in vaccines by other pathogenic microorganisms (Guan et al. 2013). Genetically engineered plants grow quickly and can be utilized for the production of viral/bacterial proteins or peptides, as well as virus-like particles, in less time. Thus, plant molecular pharming is an economical, less energy-consuming, efficient, and sustainable approach for the production of vaccines. Cold storage is not required for oral vaccines that have been developed using edible plants. Most of these vaccines remain stable at room temperature, making the transportation process easy. Protein vaccines taken in the form of edible plant proteins activate the immune system in a natural way after being absorbed from the oral route (Guan et al. 2013). Plant-based vaccines are especially beneficial against pathogens entering from the mucosa and provide stronger resistance against

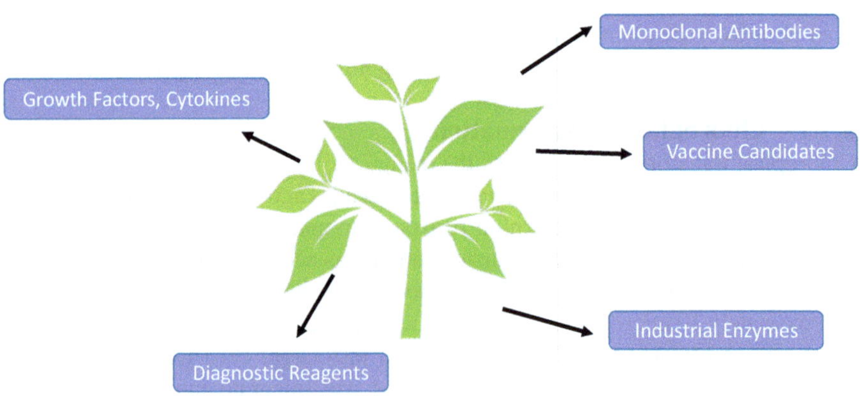

Fig. 2 Genetically engineered plants are used as bioreactors for the production of recombinant proteins, which is a novel approach and a cost-effective process

mucosal pathogens. Major challenges associated with plant-based vaccines are uniformity in productivity and complex regulatory processes associated with the approval of plant-based vaccines.

4.1 Cost-Effectiveness, Scalability, and Flexibility

Plant-based vaccines are favourable over traditional vaccines due to their cost-effectiveness, scalability, and flexibility. Plant-based vaccines are favoured over traditional vaccines due to their cost-effectiveness. Traditional vaccine production is done in bioreactors by using cell cultures or egg cultures. Bioreactors require specialized infrastructure, equipment, and facilities. Cultivation of genetically engineered plants requires low input costs as compared to animal cell culture, which requires highly technical processes. Purification of proteins from culture media is a relatively complex process compared with plants, which accumulate proteins in leaves, seeds, and fruits. Thus, minimal processing is required for the isolation and purification of plant-based vaccines, which can be grown in fields or greenhouses. Several plant varieties have been used to express recombinant proteins containing vaccine antigens in various plant tissues. Oral vaccines produced using plant-expressed antigens provide both mucosal and systemic immunity. Encapsulation of antigens expressed in plant cells prevents breakdown of antigen in the digestive system. Trans-mucosal carriers are fused to plant-expressed antigens to increase the efficiency of antigen delivery to the immune system. Fusion of antigen to trans-mucosal carriers, which assist in the delivery of antigen, is used for the development of more efficient oral vaccines (Yadav et al. 2023).

Plant vaccines have many advantages compared with traditional vaccines, as they are heat-stable, have no pathogen contamination, can be engineered to express multiple antigens, and provide protection against multiple infections. Recently developed vaccines using genetically engineered plants include the COVID vaccine and vaccines against Ebola and Zika virus. During the Ebola outbreak, researchers used tobacco plants (*Nicotiana benthamiana*) to produce the virus-like particles (VLPs) needed for an Ebola vaccine. EDIII antigen present in dengue virus has been fused with cholera toxin-B, which is expressed and assembled into a functional form in potato tubers. CTB–EDIII fusion protein is a potential oral vaccine with immunogenic potential due to its binding to specific receptors (mono-sialo-tetra-hexosyl ganglioside; GM1) on the intestinal walls (Kim et al. 2017).

A virus-like particle has been expressed in *Nicotiana benthamiana* in large quantities to provide immunity against the Zika virus. Hepatitis B core antigen (HBcAg) has been fused with ZIKV E protein domain III in this virus-like particle. Immunization of mice with this vaccine has proved it to be more efficient against multiple strains of Zika virus as well as cost-effective (Yang et al. 2017). Genetically engineered edible plants can express and assemble one or more antigens having both T- and B-cell epitopes, which give rise to both systemic and mucosal immune responses. Plant-based vaccines immunize both mice and humans efficiently. Nasal and oral vaccines are more effective in providing immunity against pathogens that enter through the mucous membrane (Bhatia and Dahiya 2015).

Genetically engineered plants can be modified to synthesize virus-like particles (VLPs) or adjuvants to enhance immunization by a plant vaccine, improving their effectiveness (Evtushenko et al. 2020). There are multiple strains of coronavirus that can cause COVID. To provide protection against COVID, immunization with multiple antigens from different strains of coronavirus needs to be done. High biomass-producing and fast-growing plant *N. benthamiana* has been genetically engineered using viral vectors for rapid production of a vaccine (Maharjan and Choe 2021). Thus, vaccine production from genetically modified plants is the best solution for rapidly emerging diseases such as COVID, Ebola, and the Zika virus. Genetically modified plants can produce more effective vaccines in large quantities against a diverse variety of pathogens and antigens.

4.2 Rapid Response to Emerging Diseases

Vaccine production from genetically modified plants is less time-consuming. Transgenic plants are produced by inserting the corresponding gene responsible for producing the antigenic protein, and vaccine production does not involve any live pathogen. Vaccine production from genetically modified plants is a cost-effective process; large-scale production of vaccines can be done using genetically modified plants with a low-cost setup. Vaccines produced using cereal grains are produced in large quantities due to the presence of a large number of seeds. For example, corn is used for the production of vast quantities of vaccines against emerging diseases. Plant-based vaccines are considered safer, and there is no risk of contamination by animal pathogens. Various studies have been successfully conducted to show that transgenic plants can express epitopes present on viruses and toxins produced by bacteria. These recombinant proteins, when correctly expressed and processed, can induce an immune response. Vaccines produced using genetically modified plants are thermally stable. These can be transported long distances at environmental temperatures without any damage, which allows for global vaccination. Streatfield et al. in 2001, produced a plant-based vaccine in maize plants by expressing the B-subunit of *Escherichia coli*, which is a heat-sensitive enterotoxin, and spike proteins from swine transmissible gastroenteritis virus in corn seeds. These vaccine proteins are expressed at a high level in corn seeds and induce a protective immune response. Several other investigations have already demonstrated transgenic plants to be a suitable replacement for producing and delivering vaccines (Gomez et al. 1998).

These vaccines can successfully induce an immune response when taken orally as feed or administered by ingesting genetically modified edible plants like bananas, tomatoes, and potatoes, which have already been engineered to produce viral or bacterial antigen proteins. After consuming these edible plants as feed, viral or bacterial proteins present in them show immunogenicity. The immune system recognizes these antigens to generate cellular and humoral immune responses. Various banana-based vaccines, such as hepatitis, cholera, and HIV, have been successfully produced. Edible vaccines stimulate the immune system in the gut. Tomato and potato plants have been genetically engineered for the production of oral vaccines against enteric viruses and Norovirus. Various non-edible plants have also been genetically modified for vaccine production on a large scale. Plant tissues (leaves, fruits, etc.) are collected, and the antigen is purified from these proteins. Proteins isolated and purified are used as injectable vaccines for immunization. Genetically modified *Nicotiana benthamiana* has been a key model for producing vaccines and has been successfully used for the production of a vaccine against the Ebola virus on a large scale (Rosales-Mendoza et al. 2017).

In the case of oral vaccines, bananas have been engineered to produce antigens against diseases like Hepatitis B, cholera, and even HIV. As an edible vaccine, the antigen is delivered directly through the food, stimulating the immune system in the gut. Various studies have also been conducted during COVID-19 for the production of plant-based vaccines on a large scale (Marwal et al. 2021). In this context, *Nicotiana benthamiana* was genetically engineered for the production and purification of the spike protein antigen present on SARS-CoV-2 (Chattopadhyay et al. 2023). Plant-based vaccines have also been found to work efficiently against the Zika virus. Research on plant-based vaccines for the Zika virus has been progressing rapidly, with plants being engineered to produce Zika virus antigens to elicit an immune response (Yang et al. 2017). The rapid production capability of plant-based platforms holds great promise for emerging viruses like Zika, which require swift vaccine development. Plant-based vaccines offer significant opportunities for the rapid and mass production of vaccines against emerging infectious diseases.

5 Challenges and Limitations of Plant Molecular Farming in Vaccine Production

5.1 Regulatory Hurdles and Low Expression Levels

Although plant-based vaccines are highly promising and efficient for large-scale production, various challenges are also associated with them. Clinical trials and regulatory analysis should be done before approval for production on a large scale, which is a complex process. Public acceptance regarding safety and efficacy is a very important aspect before vaccine production using edible and non-edible genetically modified plants. Still, further research needs to be done for vaccine optimization and quality assurance. Production of plant-based vaccines using. The process of producing plant-based vaccines in plants that are responsive to genetic transformation is challenging. *Nicotiana benthamiana* is comparatively responsive to genetic transformation and widely used for the production of plant-based vaccines. After successful transformation of plant by gene against vaccine antigen, it should be expressed at high rate for production of enough amount of vaccine antigen. Even when plants are successfully transformed, the introduced gene may not always be expressed at high levels to produce the desired amount of vaccine antigen. Consistency and stability of gene expression in vaccine-producing plants are also important factors while using plants as bioreactors (De Wilde et al. 2000). The variability of transgene expression in different plants of the same batch, as well as batch-to-batch variation, also poses a challenge, making the evaluation of vaccine doses difficult.

Variability in transgene expression from plant-to-plant and from batch-to-batch disturbs the consistency in product quality, which is necessary for vaccine efficacy and safety. The purification of antigen proteins required for vaccine production also poses a challenge for the rapid and mass production of the vaccine. Sometimes antigen protein expressed in plant cell is misfolded due to which the protein may become inactive and unable to activate an immune response. This presents another challenge for plant-based vaccines that can work efficiently. To ensure that protein antigens are correctly expressed and folded, growth conditions, a suitable time for protein harvesting, and post-harvest conditions for processing need to be optimized. Vaccine production process may become less efficient and more costly if protein production is very low compared to the plant biomass. Higher protein yield is achievable by introducing a strong promoter in the plant genome and optimizing growth conditions (Venkataraman et al. 2023).

Genetically modified edible and non-edible plants used for vaccine production fall under GMO (Genetically Modified Organisms) regulations, which may delay the approval process for vaccine production. Regulatory bodies approve vaccine production after considerable testing for allergenicity, toxicity, and the collection of data regarding safety, quality, and efficiency of the vaccine. WHO organized a meeting in 2005 to evaluate and frame guidelines regarding the development and safety of plant-based edible vaccines. It was concluded that plant-based edible vaccines must be tested according to a US investigational new drug application. It must follow GMP requirements (Kurup and Thomas 2020). There are ethical concerns about alterations in the genomes of food crops. Although plant-based vaccines are generally considered safe, there are potential risks related to cross-contamination in agricultural settings. For example, recombinant proteins might inadvertently be transferred to non-target crops or wild relatives, posing a risk to biodiversity or the human food supply. Growing genetically modified crops on a large scale for vaccine production may affect the environment. Crop yield is limited by seasons and climate throughout the year, and controlled-environment agriculture is required for production throughout the year. The yield of plant-based vaccines is significantly affected by pest attack, microbial infections, and weather conditions. Some plant-based vaccines are stable at ambient temperature, but some plant-based vaccines require storage at low temperatures, which affects the distribution of the vaccine in regions with limited resources, and storage facilities are not sufficient. Sometimes, vaccine proteins produced in plants may degrade when exposed to heat, and this affects the efficacy of the vaccine (Sahoo et al. 2020).

Vaccine developers involved in plant molecular pharming face difficulties in obtaining patent rights for newly developed plant varieties and genetic techniques involved in developing such varieties (Frisio and Ventura 2021). Despite various applications and benefits of plant-based vaccines, different challenges associated with vaccine development include regulatory approval, public acceptance, environmental issues, the complexity of genetic engineering methods, and protein purification techniques. Gene editing technology and improvement in agricultural practices with the help of plant biotechnology can overcome these challenges and may provide a better product to meet global demands (Zahedipour et al. 2024).

5.2 Glycosylation Differences

Plants, when used as bioreactors, synthesize biopharmaceuticals, including antigen proteins or vaccine proteins, on a large scale at very low cost; also, the final product is free from pathogen contamination. Plants, like all other organisms, perform protein glycosylation. Glycosylation, or the pattern in which sugars are added to plant proteins, differs from the way in which glycosylation is performed in animal or mammalian cells. This difference in glycosylation pattern is a major barrier in vaccine production through plant molecular pharming. This difference affects the immunogenicity or activity of the antigen protein expressed in the transgenic plant. Thus, the vaccine product is not efficient or effective for immunization (Singh et al. 2021). Plants and mammals, including human beings, differ in protein glycosylation pathways and in the types of glycan structure added to proteins. As a result, final proteins synthesized in plants cannot work as a vaccine. N-linked glycosylation in mammals includes sialic acid and fucose. These sugars are important in immune recognition and response. N-linked glycosylation in plants produces mannose-rich oligosaccharides. Plant proteins do not contain sialic acid and differ in folding pattern, which affects the final structure of the protein. Xylose and core fucose are present in proteins expressed in plants. The presence of these sugars in glycoproteins in plant products may produce a different immunogenic response in human beings. O-linked glycosylation is simpler in plant proteins and may not function to initiate immune response when used as vaccine. Altered glycosylation in plants may lead to undesired immune responses. Plant proteins show mannose-rich glycosylation, which is beneficial for antigen presentation by dendritic cells and for the initiation of an immune response. Mannosylated antigens are presented 1000 times more effectively than non-mannosylated antigens by macrophages or dendritic cells (Engering et al. 1997).

At the same time, incorrect glycosylation of antigenic protein in a plant cell can diminish antibody recognition or T-cell activation by binding to T-cell receptors. Glycosylation is a key regulator in T-cell development, thymocyte selection, T-cell activation, differentiation, and proliferation. If antigen proteins are not glycosylated, they will misfold in the form of aggregates or be degraded. Glycosylation plays a very important role in innate as well as adaptive immunity and has major implications in diseases like autoimmunity and cancer. There is a gap between the production and evaluation of plant-based vaccines. Evaluation by regulatory agencies requires precise and careful testing of vaccines to assess the efficacy and safety before commercialization. An understanding of glycol-engineering and the effect of glycans on structures and functions of vaccines can be utilized for the development of efficient and safer vaccines (Lembo et al. 2024).

6 Case Studies in Plant Molecular Farming for Vaccine Production

Recombinant proteins are vital across various sectors of our lives. For instance, recombinant enzymes find applications in the food and chemical industries, and also serve as high-quality proteins for research, diagnostics, and therapeutic purposes. While microbial and mammalian cell expression systems currently dominate the production of recombinant proteins, molecular farming, which involves producing these proteins in plants, has been advocated as a cost-effective alternative for the past several years. Although several products derived from molecular farming have entered the market, the number of success stories remains limited due to industrial inertia, which stems from perceptions of low productivity, high costs associated with downstream processing, and regulatory challenges that hinder effective translation (Schillberg and Finnern 2021). The demand for recombinant proteins is steadily increasing in terms of quality, quantity, and variety, leading to a global focus on developing innovative production technologies and improving traditional expression systems that utilize bacteria or mammalian cell cultures. Since the rise of plant genetic engineering in the 1980s, plants have become a valuable source for creating economically significant, biologically active non-native proteins and biopharmaceuticals, a practice known as plant molecular farming (PMF). This approach is seen as a cost-effective solution that has advanced significantly over the last 2 decades. Notably, enhancements in transient expression systems have shortened the timeline for protein production and boosted yields in plant systems. The key advantages driving the adoption of plant-based platforms as competitors to conventional expression methods include cost-effectiveness, scalability, flexibility, versatility, and robustness. Numerous biopharmaceuticals, such as recombinant vaccine antigens, monoclonal antibodies, and other commercially relevant proteins, are being produced in plants, with some already in preclinical and clinical development stages (as shown in Table 1).

6.1 Hepatitis B Vaccine in Plants

Virus-like particles (VLPs) are structures created by the self-assembly of viral capsid protein subunits, lacking any infective viral genetic material. The Hepatitis B core (HBc) antigen can form VLPs that trigger robust immune responses and has been approved as a commercial vaccine for Hepatitis B. Additionally, HBc VLPs have been utilized as a platform to present foreign epitopes to the immune system, leading to the development of vaccines for ailments such as influenza A and Foot-and-mouth disease. Plant expression systems offer a rapid, scalable, and safe method for production, ensuring proper post-translational modifications while significantly lowering upstream production costs. Consequently, producing HBc-based virus-like particles in plants could greatly enhance the efficiency of vaccine production (Vahdat et al. 2021). Plant molecular farming has made significant strides in producing vaccines, with the Hepatitis B surface antigen (HBsAg) being one of the earliest

Table 1 The significance of plant-based production systems for recombinant proteins and explores their potential in biopharmaceutical production (Shanmugaraj et al. 2020)

Applications	Plant molecular farming	Hepatitis B vaccine in plants	Norovirus-like particle (VLP) vaccine in plants	Conventional plant-based vaccines	Modern vaccines
Manufacturing framework	Employs plants, such as tobacco and lettuce, as bioreactors for the production of recombinant proteins used in vaccines	Generates Hepatitis B surface antigen (HBsAg) in plants such as lettuce and potatoes	Synthesizes Norovirus capsid proteins in plants, resulting in the formation of virus-like particles (VLPs)	Involves the extraction of antigens from plants infected with pathogens, resulting in reduced precision	Utilizes engineered cells, such as bacteria, yeast, or mammalian cells, as well as viral vectors, necessitating specialized facilities
Cost-effectiveness	Reduced costs resulting from less infrastructure needed and the ability to scale rapidly	A cost-effective alternative to conventional production methods for Hepatitis B	Economical since it bypasses the need for costly bioreactor systems	Potentially reduced costs, though it does not achieve the advanced efficiency of molecular farming	Increased costs stemming from strict requirements and intricate fermentation or cell culture systems
Expandability	Excellent scalability resulting from the simplicity of plant growth and rapid biomass accumulation	Extremely scalable for the production of the Hepatitis B vaccine	Readily scalable for the production of virus-like particles (VLPs) in high demand	Scalability is restricted when compared to molecular farming	Scalable, but necessitates infrastructure and requires high maintenance
Security	Minimizes the risk of contamination by animal pathogens, as plants do not harbour human pathogens	A safe production platform that eliminates the risk of contamination by animal and human pathogens	Virus-like particles (VLPs) are non-infectious and non-replicating, providing a guarantee of safety	Conventional methods may pose contamination risks based on the specific pathogen involved	While stringent quality control measures promote safety, risks still exist within animal-based systems

(continued)

Table 1 (continued)

Applications	Plant molecular farming	Hepatitis B vaccine in plants	Norovirus-like particle (VLP) vaccine in plants	Conventional plant-based vaccines	Modern vaccines
Immune response potential	Generates antigens that can trigger targeted immune responses	Triggers an immune response comparable to that of traditional Hepatitis B vaccines	VLPs replicate the structure of the virus, boosting the immune response while eliminating the risk of infection	Immunogenicity can vary based on the methods used for purification	Recombinant antigens combined with adjuvants demonstrate strong immunogenicity
Ecological consequences	Eco-friendly, since plants are biodegradable and produce minimal waste	Lower environmental impact by utilizing plant-based processes	Decreased environmental impact through renewable plant-based processes	Reliant on the sources of pathogens and the methods of extraction	The impact varies, being greater for animal-based or resource-intensive cell culture systems
Regulatory approval	An emerging field characterized by developing regulatory frameworks	Restricted commercial application, currently under review	Gaining traction, though additional regulatory approval is required	Restricted in scope, necessitating significant validation and regulation	Well-established regulatory guidelines, widely accepted for use in humans
Examples	ZMapp (for Ebola) and influenza vaccines	Hepatitis B prototypes developed in lettuce and potatoes	Norovirus vaccines engineered in tobacco and lettuce	Few examples exist, mostly consisting of proof-of-concept studies	Pfizer-BioNTech (for COVID-19), Gardasil (for HPV), and recombinant vaccines for Hepatitis B

and most extensively studied examples. The production of HBsAg in plants like *Nicotiana tabacum* (tobacco) and *Solanum tuberosum* (potato) involves the genetic transformation of plants to express the viral antigen. Transgenic plants expressing HBsAg can produce virus-like particles (VLPs), which are non-infectious yet immunogenic structures that resemble the hepatitis B virus, thereby triggering an immune response when administered as a vaccine (Saxena et al. 2011). The advantages of producing the hepatitis B vaccine in plants include reduced production costs, scalability, and ease of storage and distribution, especially in developing regions where cold-chain requirements pose logistical challenges. For instance, a study conducted on transgenic potatoes expressing HBsAg showed that oral administration of the plant-based vaccine induced a measurable immune response in animal models, indicating its potential as an edible vaccine (Kong et al. 2001). Nevertheless, challenges persist, particularly in maintaining consistent levels of antigen expression and securing regulatory approval for vaccines derived from plants (Ma et al. 2005).

6.2 Norovirus-Like Particle Vaccine

Noroviruses (NoV), which belong to the Caliciviridae family, are the primary global cause of acute gastroenteritis, leading to approximately 200,000 deaths each year. For about one-third of immunocompetent adults who are exposed, NoV infections result in gastrointestinal symptoms that resolve on their own. However, the impact is more severe for young children, the elderly, and those with weakened immune systems, where infections can become chronic or even life-threatening. Additionally, malnutrition has been found to play a role in the pathogenesis of NoV, increasing the risk of complications for individuals in low-resource environments (Shapiro et al. 2023). Norovirus, a highly contagious virus responsible for gastroenteritis outbreaks, has also been targeted through plant molecular farming. Norovirus VLPs can be produced in plants by expressing the capsid protein, which self-assembles into VLPs that mimic the native virus structure. The plant-based production of Norovirus-like particles has been achieved in several plant species, including *Nicotiana benthamiana* and *Lactuca sativa* (lettuce). The transient expression system using viral vectors, such as the MagnICON® system, has proven particularly effective for high-level production of Norovirus VLPs in plants. Plant-produced Norovirus VLP vaccines have been evaluated in preclinical studies, showing their potential to induce protective immunity against the virus. The immunogenicity of plant-derived VLPs has been demonstrated in animal models, where they elicited both humoral and mucosal immune responses, suggesting a possible route for oral or intranasal administration (Shapiro et al. 2023). The use of plants as biofactories for Norovirus vaccine production offers several benefits, including reduced risk of contamination with animal pathogens, cost-effectiveness, and the capacity for rapid production scaling during outbreak situations.

6.3 Conventional Plant-Based Vaccine Methods and Their Comparison with Modern Vaccines

The production of plant-based vaccines primarily involves incorporating foreign genes into plant cells. The process begins with integrating the target antigen sequence into a vector before transferring it to the expression system. The transgene can then be expressed in plants through either stable or transient transformation methods, depending on where it is inserted within the cells. Stable transformation can be achieved via nuclear or plastid integration, resulting in permanent genetic changes in the recipient cells as the target transgene becomes part of the host plant cell genome. This can be accomplished using biolistic methods or genetically modified *Agrobacterium* strains. However, the natural inability of *Agrobacterium tumefaciens* to infect many plant species restricts its use for stable transformation of desired genes. Generally, stably transgenic plant cells produce a relatively low amount of subunit antigen, ranging from 0.01 to 0.30% of total soluble plant protein (Laere et al. 2016).

Over the past decades, scientists have developed techniques for genetically modifying plants to produce foreign proteins. Initially, this was achieved through *Agrobacterium*-mediated nuclear transformation, followed by chloroplast transformation using the biolistic method (Mahmood et al. 2017; Sahu et al. 2014b; Lakra et al. 2019). These foundational approaches have paved the way for numerous transformation protocols now employed in creating various plant-based vaccines, including those produced in whole organisms, specific tissues, cell cultures, and microalgae (Monreal-Escalante et al. 2022). A particularly promising strategy involves designing and manufacturing virus-like particles (VLPs) in plants as subunit vaccines, which have proven effective in combating infectious diseases, even at industrial scales. The selection of antigens is crucial for plant-made vaccines. Researchers often choose experimental antigens that have demonstrated protective efficacy in preclinical or clinical studies, viewing plants as a potential low-cost production platform. Additionally, antigens can be identified through immune-informatic (in silico) methods (Prajapat et al. 2011a, b) such as reverse vaccinology, utilizing computational tools to predict the potential immunogenicity of pathogen proteins (Kumar et al. 2019; Ranjith et al. 2022; Sachdeva et al. 2005).

Subunit vaccines (SUVs) are a product of recombinant DNA and genetic engineering technologies, allowing for the targeted selection and production of one or more immune-protective antigens in a different organism. The rapid expansion of genomic information in databases and the ability to sequence entire genomes enable researchers to swiftly assess a pathogen protein's potential as an SUV using bioinformatics tools. These tools can be based on (a) reverse vaccinology (RV) for evaluating multiple antigens within a genome; (b) immune informatics for selecting immunogenic peptides from chosen antigens; and (c) structural vaccinology for identifying optimal three-dimensional conformations of vaccine proteins. SUVs typically have low immunogenicity, necessitating the use of adjuvants, higher doses, or booster schedules (Trott and Olson 2010).

Virus-like particles (VLPs) represent a subunit vaccine model that offers greater immunogenicity than individual antigenic proteins in three-dimensional formations. VLPs consist of repeated structural viral proteins that can assemble into particles ranging from 22 to 150 nm, with 40 nm considered optimal for dendritic cell recognition. VLPs are regarded as a safer alternative to killed or inactivated pathogen vaccines due to their lack of genetic material. They can have up to three different protein-forming layers derived from a single structural antigen. VLPs can be categorized as enveloped or non-enveloped. Non-enveloped VLPs, which do not incorporate components from the producer cell, are more commonly used in clinical trials. Enveloped VLPs, on the other hand, are more complex and include membrane components from the producer cell, with antigens exposed on the outer surface (Zeltins 2013). Conversely, transient transformation systems involve the production of desired proteins or antigens shortly after the foreign gene temporarily resides in the host cells, without being incorporated into the plant cell genome. This expression system does not require whole plant regeneration and occurs more frequently, addressing the drawbacks associated with stable integration. The two most common methods for achieving transient expression of desired proteins in plants are Agrobacterium-mediated transformation using genetically modified plant viruses and particle bombardment (Laere et al. 2016).

Molecular pharming, the technique that utilizes plants and plant cell cultures to produce high-value recombinant proteins, has made significant advancements over the past thirty years. Whether produced through stable expression in transgenic plants or using transient expression systems based on plant viruses, biopharmaceuticals have been developed to address numerous human viral diseases that have posed global pandemic threats. Plants have been effectively utilized to express various viral antigens and monoclonal antibodies. Many of these biopharmaceuticals have demonstrated considerable potential in animal studies, with several achieving success in clinical trials. The strategies and successes in developing plant-derived vaccines have been targeted at various virus-related health challenges, particularly highly contagious viral infections. Notably, it highlights the development of plant-derived biopharmaceuticals targeting Hepatitis B virus (HBV), Hepatitis C virus (HCV), the cancer-related human papillomavirus (HPV), human immunodeficiency virus (HIV), influenza virus, Zika virus, and the novel respiratory virus, severe acute respiratory syndrome coronavirus-2 (SARS-CoV-2). Additionally, the use of plant virus-derived nanoparticles (VNPs) and virus-like particles (VLPs) in the creation of plant-based vaccines is thoroughly examined (Venkataraman et al. 2021; Marwal and Gaur 2019, 2021). Comparing plant-based vaccines to traditional vaccines from mammalian cell cultures or egg-based systems reveals clear advantages. These innovative solutions offer safer and more efficient production methods that could revolutionize vaccination. One significant advantage is the lower production cost associated with plant-based systems, which do not require expensive fermentation equipment or stringent aseptic conditions. This cost reduction is especially relevant for producing vaccines for developing countries or responding to global health emergencies (Fischer et al. 2004).

Plant-derived vaccines also offer a safer production platform since there is no risk of contamination with human or animal pathogens. Unlike conventional methods, where the use of mammalian cells or chicken eggs poses a risk of contamination with animal viruses, plant systems are free from such concerns. Moreover, plant-based systems can be rapidly scaled up, which is crucial in responding to pandemics or sudden disease outbreaks. The expression of viral antigens in plants can be achieved within weeks, compared to the months required for conventional systems (Tusé and Tu 2014). However, there are limitations to plant molecular farming. Achieving consistent antigen yield and purity remains a challenge, with variability in expression levels across different batches of transgenic plants. Additionally, plant-derived vaccines (Influenza virus, Norovirus, Newcastle Disease Virus) must undergo rigorous regulatory approval to ensure their safety and efficacy, a process that can be more complex than that for traditional vaccines due to the novel nature of plant-based systems (Rybicki 2010). There are also concerns regarding the stability of plant-derived antigens (SARS-CoV-2 spike protein) during storage and transport, although lyophilization and other preservation methods have been investigated to address these issues (Ward et al. 2020).

7 Future Prospects and Innovations

Recombinant vaccines represent a significant advancement in biopharmaceutical development, offering the potential for new benchmarks and regulatory frameworks to guide the creation, approval, authorization, licensing, distribution, and marketing of plant-based vaccines. The successful approval of taliglucerase-alpha, produced in plants for enzyme replacement therapy, in conjunction with the development of a COVID-19 vaccine, illustrates the considerable promise of various plant systems—be it whole plants or plant cells—in the production of glycosylated biopharmaceuticals. Despite the discontinuation of COVIFENZ® manufacturing presenting a setback for molecular farming, recent studies highlighting successful vaccine candidate production in low- and middle-income countries (LMICs) indicate forthcoming progress in the field. The on-site production of plant-based vaccines in LMICs has the potential to mitigate challenges such as unequal vaccine distribution, prohibitive costs per dose, and complications associated with maintaining cold-chain logistics (Margolin et al. 2023; Chung et al. 2022).

Although substantial advancements have been achieved in glycol-engineering plants over the last 15 years, numerous targets remain that require modification to decrease N-glycan heterogeneity and eliminate potential immunogenic glycan epitopes. An essential objective in N-glycan engineering is the concurrent knockout of β-hexosaminidases, β-galactosidases, and Lewis A-type glycosyltransferases within a framework that incorporates XYLT/FUT knockdown or knockout strategies. Knockouts facilitated by CRISPR/Cas9 technology have achieved the removal of XYLT/FUT and P4H (Göritzer et al. 2022), laying the groundwork for the elimination of other interfering activities from expression systems such as *Nicotiana benthamiana* (Kriechbaum et al. 2020) or tobacco BY-2 cells (Mercx et al. 2017).

Current knowledge suggests that neither engineered *Nicotiana benthamiana* plants nor the derived plant cells are expected to display adverse phenotypes, hence rendering them suitable to produce recombinant glycoproteins featuring elevated levels of GnGn structures. The implementation of highly efficient gene targeting through homologous recombination in *Physcomitrella patens* will further enhance the *P. patens* expression platform by rectifying all unwanted glycan modifications (Bohlender et al. 2022).

Furthermore, to address variability and suboptimal expression levels associated with underglycosylation, it is imperative to adopt strategies that enhance N-glycosylation efficiency in commonly utilized expression systems. The stable expression of LmSTT3D, among other individual oligosaccharyltransferase (OST) subunits or engineered plant OST subunits that mimic mammalian variants, is anticipated to significantly improve existing plant-based expression systems. Progress has also been made in identifying and mitigating undesirable O-glycosylation-related modifications in plants. Given that therapeutic proteins such as erythropoietin (EPO), IgA1, etanercept, and various viral antigens contain both N- and O-glycans, a synergistic approach combining N- and O-glycan engineering methodologies is highly desirable. In addition, the presence of unwanted hydroxyproline residues at sites distinct from mammalian type O-glycosylation sites could lead to the formation of immunogenic epitopes. Transient engineering methods in *Nicotiana benthamiana* have shown that simultaneous modification of N- and O-glycans can be successfully accomplished with minimal interference from the individual glycosylation pathways. Given the advent of powerful genome editing techniques, it is likely that such multiply engineered plants or cell lines will be available in the foreseeable future (Bagdonaite and Wandall 2018; Castilho et al. 2021).

To further decrease micro-heterogeneity, there is a necessity for enhanced control over glycan processing reactions occurring in the Golgi apparatus of plants. Future initiatives will thus focus on improved regulation of the sub-compartmentalization of glycosyltransferases within the Golgi and on the engineering of glycosyltransferases to reduce substrate promiscuity. The Golgi assembly line may be refined through modifications in Golgi morphology, such as adjustments in the quantity and size of Golgi cisternae, as well as using precisely defined enzyme complexes that perform sequential modification processes, as demonstrated for plant Golgi-α-mannosidase I and GNTI. Additionally, like endeavours aimed at engineering donor nucleotide-sugar substrates, the binding site for acceptor substrates can be customized in specific glycosyltransferases to prevent the occurrence of interfering activities. A focal point for innovative protein engineering would be the development of β1,4-galactosyltransferases characterized by enhanced substrate specificity or improved catalytic efficiency, enabling the enzyme to exclusively utilize GnGn as a substrate without interference from hybrid or other intermediate N-glycan substrates. Alongside robust genome editing methodologies, strategies focused on organelle and enzyme-specific modifications are anticipated to make a substantial impact on glycol-engineering, thereby resulting in a well-defined set of plant-based expression systems conducive for recombinant glycoprotein production. Such advancements will facilitate progress towards the development of effective therapeutics (Jaiman and Thattai 2020; Cioce et al. 2021; Tan et al. 2019).

8 Conclusion

As expression and purification technologies continue to progress, it is anticipated that the usage of biologics in pharmaceutical applications will increase dramatically. Particularly, plant molecular farming has enormous potential to influence a variety of approaches and applications. Even though this technology is now mostly used for specialized applications, like the manufacturing of millions of influenza vaccine doses, it has demonstrated its potential. These achievements imply that molecular farming is in a strong position to grow its market share in the pharmaceutical sector. Prior epidemics such as Zika and Ebola, as well as the worldwide reaction to the COVID-19 pandemic, have shown that laboratories and industries may shift their platforms and tools to focus on developing vaccinations, treatments, and diagnostic reagents. Compared to traditional vaccine production, plant molecular farming requires less complex infrastructure and has the benefit of being locally deployable in many locations, which helps to overcome issues with vaccine rollout and distribution. This paradigm change has made it easier to move from studying model plant species at the laboratory scale to investigating a wider range of plants to produce secondary metabolites and recombinant proteins. Plants have successfully produced a vast range of industrial, technical, and medicinal proteins, some of which have achieved commercial viability. The development of molecular farming has largely depended on efforts to get beyond technical obstacles, especially the low yields seen in some expression systems. However, several obstacles still need to be overcome, such as resolving the environmental effects, biosafety issues, and risk assessment related to the extensive production of transgenic plants and the safety assessment of goods generated from plants. In the end, industry stakeholders oversee the promotion and delivery of commercial products. Molecular farming has encountered both technical and regulatory challenges, like any new area. Nonetheless, the industry is now positioned for faster development and more profitability in the years to come due to substantial advancements influenced by lessons learned from previous experiences.

Acknowledgements The authors are thankful to Mohanlal Sukhadia University, Udaipur, for providing the necessary facilities during the course of study.

Funding This research did not receive any specific grant from funding agencies in the public, commercial, or not-for-profit sectors.

Declaration of Competing Interests The authors declare that there are no conflicts of interest. They agree to be accountable for all aspects of the work and to ensure that questions related to the accuracy or integrity of any part of the work are appropriately investigated and resolved.

References

Altindis E, Iz SG, Ozen MO, Nartop P, Gurhan ID, Gurel A (2014) Plant derived edible vaccines and therapeutics. Front Clin Drug Res Anti Infect 1(200):10–2174

Arevalo-Villalobos JI, Alonso DO, Rosales-Mendoza S (2020a) Using carrot cells as biofactories and oral delivery vehicles of LTB-Syn: a low-cost vaccine candidate against synucleinopathies. J Biotechnol 309:75–80

Arevalo-Villalobos JI, Govea-Alonso DO, Bañuelos-Hernández B, González-Ortega O, Zarazúa S, Rosales-Mendoza S (2020b) Inducible expression of antigens in plants: a study focused on peptides related to multiple sclerosis immunotherapy. J Biotechnol 318:51–56

Bagdonaite I, Wandall HH (2018) Global aspects of viral glycosylation. Glycobiology 28(7):443–467

Balke I, Zeltins A (2019) Use of plant viruses and virus-like particles for the creation of novel vaccines. Adv Drug Deliv Rev 145:119–129

Barros GO, Woodard SL, Nikolov ZL (2011) Phenolics removal from transgenic Lemna minor extracts expressing mAb and impact on mAb production cost. Biotechnol Prog 27(2):410–418

Bevan M (1984) Binary Agrobacterium vectors for plant transformation. Nucleic Acids Res 12(22):8711–8721

Bevan MW, Flavell RB, Chilton MD (1983) A chimaeric antibiotic resistance gene as a selectable marker for plant cell transformation. Nature 304(5922):184–187

Bhatia S, Dahiya R (2015) Edible Vaccines. In: Modern Applications of Plant Biotechnology in Pharmaceutical Sciences, Academic Press, pp 333–343

Bock R (2014) Genetic engineering of the chloroplast: novel tools and new applications. Curr Opin Biotechnol 26:7–13

Bohlender LL, Parsons J, Hoernstein SN, Bangert N, Rodríguez-Jahnke F, Reski R, Decker EL (2022) Unexpected arabinosylation after humanization of plant protein N-glycosylation. Front Bioeng Biotechnol 10:838365

Breitling J, Aebi M (2013) N-linked protein glycosylation in the endoplasmic reticulum. Cold Spring Harb Perspect Biol 5(8):a013359

Buyel JF (2023) Product safety aspects of plant molecular farming. Front Bioeng Biotechnol 11:1238917

Castilho A, Schwestka J, Kienzl NF, Vavra U, Grünwald-Gruber C, Izadi S, Hiremath C, Niederhöfer J, Laurent E, Monteil V, Mirazimi A (2021) Generation of enzymatically competent SARS-CoV-2 decoy receptor ACE2-Fc in glycoengineered *Nicotiana benthamiana.* Biotechnol J 16(6):2000566

Chattopadhyay A, Jailani AA, Mandal B (2023) Exigency of plant-based vaccine against COVID-19 emergence as pandemic preparedness. Vaccine 11(8):1347

Chen Q, He J, Phoolcharoen W, Mason HS (2011) Geminiviral vectors based on bean yellow dwarf virus for production of vaccine antigens and monoclonal antibodies in plants. Hum Vaccin 7(3):331–338

Chen TH, Hu CC, Liao JT, Lee YL, Huang YW, Lin NS, Lin YL, Hsu YH (2017) Production of Japanese encephalitis virus antigens in plants using bamboo mosaic virus-based vector. Front Microbiol 8:788

Chung YH, Church D, Koellhoffer EC, Osota E, Shukla S, Rybicki EP, Pokorski JK, Steinmetz NF (2022) Integrating plant molecular farming and materials research for next-generation vaccines. Nat Rev Mater 7(5):372–388

Cioce A, Malaker SA, Schumann B (2021) Generating orthogonal glycosyltransferase and nucleotide sugar pairs as next-generation glycobiology tools. Curr Opin Chem Biol 60:66–78

Conley AJ, Zhu H, Le LC, Jevnikar AM, Lee BH, Brandle JE, Menassa R (2011) Recombinant protein production in a variety of Nicotiana hosts: a comparative analysis. Plant Biotechnol J 9(4):434–444

Dandekar AM, Fisk HJ (2004) Plant transformation: agrobacterium-mediated gene transfer. In: Transgenic plants: methods and protocols, Methods in Molecular Biology™, vol 286. pp 35–46 Humana Press. https://doi.org/10.1385/1-59259-827-7:035.
De Wilde C, Van Houdt H, De Buck S, Angenon G, De Jaeger G, Depicker A (2000) Plants as bioreactors for protein production: avoiding the problem of transgene silencing. Plant Mol Biol 43, 347–359 (2000). https://doi.org/10.1023/A:1006464304199
Dubey KK, Luke GA, Knox C, Kumar P, Pletschke BI, Singh PK, Shukla P (2018) Vaccine and antibody production in plants: developments and computational tools. Brief Funct Genomics 17(5):295–307
Engering AJ, Cella M, Fluitsma DM, Hoefsmit EC, Lanzavecchia A, Pieters J (1997) Mannose receptor mediated antigen uptake and presentation in human dendritic cells. In: Dendritic cells in fundamental and clinical immunology, vol 3, Boston, MA: Springer US. pp 183–187
Evtushenko EA, Ryabchevskaya EM, Nikitin NA, Atabekov JG, Karpova OV (2020) Plant virus particles with various shapes as potential adjuvants. Sci Rep 10(1):10365
Fifis T, Gamvrellis A, Crimeen-Irwin B, Pietersz GA, Li J, Mottram PL, McKenzie IFC, Plebanski M (2004) Size-dependent immunogenicity: therapeutic and protective properties of nano-vaccines against tumors. J Immunol 173(5):3148–3154
Fischer N, Stampacchia O, Redding K, Rochaix J (1996) Selectable marker recycling in the chloroplast. Mol Gen Genet MGG 251:373–380
Fischer R, Stoger E, Schillberg S, Christou P, Twyman RM (2004) Plant-based production of biopharmaceuticals. Curr Opin Plant Biol 7(2):152–158
Fischer R, Schillberg S, Twyman RM (2009) Molecular farming of antibodies in plants. In: Recent advances in plant biotechnology. Springer US, Boston, MA
Frisio DG, Ventura V (2021) Global innovation trends for plant-based vaccines production: a patent analysis. Plants 10(12):2558
Gallois P, Marinho P (1995) Leaf disk transformation using *Agrobacterium tumefaciens*-expression of heterologous genes in tobacco. In: Plant gene transfer and expression protocols, Methods in Molecular Biology™, vol 49. Springer, Totowa, NJ. pp 39–48
Ganapathi TR, Suprasanna P, Rao PS, Bapat VA (2004) Tobacco (*Nicotiana tabacum* L.)-a model system for tissue culture interventions and genetic engineering. Indian J Biotechnol 3:171
Gao X, Wang G, Su Q, Wang Y, An L (2007) Advances in plant transformation and transgenic crops. Biotechnol Lett 29(12):1781–1790
Gomez N, Carrillo C, Salinas J, Parra F, Borca MV, Escribano JM (1998) Expression of immunogenic glycoprotein S polypeptides from transmissible gastroenteritis coronavirus in transgenic plants. Virology 249(2):352–358
Göritzer K, Grandits M, Grünwald-Gruber C, Figl R, Mercx S, Navarre C, Ma JK, Teh AY (2022) Engineering the N-glycosylation pathway of *Nicotiana tabacum* for molecular pharming using CRISPR/Cas9. Front Plant Sci 13:1003065
Greco R, Michel M, Guetard D, Cervantes-Gonzalez M, Pelucchi N, Wain-Hobson S, Sala F, Sala M (2007) Production of recombinant HIV-1/HBV virus-like particles in *Nicotiana tabacum* and *Arabidopsis thaliana* plants for a bivalent plant-based vaccine. Vaccine 25(49):8228–8240
Guan ZJ, Guo B, Huo YL, Guan ZP, Dai JK, Wei YH (2013) Recent advances and safety issues of transgenic plant-derived vaccines. Appl Microbiol Biotechnol 97:2817–2840
Holtz BR, Berquist BR, Bennett LD, Kommineni VJ, Munigunti RK, White EL, Wilkerson DC, Wong KY, Ly LH, Marcel S (2015) Commercial-scale biotherapeutics manufacturing facility for plant-made pharmaceuticals. Plant Biotechnol J 13(8):1180–1190
Hood EE, Howard JA (2009) Over-expression of novel proteins in maize. In: Molecular genetic approaches to maize improvement. Springer Berlin Heidelberg, Berlin, Heidelberg, pp 91–105
Horn ME, Woodard SL, Howard JA (2004) Plant molecular farming: systems and products. Plant Cell Rep 22(10):711–720
Jaiman A, Thattai M (2020) Golgi compartments enable controlled biomolecular assembly using promiscuous enzymes. elife 9:e49573

Jain K, Desai N, Sharma K, Marwal A (2021) Development and screening of byproduct for its secondary metabolites, antioxidant and anti-diabetic potential from anthracnose-infected fruits of pomegranate: a sustainable approach. 3 Biotech 11:1–11

Khan I, Daniell H (2021) Oral delivery of therapeutic proteins bioencapsulated in plant cells: pre-clinical and clinical advances. Curr Opin Colloid Interface Sci 54:101452

Kim MY, Li JY, Yang MS (2017) Expression and assembly of cholera toxin B subunit and domain III of dengue virus 2 envelope fusion protein in transgenic potatoes. Protein Expr Purif 139:57–62

Kong Q, Richter L, Yang YF, Arntzen CJ, Mason HS, Thanavala Y (2001) Oral immunization with hepatitis B surface antigen expressed in transgenic plants. Proc Natl Acad Sci 98(20):11539–11544

Koropatkin NM, Cameron EA, Martens EC (2012) How glycan metabolism shapes the human gut microbiota. Nat Rev Microbiol 10(5):323–335

Kriechbaum R, Ziaee E, Grünwald-Gruber C, Buscaill P, van Der Hoorn RA, Castilho A (2020) BGAL1 depletion boosts the level of β-galactosylation of N-and O-glycans in N. benthamiana. Plant Biotechnol J 18(7):1537–1549

Kumar R, Bhanti P, Marwal A, Gaur RK (2019) Gene expression-based supervised classification models for discriminating early-and late-stage prostate cancer. Proc Natl Acad Sci India Sect B Biol Sci 90:541–565

Kurup VM, Thomas J (2020) Edible vaccines: promises and challenges. Mol Biotechnol 62(2):79–90

Kusnadi AR, Nikolov ZL, Howard JA (1997) Production of recombinant proteins in transgenic plants: practical considerations. Biotechnol Bioeng 56(5):473–484

Laere E, Ling AP, Wong YP, Koh RY, Mohd Lila MA, Hussein S (2016) Plant-based vaccines: production and challenges. J Bot 1:4928637

Lai H, Paul AM, Sun H, He J, Yang M, Bai F, Chen Q (2018) A plant-produced vaccine protects mice against lethal West Nile virus infection without enhancing Zika or dengue virus infectivity. Vaccine 36(14):1846–1852

Lakra N, Mahmood S, Marwal A, Sudheep NM, Anwar K (2019) Bioengineered plants can be an alternative source of omega-3 fatty acids for human health. In: Plant and human health, volume 2: Phytochemistry and molecular aspects, Cham: Springer International Publishing. pp 361–382

Lei X, Cai X, Yang Y (2020) Genetic engineering strategies for construction of multivalent chimeric VLPs vaccines. Expert Rev Vaccines 19(3):235–246

Lembo A, Molinaro A, De Castro C, Berti F, Biagini M (2024) Impact of glycosylation on viral vaccines. Carbohydr Polym 342:122402

Ma JK, Barros E, Bock R, Christou P, Dale PJ, Dix PJ, Fischer R, Irwin J, Mahoney R, Pezzotti M, Schillberg S (2005) Molecular farming for new drugs and vaccines: current perspectives on the production of pharmaceuticals in transgenic plants. EMBO Rep 6(7):593–599

Ma JK, Drossard J, Lewis D, Altmann F, Boyle J, Christou P, Cole T, Dale P, van Dolleweerd CJ, Isitt V, Katinger D (2015) Regulatory approval and a first-in-human phase I clinical trial of a monoclonal antibody produced in transgenic tobacco plants. Plant Biotechnol J 13(8):1106–1120

Ma F, Zhang E, Li Q, Xu Q, Ou J, Yin H, Li K, Wang L, Zhao X, Niu X, Li X (2020) A plant-produced recombinant fusion protein-based new castle disease subunit vaccine and rapid differential diagnosis platform. Vaccines (Basel) 8(1):122

Maharjan PM, Choe S (2021) Plant-based COVID-19 vaccines: current status, design, and development strategies of candidate vaccines. Vaccine 9(9):992

Mahmood S, Lakra N, Marwal A, Sudheep NÁ, Anwar K (2017) Crop genetic engineering: An approach to improve fungal resistance in plant system. In: Plant-microbe interactions in agro-ecological perspectives: volume 2: microbial interactions and agro-ecological impacts, Springer Singapore. pp 581–591

Margolin E, Oh YJ, Verbeek M, Naude J, Ponndorf D, Meshcheriakova YA, Peyret H, van Diepen MT, Chapman R, Meyers AE, Lomonossoff GP (2020) Co-expression of human calreticulin

significantly improves the production of HIV gp140 and other viral glycoproteins in plants. Plant Biotechnol J 18(11):2109–2117
Margolin E, Schäfer G, Allen JD, Gers S, Woodward J, Sutherland AD, Blumenthal M, Meyers A, Shaw ML, Preiser W, Strasser R (2023) A plant-produced SARS-CoV-2 spike protein elicits heterologous immunity in hamsters. Front Plant Sci 14:1146234
Marwal A, Gaur RK (2019) Nanophytovirology: an emerging field for disease management. In: Plant diseases-current threats and management trends. IntechOpen. UNITED KINGDOM https://doi.org/10.5772/intechopen.86653
Marwal A, Gaur RK (2021) Plant viruses as an engineered nanovehicle (PVENVs). In: Plant virus-host interaction: molecular approaches and viral evolution. Academic Press, pp 525–536
Marwal A, Sahu A, Prajapat R, Gaur RK (2013a) First report of Begomovirus infecting two ornamental plants: *Jasminum sambac* and *Millingtonia hortensis*. Indian Phytopathol 66(1):115–116
Marwal A, Sahu AK, Gaur RK (2013b) First report on the association of a begomovirus with Chrysanthemum indicum exhibiting yellowing of leaf vein disease characterized by molecular studies. J Hortic Res 21(2):17–21
Marwal A, Sahu AK, Gaur RK (2013c) Molecular characterization of begomoviruses and DNA satellites associated with a new host Spanish Flag (*Lantana camara*) in India. Int Schol Res Notice 2013(1):915703
Marwal A, Meena M, Gaur RK (2021) Molecular docking studies of coronavirus proteins with medicinal plant-based phytochemicals. Def Life Sci J 6(1):57–63
Massa S, Presenti O, Benvenuto E (2018) Engineering plants for the future: farming with value-added harvest. Prog Botany 80:65–108
Mbewana S, Mortimer E, Pera FFPG, Hitzeroth II, Rybicki EP (2015) Production of H5N1 influenza virus matrix protein 2 ectodomain protein bodies in tobacco plants and in insect cells as a candidate universal influenza vaccine. Front Bioeng Biotechnol 3:197
Mercx S, Smargiasso N, Chaumont F, De Pauw E, Boutry M, Navarre C (2017) Inactivation of the β(1,2)-xylosyltransferase and the α(1,3)-fucosyltransferase genes in Nicotiana tabacum BY-2 cells by a multiplex CRISPR/Cas9 strategy results in glycoproteins without plant-specific glycans. Front Plant Sci 8:403
Monreal-Escalante E, Ramos-Vega A, Angulo C, Bañuelos-Hernández B (2022) Plant-based vaccines: antigen design, diversity, and strategies for high level production. Vaccine 10(1):100
Muthamilselvan T, Lee CW, Cho YH, Wu FC, Hu CC, Liang YC, Lin NS, Hsu YH (2016) A transgenic plant cell-suspension system for expression of epitopes on chimeric Bamboo mosaic virus particles. Plant Biotechnol J 14(2):231–239
Nehra C, Sahu AK, Marwal A, Gaur RK (2014) Natural occurrence of Clerodendron yellow mosaic virus on Bougainvillea in India. New Dis Rep 30(19):2044–0588
Nehra C, Marwal A, Verma RK, Mishra M, Sharma P, Gaur RK (2019) Papaya yellow leaf curl virus: a newly identified begomovirus infecting *Carica papaya* L. from the Indian Subcontinent. J Hortic Sci Biotechnol 94(4):475–480
Nikolov Z, Hammes D (2002) Production of recombinant proteins from transgenic crops. In: Hood EE, Howard JA (eds) Plants as factories for protein production. Springer, Netherlands, pp 111–130
Nosaki S, Hoshikawa K, Ezura H, Miura K (2021) Transient protein expression systems in plants and their applications. Plant Biotechnol (Tokyo) 38(4):297–304
Ponndorf D, Meshcheriakova Y, Thuenemann EC, Dobon Alonso A, Overman R, Holton N, Dowall S, Kennedy E, Stocks M, Lomonossoff GP, Peyret H (2021) Plant-made dengue virus-like particles produced by co-expression of structural and non-structural proteins induce a humoral immune response in mice. Plant Biotechnol J 19(4):745–756
Prajapat R, Marwal A, Sahu A, Gaur RK (2011a) Phylogenetics and in silico docking studies between coat protein of Mimosa yellow vein virus and whey α-lactalbumin. Am J Biochem Mol Biol 1(3):265–274
Prajapat R, Gaur RK, Marwal A (2011b) Homology modeling and docking studies between AC1 rep protein of Begomovirus and Whey á-lactalbumin. Asian J Biol Sci 4(4):352–361

Ranjith S, Sathyan A, Duraisamy S, Peter AS, Marwal A, Jain K, Chidambaram P, Kumarasamy A (2022) Exploring a computational method for evaluating the Epinecidin-1 and its variants binding efficacy with breast Cancer receptor (HER-2). Int J Pept Res Ther 28(4):118

Rebelo BA, Folgado A, Ferreira AC, Abranches R (2022) Production of the SARS-CoV-2 Spike protein and its Receptor Binding Domain in plant cell suspension cultures. Front Plant Sci 13:995429

Rosales-Mendoza S, Nieto-Gómez R, Angulo C (2017) A perspective on the development of plant-made vaccines in the fight against Ebola virus. Front Immunol 8:252

Ruocco V, Strasser R (2022) Transient expression of glycosylated SARS-CoV-2 antigens in *Nicotiana benthamiana*. Plants (Basel) 11(8):1093

Rutkowsk DA, Mokoena N, Tsekoa TL, Dibakwane VS, O'Kennedy MM (2019) Plant-produced chimeric virus-like particles—a new generation vaccine against African horse sickness. BMC Vet Res 15:1–15

Rybicki EP (2010) Plant-made vaccines for humans and animals. Plant Biotechnol J 8(5):620–637

Rybicki EP, Martin DP (2011) Virus-derived ssDNA vectors for the expression of foreign proteins in plants. Curr Top Microbiol Immunol 375:19–45

Sachdeva G, Kumar K, Jain P, Ramachandran S (2005) SPAAN: a software program for prediction of adhesins and adhesin-like proteins using neural networks. Bioinformatics 21(4):483–491

Sack M, Rademacher T, Spiegel H, Boes A, Hellwig S, Drossard J, Stoger E, Fischer R (2015) From gene to harvest: insights into upstream process development for the GMP production of a monoclonal antibody in transgenic tobacco plants. Plant Biotechnol J 13(8):1094–1105

Sahoo A, Mandal AK, Dwivedi K, Kumar V (2020) A cross talk between the immunization and edible vaccine: current challenges and future prospects. Life Sci 261:118343

Sahu AK, Marwal A, Nehra C, Choudhary DK, Sharma P, Gaur RK (2014a) RNAi mediated gene silencing against betasatellite associated with Croton yellow vein mosaic begomovirus. Mol Biol Rep 41:7631–7638

Sahu AK, Marwal A, Shahid MS, Nehra C, Gaur RK (2014b) First report of a begomovirus and associated betasatellite in *Rosa indica* and in India. Aust Plant Dis Notes 9:1–4

Satheeshkumar PK (2020) Expression of single chain variable fragment (scFv) molecules in plants: a comprehensive update. Mol Biotechnol 62(2):151–167

Saxena P, Hsieh YC, Alvarado VY, Sainsbury F, Saunders K, Lomonossoff GP, Scholthof HB (2011) Improved foreign gene expression in plants using a virus-encoded suppressor of RNA silencing modified to be developmentally harmless. Plant Biotechnol J 9(6):703–712

Schillberg S, Finnern R (2021) Plant molecular farming for the production of valuable proteins–critical evaluation of achievements and future challenges. J Plant Physiol 258:153359

Schillberg S, Raven N, Fischer R, Twyman M, Schiermeyer A (2013) Molecular farming of pharmaceutical proteins using plant suspension cell and tissue cultures. Curr Pharm Des 19(31):5531–5542

Scotti N, Rybicki E (2013) Virus-like particles produced in plants as potential vaccines. Expert Rev Vaccines 12(2):211–224

Shahid N, Samiullah TR, Shakoor S, Latif A, Yasmeen A, Azam S, Shahid AA, Husnain T, Rao AQ (2020) Early stage development of a Newcastle disease vaccine candidate in corn. Front Veterin Sci 7:499

Shanmugaraj B, Bulaon CJI, Phoolcharoen W (2020) Plant molecular farming: a viable platform for recombinant biopharmaceutical production. Plants 9(7):842

Shapiro JR, Andreani G, Dubé C, Berubé M, Bussière D, Couture MM, Dargis M, Hendin HE, Landry N, Lavoie PO, Pillet S (2023) Development and characterization of a plant-derived norovirus-like particle vaccine. Vaccine 41(41):6008–6016

Shi Y, Habibi P, Haq AN, Saeed M, Gulghutay Amjad N, Khan I (2023) Seed-based system for cost-effective production of vaccine against chronic respiratory disease in chickens. Mol Biotechnol 65:570–580

Singh A, Rajput V, Singh AK, Sengar RS, Singh RK, Minkina T (2021) Transformation techniques and their role in crop improvements: a global scenario of GM crops. In: Policy Issues in Genetically Modified Crops, Elsevier, pp 515–542, Academic Press.

Spiegel H, Stöger E, Twyman RM, Buyel JF (2018) Current status and perspectives of the molecular farming landscape. In: Kermode AR, Jiang L (eds) Molecular farming: applications, challenges, and emerging areas. John Wiley & Sons, Inc., pp 3–22

Spok A, Karner S (2008) Plant molecular farming: opportunities and challenges. In: Stein AJ, Rodriguez-Cerezo J (eds) Plant molecular farming: opportunities and challenges. Institute for Prospective Technological Studies, JRC Scientific and Technical Reports. Spain, pp 1–22

Srivastava A, Pandey V, Singh N, Marwal A, Shahid MS, Gaur RK (2024) In silico identification of papaya genome-encoded microRNAs to target begomovirus genes in papaya leaf curl disease. Front Microbiol 15:1340275

Stieger M (1987) Versuche zur Integration und Expression chimärer Immunglobuline in Pflanzen [Attempts at integration and expression of chimeric immunoglobulins in plants] (Doctoral dissertation, University of Cologne)

Stieger M, Neuhaus G, Momma T, Schell J, Kreuzaler F (1991) Self assembly of immunoglobulins in the cytoplasm of the alga *Acetabularia mediterranea*. Plant Sci 73(1–2):181–190

Stöger E, Ma JKC, Fischer R, Christou P (2005) Sowing the seeds of success: pharmaceutical proteins from plants. Curr Opin Biotechnol 16(2):167–173

Streatfield SJ, Jilka JM, Hood EE, Turner DD, Bailey MR, Mayor JM, Woodard SL, Beifuss KK, Horn ME, Delaney DE, Tizard IR (2001) Plant-based vaccines: Unique advantages. Vaccine 19(17–19):2742–2748

Swapnil P, Meena M, Singh SK, Dhuldhaj UP, Marwal A (2021) Vital roles of carotenoids in plants and humans to deteriorate stress with its structure, biosynthesis, metabolic engineering and functional aspects. Curr Plant Biol 26:100203

Takeyama N, Kiyono H, Yuki Y (2015) Plant-based vaccines for animals and humans: recent advances in technology and clinical trials. Therap Adv Vaccines 3(2):139–154

Tan Y, Zhang Y, Han Y, Liu H, Chen H, Ma F, Withers SG, Feng Y, Yang G (2019) Directed evolution of an alpha1,3-fucosyltransferase using a single-cell ultrahigh-throughput screening method. Sci Adv 5(10):eaaw8451

Tekoah Y, Shulman A, Kizhner T, Ruderfer I, Fux L, Nataf Y, Bartfeld D, Ariel T, Gingis-Velitski S, Hanania U, Shaaltiel Y (2015) Large-scale production of pharmaceutical proteins in plant cell culture the protalix experience. Plant Biotechnol J 13:1199–1208

Trott O, Olson AJ (2010) AutoDock Vina: improving the speed and accuracy of docking with a new scoring function, efficient optimization, and multithreading. J Comput Chem 31(2):455–461

Tusé D, Tu T, McDonald KA (2014) Manufacturing economics of plant-made biologics: case studies in therapeutic and industrial enzymes. Biomed Res Int 1:256135

Tzfira T, Citovsky V (2006) Agrobacterium-mediated genetic transformation of plants: biology and biotechnology. Curr Opin Biotechnol 17(2):147–154

Vahdat MM, Hemmati F, Ghorbani A, Rutkowska D, Afsharifar A, Eskandari MH, Rezaei N, Niazi A (2021) Hepatitis B core-based virus-like particles: a platform for vaccine development in plants. Biotechnol Rep 29:e00605

Valenzuela P, Medina A, Rutter WJ, Ammerer G, Hall BD (1982) Synthesis and assembly of hepatitis B virus surface antigen particles in yeast. Nature 298(5875):347–350

van Eerde A, Gottschamel J, Bock R, Hansen KE, Munang'andu HM, Daniell H, Liu Clarke J (2019) Production of tetravalent dengue virus envelope protein domain III-based antigens in lettuce chloroplasts and immunologic analysis for future oral vaccine development. Plant Biotechnol J 17(6):1408–1417

Venkataraman S (2022) Plant molecular pharming and plant-derived compounds towards generation of vaccines and therapeutics against Coronaviruses. Vaccines (Basel) 10(11):1805

Venkataraman S, Hefferon K, Makhzoum A, Abouhaidar M (2021) Combating human viral diseases: will plant-based vaccines be the answer? Vaccine 9(7):761

Venkataraman S, Khan I, Habibi P, Le M, Lippert R, Hefferon K (2023) Recent advances in expression and purification strategies for plant made vaccines. Front Plant Sci 14:1273958

Verma RK, Mishra M, Marwal A, Akhtar A, Manchev N, Petrov RK (2024) CRISPR/Cas9 protection system for Chilli Leaf Curl Virus. Acta Microbiol Bulg 40(1):112–115

Vighi G, Marcucci F, Sensi L, Di Cara G, Frati F (2008) Allergy and the gastrointestinal system. Clin Exp Immunol 153(Suppl 1):3–6

Ward BJ, Landry N, Trépanier S, Mercier G, Dargis M, Couture M, D'Aoust MA, Vézina LP (2014) Human antibody response to N-glycans present on plant-made influenza virus-like particle (VLP) vaccines. Vaccine 32(48):6098–6106

Ward BJ, Gobeil P, Séguin A, Atkins J, Boulay I, Charbonneau PY, Couture M, D'Aoust MA, Dhaliwall J, Finkle C, Hager K (2020) Phase 1 trial of a candidate recombinant virus-like particle vaccine for Covid-19 disease produced in plants. medRxiv

Ward BJ, Seguin A, Couillard J, Trepanier S, Landry N (2021) Phase III: randomized observer-blind trial to evaluate lot-to-lot consistency of a new plant-derived quadrivalent virus-like particle influenza vaccine in adults 18-49 years of age. Vaccine 39(8):1528–1533

Xing H, Lawrence C, Chambers O, Davies HM, Everett N, Li Q (2006) Transformation of cereal plants with *Agrobacterium tumefaciens*. Planta 223:1024–1034

Xu JF, Ge XM, Dolan MC (2011) Towards high-yield production of pharmaceutical proteins with plant cell suspension cultures. Biotechnol Adv 29:278–299

Yadav J, Phogat S, Chaudhary D, Jaiwal R, Jaiwal PK (2023) Synthesis of plant-based, self-adjuvanted, dual antigen specific to *Mycobacterium tuberculosis* as a novel tuberculosis subunit vaccine that elicits immunogenicity in rabbit. Biotechnol Lett 45(4):703–717

Yang M, Sun H, Lai H, Hurtado J, Chen Q (2017) Plant-produced Zika virus envelope protein elicits neutralizing immune responses that correlate with protective immunity against Zika virus in mice. Plant Biotechnol J 16(2):572–580

Yusibov V, Rabindran S (2008) Recent progress in the development of plant-derived vaccines. Expert Rev Vaccines 7(8):1173–1183

Yusibov V, Streatfield SJ, Kushnir N, Roy G, Padmanaban A (2013) Hybrid viral vectors for vaccine and antibody production in plants. Curr Pharm Des 19(31):5574–5586

Zahedipour F, Zahedipour F, Zamani P, Jaafari MR, Sahebkar A (2024) Harnessing CRISPR technology for viral therapeutics and vaccines: from preclinical studies to clinical applications. Virus Res 341:199314

Zeltins A (2013) Construction and characterization of virus-like particles: a review. Mol Biotechnol 53(1):92–107

Zhang H, Liu M, Li Y, Zhao Y, He H, Yang G, Zheng C (2010) Oral immunogenicity and protective efficacy in mice of a carrot-derived vaccine candidate expressing UreB subunit against *Helicobacter pylori*. Protein Expr Purif 69(1):127–131

Biopharmaceutical Production in Plants with Plant Viruses

Ganesa Moorthy Murugan and Jebasingh Tennyson

Abstract

Plant molecular farming has the potential to alter the biopharmaceuticals sector by providing novel options for vaccine manufacturing and therapeutic protein production. Plant viruses have emerged as a strong tool for biopharmaceutical manufacturing, with significant benefits over traditional approaches, and have the potential to transform biologics manufacturing and contribute considerably to the future of medicine and agriculture. Plant viruses are expression vectors that can be used to produce high-value pharmaceutical proteins such as vaccines and antibodies. Virus-like particles (VLPs) and virus nanoparticles (VNPs) act as carrier templates and target vehicles that display peptides and synthetic molecules, making ideal platforms for drug delivery. This chapter provides an overview of biopharmaceutical production in plants with plant viruses, with special emphasis on biotechnological advances in the field of viral vectors and agroinfection in molecular farming.

Keywords

Biopharmaceuticals · Molecular farming · Plant · Virus · Vaccine · Therapeutic · Nanoagents · Drug · Virus-like particles (VLPs) · Virus nanoparticles (VNPs)

G. M. Murugan · J. Tennyson (✉)
Department of Plant Sciences, School of Biological Sciences, Madurai Kamaraj University, Madurai, Tamil Nadu, India
e-mail: jebasingh.biological@mkuniversity.ac.in

R. K. Gaur et al. (eds.), *Plant Molecular Farming*,
https://doi.org/10.1007/978-981-95-3823-2_11

1 Introduction

Biopharmaceuticals are medicinal products produced or derived from biological sources (Ecker and Seymour 2020). Biopharmaceuticals provide several benefits in therapeutic applications, including illness prevention, treatment, and diagnostics. The present spectrum of clinical biopharmaceutical products includes mAbs, enzymes, vaccines, human growth hormones, cytokines, nucleic acids, and carbohydrates (Ghanemi 2017; Loke et al. 2024). Commonly, a fermentor or bioreactor is used to generate therapeutic human proteins from mammalian cell cultures. This does not meet the projected yearly requirement for biotherapeutics, which ranges in double-digit tons, notwithstanding a growth in production facilities and biopharmaceutical expression levels. Alternatively, molecular farming can be tailored to utilize plants which are capable of producing multi-component, functionally active human proteins that have intricate posttranslational modifications (Eidenberger et al. 2023). The ability to produce sophisticated, functional biopharmaceuticals in plants is becoming more feasible and has the potential to transform biopharmaceutical production by providing a scalable and efficient platform to meet global health demands (Amanat et al. 2020; Kallolimath et al. 2021; Sun et al. 2021). Plant systems, unlike mammalian cell cultures, are devoid of human diseases, which reduces contamination hazards and improves the safety profile of final products.

Plant-based viruses have emerged as a viable platform for producing biopharmaceuticals, including vaccines, antibodies, enzymes, and therapeutic proteins. Plant viral vectors are customized forms of viruses that employ their inherent capacity to enter and proliferate in plant cells. This allows for transitory production of recombinant proteins in plants, making them useful instruments for generating biopharmaceuticals. The procedure normally entails the cloning of the desired gene in the viral genome, infecting plant cells by agroinfiltration or mechanical inoculation, and collecting the target protein. Plant-based viral systems have various advantages, including scalability, cost-effectiveness, quick production, and protection against contamination problems associated with mammalian systems. While there are limitations, such as variances in glycosylation patterns, regulatory barriers, and purifying complications in plant-based systems, advances in glycoengineering and synthetic biology are increasing the therapeutic relevance and efficiency of plant-based manufacturing systems. Despite these challenges, plant-based viral vectors have been successfully used to create vaccines (e.g., against COVID-19), monoclonal antibodies (e.g., ZMapp for Ebola), and therapeutic enzymes for disorders such as Gaucher's (Jugler et al. 2021). Plant viral vectors, such as those generated through *Tobacco mosaic virus* (TMV) and *Potato virus X* (PVX), have allowed for fast, transitory production of recombinant proteins, making them particularly useful in reacting to pandemics or emerging infectious illnesses. Thus, plant-based viral vector systems are a potent and diverse tool for biopharmaceutical manufacturing, with substantial implications for global health and the availability of cheap treatments (Abrahamian et al. 2020; Hefferon 2017; Yusibov et al. 2013).

Plant viruses have remarkable properties, which, combined with emerging genetic engineering and synthetic biology technology, make these approaches attractive for the mass and safe manufacture of biopharmaceuticals. The use of plant viruses as a vehicle for the localization of new strategies to treat and prevent diseases remains a core area of research and design as epidemiological trends continue to evolve (Abrahamian et al. 2020; Gleba et al. 2007; Gomord and Faye 2004; Lico et al. 2008).

Over the past few years, the introduction of synthetic biology and genome editing technologies like CRISPR/Cas9 has innovated plant virology. This has opened up new possibilities to precisely engineer viral genomes at a position, enabling the introduction of multiple foreign genes or specific mutations that would improve viral vector properties. Such improvements have now opened up the potential for far more biopharmaceuticals, helped by plant viruses, even to complicated proteins and multivalent vaccines (Musiychuk et al. 2007). The platform itself lets researchers more quickly create and test a new viral vector, allowing them to respond as needed when facing the next health crisis, which is especially appealing in light of emerging infectious diseases (Eidenberger et al. 2023; Venkataraman and Hefferon 2021).

Using plant viruses to make biopharmaceuticals comes with its share of problems. One big issue is how much protein gets made, which can change based on the type of plant used, the growing conditions, and the specific virus chosen. Scientists are working hard to fine-tune these factors to make sure they can produce lots of foreign proteins in different plant setups. In producing plant viruses- based biopharmaceuticals, there is an interesting mechanism of action that is both sophisticated and quite brilliant. This mechanism employs the viral life cycle in the expression of alien genes in plants. There are various steps involved in the attachment, entry, replication, assembly, movement, and release, where they operationalize the benefits of plant viruses as delivery systems for the production of therapeutics (Abrahamian et al. 2020; Gleba et al. 2007; Gomord and Faye 2004; Lico et al. 2008).

Recent developments and trends in the use of plant-based viruses to produce biopharmaceuticals have concentrated on viral vector optimization and development of hybrid viral vectors to increase the effectiveness and scalability of protein expression in plants, and agroinfiltration methodologies to enhance the viral vector delivery. Another important development is the ability to genetically modify plants to imitate human-like posttranslational modifications, offering a scalable and adaptable platform (Yusibov et al. 2013). These developments are establishing plant-based viral vector systems as a viable alternative to traditional cell-based biopharmaceutical manufacturing, providing faster, safer, and more cost-effective answers to global healthcare concerns. The current chapter explores biopharmaceutical production through plant-based viruses, covering plant-derived vectors, mechanisms of action, production methodologies, biopharmaceutical product development, upstream and downstream processing, economic and environmental considerations, as well as challenges and future prospects.

2 Plant Viruses as Expression Vectors

Plant viral expression vectors are developed from the genomes of single-stranded DNA viruses and positive-sense RNA viruses (Gleba et al. 2007; Kagale et al. 2012; Klimyuk et al. 2012). The entire viral genome is included in first-generation virus expression vectors. The foreign gene of interest is expressed either independently from a strong subgenomic promoter that is incorporated into the viral genome or as part of a fusion protein. Through insertion into exterior regions of the viral capsid protein, vaccine epitopes are displayed on the surface of virus particles. The full-virus approach has a number of drawbacks, including constraints on the size of the foreign protein that may be produced.

Conversely, second-generation viral expression vectors surpass tissue and host plant constraints, have higher production levels, and do not impose size limitations on foreign genes. The foreign gene of interest and the minimal number of viral components needed for multiplication make up these novel "deconstructed vectors" (Gleba et al. 2005; Yusibov et al. 2013). Due to the elimination of genes necessary for virus assembly and movements, deconstructed vectors need to be transferred into the host plant through vacuum infiltration of the *Agrobacterium* suspension containing the expression vector (Leuzinger et al. 2013). All plant tissues can produce the necessary medicinal protein, synchronously increasing protein production in a reduced time period.

A third generation plant virus approach that employs viruses as nanoparticles. Plant virus nanoparticle-based immunotherapy has been shown to elicit a stable, nontoxic, and biodegradable localized immune response against cancer cells (Shukla et al. 2013). Plant viruses with a rod-like shape, like potexviruses, can have their exterior surfaces functionalized to target and deliver antiviral agents into tumors. Icosahedral plant virus, such as the *Cowpea mosaic virus*, can be modified to carry cargo molecules, such as drugs, within their internal cavities. Plant viruses as nanoparticles can be used for diagnosis as well as for immunotherapy (Kagale et al. 2012; Lico et al. 2013).

3 Modifying the Virus as the Vector

The utilization of many significant alterations to optimize the efficiency and yield of foreign protein synthesis is a fundamental characteristic of the widely utilized viral systems for plant gene expression. First, potent transgenic expression depends on active viral promoters. Through the synthesis of mRNA and subsequently the functional proteins, these promoters facilitate the transcription and translation of foreign genes, leading to protein expression levels that greatly exceed those of conventional proteins (Chen and Gonçalves 2016).

Second, the foreign gene is amplified within the host plant during the viral replication processes. Viral structural proteins interact with specific viral RNA sequences that regulate replication, ensuring the efficient recognition and packaging of the persistent viral RNA after infection. Progeny virions are generated very effectively, which raises transgene expression to a high degree of protein synthesis. However, in certain cases, it is feasible to co-express heterologous proteins with viral structural proteins, or they may be produced independently to create virus-like particles (VLPs), which are very helpful for vaccine development since they are immunogenic and safe (Goldstein and Thomas 2004).

The ability of the plant virus to spread systemically throughout the host plant is an essential factor in the foreign gene's transmission. Viral movement proteins enable the modified virus to exploit plasmodesmata, the connections that exist between plant cells, enabling it to travel from cell to cell and infect the majority of the plant's tissue. The foreign gene is widely expressed thanks to this systemic mobility, which increases the biopharmaceutical product's total yield. Key benefits of plant virus-based expression systems include the systemic and virulent nature of the viral infection cycle, in contrast to other expression systems that restrict infectivity to specific cells (Gerlach et al. 2010; Loh et al. 2017; Yusibov et al. 2006). Using single-stranded RNA viruses and *Agrobacterium tumefaciens* to introduce viral genes into plant cells was a significant advancement in the creation of plant virus-based expression systems (Grimsley et al. 1986). Studies have examined the possibility that a variety of plant viruses might serve as beneficial vectors for plant genome engineering (Zahmanova et al. 2023) (Table 1). TMV (Gleba et al. 2005), PVX (Mardanova et al. 2017), *Cowpea mosaic virus* (CPMV) (Sainsbury et al. 2008), and several others are the most well-known of these.

4 RNA Virus-Based Vector Systems

4.1 Tobamovirus-Based Vectors

The genome of TMV RNA is 6.4 kb. There are four main proteins encoded by TMV RNA. The 126-kDa and 183-kDa replicase proteins are translated from the genomic RNA's initial open reading frame (ORF), with the latter being translated by sporadic readthrough of the 126-kDa protein's amber stop codon. The 3-coterminal subgenomic (sg) RNAs from the 3-proximal ORFs express the 30-kDa movement protein (MP) and the 17.4-kDa coat protein (CP) (Goelet et al. 1982). The 126- and 183-kDa replicase proteins are necessary for viral RNA replication, although the MP and CP are permissive. While the 3′-proximal CP gene of I2 RNA is translationally silent, the MP is produced by translation of the sg bicistronic intermediate-length RNA-2, also known as I2 sgRNA (Palukaitis and Zaitlin 1986). Low-molecular component (LMC), a tiny monocistronic sgRNA, is the source of this gene's expression. The sg promoter of the CP gene is exploited by the vectors based on the genomes of *Tomato mosaic virus* (ToMV) (Dohi et al. 2008), crucifer-infecting Tobamovirus (crTMV), *Turnip vein clearing virus* (TVCV) (Marillonnet et al. 2004), and TMV (Chapman 2008). This allows the synthesis of the desired protein directly either from its own or an additional CP sg promoter (Fig. 1a). Its activity is best in vectors where the CP is coupled with a tiny immunogenic epitope (Smith et al. 2009). The CP sgRNA promoter in RNA plant viruses extends downstream of

Table 1 List of viruses used for the design of viral vectors (Zahmanova et al. 2023)

Virus classification/genome organization	Virus	Vector	Achieved high yield	Recombinant proteins	References
Tobamovirus + ssRNA	*Tobacco mosaic virus* (TMV)	magnICON	5 mg/g Fresh weight (FWT)	Green Fluorescent Protein (GFP)	Marillonnet et al. (2004)
		TMV RNA-based overexpression (TRBO)	5.5 mg/g FWT	GFP	Wang et al. (2012)
Geminivirus circular DNA	*Tobacco yellow dwarf mastrevirus* (TYDV)	INPACT (In Plant Activation)	0.1 mg/g FWT	Human vitronectin	Dugdale et al. (2013)
	Bean yellow dwarf virus (BeYDV)	pBYRp19	0.5 mg/g FWT	mAb 6D8 against *Ebola virus*	Chen et al. (2011)
		pRIC	0.55 mg/g FWT	Human papillomavirus (HPV) coat protein (CP) L1	Regnard et al. (2010)
Comovirus + ssRNA	*Cowpea mosaic virus* (CPMV)	delRNA-2	–	GFP	Cañizares et al. (2006)
		pEAQ-HT	1.5 mg/g	GFP	Sainsbury et al. (2009)

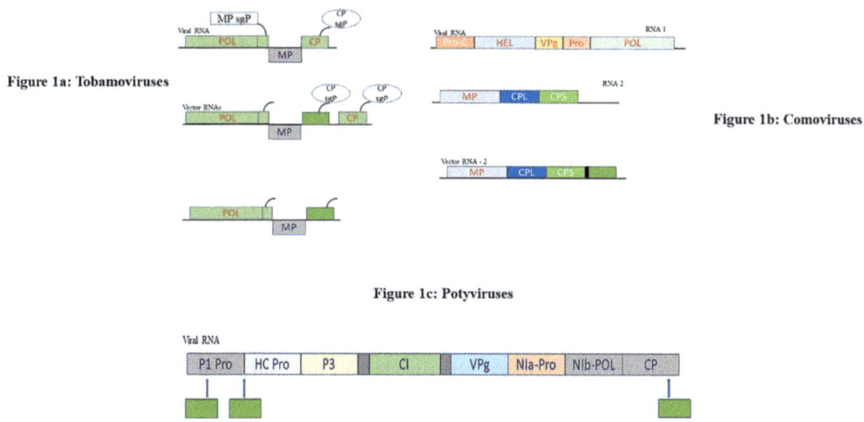

Fig. 1 Schematic representation of viral genomes and vectors used to express heterologous peptides and proteins in plants. Target protein genes are marked with a green box, and their positions in the viral vectors are indicated. The dark green box represents a leaky termination codon. Cl-cylindrical inclusion protein, CP-coat protein, CPL-large coat protein, CPS-small coat protein, HC Pro-helper component proteinase, Hel-helicase, MP-movement protein, NIa-Pro-viral proteinase, NIb-RNA-dependent RNA polymerase, P3-protein 3, POL-RNA polymerase, Pro- proteinase, sgP subgenomic promoter (marked with arrows), VPg-virus protein genome-linked (Komarova et al. 2010)

the transcription beginning point. A further method of vector-directed foreign gene expression is provided by the internal ribosome entry site (IRES) sequence.

CP mRNA of TMV is translated by conventional cap-dependent ribosome scanning. Still, vectors derived from the crTMV genome offer an additional expression approach that makes use of the IRES. Although IRES-mediated translation may produce polycistronic templates, it is inherently less efficient than cap-dependent translation (Dorokhov et al. 2006).

The first virus to be designed as a deconstructed vector system was *Tobacco mosaic virus* (TMV), a type member of Tobamovirus (Gleba et al. 2005). A total of 2130 copies of the capsid proteins cover the positive-sense RNA genome of the hard rod-shaped viral particle of TMV. The replicase, the part of the genome in charge of viral replication, is found in one of the two modules that make up the expression system. The location of foreign gene expression is contained in the other module. Full-length proteins and epitopes in the form of fusion proteins were both expressed using TMV (Huang et al. 2010; Musiychuk et al. 2007; Noris et al. 2011). Magnification is a method for transfecting plants with these recombinant viral vector modules that was created by the German biotechnology company, Icon Genetics (Gleba et al. 2005). Magnification is the delivery of a deconstructed vector that is incapable of spreading to other plants in conjunction with agroinfiltration. To produce Norwalk virus-like particles (VLPs) in plants and exhibit Influenza M2e and human papillomavirus (HPV) E7 epitopes, a deconstructed TMV vector has been used (Massa et al. 2007; Musiychuk et al. 2007; Santi et al. 2008). A TMV-based "launch vector" that is contained within an *Agrobacterium tumefaciens* binary vector was developed to create vaccines against cholera, the influenza virus, and the plague (Musiychuk et al. 2007) (Chichester et al. 2009; Hamorsky et al. 2013; Petukhova et al. 2014).

The co-expression of the *Tomato bushy stunt virus* RNA silencing suppressor gene P19 can also greatly increase the expression of foreign proteins in TMV-infected plants (Voinnet et al. 2015). Other TMV-based vectors have been developed, such as the TMV RNA-based overexpression (TRBO) vector, which removes the coat protein gene and positions the foreign gene open reading frame (ORF) nearer the 3′ end of the TMV RNA, enhancing the production of foreign protein (Lindbo 2007). An influenza vaccine developed with TMV has been well characterized. For instance, the temperature at which the viral vector was inoculated affected the spatial expression of the hemagglutinin (HA) concentration in *Nicotiana benthamiana* leaves (Matsuda et al. 2017). The ideal temperature needed to get a maximal and consistent production of HA is 20 °C. The migration of viral gene products can also be impacted by light intensity, as demonstrated by Patil and Fauquet (2015). Based on the *Sun hemp mosaic virus* (SHMV), Liu and Kearney (2010) have created a *Tobamovirus* that targets legumes and consists of some of the features of the strategies created by Icon and TRBO vectors. Their "SHEC" (SHMV elimination coat protein gene) vector replicates very poorly without the silencing suppressor P19 and is unable to generate virions due to its lack of a capsid protein. These Tobamovirus-based technologies can further control the manufacture of vaccines in highly confined and inducible environments. The recent finding of adjuvant capabilities of TMV has spurred a renewed interest in the use of this virus as a delivery vehicle for immunotherapy. Strong CD8+ T cell responses have been shown to be produced by TMV particles' uptake by dendritic cells (Kemnade et al. 2014; Liu et al. 2013). TMV particles as an adjuvant and an epitope display system have been developed to create a vaccine against the facultative intracellular pathogen *Francisella tularensis* in mice (Banik et al. 2015). The mice did not experience any negative reactions with very high doses of *F. tularensis* Live Vaccine Strain (LVS) and are shielded from respiratory difficulties. Similarly, Jones et al. (2015) developed a malaria transmission blocking vaccine (TBV) that targets Pfs25 proteins produced in the mosquito midgut during *Plasmodium falciparum* development by using this virus-based delivery platform, to lessen the spread of malaria. Additionally, HPV vaccines are based on the VLPs with the L1 protein produced by the TMV deconstructed vector. Personalized medicine attempts against non-Hodgkin's lymphoma (NHL) are arguably the most well-known application of the MagnICON deconstructed vector (McCormick et al. 2008).

In deconstructed MagnICON vectors, TMV constructs consisting of the scFv subunit and full-length idiotype IgG molecules were produced as heavy and light chains, which assembled into complete immunoglobulins in the plant (Bendandi et al. 2010; McCormick 2011). Using the tumor-homing peptide-cRGD, which has been functionalized to the virus's surface, the *Tobacco mosaic virus* has been modified to be tumor-specific. When TMV particles exhibit a tumor-homing peptide-cRGD, they have the ability to quickly integrate into tumor cells. In order to be absorbed by cancer cells and released via endocytosis, doxorubicin and other anticancer medications can also be conjugated to TMV (Bruckman et al. 2016; Hefferon 2017).

4.2 Cowpea Mosaic Virus-Based Vectors

Cowpea mosaic virus (CPMV) infects the leguminous *Vigna unguiculata* (cowpea). CPMV-based vectors have recently been recognized as a promising biopharmaceutical tool that can utilize the benefits of plant viruses to produce complex proteins, vaccines, and therapeutic agents rapidly at larger scales (Steele et al. 2017). CPMV is a member of the family Comoviridae, which has a bipartite genome with RNA-1 and RNA-2 that encode proteins involved in replication and coat protein assembly, respectively. These structural properties have allowed the development of CPMV as a versatile platform for the expression of foreign genes, making it increasingly widely applied in the production of vaccines, monoclonal antibodies, and other biopharmaceuticals (Lomonossoff 2008; Sainsbury and Lomonossoff 2008).

Delivering high-level protein expression in plants is one of the key benefits of using vectors based on CPMV. Other plant virus-based systems require integration into the plant genome and would thus be somewhat more burdensome for the regulatory process. Generally, they are not as concerned about insertional mutagenesis. On the other hand, CPMV will replicate episomally within plant cells, providing a transient yet robust expression system (Peyret and Lomonossoff 2015). This transient expression is useful for biopharmaceutical production, as it facilitates rapid scaling to meet emergency needs, such as during a pandemic or an emerging infectious disease. In addition, CPMV can infect a wide range of host plants, although most commonly it is applied in the model plant of *N. benthamiana*, which is used for transient gene expression systems. Such a broad host range increases the flexibility of the CPMV vector and makes it useful in many different production scenarios (Chen et al. 2011).

The most common approach for the engineering of CPMV-based vectors is by exchanging part of the RNA-2 sequence with the gene of interest, allowing for the expression of foreign proteins without interfering with the viral replication machinery. This approach has proven very effective in the production of a variety of therapeutic proteins, such as monoclonal antibodies and subunit vaccines. For example, CPMV vectors have been used to express monoclonal antibodies against the *Ebola virus,* providing a rapid and cost-effective alternative to traditional production methods in mammalian cell cultures (Eidenberger et al. 2023). The rapid replication of CPMV in plants enables the production of significant quantities of antibodies within weeks, an asset for solving public health emergencies (Kost and Condreay 2002).

CPMV has come into prominence in the field of vaccine development. The application areas in this direction are mainly about the production of virus-like particles, which are a group of noninfectious particles resembling the structure of the virus but lacking the viral genome, hence making them safe for application as vaccines. Immune responses induced by CPMV vectors are significantly stronger because they produce extremely repetitive and ordered presentation of antigens, which provides significant immunogenicity (Affonso de Oliveira et al. 2022). A number of studies have confirmed the efficacy of VLPs derived from CPMV as a vaccine in various infectious diseases such as influenza, human immunodeficiency virus (HIV), and HPV (Balke and Zeltins 2020). The ability to produce VLPs in plants, using CPMV-based vectors as expression vehicles, thus presents a cost-effective and scalable alternative to the traditional means of vaccine production, often relying on more expensive and time-consuming processes, such as yeast or mammalian cell fermentation (Carrillo-Tripp et al. 2021).

Another important application of CPMV vectors is in the area of cancer immunotherapy. CPMV has been seen to induce innate immunity, especially through the activation of macrophages and dendritic cells, which are key players in orchestrating tumor-specific immune responses (Wang et al. 2019). Tumor antigens can then be displayed on the surface of CPMV particles, enabling researchers to develop plant-based vaccines that can stimulate an immune response against cancer cells. This approach has been investigated in many preclinical studies, and it shows that CPMV-based vectors have great potential as a platform for personalized cancer vaccines. In these studies, CPMV vectors were engineered to express antigens specifically derived from the tumor, and the vectors will be used to vaccinate patients, thus guiding the immune system to recognize and destroy cancer cells. Rapid engineering and production of these vaccines by simply employing CPMV vectors could provide a new powerful tool to combat cancer (Chen et al. 2011; Leuzinger et al. 2013; Lomonossoff 2008; Mahmood et al. 2023).

CPMV-based vectors are thus a powerful, flexible platform for the production of biopharmaceuticals and an efficient, cost-effective, and scalable option over traditional production systems. The production capabilities of CPMV, which produce high yields of recombinant proteins, and the ease of making scalable plant-based expression systems, make this a powerful tool for producing vaccines, monoclonal antibodies, and other therapeutic proteins. Further glycoengineering and viral vector optimizations in CPMV-based systems will be able to assume significantly more important roles within the industry of biopharmaceuticals, mainly concerning personalized medicine, cancer immunotherapy, and emerging infectious diseases. CPMV is particularly attractive for the production of biosimilars and other complex biopharmaceuticals requiring sensitive posttranslational modifications (Liu et al. 2005). As the area of research unfolds, with CPMV vectors possibly becoming the basis for modern biopharmaceutical production, they may become the cornerstones for developing innovative treatments for many diseases.

Despite its numerous advantages, there are some challenges associated with the use of CPMV vectors for biopharmaceutical production. One of the primary concerns is the potential for immune responses against the viral components themselves, which could limit the effectiveness of the therapeutic proteins or vaccines produced. However, this problem can be at least partially alleviated by the use of chimeric or deconstructed CPMV particles that minimize or delete the immunogenicity of the viral coat proteins. Use of CPMV-based vectors for human therapeutics will also need to undergo rigorous regulatory evaluation regarding safety and efficacy, specifically regarding purity, consistency, and absence of contaminating components such as endotoxins or plant-derived allergens (Yusibov et al. 2000).

The RNA1 component is often left unaltered when creating CPMV-based vectors since it contains genes necessary for replication. Regarding RNA2, two methods are employed. According to the first method, RNA2 undergoes foreign gene insertion, increasing in size while maintaining its ability to disseminate throughout the body. Inserting the foreign gene as an in-frame C-terminal fusion to the RNA2-encoded polyprotein is the second and most effective method (Fig. 1b). An inducible proteolytic shunt that is added upstream of the foreign protein sequence via the Foot and mouth disease virus (FMDV) 2A catalytic peptide rescues the target protein (Liu et al. 2013; Sainsbury et al. 2009). The viral (or virus-like) particles are created using the first method mentioned above. The second approach permits the incorporation of bigger inserts but eliminates the virus's capacity to propagate both within and across plants. The mechanism is based on the finding that the RNA1-encoded replicase can only replicate RNA2 if the sequences required for this replication are found at the 5′ and 3′ ends of the RNA. This makes it possible to remove the majority of the RNA2 ORF without disrupting RNA2's capacity for replication. Since there is no natural CPMV silencing suppressor in such a vector system (Liu et al. 2004), another component—specifically, HC-Pro from *Potato virus Y*—should be included. Agroinfiltration, which enables the simultaneous delivery of several genetic expression constructs in plant cells, is typically used for this. Using the "deleted" form of RNA2 results in a greater yield of the target protein and is obviously beneficial for producing antibodies. Additionally, the CPMV system was created to simplify the cloning and expression stages using binary vectors. By infiltrating a single construct into the T-DNA region of the resulting vector, CPMV expression cassettes and the P19-encoding sequence enable high-level production of numerous polypeptides (Liu et al. 2013; Sainsbury et al. 2009).

4.3 *Potyvirus*-Based Vectors

Potyvirus-based vectors have recently received considerable attention for use in biopharmaceutical production in plants because they can transfer and express foreign genes in a large number of plant species efficiently, making them an ideal system for molecular farming. *Potyviruses* are single-stranded, positive-sense RNA viruses that replicate in the cytoplasm of infected cells and produce a polyprotein, which is processed by virus-encoded proteases into functional proteins (Martí et al. 2022). The mechanism of this potyvirus polyprotein processing is central to the design of *potyvirus*-based vectors, where foreign genes can be inserted for co-expression with viral proteins. Among the most studied potyviruses for this purpose, *Turnip mosaic virus* is engineered for the expression of therapeutic proteins, vaccines, and monoclonal antibodies. TuMV has been engineered to express epitopes from viral pathogens, including hepatitis B and human papillomavirus (HPV), on the coat proteins of TuMV (Rybicki 2020). These applications show the potential of producing vaccines in plants. Moreover, TuMV-based vectors have proven useful for the production of monoclonal antibodies, such as those directed toward the hepatitis B surface antigen and demonstrate capabilities for producing complex therapeutic proteins. Potyvirus vectors offer several advantages, including systemic infection, which allows the virus to spread throughout the plant and achieve high levels of recombinant protein expression in multiple tissues (Lico et al. 2015). This is particularly valuable for large-scale production, as the entire plant biomass can be utilized to maximize yield. In addition, potyvirus-based vectors are easily used for the rapid production of vaccines, since transient expression systems can produce therapeutic proteins within weeks. This means that the technology is very convenient for epidemic conditions when it becomes necessary to scale up vaccine production rapidly (Peyret and Lomonossoff 2015). However, some challenges are associated with-potyvirus-based vectors. One of these problems is genetic stability in a foreign gene that is inserted, and it is also prone to recombination or mutations in the viral RNA, which may cause deletion or inactivation of the foreign gene, thus lowering protein efficiency. Potyvirus-based vectors are currently being improved in stability by modification of the viral genome and use of helper components such as specific proteins or cofactors, which enhance retention of the gene during replication (Martí et al. 2022). There is also variability in the yield of proteins, as this can be affected by the species of plants used, growth conditions, and possibly the nature of the recombinant protein itself. These issues can be controlled by optimizing the expression system through choices of plant species that produce high biomass and the use of better viral constructs in order to achieve high-yield production. *Potyvirus*-based vectors are also being explored for the development of edible vaccines, where plants are engineered to produce vaccine antigens that can be consumed directly, potentially simplifying vaccine distribution and administration in regions with limited healthcare infrastructure (Kannan et al. 2020). As technology advances and our understanding of viral vector biology improves, potyvirus-based vectors are expected to play a crucial role in the production of a wide range of biopharmaceuticals, from vaccines to therapeutic antibodies, making them an important tool in the future of plant-based biopharmaceutical production (Gleba et al. 2007; Tremblay et al. 2010).

Tobacco etch virus (TEV), which was designed to produce the β-glucuronidase (GUS) reporter enzyme fused to the N-terminus of helper component proteinase (HC-Pro), was the first potyvirus to be tagged with a foreign gene. A vector based on the *Plum pox potyvirus* (PPV) was created to express full-length vaccine proteins. The NIb replicase and CP cistrons (Fig. 1c) were used to clone the foreign sequences. By cleaving at the location where the NIb and CP proteins were first separated, as well as at an extra NIa protease recognition site designed at the N-terminus, the heterologous protein is isolated from the remainder of the potyviral polyprotein. A number of additional group members are employed in the creation of foreign proteins and vectors. Two or more target proteins can be expressed simultaneously from a single vector, thanks to the next generation of potyvirus-based vectors (Fig. 1c) (Komarova et al. 2010).

5 DNA Virus-Based Vector Systems

5.1 Geminivirus-Based Vectors

Geminiviruses are capable of undergoing replication in a broad range of host plants at a high copy number. These viruses are characterized by a unique twinned (geminate) icosahedral virion morphology composed of two fused capsids that contain a circular, single-stranded DNA genome typically between 2.5 and 3.0 kb long (Hefferon 2014). *Geminiviruses* have two strands of their genomes; one is the complementary strand and the other is the virion strand, which is sometimes called the viral strand (Huang et al. 2009). The complementary strand of the virus encodes essential replication and transcriptional regulatory proteins such as transcriptional activator protein (TrAP) and replication-associated protein (Rep), while the virion strand includes the viral open reading frames for movement and capsid proteins. As a result, geminiviruses are suitable for developing as expression vectors (Hefferon 2017). *Geminiviruses* have emerged as powerful tools for producing biopharmaceuticals in plants. *Geminivirus* vectors are particularly suited for transient expression systems, offering high expression levels and large-capacity cloning regions for accommodating foreign genes. These vectors have been adapted for use in plants to efficiently express therapeutic proteins, antibodies, and vaccines (Regnard et al. 2010).

Geminiviruses replicate by rolling circle replication, the most important feature that makes them excellent for biopharmaceutical production. This allows the viral genome to be amplified to high copy numbers within a plant cell, leading to robust protein expression. Their rapid replication makes it possible to produce large amounts of recombinant proteins within a few days after the plants have been inoculated and with a minimum of manipulation. This can be particularly beneficial during times of pandemics for the rapid development of vaccines (Mahmood et al. 2023).

Among the most commonly used geminiviruses for biopharmaceutical production are *Bean yellow dwarf virus* (BeYDV) and *African cassava mosaic virus* (ACMV). To develop these viruses for the expression of therapeutic proteins, the gene of interest is usually inserted in certain locations in the genome, such as the intergenic region (IR), and replaces the nonessential genes, such as the coat protein (CP) gene, without disturbing important viral elements. The intergenic region acts as a control sequence and includes the replication origin and the promoter sequences that are bound by Rep protein (Lamprecht et al. 2016). The therapeutic DNA is usually integrated into the genome on the complementary strand, in the region where Rep and transcription activator protein (TrAP) are found. Rep plays an important role in the initiation of rolling circle replication with host DNA polymerase, which produces high amounts of replicative forms within the host cell, thus ensuring a high level of transient expression of the gene of interest. Vectors based on BeYDV, in particular, have also been successful in the production of functional monoclonal antibodies in a commercially viable manner, which attests to the viability of this approach in large-scale biopharmaceutical production (Baltes et al. 2014; Chung et al. 2011; Dugdale et al. 2014; Lamprecht et al. 2016).

Apart from monoclonal antibodies, geminivirus vectors are applied to the production of vaccines. A remarkable example was the expression of viral antigens of HPV in *Nicotiana benthamiana* using a system derived from geminivirus, which proved to be a promising candidate for vaccine production in plants (Chen et al. 2011). The potential of inserting large sequences in *geminiviruses* makes them useful for the production of large, complex, and multi-domain proteins, which further boosts their utility for therapeutic purposes.

The application of geminivirus vectors has its own set of problems. Transient expression does not mean that long-term or continuous production is possible and may necessitate repeated inoculation or the placement of the viral genome in the plant's genome. Furthermore, plant viral vector systems, including geminivirus-based platforms, must face strict regulatory policies to ensure safety and efficacy, especially with respect to the risk of recombination or environmental contamination (Rybicki 2014).

Geminivirus vectors represent a highly efficient and versatile platform for the production of biopharmaceuticals in plants. With continuous advances in molecular biology and viral vector engineering, these systems are becoming increasingly refined, offering a promising avenue for large-scale, cost-effective production of therapeutic proteins and vaccines (Dugdale et al. 2014).

6 Comparison of Various Viral Vectors

6.1 Efficiency of Protein Expression

Comparative analyses show that although high-level expression may be achieved by all four viral vectors, there are differences in the yield and quality of the proteins generated. G*eminivirus* vectors are excellent at generating complex proteins, whereas CPMV, *Potyvirus,* and *Tobamovirus* vectors often show higher yields.

6.2 Host Range and Stability

Potyvirus vectors exhibit a broad host range, making them versatile for various plant species. In contrast, geminivirus vectors, while capable of generating complex proteins, tend to have a more restricted host range, which can limit their application. CPMV and *Tobamovirus* vectors provide stability and high yields but may also have restrictions based on host plant availability.

6.3 Commercial Applications

The commercial potential of each viral vector depends on its scalability and regulatory compliance. *Potyvirus* and *Tobamovirus* vectors are well-established in research and have potential for commercial production, while geminivirus and CPMV require further optimization to meet industrial demands. The companies that use plant viruses as a vector for the production of biopharmaceuticals are listed in Table 2.

7 Techniques for Biopharmaceutical Production

In plant biotechnology, transformation systems are essential for creating a variety of beneficial products, especially in the biopharmaceuticals industry. By introducing and expressing foreign genes in plants, these systems enable the production of high-value products such as therapeutic proteins, vaccines, and antibodies (Fischer et al. 2004). Stable transformation and transient transformation are two major categories into which these systems may be divided.

1. *Stable Transformation System*

   Stable transformation systems are important in plant biotechnology. This is especially suited for the production of biopharmaceuticals. These systems allow the integration of foreign genes into the plant genome. Horsch et al. (1985) define stable genetic transformation as the insertion of recombinant genes into plant cell genomes, resulting in stable protein synthesis. This approach has various advantages; it permits stable and large-scale manufacture of recombinant proteins, including numerous proteins. However, there are certain drawbacks to this approach, including its time-consuming nature and the need for modifications to achieve mass expression. This allows for stable protein expression throughout generations, while transient expression systems, such as virus replication, provide fast production capability, and static changes offer a more sustainable solution for ongoing protein production.

2. *Transient Expression System*

   Transient expression systems using plant-based viruses have proven to be an efficient and cost-effective way of producing biopharmaceuticals. The method harnesses the natural propensity of plant viruses to invade host plant cells and induce the expression of exogenous genes. Transient expression systems can be an ideal tool for biopharmaceutics due to their multitude of advantages such as being quick, flexible, scalable, and affordable (Komarova et al. 2010).

   Transient expression is the process of expressing a foreign gene in host cells for brief intervals of time without requiring genome integration. The most popular technique for plant-based systems is the use of viral vectors, which carry a gene encoding for the target protein into plant cells and allow protein production to occur within a week to 2 weeks or days by tissue infiltration (Nosaki et al. 2021).

   Virus vectors derived from plant viruses such as *Tobacco mosaic virus* (TMV), *Cowpea mosaic virus* (CPMV), and the geminiviruses are most commonly employed for transient expression. These viruses have developed strategies that allow them to replicate and spread rapidly locally in planta to drive the expression of heterologous genes.

Table 2 Companies using plant-based platforms for pharmaceutical production (Hemmati et al. 2022)

Company	Host	Technology	Products	Website
Mapp biopharmaceutical	Tobacco leaves	magnICON- based expression Vector	ZMapp™ for *Ebola virus* disease	https://mappbio.com/
Cape bio pharms	*Nicotiana benthamiana*	Transient expression vector	PtXTM SARS-CoV-2 spike protein (S1, rabbit Fc); PtXTM SARS-CoV-2 spike protein (RBD, rabbit FC); CB_0002.5 PtXTM SARS-CoV-2 spike protein (S1- His); anti-HIV-1 p24 humanized IgG1 antibody conjugated to horseradish peroxidase (HRP); tXTM anti-HIV-1 p24 human IgG1 antibody	https://www.capebiopharms.com/
Leaf expression systems	*Nicotiana benthamiana*	Hypertrans-protein expression system	*Cowpea Mosaic virus* VLPs; recombinant SARS-CoV N-protein (nucleoprotein, His-Tag); SARS-CoV-2 nucleocapsid (N) protein	https://www.leafexpressionsystems.com/
Medicago	*Nicotiana benthamiana*	VLP*Express*™	Influenza VLP vaccine candidate; SARS-CoV-2 VLP vaccine candidate	https://www.medicago.com/en/
Planetbiotechnology	Tobacco leaves	Transient and stable expression	CMG2-Fc; TEM8-Fc; BSG-Fc;AtCry1	https://www.planetbiotechnology.com/
Healthgen biotechnology	*Oryza sativa* (rice)	$\text{Oryz}^{\text{HiExp}}$,platform	Recombinant human serum albumin (OsrHSATM); recombinant epidermal growth factor (OsrEGF)	https://www.oryzogen.net/
Kentucky bioprocessing	*Nicotiana benthamiana*	magnICON- based expression vect	ZMapp™ *Ebola virus* disease; vaccine candidate for SARS-CoV-2 (preclinical phase)	https://kentuckybioprocessing.com/
Icon genetics	*Nicotiana benthamiana*	magnICON-based expression vector	Vaccine candidate for non-Hodgkin's lymphoma; ZMapp™ for *Ebola virus* disease	https://www.icongenetics.com/

8 Virus Nanoparticles in Biopharmaceutical Production

In the field of biopharmaceutical production, viral nanoparticles (VNPs) have emerged as an innovative tool, especially when derived from plant viruses. These nanoparticles mimic the structure and characteristics of viruses but do not cause infection, making them ideal for therapeutic applications. They have unique properties that can be used in a variety of biomedical applications, including vaccine development and drug delivery (Steinmetz 2010).

Viral nanoparticles provide various benefits over regular protein expression platforms. Viral nanoparticles may be produced in huge quantities by efficient viral genome replication within plant cells. This produces a large number of nanoparticles that can be recovered from very small amounts of plant material. Because of their structural properties, viral nanoparticles can effectively activate the immune system. Their resemblance to infectious viruses allows them to provoke a strong immune response. This makes them a highly effective vaccine candidate. Viral nanoparticles can be engineered to deliver many antigens simultaneously. To improve efficacy, various vaccinations targeting different diseases have been developed. Because viral nanoparticles do not induce infection, they represent a low danger to human health. Because of this, it is a safer choice than current vaccine platforms. It is possible to design viral nanoparticles so that they can directly transport therapeutic proteins to particular tissues or cells. Boost the effectiveness of treatment and reduce side effects. For the treatment of cancer when topical therapy is required, this specialized delivery system is highly helpful. Some of the plant VNPs tested for cancers are tabulated below (Table 3) (Bamogo et al. 2019; Chen et al. 2011; Chung et al. 2020; Gleba et al. 2007; Yusibov and Rabindran 2008).

9 Virus-Like Particles (VLPs) in Biopharmaceutical Production

Plant-based systems have emerged as a new platform for VLP production. Plants can undergo complex posttranslational modifications similar to those in mammalian systems. This makes them ideal for producing functional VLPs using techniques such as culture and viral vectors. These systems have been used to introduce genes encoding viral structural proteins into plant tissues. This helps with safety, scalability, and the ability to produce VLP without the need for expensive fermentation equipment (Chen and Lai 2013).

VLPs are synthesized in eukaryotic systems, such as plant or insect cells. It is possible to make changes after the translation, which involves glycosylation. These changes are frequently necessary for the stability and function of therapeutic proteins. VLPs can be created to contain numerous antigens. This resulted in the development of multivalent vaccinations. This adaptability is especially important in coping with complex diseases that necessitate a broad immune response (Zahin et al. 2016).

Table 3 Selected examples of plant virus nanoparticles tested on some cancers (Bamogo et al. 2019)

Plant virus employed	Application examined	Carried drug	References
CPMV	Solid tumors	Ag	Gonzalez et al. (2009), Saunders et al. (2009)
TMV	Ovarian cancer	Cisplatin	Franke et al. (2018)
TMV	Cancer	Phenanthriplatin	Vernekar et al. (2018)

The composition of VLPs can vary greatly depending on the source virus. For example, VLPs derived from plant viruses such as *Cowpea mosaic virus* (CPMV) display icosahedral shapes, and *Tobacco mosaic virus* (TMV) displays filamentous, rod-like structures. In contrast, VLPs derived from animal viruses such as human immunodeficiency virus (HIV) or hepatitis B virus (HBV) can display more complex structures. It includes an encapsulated form that has two layers of lipids. The structural integrity of VLPs is critical to their function. This is because the structure of a protein affects its ability to stimulate an immune response (Chen and Lai 2013; Yusibov and Rabindran 2008). The recent development of plant virus-based transient expression systems, as shown in Fig. 2, has greatly increased the speed and yield of VLP production (Marsian and Lomonossoff 2016).

1. *Mechanical Transformation*

In plant biotechnology, mechanical transformation is a physical way of introducing genetic material into plant cells. Mechanical transformation uses force to break down plant cell walls or membranes so that foreign DNA or viral vectors may be absorbed, in contrast to more conventional genetic transformation techniques like Agrobacterium-mediated transformation or electroporation (Gelvin 2003). For transferring plant-based viruses with recombinant genes for the production of therapeutic proteins, antibodies, vaccines, and enzymes, mechanical transformation is advantageous for plant virus-based systems (Lico et al. 2008).

1. *Leaf Bombardment (Gene Gun or Biolistics):* This technique shoots metal particles coated with viral DNA into plant cells using high-pressure acceleration. The viral genetic material is inserted within plant cells by the accelerated particles when they enter plant tissues. A gene gun is frequently used for this, and it works especially well on tissues like leaves. Once within the cell, the viral vector may multiply and create the required recombinant proteins (Kikkert et al. 2005).
2. *Agitation or Homogenization:* This technique involves grinding or vortexing plant tissues, usually leaves or other plant parts, with viral vectors. Viral vectors can more easily enter cells when the mechanical force helps to crack open cell barriers. Although this approach is usually less intrusive than gene gun bombardment, the transformation process is still made possible by physical damage.

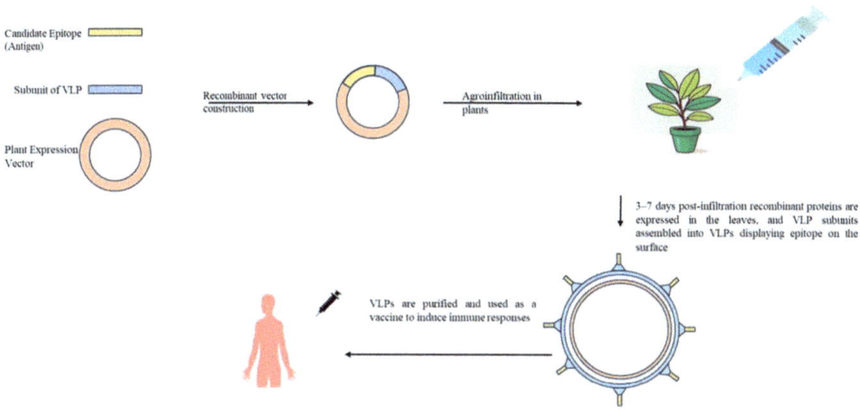

Fig. 2 Schematic representation of recombinant viral protein (VLP) expression in plants Methods of Transformation of Viral Vectors

By physically rupturing plant tissues, these techniques expose cells to viral vectors, which subsequently have the ability to infect the cells. The appropriate proteins are produced as a result of the plant virus replication and transcription of the recombinant genes after it has entered the cell (Kikkert et al. 2005).

2. *Agroinfiltration*

Agroinfiltration is a widely accepted method for the transient expression of proteins in plants, especially in herbaceous species such as *N. benthamiana*. This technique takes advantage of the natural infection process of *Agrobacterium tumefaciens*, a soil bacterium that transfers part of its DNA (T-DNA) into plants, and has become a valuable tool in the production of biopharmaceuticals, including antibodies, vaccines, and enzymes (Leuzinger et al. 2013).

The first step in the technique is to create plant expression vectors, usually utilizing a binary vector system. In these vectors, a selectable marker gene and a potent plant promoter govern a T-DNA region containing the gene of interest, including those producing therapeutic proteins or vaccination antigens. The target gene is cloned into the vector and transferred into *Agrobacterium tumefaciens* (Fig. 3) by means of triparental mating. This method involves introducing modified plant viruses into the plant tissues once they have been altered to carry the desired gene. Effective biopharmaceutical manufacturing is made possible by the viral system's ability to replicate quickly and produce the target protein at a high level. Viral vectors are frequently introduced into plants using methods like agroinfiltration, where they proliferate and stimulate the production of the therapeutic protein (Lee and Gelvin 2008).

The next step is to prepare the plant material. Young leaves of *N. benthamiana* are usually used. This increases the efficiency of T-DNA transfer and is followed by a syringe or vacuum infiltration (Fig. 3) to introduce the bacteria into the intercellular spaces of the leaf. *A. tumefaciens* transfers T-DNA into the host cell. It is

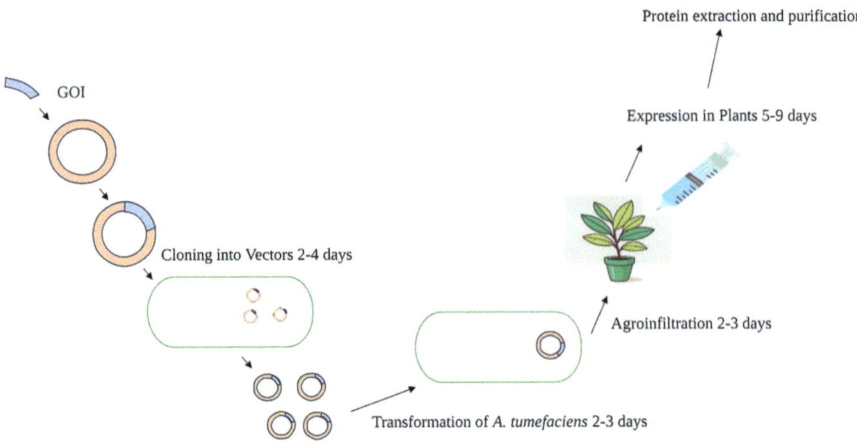

Fig. 3 Schematic representation of agroinfiltration

integrated into the plant genome or present as episomes. This leads to the expression of foreign genes (Zhang et al. 2020).

Optimizing agroinfiltration efficiency involves balancing bacterial concentration, pressure, and infiltration time. High concentrations can increase protein production but can cause plant cell death. Infiltration methods like vacuum penetration can limit *A. tumefaciens* transformation. Adjusting infiltration time based on plant growth stage can enhance total production. Interfering compounds like acetosyringone can promote virulence gene expression and improve T-DNA transfer efficiency (Chen and Lai 2013; Chincinska 2021; Tzfira and Citovsky 2006).

The agroinfiltration process provides rapid, efficient transformation, allowing crops to produce significant protein in just a few days, as opposed to static transformation methods, which can take weeks or months in the production of biopharmaceuticals. Moreover, transient expression allows researchers to rapidly evaluate multiple constructs for optimal protein expression (Kaur et al. 2021).

Another advantage is the high levels of protein expression. According to a study, agricultural drills can produce between milligrams and grams of protein per kilogram of plant material. This makes it a useful method for developing high-quality biopharmaceuticals (Chen et al. 2013). Additionally, the temporary character of agricultural introductions typically reduces the risk of genetic instability produced by static variation because foreign genes are not permanently integrated into the plant genome.

Additionally, agricultural access is highly flexible and can be used across a wide range of plant species. *N. benthamiana* is the most common host due to its susceptibility to *A. tumefaciens*. However, other species such as tomato, potato, and various members of the *Brassicaceae* family have also been explored. This versatility allows researchers to tailor agricultural entry processes to meet specific biopharmaceutical needs (Nosaki et al. 2021).

There are difficulties that might compromise the effectiveness of agroinfiltration, especially when it comes to plant viruses. Stringent management of conditions during the infiltration process is necessary because of the variability in protein expression levels caused by environmental variables such as temperature, light, and humidity, as well as host plant responses to *A. tumefaciens* infection (Leuzinger et al. 2013).

It is also important to optimize conditions for minimal plant damage and maximum expression because incorrect infiltration parameters, like an excessive bacterial concentration or infiltration pressure, can result in plant cell damage, necrosis, and decreased protein yield (Mad' Atari and Folta 2019).

Finally, the changing regulatory environment for goods produced from plants, particularly biopharmaceuticals, highlights the significance of quality assurance, safety, and efficacy in reaching successful commercialization (Huang et al. 2010).

10 Biopharmaceutical Proteins Produced from Viral Vectors

Plant viruses have been extensively researched and documented for producing foreign proteins, including pharmaceuticals and non-pharmaceuticals. Plant-produced antigens have been shown to provide vaccine candidates for both veterinary and human uses, eliciting immune responses in animal models and providing protection in challenge studies. Tables 4 and 5 show examples of several therapeutic proteins, antigens, and vaccines expressed through plant viral systems (Bamogo et al. 2019; Grill et al. 2005; Hefferon 2014; Lico et al. 2008).

11 Upstream and Downstream Processes in Biopharmaceutical Production

Biopharmaceutical production using plant-based viral vectors requires carefully coordinated upstream and downstream processes to ensure the efficient and high-quality production of therapeutic proteins, vaccines, antibodies, and other biologics. The integration of plant-based viral expression systems into the biopharmaceutical pipeline involves several key steps in both upstream and downstream processing to scale up production and meet regulatory standards for safety and efficacy (Cañizares et al. 2005).

11.1 Upstream Process

Upstream processing encompasses all operations involving the growth and development of biological systems that result in the intended biopharmaceutical output. For plant virus-based biopharmaceuticals, this comprises creating the viral vector, selecting and cultivating plants, introducing viral vectors into the plant system, and determining the conditions necessary for the optimum expression of target biopharmaceutical products (Schillberg et al. 2013).

Table 4 List of representative antigens and vaccines produced through viral vectors (Grill et al. 2005; Lico et al. 2008)

Antigen/vaccine	Viral vector	Pathogen/disease	References
FMDV structural protein, VP1	TMV	Foot and mouth disease	Wigdorovitz et al. (1999)
BRV VP8 protein	TMV	Bovine rotavirus diarrhea	P'erez-Filgueira et al. (2004)
BHV-1 glycoprotein D	TMV	Infectious bovine rhinotracheitis	Perez Filgueira et al. (2003)
HIV-1 p24 protein	TMV	Human immunodeficiency virus disease	P'erez-Filgueira et al. (2004)
Patient-specific B-cell Ig	TMV	Non-Hodgkin's lymphoma	McCormick et al. (1999, 2003)
HCV envelope protein hypervariable region + cholera toxin B subunit sequence	TMV	Hepatitis C	Nemchinov et al. (2000)
HRV coat protein	CPMV	Common cold	Porta et al. (1994)
HIV-1 coat protein	CPMV	Human immunodeficiency virus disease	Porta et al. (1994)
HIV gp41 protein	CPMV	Human immunodeficiency virus disease	McLain et al. (1996), Durrani et al. (1998)
P. aeruginosa outer membrane protein F	CPMV	Chronic pulmonary infection	Brennan et al. (1999), Gilleland et al. (2000), Staczek et al. (2000)
Epitopes from the circumsporozoite protein	TMV	Malaria	Turpen et al. (1995)
MEV VP2 capsid protein	CPMV	Mink enteritis	Dalsgaard et al. (1997)
FMDV	TMV	Foot and mouth disease virus	Wu et al. (2003)
FPV VP2 capsid protein	TMV	Feline panleukopenia virus	Pogue et al. (2004)
CPV VP2 capsid protein	CPMV	Canine parvovirus	Langeveld et al. (2001)
S. aureus fibronectin-binding protein	CPMV	Bacteremia	Rennermalm et al. (2001)
HCV envelope protein hypervariable region	CMV	Hepatitis C	Natilla et al. (2004)
Capsid epitopes	CPMV	Human immunodeficiency virus-1	Porta et al. (1994)
Several epitopes	TMV	*Plasmodium falciparum*	Turpen et al. (1995)
HA epitope	TMV	Influenza virus	Sugiyama et al. (1995)

(continued)

Table 4 (continued)

Antigen/vaccine	Viral vector	Pathogen/disease	References
VP2 protein peptide	CPMV	Canine parvovirus	Fernandez-Fernandez et al. (1998)
FnBP epitope	PPV	*Staphylococcus aureus*	Brennan et al. (1999)
VP1 protein	CPMV	Foot and mouth disease virus	Wigdorovitz et al. (1999)
Region 1 of E2	TMV	Hepatitis C virus	Nemchinov et al. (2000)
VP6 protein	TMV	Rotavirus	O'Brien et al. (2000)
Capsid epitopes	TMV	Human immunodeficiency virus-1	Marusic et al. (2001)
Chimeric peptide	PPV	Rabies virus	Yusibov et al. (2002)
Mimotope	TMV	Hepatitis C virus	Natilla et al. (2004)
Tat protein	TMV	Human immunodeficiency virus-1	Karasev et al. (2005)
G protein epitope	TMV	Respiratory syncytial virus	Yusibov et al. (2005)
p15e-Trp2 epitopes	TMV	Melanoma	McCormick et al. (2006b)
F1 and V proteins	TMV	*Yersinia pestis*	Santi et al. (2006)
M2 protein ectodomain	TMV	Influenza A virus	Nemchinov and Natilla (2007)
ESAT6—Ag85B antigens	TMV	*Mycobacterium tuberculosis*	Dorokhov et al. (2007)
E7 protein	TMV	Human papilloma virus	Massa et al. (2007)
pB5 antigenic domain	TMV	Smallpox	Golovkin et al. (2007)
F1 and V proteins	TMV	*Yersinia pestis*	Mett et al. (2007)
Protective Antigen (PA) peptide	CPMV	*Bacillus anthracis*	Phelps et al. (2007)
Protective Antigen and Lethal Factors domains	TMV	*Bacillus anthracis*	Chichester et al. (2007)

PPV Plum pox virus, *CPMV* Cowpea mosaic virus, *TMV* Tobacco mosaic virus

11.1.1 Viral Vector Design and Optimization

An initial step in upstream processing is the development and optimization of viral vectors such as *Tobacco mosaic virus* (TMV), *Cowpea mosaic virus* (CPMV), and *Geminiviruses*. They are modified to carry the genetic code of a protein, antigen, or antibody for the desired treatment. This modification involves the introduction of foreign genes into the viral genome and the use of a strong promoter to induce high levels of expression, ensuring appropriate posttranslational modifications and stability of proteins expressed in tissues (Banik et al. 2015; Regnard et al. 2010).

Table 5 Vaccines and other therapeutic proteins produced transiently with the help of plant viruses (Bamogo et al. 2019; Hefferon 2014)

Recombinant protein	Viral vectors	Expression level	Plant host	References
HPV16 L1	magnICON	0.25 mg/g fresh leaf weight (FLW)	*N. benthamiana*	Zahin et al. (2016)
M2eHBc	pEff	5–10% total soluble protein (TSP) FLW	*N. benthamiana*	Mardanova et al. (2017)
SEB	BeYDV	n/a	*N. benthamiana*	Shen and Hohn (1995)
Norwalk Virus VLPs	BeYDV	0.34 mg/g FLW	Tobacco, lettuce	Lai et al. (2012)
HBVcAg	BeYDV	0.8 mg/g FLW	*N. benthamiana*	Huang et al. (2009)
WNV E protein Mab	BeYDV	0.23–0.27 mg/g LFW	Tobacco, lettuce	Phoolcharoen et al. (2011)
Ebola Virus GP1Mab	BeYDV	0.23–0.27 g/g LFW	Tobacco, lettuce	Phoolcharoen et al. (2011)
HPV-1 L1 protein	BeYDV, mild strain	n/a	*N. benthamiana*	Regnard et al. (2010)
HIV-1 type C p24	BeYDV, mild strain	n/a	*N. benthamiana*	Regnard et al. (2010)
HAV VP1	BCTV	n/a	*N. benthamiana*	Chung et al. (2011)
Vitronectin	TYDV	2.3% TSP	*N. benthamiana*	Dugdale et al. (2014)

FLW fresh leaf weight, *n/a* not applicable, *TSP* total soluble protein

11.1.2 Plant Selection and Cultivation

Selecting plant varieties for biopharmaceutical production is essential for successful upstream processing. Common plant groups include *N. benthamiana* and *N. tabacum* (tobacco varieties), which are widely used for high biomass. They are easy to manipulate genetically and have the ability to express a wide range of foreign proteins. These plants are grown in greenhouses or specific growing chambers under regulated circumstances, which reduces the risk of contamination. Plants for biopharmaceutical manufacturing are grown in Controlled Environment Agriculture (CEA) settings under strict control of environmental factors such as light, temperature, humidity, and fertilizer levels to maximize growth and protein expression (Buyel 2019).

11.1.3 Expression and Accumulation of Biopharmaceuticals

After transferring the viral vector, the plant will operate as a biological factory. The viral genome hijacks the plant's biological machinery, causing it to express foreign proteins. Within days or weeks, the intended biopharmaceutical products accumulate in plant tissues, mostly on the leaves.

11.2 Downstream Processing

Expressed biopharmaceutical products go through downstream processing, which includes purification, formulation, quality control, and regulatory compliance, to make sure that the finished product retains its biological activity and complies with laws and regulations.

11.2.1 Harvesting

When plants reach their highest expression level of the desired product, they are harvested. Harvest time is critical to achieving maximal protein output while preserving biopharmaceutical quality. Harvested plant tissue can be processed mechanically or manually. The focus is often on leaves, where the viral vector generates the maximum degree of protein expression (Buyel 2019).

11.2.2 Extraction of Biopharmaceuticals

The next step in the downstream processing is the extraction of the biopharmaceutical product from harvested plant tissue. Plant tissue is generally homogenized to release cellular material, including expressed proteins. A variety of extraction buffers were used to ensure that the protein of interest remains stable during this step.

11.2.3 Sonication and Grinding

Physical disruption methods such as sonication or grinding are often used to increase protein release from plant cells.

11.2.4 Homogenization and Filtration

After mechanical homogenization of plant leaves, the material is filtered or centrifuged to remove large plant impurities. This results in a crude extract containing the target biopharmaceutical protein along with other plant impurities.

11.2.5 Protein Purification

After protein extraction, purification of the plant viral proteins, enzymes, and biological components is needed in order to ensure certain characteristics and eliminate contaminants for therapeutic-grade applications.

Affinity chromatography is one of the most commonly used purification procedures. This is especially true when the target protein is tagged with an affinity tag (e.g., His-tag). Affinity chromatography will select the target recombinant protein, making it possible to separate biopharmaceuticals from other ingredients effectively.

Ion exchange chromatography and size-exclusion chromatography are both used to filter proteins depending on their charge and size. This permits the protein extract to be processed to the required purity level.

Ultrafiltration and diafiltration are procedures for combining protein solutions and removing small impurities. This ensures that the finished product contains the appropriate protein concentration for medicinal purposes (Buyel 2019; Gleba et al. 2007; Schillberg et al. 2013; Tusé et al. 2020).

11.2.6 Formulation

After purification, biopharmaceutical products should be formulated as the final therapeutic form. Adjuvants that stabilize the protein, adjust pH and protein activity and shelf life are added at this stage. For vaccines, this may include the addition of adjuvants to increase immunity.

11.2.7 Lyophilization or Freeze-Drying

In many cases, purified proteins are freeze-dried to increase stability during storage and transport. This is especially true for vaccines or enzymes intended for use in remote or resource-limited situations.

11.2.8 Aseptic Filtration

It ensures sterility: Formulated biopharmaceuticals are passed through aseptic filtration to eliminate microbial contaminants that may affect product safety and effectiveness.

11.2.9 Quality Control and Testing

One of the most important aspects of downstream processing is strict quality control (QC) of the final product. QC includes verification that the purified biopharmaceutical meets all required specifications, including purity, potency, safety, and stability. This is especially important for plant-based systems, since plant-specific contaminants, such as proteases or glycosylation patterns, can affect product performance.

Quality testing can be done by the following methods:

SDS-PAGE and Western blotting are commonly used techniques to detect the presence and size of target proteins in purified products.

Mass Spectrometry and HPLC: These are used to confirm purity and to identify posttranslational modifications that may occur during expression in plants.

Bioactivity Testing: Functional testing is performed to ensure that a biopharmaceutical retains its intended therapeutic activity. For vaccines, immune studies can be conducted in animal models to determine efficacy.

11.2.10 Regulatory Compliance and Scale Up

In order to expand production scale for upstream commercial use, including improving the efficiency of downstream processes, the final product must adhere to regulatory guidelines set by organizations such as the US Food and Drug Administration (FDA) or the European Medicines Agency (EMA), in accordance with Good Manufacturing Practice (GMP) standards, to confirm the dependability and safety of large-scale manufacturing.

In order to ensure efficiency while producing plant-based biopharmaceuticals on an industrial scale from a laboratory environment, upstream processes, including plant culture, cell transfection, and expression, must be improved. During downstream purification, the products must be altered to accommodate higher quantities without sacrificing quality.

Producers must demonstrate the stability and security of their plant virus-based production methods in order to comply with regulations. Information regarding the stability of viral vector spread must be provided for protein expression and the downstream purification efficiency.

12 Economic Impact

The potential for scalable, economical, and quick production processes makes the use of plant viruses for biopharmaceutical manufacturing economically relevant (Marsian and Lomonossoff 2016). The utilization of low-cost growing techniques in host plants, as opposed to the expensive fermentation and mammalian cell culture systems, makes plant viral vectors beneficial (Gleba et al. 2007). According to Yusibov and Rabindran (2008), plants are also devoid of human pathogens, which simplifies regulations and improves production safely. These factors can result in a large reduction in manufacturing costs. According to Lai et al. (2012), this method also makes it possible to respond quickly to new diseases, since plants may be genetically modified to generate specific proteins in a couple of weeks, which can be profitable during pandemics. Plant viral systems' modular design also makes it possible to produce a variety of biologics, such as vaccines, antibodies, and therapeutic proteins, meeting consumer needs and generating new business prospects (Gleba et al. 2005). However, challenges such as obtaining regulatory permission, the expense of downstream processing, and maintaining consistent product quality still exist and might affect overall cost-effectiveness (Rybicki 2014). It is anticipated that further developments in viral vector engineering would lower production costs and increase scalability, improving the platform for the economic feasibility of producing biopharmaceuticals (Rybicki 2020).

13 Environmental Impact

Plant virus-based biopharmaceutical manufacturing effects on the environment are gaining consideration, mainly because they promise more effective and sustainable substitutes for conventional production techniques. According to Gleba et al. (2005), plant viruses provide a more economical and scalable method of creating recombinant proteins, eliminating the need for animal cell cultures and energy-intensive bioreactors. By using sunlight-driven photosynthesis as a key energy source, (Schillberg et al. 2013) also highlighted that using plants as biofactories reduces the carbon footprint associated with traditional pharmaceutical manufacturing. However, ecological problems have been brought up, such as the unintentional release of genetically engineered viruses that might impact natural ecosystems (Rybicki 2020).

Tusé et al. (2020) also emphasized the difficulties in striking a balance between the possibility of transgene escape into the environment, which might result in unintended gene flow, and achieving high-yield protein production. On the other hand, Daniell et al. (2015) found that, in comparison to fermentation-based systems, plant-based production methods using viruses drastically reduce the chemical and water inputs, supporting international efforts to promote sustainable development. The environmental advantages of waste management may be further enhanced by repurposing plant biomass wastes from these systems as feedstocks for bioenergy (Tusé et al. 2020). According to Fischer et al. (2013), the production of biopharmaceuticals based on plant viruses has the potential to reduce the environmental impact of pharmaceutical manufacturing; nonetheless, it requires strict containment and monitoring measures to handle any possible biosafety issues.

14 Future Prospects

Plant viral vectors provide environmental sustainability, scalability, cost-effectiveness, and speed in the production of biopharmaceuticals; nevertheless, future expansion is dependent on challenges and innovation.

15 Advancement in Viral Vector Technologies

15.1 Improved Viral Vectors for Higher Yields

The study of developing more potent viral vectors, such as TMV and CPMV, to increase protein synthesis in transient expression systems is a significant area in plant biopharmaceuticals. It has shown good signs; however, future advances should focus on increasing the ability to express larger, more complex proteins.

15.2 Multiple Expression Systems

In the future, viral vectors can enable the simultaneous expression of multiple therapeutic proteins in a single plant. The capacity to assemble multiple components will be necessary for the production of complex biological agents such as monoclonal antibodies. In order for this to happen, both heavy and short chains need to be expressed together. New viral vectors with multiple gene expression capabilities could result in notable increases in the productivity of plant virus-based manufacturing systems (Lomonossoff and D'Aoust 2016).

15.3 Self-Replicating RNA Vectors

Utilizing self-replicating RNA viral vectors is an additional intriguing choice. This could promote the expression of foreign genes in plant cells. Enhancing the stability and efficiency of RNA replication raises the method's competitiveness, allowing researchers to significantly enhance plant protein output.

15.4 Synthetic Biology and Custom Viral Vectors

Synthetic biology and plant virus-based medicine synthesis have incredible potential synergy. Scientists will be able to create customized viral vectors with the use of synthetic biology, which will enable the creation of vectors that are more stable, adaptable, efficient, and effective in expression.

15.5 Genome Editing

Developments such as CRISPR/Cas9 are likely to usher in a new golden age of precisely targeted viral genome modification, enhancing their activity in plant virus-based systems. This may result in vectored viral systems that are more efficient, give higher-dosage yields, and are less susceptible to degradation by host plant immune responses (Rybicki 2017).

15.6 Optimizing Viral Promoters

Promoter optimization in viral vectors for gene expression is also another area of research. Strong and plant-specific promoters can persistently produce therapeutic proteins, cooperating with low cost and yielding high quantities (Garabagi et al. 2012).

16 Expansion of Product Portfolio

16.1 New Classes of Biopharmaceuticals

Plant virus-based production platforms are uniquely suited to produce a wide range of biopharmaceuticals, from vaccines and therapeutic proteins and enzymes to diagnostic reagents. As technology grows, the range of products that can be produced using plants will expand greatly, as expected.

16.2 Plant Virus-Based Vaccines

One potential use for plant virus-based biopharmaceuticals is vaccinations. In pre-clinical and clinical studies, plant virus-based vaccines against influenza, SARS-CoV-2, and norovirus have demonstrated effectiveness. Furthermore, plant virus-based vaccines can be used to make additional vaccinations in the future, among these are customized vaccinations for the treatment of cancer (Su et al. 2023).

17 Overcoming Regulatory and Commercialization Challenges

17.1 Streamlining Regulatory Approvals

The regulatory approval process is a significant barrier to the widespread use of plant virus-based biopharmaceuticals. While regulations for these products are still being defined, plant virus-based solutions show potential and are also vital to maintain uniformity throughout all sectors.

17.2 Regulatory Harmonization

Biopharmaceuticals should be authorized and distributed worldwide; plant virus-derived biopharmaceuticals must be subject to uniform laws. International agencies such as the World Health Organization (WHO) and national regulatory agencies such as the FDA and EMA must work together to develop clear guidelines for plant virus-based biopharmaceuticals. This ensures that plant virus-based biopharmaceuticals comply with the same safety and efficacy standards as conventional pharmaceuticals (Tusé et al. 2020).

17.3 Intellectual Property and Market Access

Decisions about intellectual property will significantly influence the development of biopharmaceuticals made from plant viruses in the future. Enforcing intellectual property rights has become increasingly difficult due to the expanding number of companies and academic institutions developing unique plant viral vectors and expression systems.

17.4 Access to Technology

This is to ensure that plant virus-based biopharmaceuticals are available to everyone. Balancing the technology transfer to low- and middle-income countries with protection of intellectual property rights is a key role. Plant virus-based biopharmaceuticals will ensure that licensing agreements and open-access platforms are in the best position to support their benefits. They are thus widely adaptable to everybody, particularly public health measures like immunizations (Rybicki 2014).

18 Applications in Global Health and Pandemics

18.1 Rapid Response to Emerging Infectious Diseases

Plant virus-based systems have already shown the ability to respond rapidly to emerging infectious diseases. A plant virus (TMV)-based platform was used to produce the Zmap antibody cocktail in response to the crisis during the Ebola outbreak in 2014 (Dubey et al. 2018).

18.2 Pandemic Preparedness

International pandemic preparation activities may use plant virus-based biopharmaceutical technology because plant virus-based systems may be essential in reacting to upcoming pandemics due to their ability to quickly scale up production and produce vaccinations and treatments in a short period of time (Fischer et al. 2013).

18.3 Stockpiling of Vaccines and Therapeutics

Another purpose for the plant virus-based platform would be to create a library of remedies and vaccinations for any pandemic illnesses.This is because plants used for producing biopharmaceuticals through plant virus-based systems can be rapidly sown and harvested. According to Tusé et al. (2020), it therefore provides a scalable and adaptable way to ensure the security of global health.

19 Conclusion

The production of biopharmaceuticals using plant viral vectors is a novel approach in biotechnology. When made with these techniques, a wide range of biopharmaceutical products, including therapeutic antibodies, enzymes, cytokines, virus-like particles, and vaccinations, have shown great promise. Comparing plant-based systems to traditional manufacturing platforms reveals some significant benefits, including low cost, fast scalability, and advanced security. The continued development of plant virus expression systems is likely to expand their use. It not only meets the global demand for affordable and effective treatments but also the growing demand for personalized emergency response in biopharmaceuticals.

But the success of this method is in efficient post-processing, which includes the design and optimization of virus vectors, plant cultivation, extraction, purification, and the precise formulation of therapeutic proteins; each step must be carefully managed to ensure that the final product meets strict regulatory standards.

An important development in the realm of biopharmaceutical production is represented by viral nanoparticles. It provides a versatile and efficient platform for vaccine development. Protein delivery for treatment and application in diagnosis, high-throughput production, immunity, and safety makes it a valuable tool in addressing a wide range of medical challenges. As research and technology continue to develop, viral nanoparticles are also poised to play an increasingly important role in the future of biopharmaceuticals.

Finally, production of biopharmaceuticals using plant virus vectors is a promising approach to meet the global need for inexpensive treatments, efficient and sustainable, economic benefits from reduced production costs, scalability, and accessibility, combined with environmental benefits from reduced energy use, less pollution, and sustainability, establishment of factory-based systems, although challenges such as regulatory constraints and yield variability remain. Advances in viral vector technology and synthetic biology could transform the industry, making life-saving treatments more accessible.

References

Abrahamian P, Hammond RW, Hammond J (2020) Plant virus–derived vectors: applications in agricultural and medical biotechnology. Ann Rev Virol 7(1):513–535. https://doi.org/10.1146/annurev-virology-010720-054958

Affonsode Oliveira JF, Chan SK, Omole AO, Agrawal V, Steinmetz NF (2022) In vivo fate of Cowpea Mosaic virus in situ vaccine: biodistribution and clearance. ACS Nano 16(11):18315–18328. https://doi.org/10.1021/acsnano.2c06143

Amanat F, Stadlbauer D, Strohmeier S, Nguyen TH, Chromikova V, McMahon M, Jiang K, Arunkumar GA, Jurczyszak D, Polanco J (2020) A serological assay to detect SARS-CoV-2 seroconversion in humans. Nat Med 26(7):1033–1036

Balke I, Zeltins A (2020) Recent advances in the use of plant Virus-Like Particles as vaccines. Viruses 12(3):270. https://doi.org/10.3390/v12030270

Baltes NJ, Gil-Humanes J, Cermak T, Atkins PA, Voytas DF (2014) DNA replicons for plant genome engineering. Plant Cell 26(1):151–163

Bamogo PKA, Brugidou C, Sérémé D, Tiendrébéogo F, Djigma FW, Simpore J, Lacombe S (2019) Virus-based pharmaceutical production in plants: an opportunity to reduce health problems in Africa. Virol J 16:167. https://doi.org/10.1186/s12985-019-1263-0

Banik S, Mansour AA, Suresh RV, Wykoff-Clary S, Malik M, McCormick AA, Bakshi CS (2015) Development of a multivalent subunit vaccine against tularemia using tobacco mosaic virus (TMV) based delivery system. PLoS One 10(6):e0130858

Bendandi M, Marillonnet S, Kandzia R, Thieme F, Nickstadt A, Herz S, Fröde R, Inoges S, De Cerio AL-D, Soria E (2010) Rapid, high-yield production in plants of individualized idiotype vaccines for non-Hodgkin's lymphoma. Ann Oncol 21(12):2420–2427

Brennan, F. R., Jones, T.D., Gilleland, L. B., Bellaby, T., Xu, F., North, P. C.,Thompson, A., Staczek, J., Lin, T., Johnson, J. E., Hamilton W. D., andGilleland Jr., H. E. 1999. Pseudomonas aeruginosa outer-membrane proteinF epitopes are highly immunogenic in mice when expressed on a plant virus.Microbiology 145: 211–220.

Bruckman MA, Czapar AE, VanMeter A, Randolph LN, Steinmetz NF (2016) Tobacco mosaic virus-based protein nanoparticles and nanorods for chemotherapy delivery targeting breast cancer. J Control Release 231:103–113

Buyel JF (2019) Plant molecular farming – integration and exploitation of side streams to achieve sustainable biomanufacturing. Front Plant Sci 9. https://doi.org/10.3389/fpls.2018.01893

Cañizares MC, Liu L, Perrin Y, Tsakiris E, Lomonossoff GP (2006) A bipartite system for the constitutive and inducible expression of high levels of foreign proteins in plants. Plant Biotechnol J 4(2):183–193. https://doi.org/10.1111/j.1467-7652.2005.00170.x

Cañizares, M.C.; Liu, L.; Perrin, Y.; Tsakiris, E.; Lomonossoff, G.P. A Bipartite System for the Constitutive and Inducible Expression of High Levels of Foreign Proteins in Plants. Plant Biotechnol. J. 2006, 4, 183–193.

Carrillo-Tripp M, Santos RT-DL, Medina-Morales AM, Ríos-González L, Cruz-Hernández MA, Castell-González S, Sosa-Santillán GDJ, Acosta-Cruz E (2021) Biotechnological production of biomaterials and their applications. In: Handbook of research on bioenergy and biomaterials. Apple Academic Press

Chapman SN (2008) Construction of infectious clones for RNA viruses: TMV. In: Foster GD, Johansen IE, Hong Y, Nagy PD (eds) Plant virology protocols, vol 451. Humana Press, pp 477–490. https://doi.org/10.1007/978-1-59745-102-4_32

Chen X, Gonçalves MAFV (2016) Engineered viruses as genome editing devices. Mol Ther 24(3):447–457. https://doi.org/10.1038/mt.2015.164

Chen Q, Lai H (2013) Plant-derived virus-like particles as vaccines. Hum Vaccin Immunother 9(1):26–49. https://doi.org/10.4161/hv.22218

Chen Q, He J, Phoolcharoen W, Mason HS (2011) Geminiviral vectors based on bean yellow dwarf virus for production of vaccine antigens and monoclonal antibodies in plants. Hum Vaccin 7(3):331–338. https://doi.org/10.4161/hv.7.3.14262

Chen Q, Lai H, Hurtado J, Stahnke J, Leuzinger K, Dent M (2013) Agroinfiltration as an effective and scalable strategy of gene delivery for production of pharmaceutical proteins. Adv Tech Biol Med 1(1):103. https://doi.org/10.4172/atbm.1000103

Chichester JA, Haaheim LR, Yusibov V (2009) Using plant cells as influenza vaccine substrates. Expert Rev Vaccines 8(4):493–498. https://doi.org/10.1586/erv.09.3

Chincinska IA (2021) Leaf infiltration in plant science: old method, new possibilities. Plant Methods 17(1):83. https://doi.org/10.1186/s13007-021-00782-x

Chichester JA, Musiychuk K, de la Rosa P, Horsey A, Stevenson N, Ugulava N, Rabindran S, Palmer GA, Mett V, Yusibov V. 2007. Immunogenicity of a subunit vaccine against Bacillus anthracis. Vaccine 25: 3111–3114.

Chung HY, Lee HH, Kim KI, Chung HY, Hwang-Bo J, Park JH, Sunter G, Kim JB, Shon DH, Kim W, Chung IS (2011) Expression of a recombinant chimeric protein of hepatitis A virus VP1-Fc using a replicating vector based on Beet curly top virus in tobacco leaves and its immunogenicity in mice. Plant Cell Rep 30(8):1513–1521. https://doi.org/10.1007/s00299-011-1062-6

Chung YH, Cai H, Steinmetz NF (2020) Viral nanoparticles for drug delivery, imaging, immunotherapy, and theranostic applications. Adv Drug Deliv Rev 156:214–235. https://doi.org/10.1016/j.addr.2020.06.024

Dalsgaard, K., Uttenthal, A, Jones, T. D., Xu, F., Merryweather, A.,Hamilton, W. D. O., Langeveld, J. P. M., Boshuizen, R. S., Kamstrup, S.,Lomonossoff, G. P., Porta, C., Vela, C., Casal, J. I., Meloen, R. H., andRodgers, P. B. 1997. Plant derived vaccines protects target animals against aviral disease. Nat. Biotechnol. 15: 248–252.

Daniell H, Streatfield SJ, Rybicki EP (2015) Advances in molecular farming: key technologies, scaled up production and lead targets. Plant Biotechnol J 13(8):1011–1012. https://doi.org/10.1111/pbi.12478

Dohi K, Tamai A, Mori M (2008) Insertion in the coding region of the movement protein improves stability of the plasmid encoding a tomato mosaic virus-based expression vector. Arch Virol 153(9):1667–1675. https://doi.org/10.1007/s00705-008-0165-z

Dorokhov YL, Ivanov PA, Komarova TV, Skulachev MV, Atabekov JG (2006) An internal ribosome entry site located upstream of the crucifer-infecting tobamovirus coat protein (CP) gene can be used for CP synthesis in vivo. J Gen Virol 87(9):2693–2697. https://doi.org/10.1099/vir.0.82095-0

Dorokhov YL, Sheveleva AA, Frolova OY, Komarova TV, Zvereva AS, Ivanov PA, AtabekovJG. 2007. Superexpression of tuberculosis antigens in plant leaves. Tuberculosis (Edinb)87:218–224

Dubey KK, Luke GA, Knox C, Kumar P, Pletschke BI, Singh PK, Shukla P (2018) Vaccine and antibody production in plants: developments and computational tools. Brief Funct Genomics 17(5):295–307. https://doi.org/10.1093/bfgp/ely020

Dugdale B, Mortimer CL, Kato M, James TA, Harding RM, Dale JL (2014) Design and construction of an in-plant activation cassette for transgene expression and recombinant protein production in plants. Nat Protoc 9(5):1010–1027. https://doi.org/10.1038/nprot.2014.068

Durrani, Z., McInerney, T. L., McLain, L., Jones, T., Bellaby, T., Brennan,F.R., and Dimmock, N. J. 1998. Intranasal immunization with a plant virusexpressing a peptide from HIV-1 gp41 stimulates better mucosal and systemicHIV-1-specific IgA and IgG than oral immunization. J. Immuno. Methods 220:93–103.

Ecker DM, Seymour P (2020) Supply and demand trends: mammalian biomanufacturing industry overview. Pharma's year of accelerated innovation & convergence. https://www.cphi-online.com/46/resourcefile/10/93/83/CPhI%20annual%20rep%202020%20vJ.pdf#page=43

Eidenberger L, Kogelmann B, Steinkellner H (2023) Plant-based biopharmaceutical engineering. Nat Rev Bioeng 1(6):426–439. https://doi.org/10.1038/s44222-023-00044-6

Fernandez-Fernandez MR, Martinez-Torrecuadrada JL, Casal JI, Garcia JA. 1998.Development of an antigen presentation system based on plum pox potyvirus. FEBS Lett427:229–235

Fischer R, Stoger E, Schillberg S, Christou P, Twyman RM (2004) Plant-based production of biopharmaceuticals. Curr Opin Plant Biol 7(2):152–158. https://doi.org/10.1016/j.pbi.2004.01.007

Fischer R, Schillberg S, Buyel JF, Twyman RM (2013) Commercial aspects of pharmaceutical protein production in plants. Curr Pharm Des 19(31):5471–5477. https://doi.org/10.2174/1381612811319310002

Franke CE, Czapar AE, Patel RB, Steinmetz NF (2018) Tobacco Mosaic Virus-delivered Cisplatin restores efficacy in platinum-resistant ovarian cancer cells. Mol Pharm 15(8):2922–2931. https://doi.org/10.1021/acs.molpharmaceut.7b00466

Garabagi F, McLean MD, Hall JC (2012) Transient and stable expression of antibodies in Nicotiana species. Methods Mol Biol (Clifton, NJ) 907:389–408. https://doi.org/10.1007/978-1-61779-974-7_23

Gelvin SB (2003) Agrobacterium-mediated plant transformation: the biology behind the "Gene-Jockeying" tool. Microbiol Mol Biol Rev 67(1):16–37. https://doi.org/10.1128/MMBR.67.1.16-37.2003

Gerlach JQ, Kilcoyne M, McKeown P, Spillane C, Joshi L (2010) Plant-produced biopharmaceuticals. In: Kole C, Michler CH, Abbott AG, Hall TC (eds) Transgenic crop plants. Springer, Berlin Heidelberg, pp 269–299. https://doi.org/10.1007/978-3-642-04812-8_7

Ghanemi K (2017) Biopharmaceutical innovation: benefits and challenges. Open Access J Sci 1(1). https://doi.org/10.15406/oajs.2017.01.00004

Gilleland, H. E., Gilleland, L. B., Staczek, J., Harty, R. N., Garcia-Sastre, A.,Pale, P., Brennen, F. R., Hamilton, W. D., Bendahmane, M., and Beachy, R.N. 2000. Chimeric animal and plant viruses expressing eptiopes of outermembrane protein F as a combined vaccine against Pseudomonas aeruginosalung infection. FEMS Immunol. Med. Microbiol. 27: 291–297.

Gleba Y, Klimyuk V, Marillonnet S (2005) Magnifection—a new platform for expressing recombinant vaccines in plants. Vaccine 23(17–18):2042–2048

Gleba, Y., Klimyuk, V., & Marillonnet, S. (2007). Viral vectors for the expression of proteins in plants. Current opinion in biotechnology, 18(2), 134-141.

Goelet P, Lomonossoff GP, Butler PJ, Akam ME, Gait MJ, Karn J (1982) Nucleotide sequence of tobacco mosaic virus RNA. Proc Natl Acad Sci 79(19):5818–5822. https://doi.org/10.1073/pnas.79.19.5818

Goldstein DA, Thomas JA (2004) Biopharmaceuticals derived from genetically modified plants. QJM 97(11):705–716

Golovkin M, Spitsin S, Andrianov V, Smirnov Y, Xiao Y, Pogrebnyak N, Markley K, Brodzik R,Gleba Y, Isaacs SN, Koprowski H. 2007. Smallpox subunit vaccine produced in Plantaconfers protection in mice. Proc Natl Acad Sci USA 104:6864–6869

Gomord V, Faye L (2004) Posttranslational modification of therapeutic proteins in plants. Curr Opin Plant Biol 7(2):171–181. https://doi.org/10.1016/j.pbi.2004.01.015

Gonzalez MJ, Plummer EM, Rae CS, Manchester M (2009) Interaction of Cowpea Mosaic Virus (CPMV) nanoparticles with antigen presenting cells in. Citeseer. https://citeseerx.ist.psu.edu/document?repid=rep1&type=pdf&doi=2891432831bb0615e6c02914e31c54ffe6a76d4f

Grill LK, Palmer KE, Pogue GP (2005) Use of plant viruses for production of plant-derived vaccines. Crit Rev Plant Sci 24(4):309–323. https://doi.org/10.1080/07352680500253180
Grimsley N, Hohn B, Hohn T, Walden R (1986) "Agroinfection," an alternative route for viral infection of plants by using the Ti plasmid. Proc Natl Acad Sci 83(10):3282–3286. https://doi.org/10.1073/pnas.83.10.3282
Hamorsky KT, Kouokam JC, Bennett LJ, Baldauf KJ, Kajiura H, Fujiyama K, Matoba N (2013) Rapid and scalable plant-based production of a cholera toxin B subunit variant to aid in mass vaccination against cholera outbreaks. PLoS Negl Trop Dis 7(3):e2046
Hefferon KL (2014) DNA virus vectors for vaccine production in plants: spotlight on Geminiviruses. Vaccine 2(3):642–653. https://doi.org/10.3390/vaccines2030642
Hefferon K (2017) Plant virus expression vectors: a powerhouse for global health. Biomedicine 5(3):44. https://doi.org/10.3390/biomedicines5030044
Hemmati F, Hemmati-Dinarvand M, Karimzade M, Rutkowska D, Eskandari MH, Khanizadeh S, Afsharifar A (2022) Plant-derived VLP: a worthy platform to produce vaccine against SARS-CoV-2. Biotechnol Lett 44(1):45–57. https://doi.org/10.1007/s10529-021-03211-0
Horsch, R. B., Fry, J. E., Hoffmann, N. L., Wallroth, M., Eichholtz, D., Rogers, S. G., & Fraley, R. T. (1985). A simple and general method for transferring genes into plants. science, 227(4691), 1229-1231.
Huang Z, Chen Q, Hjelm B, Arntzen C, Mason H (2009) A DNA replicon system for rapid high-level production of virus-like particles in plants. Biotechnol Bioeng 103(4):706–714. https://doi.org/10.1002/bit.22299
Huang Z, Phoolcharoen W, Lai H, Piensook K, Cardineau G, Zeitlin L, Whaley KJ, Arntzen CJ, Mason HS, Chen Q (2010) High-level rapid production of full-size monoclonal antibodies in plants by a single-vector DNA replicon system. Biotechnol Bioeng 106(1):9–17. https://doi.org/10.1002/bit.22652
Jones RM, Chichester JA, Manceva S, Gibbs SK, Musiychuk K, Shamloul M, Norikane J, Streatfield SJ, Van De Vegte-Bolmer M, Roeffen W, Sauerwein RW, Yusibov V (2015) A novel plant-produced Pfs25 fusion subunit vaccine induces long-lasting transmission blocking antibody responses. Hum Vaccin Immunother 11(1):124–132. https://doi.org/10.4161/hv.34366
Jugler C, Sun H, Chen Q (2021) SARS-CoV-2 spike protein-induced interleukin 6 signaling is blocked by a plant-produced anti-interleukin 6 receptor monoclonal antibody. Vaccine 9(11):1365
Kagale S, Uzuhashi S, Wigness M, Bender T, Yang W, Borhan MH, Rozwadowski K (2012) TMV-Gate vectors: gateway compatible tobacco mosaic virus based expression vectors for functional analysis of proteins. Sci Rep 2(1):874
Kallolimath S, Sun L, Palt R, Stiasny K, Mayrhofer P, Gruber C, Kogelmann B, Chen Q, Steinkellner H (2021) Highly active engineered IgG3 antibodies against SARS-CoV-2. Proc Natl Acad Sci 118(42):e2107249118. https://doi.org/10.1073/pnas.2107249118
Kannan M, Zainal Z, Ismail I, Baharum SN, Bunawan H (2020) Application of reverse genetics in functional genomics of Potyvirus. Viruses 12(8):Article 8. https://doi.org/10.3390/v12080803
Karasev AV, Foulke S, Wellens C, Rich A, Shon KJ, Zwierzynski I, Hone D, Koprowski H,Reitz M. 2005. Plant based HIV-1 vaccine candidate: Tat protein produced in spinach.Vaccine 23:1875–1880.
Kaur M, Manchanda P, Kalia A, Ahmed FK, Nepovimova E, Kuca K, Abd-Elsalam KA (2021) Agroinfiltration mediated scalable transient gene expression in genome edited crop plants. Int J Mol Sci 22(19):10882. https://doi.org/10.3390/ijms221910882
Kemnade JO, Seethammagari M, Collinson-Pautz M, Kaur H, Spencer DM, McCormick AA (2014) Tobacco mosaic virus efficiently targets DC uptake, activation and antigen-specific T cell responses in vivo. Vaccine 32(33):4228–4233
Kikkert JR, Vidal JR, Reisch BI (2005) Stable transformation of plant cells by particle bombardment/biolistics. Methods Mol Biol (Clifton, N.J.) 286:61–78. https://doi.org/10.1385/1-59259-827-7:061
Klimyuk V, Pogue G, Herz S, Butler J, Haydon H (2012) Production of recombinant antigens and antibodies in Nicotiana benthamiana using 'Magnifection' technology: GMP-compliant

facilities for small- and large-scale manufacturing. In: Palmer K, Gleba Y (eds) Plant viral vectors, vol 375. Springer, Berlin Heidelberg, pp 127–154. https://doi.org/10.1007/82_2012_212

Komarova TV, Baschieri S, Donini M, Marusic C, Benvenuto E, Dorokhov YL (2010) Transient expression systems for plant-derived biopharmaceuticals. Expert Rev Vaccines 9(8):859–876. https://doi.org/10.1586/erv.10.85

Kost TA, Condreay JP (2002) Recombinant baculoviruses as mammalian cell gene-delivery vectors. Trends Biotechnol 20(4):173–180

Kost, T. A., & Condreay, J. P. (2002). Recombinant baculoviruses as mammalian cell gene-delivery vectors. Trends in biotechnology, 20(4), 173-180.

Lai H, He J, Engle M, Diamond MS, Chen Q (2012) Robust production of virus-like particles and monoclonal antibodies with geminiviral replicon vectors in lettuce. Plant Biotechnol J 10(1):95–104. https://doi.org/10.1111/j.1467-7652.2011.00649.x

Langeveld, J. P., Brennan, F. R., Martinez-Torrecuadrada, J. L., Jones, T. D.,Boshuizen, R. S., Vela, C., Casal, J. I., Kamstrup. S., Dalsgaard, K., Meloen.R., H., Bendig, M. M., and Hamilton, W. D. 2001. Inactivated recombinantplant virus protects dogs from a lethal challenge with canine parvovirus.Vaccine 19: 3661–3670.

Lamprecht RL, Kennedy P, Huddy SM, Bethke S, Hendrikse M, Hitzeroth II, Rybicki EP (2016) Production of human papillomavirus pseudovirions in plants and their use in pseudovirion-based neutralisation assays in mammalian cells. Sci Rep 6:20431. https://doi.org/10.1038/srep20431

Lee L-Y, Gelvin SB (2008) T-DNA binary vectors and systems. Plant Physiol 146(2):325–332. https://doi.org/10.1104/pp.107.113001

Leuzinger K, Dent M, Hurtado J, Stahnke J, Lai H, Zhou X, Chen Q (2013) Efficient agroinfiltration of plants for high-level transient expression of recombinant proteins. J Vis Exp 77. https://pmc.ncbi.nlm.nih.gov/articles/PMC3846102/

Lico C, Chen Q, Santi L (2008) Viral vectors for production of recombinant proteins in plants. J Cell Physiol 216(2):366–377. https://doi.org/10.1002/jcp.21423

Lico C, Schoubben A, Baschieri S, Blasi P, Santi L (2013) Nanoparticles in biomedicine: new insights from plant viruses. Curr Med Chem 20(28):3471–3487

Lico C, Benvenuto E, Baschieri S (2015) The Two-Faced Potato Virus X: from plant pathogen to smart nanoparticle. Front Plant Sci 6. https://doi.org/10.3389/fpls.2015.01009

Lindbo JA (2007) High-efficiency protein expression in plants from agroinfection-compatible Tobacco mosaic virus expression vectors. BMC Biotechnol 7(1):52. https://doi.org/10.1186/1472-6750-7-52

Liu Z, Kearney CM (2010) A tobamovirus expression vector for agroinfection of legumes and Nicotiana. J Biotechnol 147(3–4):151–159

Liu L, Grainger J, Cañizares MC, Angell SM, Lomonossoff GP (2004) *Cowpea mosaic virus* RNA-1 acts as an amplicon whose effects can be counteracted by a RNA-2-encoded suppressor of silencing. Virology 323(1):37–48. https://doi.org/10.1016/j.virol.2004.02.013

Liu L, Cañizares MC, Monger W, Perrin Y, Tsakiris E, Porta C, Shariat N, Nicholson L, Lomonossoff GP (2005) Cowpea mosaic virus-based systems for the production of antigens and antibodies in plants. Vaccine 23(15):1788–1792. https://doi.org/10.1016/j.vaccine.2004.11.006

Liu R, Vaishnav RA, Roberts AM, Friedland RP (2013) Humans have antibodies against a plant virus: evidence from tobacco mosaic virus. PLoS One 8(4):e60621

Loh H-S, Green BJ, Yusibov V (2017) Using transgenic plants and modified plant viruses for the development of treatments for human diseases. Curr Opin Virol 26:81–89

Loke YH, Jayakrishnan A, Mod Razif MRF, Yee KM, Kee PE, Goh BH, Helal Uddin ABM, Lakshminarayanan V, Liew KB (2024) A comprehensive review of challenges in oral drug delivery systems and recent advancements in innovative design strategies. Curr Pharm Des 30. https://doi.org/10.2174/0113816128338560240923073357

Lomonossoff GP (2008) Cowpea Mosaic Virus. Encyclopedia Virol 569. https://doi.org/10.1016/B978-012374410-4.00562-8

Lomonossoff GP, D'Aoust M-A (2016) Plant-produced biopharmaceuticals: a case of technical developments driving clinical deployment. Science 353(6305):1237–1240. https://doi.org/10.1126/science.aaf6638

Mad' Atari MFB, Folta KM (2019) Transformation improvement with the Standardized Pressure Agrobacterium Infiltration Device (SPAID). BMC Res Notes 12:144. https://doi.org/10.1186/s13104-019-4117-3

Mahmood MA, Naqvi RZ, Rahman SU, Amin I, Mansoor S (2023) Plant virus-derived vectors for plant genome engineering. Viruses 15(2):531

Mardanova ES, Blokhina EA, Tsybalova LM, Peyret H, Lomonossoff GP, Ravin NV (2017) Efficient transient expression of recombinant proteins in plants by the novel pEff vector based on the genome of potato virus X. Front Plant Sci 8:247

Marillonnet S, Giritch A, Gils M, Kandzia R, Klimyuk V, Gleba Y (2004) *In planta* engineering of viral RNA replicons: efficient assembly by recombination of DNA modules delivered by *Agrobacterium*. Proc Natl Acad Sci 101(18):6852–6857. https://doi.org/10.1073/pnas.0400149101

Marsian J, Lomonossoff GP (2016) Molecular pharming—VLPs made in plants. Curr Opin Biotechnol 37:201–206. https://doi.org/10.1016/j.copbio.2015.12.007

Martí M, Merwaiss F, Butković A, Daròs J-A (2022) Production of Potyvirus-derived nanoparticles decorated with a Nanobody in biofactory plants. Front Bioeng Biotechnol 10. https://doi.org/10.3389/fbioe.2022.877363

Massa S, Franconi R, Brandi R, Muller A, Mett V, Yusibov V, Venuti A (2007) Anti-cancer activity of plant-produced HPV16 E7 vaccine. Vaccine 25(16):3018–3021

Marusic C, Rizza P, Lattanzi L, Mancini C, Spada M, Belardelli F, Benvenuto E, Capone I. 2001. Chimeric plant virus particles as immunogens for inducing murine and human immuneresponses against human immunodeficiency virus type 1. J Virol 75:8434–8439.

Matsuda R, Abe T, Fujiuchi N, Matoba N, Fujiwara K (2017) Effect of temperature post viral vector inoculation on the amount of hemagglutinin transiently expressed in Nicotiana benthamiana leaves. J Biosci Bioeng 124(3):346–350

Mett V, Lyons J, Musiychuk K, Chichester JA, Brasil T, Couch R, Sherwood R, Palmer GA,Streatfield SJ, Yusibov V. 2007. A plant-produced plague vaccine candidate confersprotection to monkeys. Vaccine 25:3014–3017

McCormick, A. A., Kumagai, M. H., Hanley, K., Turpen, T. H., Hakim, I., Grill,L. K., Tuse, D., Levy, S., and Levy, R. 1999. Rapid production of specificvaccines for lymphoma by expression of the tumor-derived single-chain Fvepitopes in tobacco plants. Proc. Natl. Acad. Sci. USA 96: 703–708.

McCormick, A. A., Reinl, S. J., Cameron, T. I., Vojdani, F., Fronefield, M., Levy,R., and Tus´e, D. T. 2003. Individualized human scFv vaccines produced inplants: humoral anti-idiotype responses in vaccinated mice confirm relevanceto the tumor Ig. J. Immunol. Meth. 278: 95–104.

McCormick AA, Corbo TA, Wykoff-Clary S, Palmer KE, Pogue GP. 2006b. Chemicalconjugate TMV-peptide bivalent fusion vaccines improve cellular immunity and tumorprotection. Bioconjug Chem 17:1330–1338

McCormick AA (2011) Tobacco derived cancer vaccines for non-Hodgkin's lymphoma: perspectives and progress. Hum Vaccin 7(3):305–312. https://doi.org/10.4161/hv.7.3.14163

McCormick AA, Reddy S, Reinl SJ, Cameron TI, Czerwinkski DK, Vojdani F, Hanley KM, Garger SJ, White EL, Novak J, Barrett J, Holtz RB, Tusé D, Levy R (2008) Plant-produced idiotype vaccines for the treatment of non-Hodgkin's lymphoma: safety and immunogenicity in a phase I clinical study. Proc Natl Acad Sci 105(29):10131–10136. https://doi.org/10.1073/pnas.0803636105

McLain, L., Durrani, Z., Wisniewski, L. A., Porta, C., Lomonossoff, G. P.,and Dimmock, N. J. 1996. Stimulation of neutralizing antibodies to humanimmunodeficiency virus type 1 in three strains of mice immunized with a 22amino acid peptide of gp41 expressed on the surface of a plant virus. Vaccine14: 799–810.

Musiychuk K, Stephenson N, Bi H, Farrance CE, Orozovic G, Brodelius M, Brodelius P, Horsey A, Ugulava N, Shamloul A-M, Mett V, Rabindran S, Streatfield SJ, Yusibov V (2007) A

launch vector for the production of vaccine antigens in plants. Influenza Other Respir Viruses 1(1):19–25. https://doi.org/10.1111/j.1750-2659.2006.00005.x
Natilla, A., Piazzolla, G., Nuzzaci, M., Saldarelli, P., Tortorella, C., Antonaci,S., and Piazzolla, P. 2004. Cucumber mosaic virus as carrier of a hepatitis Cvirus-derived epitope. Arch. Virol. 149: 137–154.
Nemchinov, L. G., Liang, T. J., Rifaat, M. M., Mazyad, H. M., Hadidi, A., andKeith, J. M. 2000. Development of a plant-derived subunit vaccine candidateagainst hepatitis C virus. Arch. Virol. 145: 2557–2573.
Nemchinov LG, Natilla A. 2007. Transient expression of the ectodomain of matrix protein 2(M2e) of avian influenza A virus in plants. Protein Expr Purif 56:153–159.
Noris E, Poli A, Cojoca R, Rittà M, Cavallo F, Vaglio S, Matic S, Landolfo S (2011) A human papillomavirus 8 E7 protein produced in plants is able to trigger the mouse immune system and delay the development of skin lesions. Arch Virol 156(4):587–595. https://doi.org/10.1007/s00705-010-0893-8
Nosaki S, Hoshikawa K, Ezura H, Miura K (2021) Transient protein expression systems in plants and their applications. Plant Biotechnol 38(3):297–304. https://doi.org/10.5511/plantbiotechnology.21.0610a
O'Brien GJ, Bryant CJ, Voogd C, Greenberg HB, Gardner RC, Bellamy AR. 2000. RotavirusVP6 expressed by PVX vectors in Nicotiana benthamiana coats PVX rods and alsoassembles into viruslike particles. Virology 270:444–453.
Palukaitis P, Zaitlin M (1986) Tobacco Mosaic Virus infectivity and replication. In: Van Regenmortel MHV, Fraenkel-Conrat H (eds) The plant viruses. Springer US, pp 105–131. https://doi.org/10.1007/978-1-4684-7026-0_5
Patil BL, Fauquet CM (2015) Light intensity and temperature affect systemic spread of silencing signal in transient agroinfiltration studies. Mol Plant Pathol 16(5):484–494. https://doi.org/10.1111/mpp.12205
Perez Filgueira, D. M., Zamorano, P. I., Dominguez, M. G., Taboga, O., DelMedico Zajac, M. P., Puntel, M., Romera, S. A., Morris, T. J., Borca, M. V.,and Sadir, A. M. 2003. Bovine herpes virus G-D protein produced in plantsusing a recombinant tobacco mosaic virus (TMV) vector possesses authenticantigenicity. Vaccine 21: 4201–4209
P´erez-Filgueira, D. M., Mozgovoj, M., Wigdorovitz, A., Dus Santos, M. J.,Parre˜no, V., Trono, K., Fernandez, F. M., Carrillo, C., Babiuk, L. A.,Morris, T. J., and Borca, M. V. 2004. Passive protection of bovine rotavirus(BRV) infection induced by a BRV VP8 produced in plants using a TMV-based vector. Arch. Virol. 149: 2337–2348
Petukhova N, Gasanova T, Ivanov P, Atabekov J (2014) High-level systemic expression of conserved Influenza Epitope in plants on the surface of rod-shaped Chimeric particles. Viruses 6(4):1789–1800. https://doi.org/10.3390/v6041789
Peyret H, Lomonossoff GP (2015) When plant virology met agrobacterium: the rise of the deconstructed clones. Plant Biotechnol J 13(8):1121–1135. https://doi.org/10.1111/pbi.12412
Phelps JP, Dang N, Rasochova L. 2007. Inactivation and purification of cowpea mosaic virus-like particles displaying peptide antigens from Bacillus anthracis. J Virol Methods 141: 146–153.
Phoolcharoen W, Bhoo SH, Lai H, Ma J, Arntzen CJ, Chen Q, Mason HS (2011) Expression of an immunogenic Ebola immune complex in *Nicotiana benthamiana*. Plant Biotechnol J 9(7):807–816. https://doi.org/10.1111/j.1467-7652.2011.00593.x
Pogue, G. P., Lindbo, J. A., McCulloch, M. J., Lawrence, J. E., Gross, C. S., andGarger, S. J. 2004. Parvovirus vaccine as viral coat protein fusions. UnitedStates Patent #6,730,306.
Porta, C., Spall, V. E., Loveland, J., Johnson, J. E., Barker, P. J., and Lomonos-soff, G. P. 1994. Development of cowpea mosaic virus as a high-yieldingsystem for the presentation of foreign peptides. Virology 202: 949–955.
Regnard GL, Halley-Stott RP, Tanzer FL, Hitzeroth II, Rybicki EP (2010) High level protein expression in plants through the use of a novel autonomously replicating geminivirus shuttle vector. Plant Biotechnol J 8(1):38–46. https://doi.org/10.1111/j.1467-7652.2009.00462.x

Rennermalm, A., Li, Y. H., Bohaufs, L., Jarstrand, C., Brauner, A., Brennan,F. R., and Flock, J. I. 2001. Antibodies against a truncated Staphylococcusaureus fibronectin-binding protein protect against dissemination of infectionin the rat. Vaccine 19: 3376–3383
Rybicki EP (2014) Plant-based vaccines against viruses. Virol J 11(1):205. https://doi.org/10.1186/s12985-014-0205-0
Rybicki EP (2017) Plant-made vaccines and reagents for the One Health initiative. Hum Vaccin Immunother 13(12):2912–2917. https://doi.org/10.1080/21645515.2017.1356497
Rybicki EP (2020) Plant molecular farming of virus-like nanoparticles as vaccines and reagents. WIREs Nanomed Nanobiotechnol 12(2):e1587. https://doi.org/10.1002/wnan.1587
Sainsbury F, Lomonossoff GP (2008) Extremely high-level and rapid transient protein production in plants without the use of viral replication. Plant Physiol 148(3):1212–1218
Sainsbury F, Lavoie P, D'Aoust M, Vézina L, Lomonossoff GP (2008) Expression of multiple proteins using full-length and deleted versions of cowpea mosaic virus RNA-2. Plant Biotechnol J 6(1):82–92. https://doi.org/10.1111/j.1467-7652.2007.00303.x
Sainsbury F, Liu L, Lomonossoff GP (2009) Cowpea Mosaic Virus-based systems for the expression of antigens and antibodies in plants. In: Faye L, Gomord V (eds) Recombinant proteins from plants: methods and protocols. Humana Press, pp 25–39. https://doi.org/10.1007/978-1-59745-407-0_2
Santi L, Giritch A, Roy CJ, Marillonnet S, Klimyuk V, Gleba Y, Webb R, Arntzen CJ, Mason HS.2006. Protection conferred by recombinant Yersinia pestis antigens produced by a rapidand highly scalable plant expression system. Proc Natl Acad Sci USA 103:861–866
Santi L, Batchelor L, Huang Z, Hjelm B, Kilbourne J, Arntzen CJ, Chen Q, Mason HS (2008) An efficient plant viral expression system generating orally immunogenic Norwalk virus-like particles. Vaccine 26(15):1846–1854
Saunders K, Sainsbury F, Lomonossoff GP (2009) Efficient generation of cowpea mosaicvirus empty virus-like particles by the proteolytic processing of precursors in insect cells and plants. Virology 393(2):329–337
Schillberg S, Raven N, Fischer R, Twyman RM, Schiermeyer A (2013) Molecular farming of pharmaceutical proteins using plant suspension cell and tissue cultures. Curr Pharm Des 19(31):5531–5542. https://doi.org/10.2174/1381612811319310008
Shen W-H, Hohn B (1995) Vectors based on maize streak virus can replicate to high copy numbers in maize plants. J Gen Virol 76(4):965–969. https://doi.org/10.1099/0022-1317-76-4-965
Shukla S, Ablack AL, Wen AM, Lee KL, Lewis JD, Steinmetz NF (2013) Increased tumor homing and tissue penetration of the filamentous plant viral nanoparticle *Potato virus X*. Mol Pharm 10(1):33–42. https://doi.org/10.1021/mp300240m
Smith ML, Fitzmaurice WP, Turpen TH, Palmer KE (2009) Display of peptides on the surface of Tobacco Mosaic virus particles. In: Karasev AV (ed) Plant-produced microbial vaccines, vol 332. Springer, Berlin Heidelberg, pp 13–31. https://doi.org/10.1007/978-3-540-70868-1_2
Staczek, J., Bendahmane, M., Gilleland, L. B., Beachy, R. N., and Gilleland,H. E. Jr. 2000. Immunization with a chimeric tobacco mosaic virus contain-ing an epitope of outer membrane protein F of Pseudomonas aeruginosaprovides protection against challenge with P. aeruginosa. Vaccine 18: 2266–2274.
Steele, J. F., Peyret, H., Saunders, K., Castells-Graells, R., Marsian, J., Meshcheriakova, Y., & Lomonossoff, G. P. (2017). Synthetic plant virology for nanobiotechnology and nanomedicine. Wiley Interdisciplinary Reviews: Nanomedicine and Nanobiotechnology, 9(4), e1447.
Steinmetz NF (2010) Viral nanoparticles as platforms for next-generation therapeutics and imaging devices. Nanomedicine 6(5):634–641. https://doi.org/10.1016/j.nano.2010.04.005
Su H, van Eerde A, Rimstad E, Bock R, Branza-Nichita N, Yakovlev IA, Clarke JL (2023) Plant-made vaccines against viral diseases in humans and farm animals. Front Plant Sci 14. https://doi.org/10.3389/fpls.2023.1170815
Sugiyama, Y., Hamamoto, H., Takemoto, S., Watanabe, Y., and Okada, Y. 1995.Systemic production of foreign peptides on the particle surface of tobaccomosaic virus. FEBS Let. 359: 247–250
Sun L, Kallolimath S, Palt R, Stiasny K, Mayrhofer P, Maresch D, Eidenberger L, Steinkellner H (2021) Increased in vitro neutralizing activity of SARS-CoV-2 IgA1 dimers compared

to monomers and IgG. Proc Natl Acad Sci 118(44):e2107148118. https://doi.org/10.1073/pnas.2107148118
Tremblay R, Wang D, Jevnikar AM, Ma S (2010) Tobacco, a highly efficient green bioreactor for production of therapeutic proteins. Biotechnol Adv 28(2):214–221. https://doi.org/10.1016/j.biotechadv.2009.11.008
Turpen, T. H., Reinl, S. J., Charoenvit, Y., Hoffman, S. L., Fallarme, V., andGrill, L. K. 1995. Malarial epitopes expressed on the surface of recombinanttobacco mosaic virus. Nat. Biotechnol. 13: 53–57
Tusé D, Nandi S, McDonald KA, Buyel JF (2020) The emergency response capacity of plant-based biopharmaceutical manufacturing-what it is and what it could be. Front Plant Sci 11. https://doi.org/10.3389/fpls.2020.594019
Tzfira T, Citovsky V (2006) Agrobacterium-mediated genetic transformation of plants: biology and biotechnology. Curr Opin Biotechnol 17(2):147–154. https://doi.org/10.1016/j.copbio.2006.01.009
Venkataraman S, Hefferon K (2021) Application of plant viruses in biotechnology, medicine, and human health. Viruses 13(9):1697
Vernekar, A. A., Berger, G., Czapar, A. E., Veliz, F. A., Wang, D. I., Steinmetz, N. F., & Lippard, S. J. (2018). Speciation of phenanthriplatin and its analogs in the core of tobacco mosaic virus. Journal of the American Chemical Society, 140(12), 4279-4287.
Voinnet O, Rivas S, Mestre P, Baulcombe D (2015) An enhanced transient expression system in plants based on suppression of gene silencing by the p19 protein of tomato bushy stunt virus (Retration of vol 33, pg 949, 2003). Plant J 84(4):846–846
Wang L-Y, Lin S-S, Hung T-H, Li T-K, Lin N-C, Shen T-L (2012) Multiple domains of the *Tobacco mosaic virus* p126 protein can independently suppress local and systemic RNA silencing. Mol Plant-Microbe Interact 25(5):648–657. https://doi.org/10.1094/MPMI-06-11-0155
Wang C, Fiering SN, Steinmetz NF (2019) Cowpea Mosaic Virus promotes anti-tumor activity and immune memory in a Mouse Ovarian Tumor model. Adv Therap 2(5):1900003. https://doi.org/10.1002/adtp.201900003
Wigdorovitz A, Perez Filgueira DM, Robertson N, Carrillo C, Sadir AM, Morris TJ, Borca MV. 1999. Protection of mice against challenge with foot and mouth disease virus (FMDV) by immunization with foliar extracts from plants infected with recombinant tobacco mosaic virus expressing the FMDV structural protein VP1. Virology 264: 85–91.
Wu, L., Jiang, L., Zhou, Z., Fan, J., Zhang, Q., Zhu, H., Han, Q., andXu, Z. 2003. Expression of foot-and-mouth disease virus in tobaccoleaves using a tobacco mosaic virus-based vector. Vaccine 21: 4390–4398
Yusibov V, Hooper DC, Spitsin SV, Fleysh N, Kean RB, Mikheeva T, Deka D, Karasev A, CoxS, Randall J, Koprowski H. 2002. Expression in plants and immunogenicity of plant virus-based experimental rabies vaccine. Vaccine 20:3155–3164.
Yusibov V, Mett V, Davidson C, Musiychuk K, Gilliam S, Farese A, Macvittie T, Mann D.2005. Peptide-based candidate vaccine against respiratory syncytial virus. Vaccine 23:2261–2265.
Yusibov V, Rabindran S (2008) Recent progress in the development of plant derived vaccines. Expert Rev Vaccines 7(8):1173–1183. https://doi.org/10.1586/14760584.7.8.1173
Yusibov V, Shivprasad S, Turpen TH, Dawson W, Koprowski H (2000) Plant viral vectors based on Tobamoviruses. In: Hammond J, McGarvey P, Yusibov V (eds) Plant biotechnology, vol 240. Springer, Berlin Heidelberg, pp 81–94. https://doi.org/10.1007/978-3-642-60234-4_4
Yusibov V, Rabindran S, Commandeur U, Twyman RM, Fischer R (2006) The potential of plant virus vectors for vaccine production. Drugs R & D 7(4):203–217. https://doi.org/10.2165/00126839-200607040-00001
Yusibov V, Streatfield S, Kushnir N, Roy G, Padmanaban A (2013) Hybrid viral vectors for vaccine and antibody production in plants. Curr Pharm Des 19(31):5574–5586. https://doi.org/10.2174/13816128113199990335
Zahin M, Joh J, Khanal S, Husk A, Mason H, Warzecha H, Ghim S, Miller DM, Matoba N, Jenson AB (2016) Scalable production of HPV16 L1 protein and VLPs from tobacco leaves. PLoS One 11(8):e0160995

Zahmanova G, Aljabali AA, Takova K, Toneva V, Tambuwala MM, Andonov AP, Lukov GL, Minkov I (2023) The plant viruses and molecular farming: how beneficial they might be for human and animal health? Int J Mol Sci 24(2):1533

Zhang Y, Chen M, Siemiatkowska B, Toleco MR, Jing Y, Strotmann V, Zhang J, Stahl Y, Fernie AR (2020) A highly efficient agrobacterium-mediated method for transient gene expression and functional studies in multiple plant species. Plant Commun 1(5):100028. https://doi.org/10.1016/j.xplc.2020.100028

Cutting-Edge Technologies in the Production of Different Biopharmaceuticals

Sandesh Poudel and Raj Kumar Thapa

Abstract

Biopharmaceuticals have gained considerable interest since the introduction of the first recombinant therapeutic, human insulin, in 1982, which led to the development of a new branch of pharmaceutical manufacturing and therapy. The market for biopharmaceuticals has been steadily growing, as represented by the approval of >450 biopharmaceuticals in the EU and the US. The conventional bioreactors and fermentation capacities can be bottlenecks for the efficient production of biopharmaceuticals. Therefore, plants can be utilized as a viable alternative in the production of biopharmaceuticals, thereby offering a variety of options for transgene targeting and modification. This chapter will focus on the potential of plants as biological factories. The key focus will be placed on recent advancements in plant-based expression systems and the utilization of cutting-edge technologies that streamline the production of different biopharmaceuticals.

Keywords

Biology · Biopharmaceuticals · Plants · Targeting · Transgene · Therapy

S. Poudel
School of Health and Allied Sciences, Pokhara University, Pokhara, Kaski, Nepal

R. K. Thapa (✉)
Pharmacy Program, Gandaki University, Pokhara, Kaski, Nepal

Research Center, Invention and Innovation Center, Gandaki University,
Pokhara, Kaski, Nepal
e-mail: rkthapa@gandakiuniversity.edu.np

R. K. Gaur et al. (eds.), *Plant Molecular Farming*,
https://doi.org/10.1007/978-981-95-3823-2_12

1 Introduction

Therapy of a diseased condition with pharmaceuticals that improve altered physiology is the most trusted approach to treatment (Yuan et al. 2016). Over the years, the scope of the term "pharmaceuticals" has been expanding with the advancement in science and technology. In its earlier days, pharmaceuticals were chemical entities derived from different sources: plants, animals, microbes, or chemical synthesis, that altered the pathophysiology and aided in the treatment of diseases. However, the progress in science and technology brought advancement in treatment modalities, and the term pharmaceuticals now covers a wide range of substances, including chemical entities, naturally obtained biologics, and the products of biotechnology (Rader 2008).

The term "biopharmaceutical" is often interpreted differently, leading to confusion. While some define biopharmaceuticals as biologically derived substances obtained from natural sources (Yuan-Chuan and Ming-Kung 2018), others consider them as products manufactured specifically through biotechnology (Rader 2008). For the purpose of this book chapter, we adopt the second definition of biopharmaceuticals. Thus, for a pharmaceutical to be termed a biopharmaceutical, it must be partly or wholly obtained from biological sources (plants, animals, microbes, etc.) using biotechnology tools, and it must also be biological (proteins, peptides, etc.). For example, recombinant human insulin is a biopharmaceutical because its nature and source are biological and are synthesized in recombinant hosts (Goeddel et al. 1979), but steroidal hormones and non-biological antibiotics are not biopharmaceuticals because, although they are obtained from recombinant biological sources, they are chemical in nature. This definition of biopharmaceuticals distinguishes pharmaceuticals and other biologics (those that are produced by conventional biological methods) from biopharmaceuticals.

1.1 Expression Systems for Biopharmaceuticals

As previously discussed, biopharmaceuticals are the products of biotechnology and are synthesized in the biological factories of cells. Owing to the well-understood cellular physiology, prokaryotes are the most widely used production systems in the manufacture of biopharmaceuticals. This is also true because prokaryotic production systems are the pioneering expression systems in the production of biopharmaceuticals (Sodoyer 2004). However, biopharmaceuticals are often complex proteins, and prokaryotic expression systems are not always capable of efficiently producing and modifying them (i.e., posttranslational modifications). For these reasons, multiple expression systems have been developed over the years, with mammalian being the preferred expression system, according to recent reports (Estes and Melville 2014; Walsh and Walsh 2022). Below, we discuss each source and provide an example of a biopharmaceutical produced from it, as well as the pros and cons associated with the source.

1.2 Prokaryotic Expression Systems

Prokaryotic expression systems, especially *E. coli*, are widely employed for recombinant protein production owing to their rapid growth, cost-effectiveness, and ease of genetic manipulation (Baeshen et al. 2015). These are also the pioneering expression systems in protein production, with recombinant human insulin (Humulin) being the first biopharmaceutical to be successfully commercialized (Goeddel et al. 1979). Prokaryotes are known to multiply rapidly, hence speeding the production of biopharmaceuticals. Further, prokaryotic cultures are scalable and cost-effective. However, the major setback of prokaryotic production systems is their inability to undergo complex posttranslational modifications of biopharmaceuticals (Sanchez-Garcia et al. 2016).

1.3 Eukaryotic Expression Systems

1.3.1 Yeast

Yeasts, just like bacteria, retain the ability to grow and multiply rapidly, allowing rapid production of biopharmaceuticals. Apart from this, these expression systems offer an additional benefit, i.e., they are eukaryotes and can introduce some of the posttranslational modifications that are desired in complex biopharmaceuticals (Sodoyer 2004). *Saccharomyces cerevisiae* has been one of the most extensively utilized expression systems to produce recombinant proteins for these very reasons. Gardasil®, a vaccine approved against human papillomavirus, represents one example of the successful implementation of yeast as an eukaryotic expression system (Group, 2007). Although bioengineered yeasts can perform basic glycosylations, complex posttranslational modifications still remain a limitation of yeast-based eukaryotic expression systems. Further, yeast expression systems are known to hyperglycosylate proteins, reducing therapeutic efficacy and increasing the chances of immunogenicity, as in bacteria (Nevalainen and Peterson 2014).

1.3.2 Mammalian Cells

Mammalian cell expression systems are considered the gold standard for producing complex biopharmaceuticals for multiple reasons. They are capable of complex posttranslational modifications, which were major drawbacks of prokaryotic and yeast-based expression systems. Further, biopharmaceuticals from mammalian expression systems are known to be more compatible and pose a lower risk of hypersensitivity (Schmidt 2004). Herceptin®, a monoclonal antibody targeting the HER2 receptor in the management of breast cancer, is an excellent example of a biopharmaceutical obtained using a mammalian expression system produced using Chinese Hamster Ovary (CHO) cells (Wilson et al. 2018). However, mammalian expression systems face significant limitations, including their complexity and the need for extensive optimization to ensure consistent and reliable production. Apart from this, mammalian cell systems are known to have lower production rates compared to prokaryotic production systems (Schmidt 2004).

1.3.3 Plants

Plant cell cultures are viewed as an excellent substitute for mammalian cell-based expression systems. The use of plant expression systems in the manufacture of biopharmaceuticals is commonly termed molecular pharming, a portmanteau of farming and pharmaceutical (Fischer and Emans 2000). Similar to mammalian expression systems, plants possess the capability to perform complex posttranslational modifications. In addition, plant biomolecules are distinctive enough from mammalian products, and hence, biopharmaceuticals from plants are known to exhibit a lower degree of immunogenicity. Apart from these advantages, plant-based expression provides rapid scalability and a lower production cost compared to mammalian cell-based expression systems (Fischer et al. 2004; Schmidt 2004). Elelyso® (taliglucerase alfa) represents the first plant-based biopharmaceutical to achieve regulatory approval for commercialization. This recombinant form of human glucocerebrosidase was successfully produced using transgenic carrot cell cultures and received FDA approval in 2012 (Mor 2015). Although plant-based expression has shown promising results, it is slower than prokaryotic expression systems and is known to show nonhuman glycosylation patterns, affecting the safety and efficacy of the biopharmaceuticals. Further, proteases in plants are known to reduce the accumulation of biopharmaceuticals, drastically reducing their yields (Margolin et al. 2020a, b) (Table 1).

2 History of Biopharmaceuticals

The advancement of biotechnology over the past decades has brought an end to human suffering. The pioneering miracle of biotechnology and recombinant deoxyribonucleic acid (rDNA) technology in the field of medicine is the production of recombinant human insulin in *Escherichia coli* (Goeddel et al. 1979). Although the treatment of diabetes using animal-derived insulin was practiced since the 1920s (Banting and Best 1987), the production required enormous quantities of animal pancreas and often led to insulin resistance and hypersensitivity reactions in patients administered with animal insulins. The production of recombinant human insulin in *E. coli* addressed many of these challenges, leading to a groundbreaking achievement. On October 28, 1982, after just 5 months of review, the FDA granted approval for the commercialization of recombinant human insulin, marketed as Humulin, by Eli Lilly and Co. (FDA 2022).

The approval of Humulin opened a new era of biopharmaceuticals, with more than 80 biopharmaceuticals being approved within just 20 years of the approval of Humulin (FDA 2022). After this, the market growth curve of biopharmaceuticals appears steeper each year, reaching a total of 443 active licenses for individual biopharmaceuticals by 2022 in Europe and the USA (Walsh and Walsh 2022).

Although prokaryotes were the pioneering expression systems, they are not ideal. This is especially true for complex biopharmaceuticals that require posttranslational folding and glycosylation. Thus, the search for an ideal expression system

Table 1 Expression systems for biopharmaceuticals. (Shanmugaraj et al. 2020b)

Expression system		Advantages	Disadvantages
Prokaryotic		User-friendly for modifications Economical to produce Elevated protein yield Scalability for industrial production Rapid production timeline Well-defined regulatory approval processes	Misfolding of proteins, limiting functionality for complex proteins Lack of posttranslational modifications (e.g., phosphorylation, glycosylation) Accumulation of endotoxins, requiring additional purification steps
Eukaryotic	*Yeast*	Rapid growth rate and scalability Simple genetic manipulation Cost-effective media and culture conditions Capability for posttranslational modifications	Robust cell walls complicating cell disruption Hyperglycosylation of proteins, which can affect functionality and immunogenicity
	Mammalian cells	Accurate protein folding and genuine posttranslational modifications Well-defined regulatory processes and approval criteria	Expensive media and culture conditions High production costs
	Plant	Rapid and cost-effective production Optimized growth conditions for large-scale production Free from pathogen and bacterial toxin contamination Economical for developing countries Capability for posttranslational modifications resembling mammalian systems	Regulatory compliance requirements can be complex and time-consuming Limited glycosylation capacity compared to mammalian systems

to efficiently produce biopharmaceuticals has always been a matter of concern to researchers (Schmidt 2004).

The first documented use of plant expression systems for biopharmaceutical production dates back to 1989, when researchers successfully demonstrated the expression of antibodies in transgenic tobacco plants (Hiatt et al. 1989). This groundbreaking finding paved the path for the future of modern molecular pharming, and since this discovery, many transgenic plant models have been generated to express different proteins in plants. However, it was not until 1997 that the products of pharming were tested in clinical trials. In 1997, a trial was conducted to test the first-ever plant-derived biopharmaceutical, CaroRx®, an immunoglobulin A (IgA) antibody that specifically targeted *Streptococcus mutans* and was developed as a prophylactic

Table 2 Highlights of the timeline in plant-based biopharmaceutical production

Year	Product/innovation	Host plant	Significance	References
1986	Human growth hormone	Tobacco	First therapeutic protein produced using plant expression systems	Barta et al. (1986)
1997	CaroRx™ (antibody)	Tobacco	First plant-derived therapeutic tested in humans	Ma et al. (1998)
2012	Elelyso™ (Taliglucerase alfa)	Carrot cell cultures	First FDA-approved plant-based biopharmaceutical	Mor (2015)
2014	ZMapp™	*Nicotiana benthamiana*	First plant-derived monoclonal antibody for Ebola	Qiu et al. (2014)
2022	Covifenz® (COVID-19 vaccine)	*Nicotiana benthamiana*	First vaccine developed from plants for human use	Ward et al. (2021a, b)

treatment for dental caries. The biopharmaceutical was expressed in *Nicotiana tabacum*. The trial demonstrated that the product was safe, efficacious, and well-tolerated by the subjects. However, regulatory hurdles restricted the widespread therapeutic use (Thomas et al. 2011; Wycoff 2005). In 2003, trypsin produced in maize plants received regulatory approval for use in veterinary medicine. This was the first plant-based biopharmaceutical to obtain regulatory approval (Howard and Hood 2005).

However, it was only after 2012 that regulatory agencies approved plant-based biopharmaceuticals for human use. In 2012, the FDA-approved Elelyso® (taliglucerase alfa), a recombinant human glucocerebrosidase developed for the treatment of Gaucher's disease (Mor 2015). This biopharmaceutical was successfully produced using a carrot cell suspension culture system. The transformation of the *β*-glucosidase gene was mediated by *Agrobacterium tumefaciens*. The kanamycin resistance gene was also incorporated into the Ti-plasmid to selectively allow the growth of positively transformed cells. These successfully transformed cultures were subsequently propagated in large suspension systems to generate seed cultures, which were then inoculated into bioreactors designed to maintain optimal conditions for efficient protein expression (Fox 2012; Grabowski et al. 2014) (Table 2).

3 Current Advancements in the Use of Plants as a Source of Biopharmaceuticals

Plants can be genetically modified to synthesize a wide array of biologics, including vaccines, monoclonal antibodies, therapeutic proteins, enzymes, hormones, and cytokines, as well as recombinant proteins. This biotechnological process encompasses the modification of plant genomes to enable the expression of proteins and bioactive molecules that are not naturally synthesized by the plant (England et al. 2023). Genetic modification enables researchers to precisely regulate gene expression, fine-tune metabolic pathways, and customize protein characteristics to fulfill specific therapeutic needs (Van Eck 2020).

Plant-derived biologics often exhibit intricate molecular architectures that may include additional molecular components, such as glycosylated carbohydrates and fatty acid residues. These biologics fall into distinct categories, including monoclonal antibodies, vaccines, therapeutic replacement enzymes, receptor modulators, and bioactive small molecules (Venkataraman 2022). Progress in plant-derived biologics and industrial protein development has contributed significantly to the growth of the commercial biopharmaceutical market, with several examples outlined in Table 3, highlighting recent developments and innovations.

3.1 Therapeutic Enzymes: Plant-Based Production and Clinical Applications

Plants offer a highly efficient platform for the production of recombinant enzymes with clinically enhanced properties. For instance, Elelyso™ (β-glucocerebrosidase) demonstrates superior therapeutic effectiveness compared to its CHO cell-derived counterpart (Mor 2015). This advantage arises from the presence of terminal mannose residues on the plant-derived enzyme, a critical feature for the effectiveness in treating Gaucher's disease through enzyme replacement therapy.

Preclinical and clinical investigations have also demonstrated the potential of plant-based systems for producing therapeutic proteins such as insulin and pegunigalsidase alpha. A unique advantage of plant-produced enzymes is their ability to be delivered orally. The cellulose-rich cell walls of plant cells serve as a protective barrier, shielding recombinant proteins from degradation by stomach acids and digestive enzymes (Schiffmann et al. 2019). These cell wall polymers, which contain β-1,4- and β-1,6-glycosidic linkages, resist hydrolysis during gastric transit. Instead, commensal gut microbiota break down the cell wall in the intestinal epithelium, leading to the release of bioencapsulated recombinant proteins. These proteins are subsequently identified by the gut-associated lymphoid tissues (GALT), initiating an adaptive immune response (Koizumi et al. 2022).

Plant expression systems provide several benefits over conventional production platforms, such as cost-efficiency, scalability, and the capability to produce structurally intricate recombinant proteins (Feng et al. 2022). The results from preclinical and clinical studies underscore their potential to revolutionize human and veterinary health by providing a sustainable and economical alternative for the production of therapeutic biologics.

3.2 Vaccines and Virus-like Particles (VLPs): Plant-Based Advancements

Plants provide a versatile and efficient system for producing vaccines and virus-like particles (VLPs). VLPs are highly organized assemblies of virus coat proteins that

Table 3 Advancements in plant-derived biologics

S. No.	Biologic	Applications	Plant/ expression system	Status	Company	References
Therapeutics						
1.	Virus-trapping proteins CTB-ACE2 (angiotensin-converting enzyme 2 fused to the nontoxic cholera toxin subunit B)	SARS-CoV-2	*Lactuca sativa* (stable chloroplast expression)	Phase I/II clinical trial of chewing gum with CTB-ACE2 for SARS-CoV-2 prevention	University of Pennsylvania	Daniell et al. (2022)
2.	β-Glucocerebrosidase	Gaucher's disease and enzyme replacement therapy	*Daucus carota* (stable gene expression using ProCellEx®)	Elelyso™ has received approval from the United States Department of Agriculture (USDA) and European Medicines Agency (EMA) as the first plant-derived biologic available on the market	Protalix BioTherapeutics Inc., Karmiel, Israel/Pfizer	Mor (2015)
3.	Pegunigalsidase alpha	Fabry disease	Stable gene expression using ProCellEx®	ELFABRIO™ has received approval for the USA and EU for treatment of Fabry disease.	Protalix BioTherapeutics Inc., Israel	Holida et al. (2025)
4.	Uricase (PRX 115)	Severe gout	Stable gene expression using ProCellEx®	Phase I clinical trials for recombinant PEGylated uricase	Protalix BioTherapeutics Inc., Karmiel, Israel/Pfizer	Biotherapeutics (2024)
5.	Lactoferrin (VEN120)	Inflammatory bowel disease	*Oryza sativa* (cell culture media)	Products available in the market	Ventria bioscience, Junction City, KS, USA	MacManus et al. (2017)
	Lactoferrin (VEN-BETA)	*E. coli* gastroenteritis				Dierick et al. (2022)

6.	Insulin	Diabetes	*Helianthus annuus*, (stable gene expression)	Phase I/II clinical trials completed (2008)	SemBioSys genetics Inc., Calgary, Alberta, Canada; operation terminated in 2012	OTS/ PRNewswire (2008)
7.	Allergens bioparticles	Allergies and immunotherapy	*Nicotiana benthamiana* (transient expression)	High-quality allergens (natural-like) produced for pharmaceutical use	Angani Inc., Quebec, QC, Canada	Gomord et al. (2020)
Therapeutic antibodies						
1.	ZMapp™	Anti-Ebola monoclonal antibodies	*Nicotiana benthamiana* (transient expression)	Failed to meet statistical efficacy in randomized controlled trial for Ebola treatment	Mapp Biotherapeutics, Inc., San Diego, CA, USA	Qiu et al. (2014)
2.	Anti-human antibody (BLX-301)	Non-Hodgkin's lymphoma	*Lemina minor* (stable gene expression)	Phase II clinical trials; system abandoned after company closure	Biolex Inc., Pittsboro, NC, USA (company closed in 2012)	Edgue et al. (2017)
3.	Anti-spike antibody (mAb B38, H4)	SARS-CoV-2 monoclonal antibodies	*Nicotiana benthamiana* (transient expression)	Demonstrated effective neutralization of SARS-CoV-2 in vitro	Baiya Phytopharm, Bangkok, Thailand	Shanmugaraj et al. (2020c)
4.	P2G12	HIV-neutralizing human monoclonal antibody	*Nicotiana tabacum* (stable gene expression)	Phase I clinical trial underway	Pharma-planta consortium, Fraunhofer Institute for Molecular Biology and Applied Ecology (IME), Schmallenberg, Germany	Ma et al. (2015)

(continued)

Table 3 (continued)

S. No.	Biologic	Applications	Plant/ expression system	Status	Company	References
5.	Plant-derived monoclonal antibody	Against ricin exposure	VivoXPRESS® (plant-based manufacturing system)	–	AntoXa corporation, Toronto, Ontario, Canada and SwiftPharma, Belgium	AntoXa (2024)
Vaccines						
1.	Influenza virus-like particles (VLPs) vaccine	Seasonal flu	*Nicotiana benthamiana* (transient expression)	Phase III clinical trials	Medicago Inc., Quebec City, QC, Canada; (company ceased operations in 2023)	B. J. Ward et al. (2021a, b)
2.	KBP-201 with cyotsine phosphoguanine (CpG) oligonucleotides	SARS-CoV-2 vaccine	*Nicotiana benthamiana* (transient expression)	Phase I/II clinical trials for enhanced immunity to SARS-CoV-2	Kentucky BioProcessing, Owensboro, KY, USA	Shanmugaraj et al. (2021)
3.	Covifenz®	SARS-CoV-2 vaccine	*Nicotiana benthamiana* (transient expression)	Approved by Health Canada after the successful completion of clinical trials	Medicago Inc., Quebec City, QC, Canada	Brian J. Ward et al. (2021a, b)
4.	IBIO-201 and IBIO-202	SARS-CoV-2 vaccine	*Nicotiana benthamiana* (transient expression)	Preclinical studies of recombinant vaccines targeting SARS-CoV-2	iBio Inc., Bryan, TX, USA	Shanmugaraj et al. (2021)
5.	SARS-CoV-2 receptor-binding domain (RBD) vaccine	SARS-CoV-2 vaccine	*Chlamydomonas reinhardtii* (algae)	Evaluation of freeze-dried algae biomass as an oral immunogen	–	Berndt et al. (2021)
6.	Baiya SARS-CoV-2 vax 1	SARS-CoV-2 vaccine	*Nicotiana benthamiana* (transient expression)	Ongoing phase I clinical trial for SARS-CoV-2 vaccine	Baiya Phytopharm Co., Ltd., Bangkok, Thailand	Shanmugaraj et al. (2021)

7.	HERBAVAC™ CSF green marker	Classical swine fever virus (CSFV)	*Nicotiana benthamiana* (transient expression)	Registered by the World Organization for Animal Health (WOAH)	BioApplications Inc., Pohang, Republic of Korea	Oh et al. (2021)

mimic the structure of viruses without possessing genetic material, ensuring their safety as vaccine candidates and for use in nanoparticle technologies (Chen and Lai 2013). Plant-derived VLPs, as well as chimeric VLPs—comprising structural proteins or immunogenic epitopes from multiple viruses—function as potent immune modulators and self-adjuvants. These properties enable them to elicit robust immune responses targeting various diseases, such as viral infections, allergies, cancer, and autoimmune disorders (Nooraei et al. 2021).

Plant-based vaccine production offers numerous advantages, like reduced production expenses, enhanced stability, decreased reliance on cold-chain logistics, and the potential for oral administration. Among these innovations, edible vaccines have garnered particular interest. These vaccines use genetically engineered plants to express antigenic proteins, offering a transformative approach to vaccine delivery (Laere et al. 2016). The ability to administer vaccines orally—bypassing the need for injections and complex storage infrastructure—positions this technology as a game-changer for immunization efforts, particularly in resource-limited regions. However, challenges remain, notably the precise control of vaccine dosage in edible formats (Vela Ramirez et al. 2017).

In addition to their immunological applications, VLPs produced in plants show promise as nanoparticles for drug delivery systems. The ability to engineer plant-derived VLPs with high precision and scalability further underscores their potential to revolutionize vaccine development and therapeutic delivery, providing safe, cost-effective, and accessible solutions for global health challenges (Chung et al. 2020).

3.3 Antibodies and Antibody Fragments: Plant-Based Production Systems

Plant-based production systems have gained recognition as a flexible platform for manufacturing monoclonal antibodies and antibody fragments, collectively referred to as "plantibodies." These biomolecules have diverse applications in therapeutics, diagnostics, and biomedical research. Plantibodies have proven effective in targeting infectious pathogens, cancer markers, and various targets (Sindelar 2013).

A notable example is the plant-derived monoclonal antibody cocktail ZMapp™, which was developed during the 2014 Ebola outbreak. ZMapp™ significantly improved survival rates among treated patients, although the limited number of cases made a comprehensive statistical evaluation of its efficacy challenging (Qiu et al. 2014). This highlighted the feasibility and rapid scalability of plant-based antibody production in response to public health emergencies.

Plantibodies offer several key benefits compared to traditional production methods, such as reduced costs, scalability, and the potential for precise customization to meet specific therapeutic needs. These benefits position plant-based systems as a valuable alternative for antibody production, particularly for large-scale or emergent demands, paving the way for innovative solutions in global health management (Oluwayelu and Adebiyi 2016).

3.4 Small Molecules: Plant-Based Biosynthesis and Metabolic Engineering

Plants serve as an abundant natural reservoir of bioactive secondary metabolites, including triterpenoids, alkaloids, and phenolics, which are characterized by their complex multi-step biosynthetic pathways and specific stereochemical configurations (Reshi et al. 2023). These metabolites serve as critical medicinal compounds, yet their extraction from native species often presents challenges due to limited yields, the labor-intensive nature of cultivation, and high purification costs. While complete chemical synthesis offers an alternative, it is frequently impractical because of the intricate structure of these molecules (Sasidharan et al. 2011).

To address these challenges, significant efforts have focused on elucidating and characterizing the biosynthetic pathways of plant-based metabolites. This knowledge enables the reconstruction of these pathways in more tractable systems, such as yeast, bacteria, or other plant species, via the simultaneous expression of biosynthetic enzymes (Cravens et al. 2019). Importantly, this metabolic engineering approach allows for the production of novel variants of natural molecules by combining different enzymatic activities, opening pathways for the development of compounds with tailored bioactivities.

A landmark example of this approach is the reconstitution of the biosynthetic pathway for the antimalarial compound artemisinin in yeast. Afterward, the pathway enzymes were expressed in *Nicotiana benthamiana* and other more easily manipulated organisms, providing an alternative to the native source, *Artemisia annua* (Al-Khayri et al. 2022). Transient expression of biosynthetic enzymes in plants has become the preferred method over stable transformation due to its rapid implementation and flexibility. This method enables combinatorial expression strategies to generate a wide array of related compounds with diverse bioactivities (Yao et al. 2022).

The success of this approach is exemplified by the production of triterpenes, where transient combinatorial expression of up to 16 enzymes in *Nicotiana benthamiana* led to the synthesis of saponin molecules (Stephenson et al. 2018). These molecules were subsequently bioengineered to create vaccine adjuvants, showcasing the potential of this strategy in pharmaceutical development. Similar approaches have been employed to elucidate and manipulate the biosynthetic pathways of anticancer drugs such as vinblastine and paclitaxel, further demonstrating the transformative impact of metabolic engineering in advancing drug discovery and production (Changxing et al. 2020).

3.5 Receptor Modulators: Plant-Based Production of Regulatory Proteins

Plant-based systems are proven effective in the production of small polypeptides and glycoproteins that regulate mammalian cellular processes, including cytokines and hormones. For example, human growth hormone (hGH) expressed in *Nicotiana benthamiana* demonstrated functional biological activity in hypophysectomized rats, confirming its potential therapeutic utility (Rabindran et al. 2009).

Cytokines, a critical class of signaling proteins, play a pivotal role in controlling inflammation and immune responses. Their therapeutic applications extend to cancer treatment, immune modulation, and the management of various inflammatory and immune-related disorders (Bhol et al. 2024). Using plant-based systems, researchers have successfully expressed numerous cytokines and regulatory proteins, including erythropoietin, interleukins (IL-2, IL-4, IL-12, IL-13, IL-18), human granulocyte-macrophage colony-stimulating factor (GM-CSF), cardiotrophin-1, tumor necrosis factor-alpha (TNF-α), human fibroblast growth factor 8b, interferon-alpha, and insulin-like growth factor 1 (Sirko et al. 2011).

To optimize the yield and stability of these recombinant proteins, various strategies have been employed, including transient expression systems, codon optimization, and the use of molecular chaperones and stabilizing agents (Liu and Timko 2022). These advancements highlight the versatility of plant-based systems for producing bioactive receptor modulators, offering a scalable and cost-effective substitute for the development of therapeutics targeting critical regulatory pathways in human health.

3.6 Bioactive Proteins from Plants: Applications and Potential

Plant-based biologics extend to bioactive proteins, including lectins, which are natural carbohydrate-binding proteins with high specificity for sugar moieties on other molecules (Naithani et al. 2021). Lectins exhibit diverse biological activities, such as potent antimicrobial properties mediated by binding to carbohydrates on microbial surfaces, leading to alterations in cell permeability and the formation of pores (Carneiro et al. 2022). These properties make lectins valuable as microbicides, vaccine adjuvants, and antitumor agents.

One notable example is mistletoe lectins (ML-I, ML-II, and ML-III), which were transiently expressed in *Nicotiana benthamiana*. These plant-expressed lectins demonstrated significant anticancer activity, showcasing their potential as therapeutic agents (Mazalovska and Kouokam 2020).

The ability to express bioactive proteins like lectins in plant systems offers numerous advantages, including cost-effective production, scalability, and the potential for customization. These advancements highlight the role of plant-based platforms in expanding the toolkit of bioactive molecules for therapeutic and biomedical applications (Mungwari et al. 2025).

4 Plant-Based Biologics Production Methods

Advancements in plant molecular farming have addressed many of its early technical challenges, such as low recombinant-protein yields, limited recovery during downstream processing, and plant-specific glycosylation (Shanmugaraj et al.

2020a). Emerging technologies—including stable plastid transformation, transient expression, plant cell cultures, *Agrobacterium*-mediated infiltration, glycosylation pathway modeling, plant genome editing, and adherence to current Good Manufacturing Practices (cGMP)—have significantly improved the scalability, efficiency, and functionality of plant-derived biologics (Gelvin 2003). Key methods used in plant-based biologic production are outlined in Table 4. These methodologies collectively advance the scalability, efficiency, and versatility of plant-based biologic production, driving innovation in therapeutics and industrial applications.

5 Challenges and Resolving Strategies in the Production and Use of Plant-Derived Biologics

5.1 Complex Protein Expression

Producing complex recombinant proteins in plants presents considerable challenges because of the intricate requirements for proper folding, assembly, and posttranslational modifications. Many proteins necessitate sophisticated folding mechanisms and assembly processes that might be problematic to replicate in plant cellular environments (Feng et al. 2022).

Two principal strategies are employed for expressing multiple recombinant proteins in plants: (i) Co-transformation: This involves using individual *Agrobacterium* strains, each carrying a vector for a specific gene of interest. (ii) Tandem Expression Constructs: This approach uses a single expression vector containing multiple recombinant genes arranged for coordinated expression (Pinneh et al. 2022).

Higher accumulation of complex proteins is generally observed when these proteins are directed to the apoplast or endoplasmic reticulum (ER). This is attributed to: i) Facilitated Folding and Glycosylation: The secretory pathway supports proper protein folding and glycosylation. ii) Reduced Proteolytic Activity: Protease activity in the apoplast and ER is comparatively lower than in the cytosol, reducing protein degradation (Song et al. 2024).

Co-expression of recombinant proteins alongside molecular chaperones from the same species as the target protein has been demonstrated to improve folding and accumulation. For example, co-expression with calreticulin has been demonstrated to improve the yield and folding of gp140 HIV-1 antigens and other viral glycoproteins by mitigating ER stress; however, co-expression of calreticulin with the receptor-binding domain (RBD) glycoprotein of SARS-CoV-2 did not enhance protein yield or improve folding (Margolin et al. 2020a, b).

These variabilities highlight the complexity of designing plant-based expression systems tailored to the structural and functional requirements of individual proteins. The development of optimized systems capable of accommodating such intricate protein structures remains a significant technical hurdle in plant molecular farming.

Table 4 Techniques used in plant-based biologic production

S. No.	Technique	Description	Advantages	Disadvantages	References
1.	Stable nuclear transformation	Integration of the target gene into the plant genome via *Agrobacterium*- or biolistics-mediated delivery, enabling consistent and long-term protein production	Suitable for large-scale production of biologics requiring complex posttranslational modifications. Supports edible plants for oral delivery and cereals for long-term storage	Time-intensive transformation and regeneration processes. Lower protein yields, gene silencing, potential position effects, and genetically modified organism (GMO) regulatory concerns	Hwang et al. (2017)
2.	Viral vectors	Use of viral vectors (e.g., tobacco mosaic virus) for enhancement of protein expression by exploiting viral replication machinery in plants	Achieves higher protein yields and is compatible with transient as well as stable expression systems	Potential for viral contamination, biosafety concerns, and potential negative effects on plant growth. Additional steps needed for viral vector construction	Lico et al. (2008)
3.	Plant cell cultures	Recombinant protein production in suspension cell cultures under aseptic fermentation conditions	Enables scalable manufacturing with reduced contamination risks. Regulatory frameworks align with mammalian cell production systems	Reduced growth rates and lower yields in comparison to microbial or mammalian systems. Moderate production costs	Plasson et al. (2009)
4.	Gene editing	Precise genome modification using tools like CRISPR/Cas9 to enhance protein yield and minimize degradation	Allows targeted improvement of traits and multiplex editing to optimize multiple pathways simultaneously	Requires technical expertise, potential off-target effects, and regulatory scrutiny for GMOs	Cardi et al. (2023)

5.	Downstream processing	Purification methods to eradicate plant-specific contaminants, safeguarding product quality and stability	Ensures high purity and tailored processes for specific biologics	Increases costs and complexity of production, requiring customization for different biomolecules	Buyel et al. (2015)
6.	Stable chloroplast transformation	Integration of genes into chloroplast genomes, enabling high-level protein expression	High protein yields (up to 45% of total soluble proteins), absence of gene silencing, ability to express toxic proteins and complex structures, and no gene flow	Labor-intensive procedures with low transformation efficiency, development of inclusion bodies, challenges in protein purification, and regulatory complexities	Gao et al. (2012)
7.	Transient expression	Introduction of target genes using agroinfiltration or viral vectors for rapid protein production	Quick, high-yield production suitable for emergencies or rapid responses. Versatile for various biologics	Temporary protein expression requiring repeated infiltrations for continuous production	Norkunas et al. (2018)
8.	Hairy roots	Use of *Rhizobium rhizogenes* for rhizosecretion of recombinant proteins into hydroponic media	Facilitates easier protein purification and improves homogeneity of the product	Challenges include protein degradation, elevated proteolytic activity, and stringent Good Manufacturing Practice (GMP) regulations	Gaume et al. (2003)
9.	Glycoengineering	Modifying plant glycosylation pathways to eliminate unwanted glycans and produce human-compatible glycoproteins	Enables production of biologics with human-like glycosylation, eliminating plant-specific glycan modifications	Not all plants tolerate glycoengineering. N-glycan heterogenicity and GMO safety concerns remain	Castilho and Steinkellner (2012)

(continued)

Table 4 (continued)

S. No.	Technique	Description	Advantages	Disadvantages	References
10.	Formulation and delivery	Development of advanced systems to enhance the stability, bioavailability, and targeted delivery of biologics	Improves storage stability, bioavailability, and targeted efficacy in tissues or cells	Increases production costs and complexity. Challenges in achieving precise targeted delivery	Lv et al. (2024)

5.2 Proteolytic Degradation of the Recombinant Proteins

In plant cells, a diverse array of endogenous proteases is active, playing crucial roles in maintaining cellular homeostasis, protein turnover, and stress responses (Luciński and Adamiec 2023). However, these proteases pose a significant challenge in the context of plant molecular farming, as they contribute to the proteolytic degradation of recombinant proteins, reducing yield, stability, and functionality (Benchabane et al. 2008). Several specific proteases have been identified as primary contributors to this degradation. These include serine proteases, aspartic proteases, cysteine proteases, and metalloproteases, which are localized in various cellular compartments such as vacuoles, lysosomes, and the apoplast (Moloi and Ngara 2023). To mitigate proteolytic degradation and enhance the stability of recombinant proteins, several advanced strategies have been employed:

(i) Protease Gene Editing and Silencing (Karki et al. 2021):

- CRISPR/Cas9 Technology: Enables precise knockout or knockdown of specific protease genes to reduce their activity.
- Ribonucleic acid (RNA) Interference (RNAi): Effectively silences protease gene expression at the mRNA level.
- Transcription Activator-Like Effector Nucleases (TALENs): Facilitate targeted gene disruption for protease activity suppression.

(ii) Co-Expression of Protease Inhibitors (Moloi and Ngara 2023):

- Plant cells can be engineered to express endogenous or heterologous protease inhibitors that neutralize specific proteases.
- Protease inhibitors such as cystatins and serpins have been co-expressed alongside recombinant proteins to enhance stability.

(iii) pH Regulation (Coates et al. 2022):

- The intracellular environment can be modulated to less favorable pH levels for protease activity, reducing degradation.
- Targeting recombinant protein expression to specific compartments (e.g., apoplast, endoplasmic reticulum) with controlled pH conditions has shown promise in mitigating protease activity.

These approaches collectively aim to create a protective environment for recombinant proteins within plant cells, enhancing their accumulation, integrity, and functionality. The strategic combination of these technologies continues to play a pivotal role in optimizing plant-based production systems for biologics.

5.3 Posttranslational Modifications in Plant Expression Systems

Posttranslational modifications (PTMs) are critical determinants of the functionality, efficacy, and stability of biologics. However, the PTMs observed in plant expression systems often differ from those in mammalian systems, presenting challenges in replicating human-like modifications necessary for therapeutic efficacy (Gomord and Faye 2004).

Plants are capable of performing intricate N-glycosylation of recombinant glycoproteins. This process is defined by plant-specific glycan structures, including β (1,2)-xylose and core α (1,3)-fucose, which are absent in mammalian glycoproteins. Conversely, plants lack the enzymatic pathways required for galactosylation, core α (1,6)-fucosylation, sialylation, and the addition of bisecting GlcNAc residues, all of which are commonly found in human glycoproteins. These differences in glycosylation patterns significantly influence the biological properties of recombinant glycoproteins, such as their immunogenicity, stability, and bioactivity (Izadi et al. 2023).

To overcome these limitations, engineering the N-glycosylation pathways in plants has emerged as a critical area of research. Key approaches include:

(i) CRISPR/Cas9-Mediated Gene Editing (Jansing et al. 2019):

- Targeted knockout of genes encoding plant-specific glycosyltransferases (e.g., responsible for β (1,2)-xylose and α (1,3)-fucose) to eliminate undesired glycan structures.
- Introduction of mammalian glycosylation pathway genes into plant cells to enable the synthesis of human-like glycan structures, including sialylation and galactosylation.

(ii) Synthetic Biology and Glycoengineering (Strasser 2023):

- Incorporating mammalian glycosyltransferases and associated regulatory proteins into plant cells to mimic human glycosylation pathways.
- Engineering endoplasmic reticulum and Golgi apparatus compartments in plants to facilitate the addition of complex human-like glycans.

(iii) Expression Targeting (Song et al. 2024):

- Directing the expression of recombinant glycoproteins to specific subcellular compartments, such as the endoplasmic reticulum, to leverage the local glycosylation machinery and reduce unwanted modifications.

Despite these advances, achieving complete fidelity in posttranslational modifications to match mammalian systems remains a significant challenge. Future innovations in plant-based systems are expected to focus on refining glycoengineering strategies and enhancing the consistency and precision of PTMs, thereby broadening the applicability of plant-derived biologics in therapeutic contexts.

5.4 Downstream Processing

Effective downstream processing of biologics from plants remains a significant challenge, primarily attributed to the occurrence of plant-specific contaminants and the inherent structural complexity of biologics. These contaminants include phenolic compounds, pigments, polysaccharides, and other secondary metabolites, which can interfere with purification processes and compromise product quality (Buyel et al. 2015).

The structural complexity of biologics introduces additional hurdles, as they are susceptible to various forms of degradation, including (Phyo et al. 2021):

- Aggregation: Protein clumping that can reduce biological activity and increase immunogenicity.
- Isomerization: Conversion of amino acid residues, potentially altering protein function.
- Hydrolysis: Breakdown of peptide bonds, leading to fragmentation.
- Deamidation: Conversion of asparagine residues to aspartic acid, which can impact stability and efficacy.
- Oxidation: Modification of amino acid side chains, particularly methionine and cysteine, affecting protein activity and shelf life.

To address these challenges, robust and scalable purification strategies are essential. Key considerations include:

(i) Contaminant Removal (Venkataraman et al. 2023):

- Initial Clarification: Techniques such as filtration and centrifugation to remove large plant debris.
- Targeted removal of plant-specific impurities: Incorporating steps like adsorption chromatography, which binds and removes phenolic compounds and pigments.

(ii) Protein Stabilization (Benchabane et al. 2009):

- Protease Inhibitors: Addition of inhibitors during extraction to prevent proteolytic degradation of recombinant proteins.
- Buffer Optimization: Employing buffers that stabilize protein structure and minimize pH-induced aggregation or deamidation.

(iii) Advanced Chromatographic Techniques (Wingfield 2015):

- Use of affinity chromatography, hydrophobic interaction chromatography, and ion exchange chromatography to achieve high purity while preserving protein integrity.
- Integration of multi-modal chromatography to streamline purification steps and reduce processing time.

(iv) Process Analytical Technology (PAT) (Graf et al. 2024):

- Real-time monitoring of critical quality attributes (CQAs) throughout purification to ensure product consistency and minimize degradation.

(v) Scalability and Cost-Efficiency (Tripathi and Shrivastava 2019):

- Developing scalable purification methods compatible with industrial-scale production, such as single-use systems or membrane-based separations.
- Incorporating automation to enhance process reproducibility and reduce operational costs.

Downstream processing of plant-derived biologics must balance the need for cost-effectiveness with stringent quality requirements to ensure that final products maintain their structural integrity, biological activity, and therapeutic efficacy. Innovations in purification technologies and process optimization will be critical for advancing the viability of plant-based biologic production systems.

5.5 Posttranscriptional Gene Silencing and its Modulation

Posttranscriptional gene silencing (PTGS) is a naturally occurring regulatory mechanism in plants that can limit the expression of transgenes, thereby reducing the accumulation of recombinant proteins. Overcoming this challenge is critical for achieving high levels of recombinant protein production in plant-based expression systems (Lucy et al. 2000).

Several strategies have been successfully employed to mitigate PTGS and enhance recombinant protein expression:

(i) Utilization of Mutant Lines (Peragine et al. 2004):
   - Mutants of *Arabidopsis thaliana*, such as sgs2 and sgs3, which exhibit impaired PTGS, have been used to significantly increase transgene expression. For instance, these mutants facilitated the high-level production of β-glucuronidase (GUS), demonstrating the potential of PTGS repression in improving transgene expression.

(ii) RNA Interference (RNAi) Technology (Matsuo 2022):
   - RNAi-mediated knockdown of specific genes involved in the RNA silencing pathway has been effective in enhancing recombinant protein production. In *Nicotiana benthamiana*, silencing of DCL2 and DCL4 genes—key players in the RNA silencing mechanism—has shown notable success in increasing transgene expression.

(iii) Genome Editing Approaches (Matsuo and Atsumi 2019):
   - CRISPR/Cas9 technology has empowered precise and efficient knockouts of genes involved in PTGS. For instance, targeting genes such as RDR6 and AGO2 in *Nicotiana benthamiana* resulted in significant improvements in recombinant protein accumulation. These modifications reduce the plant's ability to degrade transgenic mRNA, thus enhancing protein yield.

(iv) Co-Expression with Gene Silencing Suppressors (Canto et al. 2006):

- Suppressors of gene silencing, like the P19 protein from Tomato bushy stunt virus, have been widely utilized to inhibit PTGS. Co-expression of P19 with the gene of interest stabilizes transgenic mRNA and prevents its degradation, leading to increased protein expression levels.

(v) Targeted Approaches to Modulate Host Silencing Pathways (Kryovrysanaki et al. 2022):

- Engineering plants to selectively repress components of the RNA silencing machinery without affecting their overall growth and development is a promising avenue.
- Enhancing plant host compatibility with transgenes by optimizing codon usage and introducing strong promoters can complement PTGS mitigation strategies.

By leveraging genetic, molecular, and biotechnological tools to suppress PTGS, it is possible to achieve high and stable expression of recombinant proteins in plant systems. These advancements not only increase protein yield but also enhance the scalability and cost-effectiveness of plant-based biologic production.

6 Regulatory Considerations for Plant-Derived Biologics

Regulatory frameworks are critical to ensuring the safety, efficacy, and quality of plant-derived therapeutics while accommodating their unique characteristics. These biologics differ significantly from those produced through other systems, necessitating specific regulatory guidelines tailored to their distinct features (Spök et al. 2008).

6.1 Unique Challenges and Regulatory Oversight

Regulatory agencies, the US FDA and the EMA, have a key role in evaluating plant-derived biologics. The key challenges include:

(i) Source Evaluation (McHughen 2016):

- Regulatory scrutiny extends to the plant species, expression vectors used in production, and gene modifications.
- Assessments focus on potential risks associated with genetic modifications, transgene stability, and the potential for contamination with allergens or toxins specific to the plant species.

(ii) Characterization (Ambrogelly et al. 2018):

- Plant-derived biologics require detailed characterization, including structural attributes, stability, potency, and purity.

- Advanced analytical techniques and assays are critical for the evaluation of these parameters, ensuring consistency and reproducibility.

(iii) Equivalence to Non-Plant-Derived Biologics (Webster et al. 2021):

- While comparability to biologics from other systems is considered, the unique nature of plant-derived biologics demands case-by-case evaluations rather than broad equivalence.
- This nuanced approach assesses the safety, functionality, and efficacy of each product independently.

6.2 Advances in Regulatory Frameworks

In recent years, regulatory agencies have made significant advancements in adapting their frameworks to accommodate plant-derived biologics:

(i) Case-By-Case Evaluations (Wainwright et al. 2022):

- Recognizing the diversity in plant-derived biologics, regulators employ tailored assessments for each product rather than a universal standard.
- This allows for flexibility in addressing specific characteristics, production processes, and therapeutic applications.

(ii) Global Harmonization (Niazi 2023):

- Efforts to standardize regulations across regions aim to simplify the development and approval process of biologics derived from plants.
- International guidelines and standards foster consistency while facilitating cross-border collaboration in biotechnology innovation.

(iii) Incorporating Scientific Advancements (Dey 2021):

- Regulatory frameworks increasingly integrate emerging technologies, such as genome editing (e.g., CRISPR/Cas9), to assess the genetic modifications used in plant-based expression systems.
- Advances in analytical and bioinformatics tools enhance the ability to evaluate product safety and efficacy comprehensively.

6.3 Environmental and Ethical Considerations

- Environmental Impact (Bauer-Panskus et al. 2020):
  - Regulatory guidelines assess the potential ecological risks associated with genetically modified plants, including gene flow to wild relatives and impacts on biodiversity.
  - Containment measures, such as using nonfood crops or specific growth environments, are often mandated.

- Public Perception and Ethical Issues (Bawa and Anilakumar 2013):
  - Addressing public concerns about GMOs is a priority. Transparent communication and adherence to rigorous safety standards help build trust.

6.4 Ensuring Safety, Efficacy, and Quality

Regulatory agencies ensure that plant-derived biologics meet stringent standards across multiple dimensions (Wainwright et al. 2022):

- Safety: Comprehensive preclinical and clinical evaluations assess potential immunogenicity, allergenicity, and toxicity.
- Efficacy: Robust trials are required to demonstrate the therapeutic potential of the biologic.
- Quality: Manufacturing processes must adhere to cGMP guidelines, ensuring product consistency and integrity.

Regulatory considerations for plant-derived biologics are complex and multifaceted, reflecting their distinct nature. Recent advancements in regulatory approaches, including case-by-case evaluations and global harmonization, support the efficient development and approval of these products. By addressing safety, efficacy, quality, and environmental concerns, these frameworks facilitate the integration of plant-derived biologics into the biotechnology landscape while meeting rigorous standards for therapeutic application.

7 Conclusions and Outlook

Plant-based bioengineering presents a promising solution to satisfy the growing need for innovative protein-based pharmaceuticals, particularly in light of potential capacity limitations in mammalian cell-based systems. Despite significant advancements in plant-derived protein production, the limited industrial translation is more likely due to regulatory and industrial inertia than the intrinsic inadequacy of plant-based products. A growing body of evidence supports the functional equivalence or superiority of plant-produced biologics.

Advances in modular transient expression systems in plants have demonstrated remarkable scalability, with the potential to produce up to 7.5 billion vaccine doses annually in facilities with a capacity exceeding 100 hectares. For instance, Medicago, Inc.'s 9-hectare facility, expected to open in 2023/2024, was projected to deliver up to 1 billion vaccine doses annually. Recent technological advancements, such as Plant Cell Platforms (PCPs) for high-throughput screening of over 1000 candidate variants per week, enable rapid scalability of expression results to intact plants. Additionally, techno-economic analysis (TEAs) predicts significant cost reductions in capital investment and cost of goods for plant-based transient expression platforms compared to mammalian cell systems at similar production scales. However, industrial-scale production is required to validate these projections.

7.1 Addressing Global Challenges with Plant-Based Manufacturing

The increasing frequency of viral outbreaks in the twenty-first century, such as Ebola and coronaviruses, has underscored the urgent need for rapid, large-scale pharmaceutical production. Mammalian cell-based manufacturing facilities, predominantly located in North America, Europe, and Asia, often fall short in meeting these demands. Plant-based manufacturing offers distinct advantages:

(i). Scalability and Accessibility:

- Facilities can be established in diverse geographical settings, reducing reliance on centralized production.
- Minimal cold-chain requirements facilitate distribution in low-resource settings.

(ii). Needle-Free Administration:

- Plant-derived pharmaceuticals can be formulated for aerosol delivery using nanoparticles and IgA antibodies, enhancing accessibility and compliance.

(iii). Speed and Flexibility:

- Plants enable high-throughput screening, rapid biomass scaling, and the production of virus-like particles (VLPs) and protein bodies within 12 months.
- This timeline surpasses the production speed of mammalian cell-based systems, as demonstrated by plant-derived influenza and SARS-CoV-2 vaccines.

7.2 Future Prospects

Two key applications exemplify the potential of plant-based production systems:

(i) Vaccines Against Emerging Pathogens:

- Rapid development of vaccines to address new infectious diseases, ensuring timely responses to outbreaks.

(ii) Modular Cancer Vaccines:

- Individualized cancer therapies leveraging the flexibility and scalability of plant-based platforms.

Emerging research on plant-derived self-assembling nanoparticles offers exciting opportunities for the development of self-adjuvanting vaccines and advanced drug delivery systems. These advancements could revolutionize drug development and production.

7.3 Equitable Access and Global Health

Plant-based platforms hold the potential to democratize access to biopharmaceuticals by enabling rapid, cost-effective production on a global scale. By addressing critical bottlenecks in traditional manufacturing, these systems could ensure the timely and equitable delivery of essential medicines during public health emergencies and beyond.

In conclusion, plant-based bioengineering represents a transformative technology in biopharmaceutical production. With ongoing advancements in scalability, cost-efficiency, and regulatory frameworks, plant-based systems could redefine the future of pharmaceutical manufacturing, offering a sustainable and equitable approach to global health challenges.

References

Al-Khayri JM, Sudheer WN, Lakshmaiah VV, Mukherjee E, Nizam A, Thiruvengadam M et al (2022) Biotechnological approaches for production of Artemisinin, an anti-malarial drug from Artemisia annua L. Molecules 27(9). https://doi.org/10.3390/molecules27093040

Ambrogelly, A., Gozo, S., Katiyar, A., Dellatore, S., Kune, Y., Bhat, R.,… Liu, H. (2018). Analytical comparability study of recombinant monoclonal antibody therapeutics. MAbs, 10(4), 513–538. doi: https://doi.org/10.1080/19420862.2018.1438797

AntoXa. (2024). AntoXa and SwiftPharma to present plant-made anti-ricin antibody technology to NATO expert panel. https://antoxacorp.com/antoxa-and-swiftpharma-to-present-plant-made-anti-ricin-antibody-technology-to-nato-expert-panel/

Baeshen MN, Al-Hejin AM, Bora RS, Ahmed MM, Ramadan HA, Saini KS et al (2015) Production of biopharmaceuticals in E. coli: current scenario and future perspectives. J Microbiol Biotechnol 25(7):953–962. https://doi.org/10.4014/jmb.1412.12079

Banting FG, Best CH (1987) Nutrition classics. The Journal of Laboratory and Clinical Medicine, Vol. VII, 1922: The internal secretion of the pancreas. By F. G. Banting, and C. H. Best. Nutr Rev 45(2):55–57. https://doi.org/10.1111/j.1753-4887.1987.tb07442.x

Barta A, Sommergruber K, Thompson D, Hartmuth K, Matzke MA, Matzke AJM (1986) The expression of a nopaline synthase — human growth hormone chimaeric gene in transformed tobacco and sunflower callus tissue. Plant Mol Biol 6(5):347–357. https://doi.org/10.1007/BF00034942

Bauer-Panskus A, Miyazaki J, Kawall K, Then C (2020) Risk assessment of genetically engineered plants that can persist and propagate in the environment. Environ Sci Eur 32(1):32. https://doi.org/10.1186/s12302-020-00301-0

Bawa AS, Anilakumar KR (2013) Genetically modified foods: safety, risks and public concerns-a review. J Food Sci Technol 50(6):1035–1046. https://doi.org/10.1007/s13197-012-0899-1

Benchabane M, Goulet C, Rivard D, Faye L, Gomord V, Michaud D (2008) Preventing unintended proteolysis in plant protein biofactories. Plant Biotechnol J 6(7):633–648. https://doi.org/10.1111/j.1467-7652.2008.00344.x

Benchabane M, Rivard D, Girard C, Michaud D (2009) Companion protease inhibitors to protect recombinant proteins in transgenic plant extracts. Methods Mol Biol 483:265–273. https://doi.org/10.1007/978-1-59745-407-0_15

Berndt AJ, Smalley TN, Ren B, Simkovsky R, Badary A, Sproles AE et al (2021) Recombinant production of a functional SARS-CoV-2 spike receptor binding domain in the green algae Chlamydomonas reinhardtii. PLoS One 16(11):e0257089. https://doi.org/10.1371/journal.pone.0257089

Bhol NK, Bhanjadeo MM, Singh AK, Dash UC, Ojha RR, Majhi S et al (2024) The interplay between cytokines, inflammation, and antioxidants: mechanistic insights and therapeutic potentials of various antioxidants and anti-cytokine compounds. Biomed Pharmacother 178:117177. https://doi.org/10.1016/j.biopha.2024.117177

Biotherapeutics P (2024) PRX-115. https://protalix.com/pipeline/prx-115

Buyel JF, Twyman RM, Fischer R (2015) Extraction and downstream processing of plant-derived recombinant proteins. Biotechnol Adv 33(6, Part 1):902–913. https://doi.org/10.1016/j.biotechadv.2015.04.010

Canto T, Uhrig JF, Swanson M, Wright KM, MacFarlane SA (2006) Translocation of tomato bushy stunt virus P19 protein into the nucleus by ALY proteins compromises its silencing suppressor activity. J Virol 80(18):9064–9072. https://doi.org/10.1128/jvi.00953-06

Cardi T, Murovec J, Bakhsh A, Boniecka J, Bruegmann T, Bull SE et al (2023) CRISPR/Cas-mediated plant genome editing: outstanding challenges a decade after implementation. Trends Plant Sci 28(10):1144–1165. https://doi.org/10.1016/j.tplants.2023.05.012

Carneiro DC, Fernandez LG, Monteiro-Cunha JP, Benevides RG, Cunha Lima ST (2022) A patent review of the antimicrobial applications of lectins: perspectives on therapy of infectious diseases. J Appl Microbiol 132(2):841–854. https://doi.org/10.1111/jam.15263

Castilho A, Steinkellner H (2012) Glyco-engineering in plants to produce human-like N-glycan structures. Biotechnol J 7(9):1088–1098. https://doi.org/10.1002/biot.201200032

Changxing L, Galani S, Hassan F-u, Rashid Z, Naveed M, Fang D et al (2020) Biotechnological approaches to the production of plant-derived promising anticancer agents: an update and overview. Biomed Pharmacother 132:110918. https://doi.org/10.1016/j.biopha.2020.110918

Chen Q, Lai H (2013) Plant-derived virus-like particles as vaccines. Hum Vaccin Immunother 9(1):26–49. https://doi.org/10.4161/hv.22218

Chung YH, Cai H, Steinmetz NF (2020) Viral nanoparticles for drug delivery, imaging, immunotherapy, and theranostic applications. Adv Drug Deliv Rev 156:214–235. https://doi.org/10.1016/j.addr.2020.06.024

Coates RJ, Young MT, Scofield S (2022) Optimising expression and extraction of recombinant proteins in plants [review]. Front Plant Sci 13

Cravens A, Payne J, Smolke CD (2019) Synthetic biology strategies for microbial biosynthesis of plant natural products. Nat Commun 10(1):2142. https://doi.org/10.1038/s41467-019-09848-w

Daniell H, Nair SK, Esmaeili N, Wakade G, Shahid N, Ganesan PK et al (2022) Debulking SARS-CoV-2 in saliva using angiotensin converting enzyme 2 in chewing gum to decrease oral virus transmission and infection. Mol Ther 30(5):1966–1978. https://doi.org/10.1016/j.ymthe.2021.11.008

Dey A (2021) CRISPR/Cas genome editing to optimize pharmacologically active plant natural products. Pharmacol Res 164:105359. https://doi.org/10.1016/j.phrs.2020.105359

Dierick M, Ongena R, Vanrompay D, Devriendt B, Cox E (2022) Lactoferrin decreases enterotoxigenic Escherichia coli-induced fluid secretion and bacterial adhesion in the porcine small intestine. Pharmaceutics 14(9). https://doi.org/10.3390/pharmaceutics14091778

Edgue G, Twyman RM, Beiss V, Fischer R, Sack M (2017) Antibodies from plants for bionanomaterials. Wiley Interdiscip Rev Nanomed Nanobiotechnol 9(6). https://doi.org/10.1002/wnan.1462

England C, TrejoMartinez J, PerezSanchez P, Karki U, Xu J (2023) Plants as biofactories for therapeutic proteins and antiviral compounds to combat COVID-19. Life (Basel) 13(3). https://doi.org/10.3390/life13030617

Estes S, Melville M (2014) Mammalian cell line developments in speed and efficiency. Adv Biochem Eng Biotechnol 139:11–33. https://doi.org/10.1007/10_2013_260

FDA (2022) 100 Years of Insulin. https://www.fda.gov/about-fda/fda-history-exhibits/100-years-insulin

Feng Z, Li X, Fan B, Zhu C, Chen Z (2022) Maximizing the production of recombinant proteins in plants: from transcription to protein stability. Int J Mol Sci 23(21). https://doi.org/10.3390/ijms232113516

Fischer R, Emans N (2000) Molecular farming of pharmaceutical proteins. Transgenic Res 9(4–5):279–299.; discussion 277. https://doi.org/10.1023/a:1008975123362

Fischer R, Stoger E, Schillberg S, Christou P, Twyman RM (2004) Plant-based production of biopharmaceuticals. Curr Opin Plant Biol 7(2):152–158. https://doi.org/10.1016/j.pbi.2004.01.007

Fox JL (2012) First plant-made biologic approved. Nat Biotechnol 30(6):472–472. https://doi.org/10.1038/nbt0612-472

Gao M, Li Y, Xue X, Wang X, Long J (2012) Stable plastid transformation for high-level recombinant protein expression: promises and challenges. J Biomed Biotechnol 2012:158232. https://doi.org/10.1155/2012/158232

Gaume A, Komarnytsky S, Borisjuk N, Raskin I (2003) Rhizosecretion of recombinant proteins from plant hairy roots. Plant Cell Rep 21(12):1188–1193. https://doi.org/10.1007/s00299-003-0660-3

Gelvin SB (2003) Agrobacterium-mediated plant transformation: the biology behind the "gene-jockeying" tool. Microbiol Mol Biol Rev 67(1):16–37., Table of contents. https://doi.org/10.1128/mmbr.67.1.16-37.2003

Goeddel DV, Kleid DG, Bolivar F, Heyneker HL, Yansura DG, Crea R et al (1979) Expression in Escherichia coli of chemically synthesized genes for human insulin. Proc Natl Acad Sci U S A 76(1):106–110. https://doi.org/10.1073/pnas.76.1.106

Gomord V, Faye L (2004) Posttranslational modification of therapeutic proteins in plants. Curr Opin Plant Biol 7(2):171–181. https://doi.org/10.1016/j.pbi.2004.01.015

Gomord V, Stordeur V, Fitchette AC, Fixman ED, Tropper G, Garnier L et al (2020) Design, production and immunomodulatory potency of a novel allergen bioparticle. PLoS One 15(12):e0242867. https://doi.org/10.1371/journal.pone.0242867

Grabowski GA, Golembo M, Shaaltiel Y (2014) Taliglucerase alfa: an enzyme replacement therapy using plant cell expression technology. Mol Genet Metab 112(1):1–8. https://doi.org/10.1016/j.ymgme.2014.02.011

Graf T, Naumann L, Bonnington L, Heckel J, Spensberger B, Klein S et al (2024) Expediting online liquid chromatography for real-time monitoring of product attributes to advance process analytical technology in downstream processing of biopharmaceuticals. J Chromatogr A 1729:465013. https://doi.org/10.1016/j.chroma.2024.465013

Group, F. I. S (2007) Quadrivalent vaccine against human papillomavirus to prevent high-grade cervical lesions. N Engl J Med 356(19):1915–1927. https://doi.org/10.1056/NEJMoa061741

Hiatt A, Cafferkey R, Bowdish K (1989) Production of antibodies in transgenic plants. Nature 342(6245):76–78. https://doi.org/10.1038/342076a0

Holida M, Linhart A, Pisani A, Longo N, Eyskens F, Goker-Alpan O et al (2025) A phase III, open-label clinical trial evaluating pegunigalsidase alfa administered every 4 weeks in adults with Fabry disease previously treated with other enzyme replacement therapies. J Inherit Metab Dis 48(1):e12795. https://doi.org/10.1002/jimd.12795

Howard JA, Hood E (2005) Bioindustrial and biopharmaceutical products produced in plants. In: Advances in agronomy, vol 85. Academic Press, pp 91–124. https://doi.org/10.1016/S0065-2113(04)85002-8

Hwang HH, Yu M, Lai EM (2017) Agrobacterium-mediated plant transformation: biology and applications. Arabidopsis Book 15:e0186. https://doi.org/10.1199/tab.0186

Izadi S, Kunnummel V, Steinkellner H, Werner S, Castilho A (2023) Assessment of transient expression strategies to sialylate recombinant proteins in N. benthamiana. J Biotechnol 365:48–53. https://doi.org/10.1016/j.jbiotec.2023.02.004

Jansing J, Sack M, Augustine SM, Fischer R, Bortesi L (2019) CRISPR/Cas9-mediated knockout of six glycosyltransferase genes in Nicotiana benthamiana for the production of recombinant proteins lacking β-1,2-xylose and core α-1,3-fucose. Plant Biotechnol J 17(2):350–361. https://doi.org/10.1111/pbi.12981

Karki U, Fang H, Guo W, Unnold-Cofre C, Xu J (2021) Cellular engineering of plant cells for improved therapeutic protein production. Plant Cell Rep 40(7):1087–1099. https://doi.org/10.1007/s00299-021-02693-6

Koizumi A, Miyazawa K, Ogata M, Takahashi Y, Yano S, Yoshimi A et al (2022) Cleavage of α-1,4-glycosidic linkages by the glycosylphosphatidylinositol-anchored α-amylase AgtA decreases the molecular weight of cell wall α-1,3-glucan in Aspergillus oryzae. Front Fungal Biol 3:1061841. https://doi.org/10.3389/ffunb.2022.1061841

Kryovrysanaki N, James A, Tselika M, Bardani E, Kalantidis K (2022) RNA silencing pathways in plant development and defense. Int J Dev Biol 66(1-2-3):163–175. https://doi.org/10.1387/ijdb.210189kk

Laere E, Ling APK, Wong YP, Koh RY, Mohd Lila MA, Hussein S (2016) Plant-based vaccines: production and challenges. J Bot 2016(1):4928637. https://doi.org/10.1155/2016/4928637

Lico C, Chen Q, Santi L (2008) Viral vectors for production of recombinant proteins in plants. J Cell Physiol 216(2):366–377. https://doi.org/10.1002/jcp.21423

Liu H, Timko MP (2022) Improving protein quantity and quality-the next level of plant molecular farming. Int J Mol Sci 23(3). https://doi.org/10.3390/ijms23031326

Luciński R, Adamiec M (2023) The role of plant proteases in the response of plants to abiotic stress factors [review]. Front Plant Physiol 1

Lucy AP, Guo HS, Li WX, Ding SW (2000) Suppression of post-transcriptional gene silencing by a plant viral protein localized in the nucleus. EMBO J 19(7):1672–1680. https://doi.org/10.1093/emboj/19.7.1672

Lv Y, Li W, Liao W, Jiang H, Liu Y, Cao J et al (2024) Nano-drug delivery systems based on natural products. Int J Nanomedicine 19(null):541–569. https://doi.org/10.2147/IJN.S443692

Ma, J. K., Hikmat, B. Y., Wycoff, K., Vine, N. D., Chargelegue, D., Yu, L.,...Lehner, T. (1998). Characterization of a recombinant plant monoclonal secretory antibody and preventive immunotherapy in humans. Nat Med, 4(5), 601–606. doi: https://doi.org/10.1038/nm0598-601

Ma, J. K., Drossard, J., Lewis, D., Altmann, F., Boyle, J., Christou, P.,...Fischer, R. (2015). Regulatory approval and a first-in-human phase I clinical trial of a monoclonal antibody produced in transgenic tobacco plants. Plant Biotechnol J, 13(8), 1106–1120. doi: https://doi.org/10.1111/pbi.12416

MacManus CF, Collins CB, Nguyen TT, Alfano RW, Jedlicka P, de Zoeten EF (2017) VEN-120, a recombinant human Lactoferrin, promotes a regulatory T cell [Treg] phenotype and drives resolution of inflammation in distinct murine models of inflammatory bowel disease. J Crohns Colitis 11(9):1101–1112. https://doi.org/10.1093/ecco-jcc/jjx056

Margolin E, Oh YJ, Verbeek M, Naude J, Ponndorf D, Meshcheriakova YA et al (2020a) Co-expression of human calreticulin significantly improves the production of HIV gp140 and other viral glycoproteins in plants. Plant Biotechnol J 18(10):2109–2117. https://doi.org/10.1111/pbi.13369

Margolin EA, Strasser R, Chapman R, Williamson AL, Rybicki EP, Meyers AE (2020b) Engineering the plant secretory pathway for the production of next-generation pharmaceuticals. Trends Biotechnol 38(9):1034–1044. https://doi.org/10.1016/j.tibtech.2020.03.004

Matsuo K (2022) CRISPR/Cas9-mediated knockout of the DCL2 and DCL4 genes in Nicotiana benthamiana and its productivity of recombinant proteins. Plant Cell Rep 41(2):307–317. https://doi.org/10.1007/s00299-021-02809-y

Matsuo K, Atsumi G (2019) CRISPR/Cas9-mediated knockout of the RDR6 gene in Nicotiana benthamiana for efficient transient expression of recombinant proteins. Planta 250(2):463–473. https://doi.org/10.1007/s00425-019-03180-9

Mazalovska M, Kouokam JC (2020) Transiently expressed mistletoe lectin II in Nicotiana benthamiana demonstrates anticancer activity in vitro. Molecules 25(11). https://doi.org/10.3390/molecules25112562

McHughen A (2016) A critical assessment of regulatory triggers for products of biotechnology: product vs. process. GM Crops Food 7(3–4):125–158. https://doi.org/10.1080/21645698.2016.1228516

Moloi SJ, Ngara R (2023) The roles of plant proteases and protease inhibitors in drought response: a review. Front Plant Sci 14:1165845. https://doi.org/10.3389/fpls.2023.1165845
Mor TS (2015) Molecular pharming's foot in the FDA'S door: Protalix's trailblazing story. Biotechnol Lett 37(11):2147–2150. https://doi.org/10.1007/s10529-015-1908-z
Mungwari CP, King'ondu CK, Sigauke P, Obadele BA (2025) Conventional and modern techniques for bioactive compounds recovery from plants: review. Scientific African 27:e02509. https://doi.org/10.1016/j.sciaf.2024.e02509
Naithani S, Komath SS, Nonomura A, Govindjee G (2021) Plant lectins and their many roles: carbohydrate-binding and beyond. J Plant Physiol 266:153531. https://doi.org/10.1016/j.jplph.2021.153531
Nevalainen H, Peterson R (2014) Making recombinant proteins in filamentous fungi- are we expecting too much? Front Microbiol 5:75. https://doi.org/10.3389/fmicb.2014.00075
Niazi SK (2023) A proposed global medicines agency (GMA) to make biological drugs accessible: starting with the league of Arab states. Healthcare (Basel) 11(14). https://doi.org/10.3390/healthcare11142075
Nooraei S, Bahrulolum H, Hoseini ZS, Katalani C, Hajizade A, Easton AJ, Ahmadian G (2021) Virus-like particles: preparation, immunogenicity and their roles as nanovaccines and drug nanocarriers. J Nanobiotechnol 19(1):59. https://doi.org/10.1186/s12951-021-00806-7
Norkunas K, Harding R, Dale J, Dugdale B (2018) Improving agroinfiltration-based transient gene expression in Nicotiana benthamiana. Plant Methods 14:71. https://doi.org/10.1186/s13007-018-0343-2
Oh Y, Park Y, Choi BH, Park S, Gu S, Park J et al (2021) Field application of a new CSF vaccine based on plant-produced recombinant E2 marker proteins on pigs in areas with two different control strategies. Vaccines (Basel) 9(6). https://doi.org/10.3390/vaccines9060537
Oluwayelu DO, Adebiyi AI (2016) Plantibodies in human and animal health: a review. Afr Health Sci 16(2):640–645. https://doi.org/10.4314/ahs.v16i2.35
OTS/PRNewswire (2008) SemBioSys Eligible to Proceed With Phase I/II Plant-Produced Insulin Trial After Submission of IND. https://www.ots.at/presseaussendung/OTE_20080916_OTE0007/sembiosys-eligible-to-proceed-with-phase-iii-plant-produced-insulin-trial-after-submission-of-ind
Peragine A, Yoshikawa M, Wu G, Albrecht HL, Poethig RS (2004) SGS3 and SGS2/SDE1/RDR6 are required for juvenile development and the production of trans-acting siRNAs in Arabidopsis. Genes Dev 18(19):2368–2379. https://doi.org/10.1101/gad.1231804
Phyo P, Zhao X, Templeton AC, Xu W, Cheung JK, Su Y (2021) Understanding molecular mechanisms of biologics drug delivery and stability from NMR spectroscopy. Adv Drug Deliv Rev 174:1–29. https://doi.org/10.1016/j.addr.2021.02.007
Pinneh EC, van Dolleweerd CJ, Göritzer K, Drake PMW, Ma JK, Teh AY (2022) Multiple gene expression in plants using MIDAS-P, a versatile type II restriction-based modular expression vector. Biotechnol Bioeng 119(6):1660–1672. https://doi.org/10.1002/bit.28073
Plasson C, Michel R, Lienard D, Saint-Jore-Dupas C, Sourrouille C, de March GG, Gomord V (2009) Production of recombinant proteins in suspension-cultured plant cells. Methods Mol Biol 483:145–161. https://doi.org/10.1007/978-1-59745-407-0_9
Qiu X, Wong G, Audet J, Bello A, Fernando L, Alimonti JB et al (2014) Reversion of advanced Ebola virus disease in nonhuman primates with ZMapp. Nature 514(7520):47–53. https://doi.org/10.1038/nature13777
Rabindran S, Stevenson N, Roy G, Fedorkin O, Skarjinskaia M, Ensley B, Yusibov V (2009) Plant-produced human growth hormone shows biological activity in a rat model. Biotechnol Prog 25(2):530–534. https://doi.org/10.1002/btpr.130
Rader RA (2008) (Re)defining biopharmaceutical. Nat Biotechnol 26(7):743–751. https://doi.org/10.1038/nbt0708-743
Reshi ZA, Ahmad W, Lukatkin AS, Javed SB (2023) From nature to lab: A review of secondary metabolite biosynthetic pathways, environmental influences, and in vitro approaches. Metabolites 13(8). https://doi.org/10.3390/metabo13080895

Sanchez-Garcia L, Martín L, Mangues R, Ferrer-Miralles N, Vázquez E, Villaverde A (2016) Recombinant pharmaceuticals from microbial cells: a 2015 update. Microb Cell Fact 15(1):33. https://doi.org/10.1186/s12934-016-0437-3

Sasidharan S, Chen Y, Saravanan D, Sundram KM, Yoga Latha L (2011) Extraction, isolation and characterization of bioactive compounds from plants' extracts. Afr J Tradit Complement Altern Med 8(1):1–10

Schiffmann R, Goker-Alpan O, Holida M, Giraldo P, Barisoni L, Colvin RB et al (2019) Pegunigalsidase alfa, a novel PEGylated enzyme replacement therapy for Fabry disease, provides sustained plasma concentrations and favorable pharmacodynamics: A 1-year phase 1/2 clinical trial. J Inherit Metab Dis 42(3):534–544. https://doi.org/10.1002/jimd.12080

Schmidt FR (2004) Recombinant expression systems in the pharmaceutical industry. Appl Microbiol Biotechnol 65(4):363–372. https://doi.org/10.1007/s00253-004-1656-9

Shanmugaraj B, Bulaon CJI, Phoolcharoen W (2020a) Plant molecular farming: A viable platform for recombinant biopharmaceutical production. Plants (Basel) 9(7). https://doi.org/10.3390/plants9070842

Shanmugaraj B, Malla A, Phoolcharoen W (2020b) Emergence of novel coronavirus 2019-nCoV: need for rapid vaccine and biologics development. Pathogens 9(2). https://doi.org/10.3390/pathogens9020148

Shanmugaraj B, Rattanapisit K, Manopwisedjaroen S, Thitithanyanont A, Phoolcharoen W (2020c) Monoclonal antibodies B38 and H4 produced in Nicotiana benthamiana neutralize SARS-CoV-2 in vitro. Front Plant Sci 11:589995. https://doi.org/10.3389/fpls.2020.589995

Shanmugaraj B, Siriwattananon K, Malla A, Phoolcharoen W (2021) Potential for developing plant-derived candidate vaccines and biologics against emerging coronavirus infections. Pathogens 10(8). https://doi.org/10.3390/pathogens10081051

Sindelar RD (2013) Genomics, other "Omic" technologies, personalized medicine, and additional biotechnology-related techniques. In: Pharmaceutical biotechnology. Springer Science+Business Media, New York, pp 179–221. https://doi.org/10.1007/978-1-4614-6486-0_8

Sirko A, Vaněk T, Góra-Sochacka A, Redkiewicz P (2011) Recombinant cytokines from plants. Int J Mol Sci 12(6):3536–3552. https://doi.org/10.3390/ijms12063536

Sodoyer R (2004) Expression systems for the production of recombinant pharmaceuticals. BioDrugs 18(1):51–62. https://doi.org/10.2165/00063030-200418010-00005

Song SJ, Diao HP, Guo YF, Hwang I (2024) Advances in subcellular accumulation design for recombinant protein production in tobacco. Biodes Res 6:0047. https://doi.org/10.34133/bdr.0047

Spök A, Twyman RM, Fischer R, Ma JKC, Sparrow PAC (2008) Evolution of a regulatory framework for pharmaceuticals derived from genetically modified plants. Trends Biotechnol 26(9):506–517. https://doi.org/10.1016/j.tibtech.2008.05.007

Stephenson MJ, Reed J, Brouwer B, Osbourn A (2018) Transient expression in nicotiana benthamiana leaves for triterpene production at a preparative scale. J Vis Exp 138. https://doi.org/10.3791/58169

Strasser R (2023) Plant glycoengineering for designing next-generation vaccines and therapeutic proteins. Biotechnol Adv 67:108197. https://doi.org/10.1016/j.biotechadv.2023.108197

Thomas DR, Penney CA, Majumder A, Walmsley AM (2011) Evolution of plant-made pharmaceuticals. Int J Mol Sci 12(5):3220–3236. https://doi.org/10.3390/ijms12053220

Tripathi NK, Shrivastava A (2019) Recent developments in bioprocessing of recombinant proteins: expression hosts and process development. Front Bioeng Biotechnol 7:420. https://doi.org/10.3389/fbioe.2019.00420

Van Eck J (2020) Applying gene editing to tailor precise genetic modifications in plants. J Biol Chem 295(38):13267–13276. https://doi.org/10.1074/jbc.REV120.010850

Vela Ramirez JE, Sharpe LA, Peppas NA (2017) Current state and challenges in developing oral vaccines. Adv Drug Deliv Rev 114:116–131. https://doi.org/10.1016/j.addr.2017.04.008

Venkataraman S (2022) Plant molecular pharming and plant-derived compounds towards generation of vaccines and therapeutics against coronaviruses. Vaccines (Basel) 10(11). https://doi.org/10.3390/vaccines10111805

Venkataraman S, Khan I, Habibi P, Le M, Lippert R, Hefferon K (2023) Recent advances in expression and purification strategies for plant made vaccines. Front Plant Sci 14:1273958. https://doi.org/10.3389/fpls.2023.1273958

Wainwright CL, Teixeira MM, Adelson DL, Braga FC, Buenz EJ, Campana PRV et al (2022) Future directions for the discovery of natural product-derived immunomodulating drugs: an IUPHAR positional review. Pharmacol Res 177:106076. https://doi.org/10.1016/j.phrs.2022.106076

Walsh G, Walsh E (2022) Biopharmaceutical benchmarks 2022. Nat Biotechnol 40(12):1722–1760. https://doi.org/10.1038/s41587-022-01582-x

Ward BJ, Gobeil P, Séguin A, Atkins J, Boulay I, Charbonneau P-Y et al (2021a) Phase 1 randomized trial of a plant-derived virus-like particle vaccine for COVID-19. Nat Med 27(6):1071–1078. https://doi.org/10.1038/s41591-021-01370-1

Ward BJ, Séguin A, Couillard J, Trépanier S, Landry N (2021b) Phase III: randomized observer-blind trial to evaluate lot-to-lot consistency of a new plant-derived quadrivalent virus like particle influenza vaccine in adults 18-49 years of age. Vaccine 39(10):1528–1533. https://doi.org/10.1016/j.vaccine.2021.01.004

Webster CJ, George KL, Woollett GR (2021) Comparability of biologics: global principles, evidentiary consistency and unrealized reliance. BioDrugs 35(4):379–387. https://doi.org/10.1007/s40259-021-00488-5

Wilson FR, Coombes ME, Brezden-Masley C, Yurchenko M, Wylie Q, Douma R et al (2018) Herceptin® (trastuzumab) in HER2-positive early breast cancer: a systematic review and cumulative network meta-analysis. Syst Rev 7(1):191. https://doi.org/10.1186/s13643-018-0854-y

Wingfield PT (2015) Overview of the purification of recombinant proteins. Curr Protoc Protein Sci 80:6.1.1-6.1.35. https://doi.org/10.1002/0471140864.ps0601s80

Wycoff KL (2005) Secretory IgA antibodies from plants. Curr Pharm Des 11(19):2429–2437. https://doi.org/10.2174/1381612054367508

Yao X, Wuzhang K, Peng B, Chen T, Zhang Y, Liu H et al (2022) Engineering the expression of plant secondary metabolites-genistein and scutellarin through an efficient transient production platform in Nicotiana benthamiana L. Front Plant Sci 13:994792. https://doi.org/10.3389/fpls.2022.994792

Yuan H, Ma Q, Ye L, Piao G (2016) The traditional medicine and modern medicine from natural products. Molecules 21(5). https://doi.org/10.3390/molecules21050559

Yuan-Chuan C, Ming-Kung Y (2018) Introductory chapter: biopharmaceuticals. In: Ming-Kung Y, Yuan-Chuan C (eds) Biopharmaceuticals (pp. Ch. 1). IntechOpen. https://doi.org/10.5772/intechopen.79194

Harnessing Molecular Docking Approach for Nutraceutical Development from Plants: Current Knowledge and Emerging Paradigm

Varsha Rani, Kanchan Yadav, D. C. Joshi, and Dinesh Yadav

Abstract

Nutraceuticals' health-promoting and disease-preventive properties, commonly found in food sources, have gained considerable attention in recent years. A growing number of nutraceuticals are available on the market today. Their variety of applications has led to an increase in global nutraceutical prevalence. In light of this, they have an exceptional opportunity to explore this field further using innovative, reliable, low-cost methods to produce next-generation nutraceuticals. As a result, computational techniques like molecular docking have become an influential trend in academic and industrial research, predicting ligands' binding affinity to receptor proteins. A molecular target can contribute to developing disease-specific new therapies, which is the basis for developing novel nutraceuticals, followed by the necessity of establishing the reliability of scoring functions and confirming their efficacy through experimental testing. As the title suggests, this chapter will briefly introduce nutraceuticals, emphasizing the importance of chemoinformatic platforms, their libraries, and the application of molecular docking in developing new nutraceuticals. These tools enable researchers to predict the biological activity of compounds, optimize their properties, and streamline the discovery process. By leveraging these technologies, the development of effective and safe nutraceuticals can be significantly accelerated.

V. Rani
Department of Biotechnology, School of Engineering and Technology, Sandip University, Nashik, Maharashtra, India

K. Yadav · D. Yadav (✉)
Department of Biotechnology, Deen Dayal Upadhyaya Gorakhpur University, Gorakhpur, Uttar Pradesh, India
e-mail: dinesh.biotech@ddugu.ac.in

D. C. Joshi
ICAR-Vivekananda Institute of Hill Agriculture, Almora, Uttarakhand, India

R. K. Gaur et al. (eds.), *Plant Molecular Farming*,
https://doi.org/10.1007/978-981-95-3823-2_13

Keywords

Bioactive compound · Chemo libraries · In silico tools · Databases · Molecular docking · Nutraceuticals · Nano nutraceuticals

1 Introduction

The use of medicinal plants in medical research has a long-standing heritage in human civilization. The potential of these plants in contemporary medicine is substantial, and their systematic examination has progressed significantly with scientific advancement. A growing number of populations are recognizing the existence of bioactive chemicals in medicinal plants, which may have considerable pharmacological effects. In recent years, the importance of plants as a potential source of alternative medicine has gained more attention. Several plant species have the potential to be used as therapeutics (Yurdakok-Dikmen and Filazi 2021). Medicines made from botanical sources are generally easier for people to tolerate and are easily accessible, affordable, safe, and effective. Among the myriad of plants employed for medicinal purposes throughout history, those with significant potential as innovative medicines are primarily those with antibacterial properties, anticancer agents, and antihepatotoxic chemicals (Nasri et al. 2014). The World Health Organization considers medicinal plants to be the most effective source for various therapies. It is estimated that almost 80% of the population in developed nations utilizes traditional medicine, frequently using plant-based compounds for therapeutic purposes (Arunkumar and Muthuselvam 2009). Medicinal plants encompass a variety of bioactive chemicals, including tannins, alkaloids, polysaccharides, terpenoids, steroids, and flavonoids, which affect human physiology (Edeoga et al. 2005). Understanding the chemical makeup of plants is helpful when trying to create complex new molecules (Daliu et al. 2019). Enhancing research on medicinal plants is essential for the advancement of novel therapeutic opportunities. Technological advancements have brought innovative methods for nutraceutical research and development. In silico methodologies, including molecular docking, pharmacophore modeling, and molecular dynamics simulations, have become essential for the identification and optimization of bioactive substances. These computational techniques facilitate the prediction of interactions between nutraceuticals and biological targets, optimizing the development process and diminishing dependence on lengthy laboratory experiments. Molecular docking has been employed to forecast the binding affinity of ligands (nutraceuticals) to certain proteins, facilitating the discovery of prospective therapeutic agents (Rani et al. 2024). This technique not only confirms current use but also investigates emerging possibilities to create sustainable and effective healthcare solutions (Agu et al. 2023).

In drug development, in silico modeling has emerged as an essential technique, significantly accelerating drug discovery. However, most molecular docking algorithms operate under the assumption that proteins are rigid entities, which often leads to discrepancies in docking outcomes (Kumar et al. 2019). Notably, no existing docking technique can reliably predict ligand binding affinities at the molecular interface (Lakshmi Ranganatha et al. 2013). Despite these limitations, in silico approaches offer a substantial advantage by expediting drug design, screening, and manufacturing. They achieve this by analyzing molecular properties, predicting potential therapeutic targets, and identifying safety profiles. Consequently, these computational techniques help minimize reliance on costly and time-consuming laboratory procedures, such as in vitro assays and animal testing, thereby enhancing efficiency in the drug development pipeline. This chapter will explore nutraceuticals, their role as therapeutic compounds, and their increasing market demand. It will also shed light on emerging computational methodologies, including databases, molecular docking, and simulation approaches for determining the stability of nutraceuticals with target proteins. By profiling nutraceuticals according to Absorption, Distribution, Metabolism, Excretion, and Toxicity (ADMET) criteria, we will gain insight into their safety and sustainability as therapeutic solutions.

2 Concept of Nutraceuticals

The term "nutraceutical" was coined in 1989 by Dr. Stephen De Felice, deriving from the amalgamation of nutrients and pharmaceuticals. The current definition of nutraceutical denotes a food (or component of a food) that offers medical or health advantages, including disease prevention and/or treatment (Tsiaka et al. 2022). With the evolution of the language over time, "nutraceutical" is now defined as a substance extracted or refined from foods frequently offered in pharmaceutical forms but not associated with the food (DeFelice 1995). According to the literature, numerous definitions characterize nutraceuticals as food, dietary components, or nutrients that confer health advantages beyond their nutritional value (Pandey et al. 2010). The precise definition of nutraceutical supplements based on their intentional use is still debatable. Although "nutraceuticals" lacks a universally accepted meaning, it encompasses the nutritional content of goods with health-promoting and disease-preventing properties (Andrew and Izzo 2017; Tsiaka et al. 2022). Based on origin, composition, and method of development, nutraceuticals can be classified into traditional nutraceuticals and nontraditional nutraceuticals (Fig. 1). Traditional nutraceuticals are naturally occurring compounds derived from plants, animals, or microorganisms, offering significant physiological and therapeutic benefits. These can be categorized into chemical constituents, probiotic organisms, and nutraceutical enzymes (Dudeja and Gupta 2017). Chemical constituents include essential nutrients such as vitamins (e.g., vitamin C, vitamin D, and vitamin B-complex), minerals (e.g., calcium, magnesium, iron, and zinc), amino acids (e.g., lysine, arginine, and glutamine), and fatty acids (e.g., omega-3 and omega-6 fatty acids). Additionally, plant-derived phytochemicals, including flavones (e.g., 6-Hydroxyluteolin and apigenin), flavonoids (e.g., quercetin, taxifolin, and catechins), lignans (e.g., Cyclolariciresinol and Sesamol), phenolic acids (e.g., ferulic acid, cinnamic acid, and gallic acid), and stilbenes (e.g., pinosylvin), exhibit antioxidant, anti-inflammatory, and anticancer properties. Probiotic microorganisms

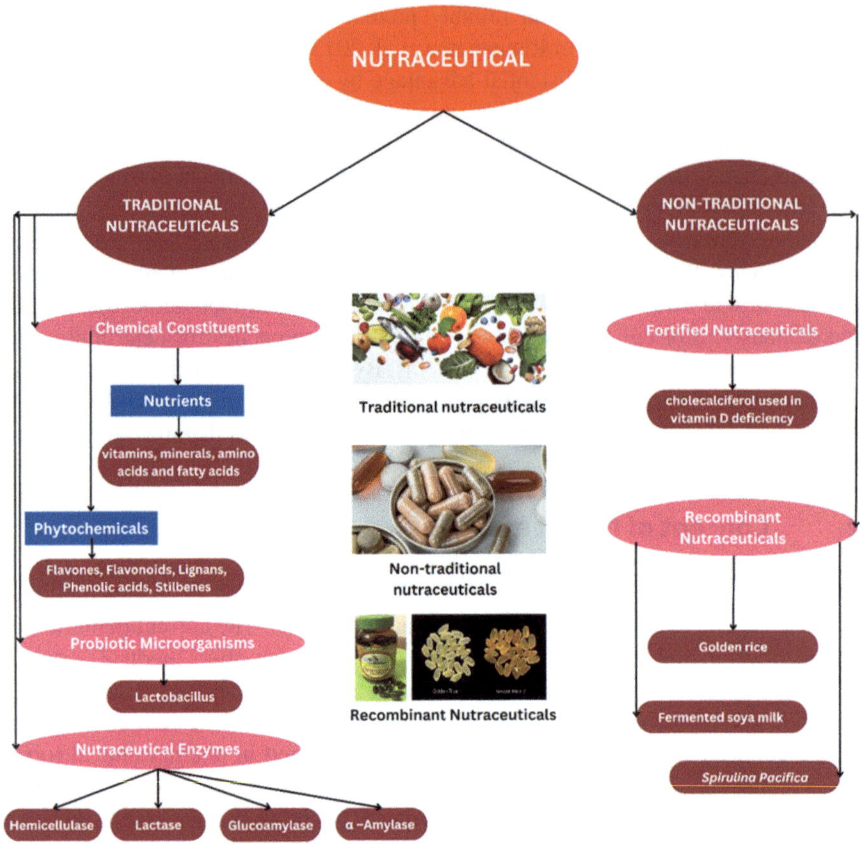

Fig. 1 Classification of Nutraceuticals. Nutraceuticals are categorized into traditional and non-traditional types. Traditional nutraceuticals are naturally derived from food sources and encompass: (i) Chemical constituents, including essential nutrients (vitamins, minerals, amino acids, and fatty acids) and bioactive phytochemicals (flavones, flavonoids, lignans, phenolic acids, and stilbenes); (ii) Probiotic microorganisms, such as Lactobacillus; and (iii) Nutraceutical enzymes, including hemicellulase, lactase, glucoamylase, and α-amylase, which support digestion and metabolism. In contrast, non-traditional nutraceuticals are modified or bioengineered to amplify health benefits and include: (i) Fortified nutraceuticals, such as cholecalciferol-enriched products for addressing vitamin D deficiency; and (ii) Recombinant nutraceuticals developed through biotechnology, examples being Golden rice (β-carotene-enriched), fermented soya milk, and Spirulina pacifica

such as *Lactobacillus* spp. are commonly found in fermented foods like yogurt, promoting gut health and immune system modulation. Nutraceutical enzymes, including hemicellulase (for fiber digestion), lactase (for lactose digestion), glucoamylase (for carbohydrate metabolism), and α-amylase (for starch breakdown), play a crucial role in enhancing digestion and nutrient absorption. On the other hand, nontraditional nutraceuticals are developed through scientific advancements such as biofortification, genetic engineering, and microbial fermentation to enhance their efficacy and health benefits. These can be classified into fortified nutraceuticals and recombinant nutraceuticals. Fortified nutraceuticals are enriched with essential nutrients to address deficiencies and improve health, with cholecalciferol (Vitamin D3) being a prominent example, commonly used in dairy products and cereals to prevent vitamin D deficiency and support bone health. Recombinant nutraceuticals, produced using genetic engineering and fermentation technology, enhance bioavailability and therapeutic properties. Examples include Golden Rice, a genetically modified rice variety enriched with beta-carotene to combat vitamin A deficiency; fermented soya milk, which contains bioactive peptides and probiotics beneficial for gut and cardiovascular health; and *Spirulina Pacifica*, a cyanobacterium rich in proteins, vitamins, antioxidants, and essential fatty acids, widely used in dietary supplements for immune and metabolic support (Singh and Sinha 2012; Bairagi and Patel 2021).

3 Databases for Nutraceuticals

Databases are essential in nutraceutical research, offering comprehensive information on bioactive chemicals, their origins, chemical structures, pharmacological characteristics, and molecular interactions. Table 1 represents numerous specialized databases that enable the identification, study, and development of nutraceuticals (Yi et al. 2018; Kumar et al. 2019). Phenol-Explorer and PhInd focus on plant metabolites and polyphenol content, respectively, facilitating the understanding of the chemistry and biogenesis of plant-derived chemicals. NutriChem analyzes drug-food interactions, providing insights into possible nutraceutical applications. Natural Product Activity and Species Source (NPASS) and Indian Medicinal Plants, Phytochemistry and Therapeutics (IMPPAT) offer essential data regarding the activity of natural products and Indian medicinal flora, facilitating both traditional and contemporary nutraceutical research. Several popular databases, such as PubChem, DrugBank, BindingDB, and Chemical Entities of Biological Interest (ChEBI), aggregate chemical, biological, and pharmacokinetic data that are essential for determining the efficacy and safety of nutraceuticals. NAPRALERT is a comprehensive bibliographic library covering pharmacological and biological data on natural compounds. Alkamid focuses on plant-derived N-alkyl amides, emphasizing their production and pharmacological importance. TCMSP provides data on more than 500 Chinese pharmacopoeial compounds, facilitating comprehensive research on nutraceuticals derived from traditional Chinese medicine. Computational resources like Timtec and ZINC offer extensive libraries of natural products and three-dimensional chemical structures, enabling in silico investigations such as molecular docking for drug discovery. Food & Nutrition-NBRI supports studies on the connections between diet and disease by acting as a vital repository of dietary and health-associated data. These databases jointly improve the investigation, verification, and utilization of nutraceuticals in illness management and functional food production.

Table 1 Nutraceutical databases of plant secondary metabolites

S.No.	Database	Description	Link
1	Phenol-explorer	Databases on the plant metabolome describe the chemistry and biosynthesis of plant chemicals	http://www.phenol-explorer.eu
2	PhInd	Comprehensive database on polyphenol content in by-products or waste generated in food processing or harvesting	https://phind.uns.ac.rs/explore/index
3	NutriChem	NutriChem 2.0 contains predicted interactions between 428 drugs and 339 foods, supported by 107 jointly targeted proteins	http://sbb.hku.hk/services/NutriChem-2.0/
4	NPASS	Natural product activity and species source database (NPASS) has served as one of the leading data sources for the NP research community since its first version release in 2017	https://bidd.group/NPASS/
5	IMPPAT	Indian medicinal plants, Phytochemistry and therapeutics database (IMPPAT) is a database of bioactive compounds from Indian medicinal plants, including their pharmacological properties and molecular interactions	https://cb.imsc.res.in/imppat/
6	PubChem	A widely used database containing chemical structures, biological activities, safety, and toxicity data for numerous bioactive compounds, including nutraceuticals	https://pubchem.ncbi.nlm.nih.gov/
7	DrugBank	Contains detailed information on bioactive compounds, including nutraceuticals, with pharmacokinetics, metabolism, and target interactions	https://go.drugbank.com/
8	BindingDB	A database of molecular interactions, focusing on small molecules, including nutraceutical compounds, and their protein targets	https://www.bindingdb.org/rwd/bind/index.jsp
9	ChEBI	Chemical entities of biological interest (ChEBI) provides data on bioactive molecules, including naturally occurring compounds used in nutraceuticals and pharmaceuticals	https://www.ebi.ac.uk/chebi/
10	NAPRALERT	A natural product databasebased on more than200,000 bibliographicalrecords, including pharmacologicaland biochemicalinformation	https://pharmacognosy.pharmacy.uic.edu/napralert/
11	Alkamid	Details of plants' sources,biosynthesis and pharmacologicalinformationrelated to N-alkylamides	https://alkamid.ugent.be/
12	TCMSP	Information about500 Chinesepharmacopoeia	https://tcmsp-e.com/tcmsp.php
13	Timtec	Database of more than10,000 natural productsand their derivatives	https://www.timtec.net/chemical-compounds/bioactive-compounds/
14	ZINC	Computational databaseof millions of compoundsand 3D structure fordocking	https://zinc.docking.org/substances/home/
15	Food & Nutrition-NBRI	Food & nutrition-NBRI is a vital national and international resource providing expertise and accurate data/information essential for high-quality research on relationships between diet and health	https://fnnbri.quadram.ac.uk/data/

4 Molecular Docking

The molecular docking method is a computational method widely used in drug discovery and development to predict interactions between biological macromolecules and small molecules (Kitchen et al. 2004; Rani et al. 2024). By utilizing this method, nutraceutical compounds can be identified as potential therapeutic agents that target specific proteins associated with various diseases, thus potentially acting as therapeutic agents. Several molecular docking tools and software are used in this procedure (Table 2). One of the most popular tools is AutoDock and AutoDock Vina (Eberhardt et al. 2021), which use genetic algorithms to predict ligand-protein interactions in a highly accurate and efficient manner. SwissDock offers a free online docking service based on the EADock DSS that allows nutraceutical compounds to be screened rapidly. (Grosdidier et al. 2011). HADDOCK docking platforms incorporate experimental data to enhance docking precision, making them ideal for analyzing protein-protein and protein-ligand interactions (De Vries et al. 2010). Shape-based docking approaches, such as PatchDock, predict molecular interactions accurately by utilizing geometric compatibility between ligands and receptors (Schneidman-Duhovny et al. 2005). In addition to clustering algorithms, ClusPro can also be used to generate reliable docking solutions for complex biomolecular interactions (Kozakov et al. 2017). Using DockThor, an advanced web server that facilitates docking small molecules into macromolecular targets, is highly beneficial for screening natural bioactive compounds from nutraceuticals (Kozakov et al. 2017). In addition to enhancing structure-based drug design by providing high-precision docking, the use of commercial software such as Glide (Schrödinger) in virtual screening applications enhances structure-based drug design (Halgren et al. 2004). As a consequence of considering ligand flexibility, iDock and FlexAID provide efficient docking simulations when dealing with a variety of nutraceutical-derived phytochemicals. A combination of molecular docking, modeling, and visualization facilitates the identification of lead compounds using MOE (Molecular Operating Environment) (Vilar et al. 2008). A high-throughput screening approach is employed by Surflex-Dock, whereas LeDock is regarded for its lightweight, efficient docking capabilities. As part of GOLD (Genetic Optimization for Ligand Docking), genetic algorithms are used to refine ligand binding conformations (Verdonk et al. 2003). In order to evaluate nutraceutical compounds as potential therapeutic candidates through large-scale virtual screening using rDock, this docking program is essential. Researchers can optimize structure-based drug design through pharmacophore modeling using advanced docking platforms such as PharmDock. BioDock, a cloud-based tool, can also be used to predict the binding affinity of nutraceutical-derived molecules. In order to develop degraders that degrade disease-associated proteins, the PROTAC-DB database facilitates docking studies for the PROTAC-based drug discovery (Weng et al. 2021). For the identification of binding pockets in protein structures without knowing their active sites, cavity-based blind docking servers such as CB-Dock2 are particularly useful. In addition to improving the accuracy of ligand-receptor interactions, enhanced docking algorithms, such as PSOVina, can improve the screening accuracy for

Table 2 Freely available molecular docking tools and softwares

S.No.	Docking server/tool/software	Description	Link
1	AutoDock	A widely used molecular docking tool that employs genetic algorithms for predicting ligand-protein interactions	https://github.com/ccsb-scripps/AutoDock-GPU
2	AutoDock Vina	An improved version of AutoDock with higher docking speed and accuracy for small molecule docking	https://github.com/ccsb-scripps/AutoDock-Vina
3	SwissDock	A free online docking service based on EADock DSS that predicts protein-ligand interactions	https://www.swissdock.ch/
4	HADDOCK	A flexible docking platform that accounts for molecular flexibility and incorporates experimental data	https://rascar.science.uu.nl/haddock2.4/
5	PatchDock	A shape-based molecular docking algorithm that efficiently predicts protein-ligand interactions	https://bioinfo3d.cs.tau.ac.il/PatchDock/
6	ClusPro	A protein-protein docking server that utilizes clustering algorithms to generate docking solutions	https://cluspro.org/help.php
7	DockThor	A web server designed for flexible docking of small molecules into macromolecular targets	https://dockthor.lncc.br/v2/
8	Glide (Schrödinger)	A commercial docking software with high precision and accuracy, used in virtual screening	https://www.schrodinger.com/platform/products/glide/
9	iDock	A fast molecular docking tool designed to predict protein-ligand interactions using efficient algorithms	istar.cse.cuhk.edu.hk/idock
10	FlexAID	A flexible docking tool that considers ligand and protein flexibility during docking simulations	http://bcb.med.usherbrooke.ca/FlexAID
11	Molecular operating environment (MOE)	A commercial software suite offering molecular docking, modeling, and virtual screening	https://www.chemcomp.com/en/Download_Request.htm
13	Surflex-dock	A molecular docking tool that uses an incremental construction approach for flexible docking	https://www.computabio.com/applications-of-surflex-dock-software.html
14	LeDock	A lightweight and efficient docking software for small molecule docking	https://www.lephar.com/software
15	Genetic optimization for ligand docking (gold)	A docking program that uses genetic algorithms for flexible ligand docking	https://www.ccdc.cam.ac.uk/solutions/software/gold/

16	rDock	A molecular docking program for high-throughput virtual screening and flexible ligand docking	https://rdock.github.io/
17	PharmDock	A docking tool optimized for structure-based drug design and pharmacophore modeling	http://people.pharmacy.purdue.edu/~mlill/software/pharmdock
20	BioDock	A cloud-based molecular docking tool designed for protein-ligand docking and structure-based drug discovery	https://www.biodock.ai/
21	PROTAC-DB	A docking tool specifically designed for PROTAC-based drug discovery	http://cadd.zju.edu.cn/protacdb/
22	CB-Dock2	A cavity-based blind docking server for predicting ligand binding interactions in proteins	https://cadd.labshare.cn/cb-dock2/index.php
23	PSOVina	An enhanced version of AutoDock Vina that uses particle swarm optimization for improved docking results	https://app.cbbio.online/psovina/home
24	Molecular docking server	A free web-based docking service that provides flexible ligand docking to protein targets	https://www.dockingserver.com/web

nutraceutical-based drugs by utilizing particle swarm optimization. It provides a free web-based platform that facilitates flexible docking of ligands to a variety of protein targets as an additional resource for researchers exploring functional foods and bioactive compounds in drug discovery (Das et al. 2020; Agu et al. 2023; Alagarsamy et al. 2025).

5 Application of Molecular Docking for Analysis of Drug-like Properties of Nutraceutical

Molecular docking can be applied to identify nutraceuticals that target key proteins involved in diabetes, such as insulin receptors or glucose transporters. By predicting the binding affinity and interactions of these natural compounds with diabetes-related proteins, we can discover potential therapeutic agents that enhance insulin sensitivity or regulate blood glucose levels (Vigneshwara et al. 2021; Qin et al. 2025). Polyphenols possess antidiabetic properties by reducing insulin resistance or secretion through various signaling pathways (Devi et al. 2014). Polyphenols found in different plant extract fractions have been shown to decrease starch digestibility and absorption, lowering the glycemic response (Lakshmi Kumari and Sumathi 2002). This approach allows for a systematic exploration of dietary components and natural products, offering promising avenues for developing new treatments for diabetes. Figure 2 represents the molecular docking interactions between human pancreatic amylase (HPA) and two different ligands, demonstrated using 3D and 2D visualization tools, to illustrate various binding interactions such as van der Waals forces, conventional hydrogen bonds, and π-π T-shaped interactions. In panel (a), Montbretin A, a natural α-amylase inhibitor, binds to the active site of HPA through stabilizing van der Waals forces, strong conventional hydrogen bonds, and π-π T-shaped interactions, contributing to ligand stability. The 2D diagram highlights specific residues involved in binding, such as $Asp^{300,}$ Trp^{58}, and Tyr^{62}, which play a critical role in α-amylase inhibition. In panel (b), taxifolin, a naturally occurring polyphenol, docks into the same active site of HPA, interacting with crucial residues including Asp^{300}, Trp^{58}, Tyr^{62}, and His^{299}. The interaction pattern closely matches that of Montbretin A, indicating taxifolin's potential to effectively inhibit α-amylase. The presence of hydrogen bonds, van der Waals forces, and π-π stacking interactions suggests a strong binding affinity, further aligning with experimentally validated inhibitory residues. Given that α-amylase is a key enzyme in carbohydrate digestion, its inhibition reduces glucose absorption, which is beneficial for managing postprandial hyperglycemia in diabetes. Docking results confirm that taxifolin interacts with α-amylase in a similar manner to Montbretin A, suggesting it as a promising natural inhibitor for diabetes treatment. This study highlights molecular docking as a powerful tool for identifying phytochemicals with antidiabetic potential, supporting further in vitro and in vivo validation of taxifolin as a natural α-amylase inhibitor.

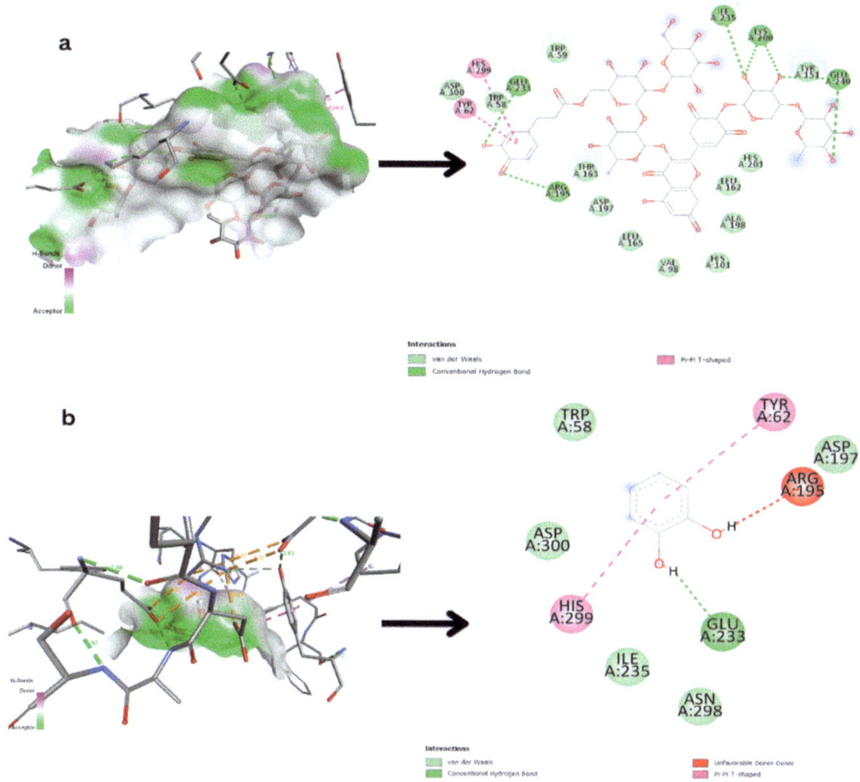

Fig. 2 Interaction between ligand and target: (**a**) control: human pancreatic alpha-amylase in complex with Montbretin A and (**b**) human pancreatic alpha-amylase in complex with taxifolin

6 Molecular Dynamics Simulations

The Molecular Dynamics Simulation (MD) is a widely used computational method for analyzing the behavior of proteins, ligand-protein interactions, and biomolecules with high temporal resolution and atomic-level precision (Hollingsworth and Dror 2018). By utilizing interatomic potentials or molecular mechanics force fields, this technique, which is based on Newton's equations of motion, allows for predicting atomic and molecular movements. MD simulations provide a detailed molecule-level perspective that enables the analysis of molecular systems on an atomic level to understand their dynamics. Numerous force fields and software tools have been developed to facilitate MD simulation (Tsiaka et al. 2022). The most used tools include AMBER, GROMACS, CHARMM-GUI, DESMOND, and NanoscaleMD (NAMD). MD simulations have significantly enhanced molecular biology and drug discovery to understand protein functions and disease mechanisms and optimize small molecules structurally (Chen et al. 2019). MD simulations play a crucial role in drug design by allowing researchers to visualize and predict how potential drug candidates interact with target proteins at the molecular level. This insight helps optimize small molecules' binding affinity and specificity, ultimately improving their efficacy as therapeutic agents. Furthermore, MD simulations can identify potential off-target effects and guide the design of compounds with reduced side effects.

7 ADMET Profiling

ADMET (Absorption, Distribution, Metabolism, Excretion, and Toxicity) represents its key pharmacokinetic and toxicity properties among a molecule's pharmacokinetics and toxicity characteristics (Guan and Yang 2018). Optimizing these characteristics early in the drug discovery process can reduce potential ADMET-related challenges later in the development process, thus playing a vital role in early drug discovery. As a strategic approach to assessing these factors early on, ADMET analysis allows researchers to identify and exclude compounds that have undesirable ADMET profiles in advance of further development. Adopting a proactive approach will require fewer expensive modifications and reformulations in later stages of development. Performing structural refinement evaluations before committing resources to synthesize new compounds assists in evaluating proposed modifications to enhance ADMET properties. This prescreening process assists in prioritizing candidates with a higher probability of success in advanced stages of drug development. Early integration of ADMET considerations during drug discovery contributes to creating safer and more effective therapeutic compounds by assisting researchers in making informed decisions. ADMET analysis significantly impacts cost efficiency by identifying and eliminating compounds with poor pharmacokinetic and toxicity profiles early in the drug discovery process (Biswas et al. 2023; Nagre et al. 2023). This preemptive screening reduces the likelihood of costly failures in later stages of development, where more resources have already been invested. By focusing on promising candidates from the outset, research teams can allocate their budgets more effectively, minimizing financial risks associated with drug development.

8 Conclusion

A molecular docking approach is an effective means of identifying the molecular targets of nutraceuticals in treating diseases. A potential treatment target can be identified by predicting nutraceuticals' binding affinity and conformation with target proteins. As databases become increasingly available and computational tools become more advanced, molecular docking has become an indispensable tool for drug discovery. This technology has made drug discovery more effective and efficient by cutting the time and costs associated with conventional experimental procedures. As such, molecular docking can be utilized in dietary supplement research to identify novel therapeutic targets and to develop secure and efficient nutritional supplements to treat diseases in several ways. By leveraging molecular docking, researchers can gain insights into the mechanisms of action of nutraceuticals, optimize their efficacy, and minimize potential side effects. This approach paves the way for the development of innovative dietary supplements that can address various health concerns and improve overall well-being.

References

Agu PC, Afiukwa CA, Orji OU, Ezeh EM, Ofoke IH, Ogbu CO, Ugwuja EI, Aja PM (2023) Molecular docking as a tool for the discovery of molecular targets of nutraceuticals in diseases management. Sci Rep 13:1–18. https://doi.org/10.1038/s41598-023-40160-2

Alagarsamy V, Sulthana MT, Solomon VR, Satishchandra A, Kulkarni VS, Narendhar B, Murugesan S, Muzaffar-Ur-Rehman M, Chandu A (2025) Identification of potential inhibitors from medicinal plant-based phytochemicals for the influential C4 target of diabetic retinopathy by molecular docking studies. Curr Pharm Des 31:307–319

Andrew R, Izzo AA (2017) Principles of pharmacological research of nutraceuticals. Br J Pharmacol 174:1177

Arunkumar S, Muthuselvam M (2009) Analysis of phytochemical constituents and antimicrobial activities of Aloe vera L. against clinical pathogens. World J Agric Sci 5:572–576

Bairagi GR, Patel VP (2021) Nutraceutical a review on: basic need, classification, recent trends in industry and delivery systems. Jetir 8:183–199

Biswas B, Golder M, Devnath HS, Mazumder K, Sadhu SK (2023) Comparative antihyperglycemic, analgesic and anti-inflammatory potential of ethanolic aerial root extracts of Ceriops decandra and Ceriops tagal: supported by molecular docking and ADMET analysis. Heliyon 9:e14254. https://doi.org/10.1016/j.heliyon.2023.e14254

Chen G, Huang K, Miao M, Feng B, Campanella OH (2019) Molecular dynamics simulation for mechanism elucidation of food processing and safety: state of the art. Compr Rev Food Sci Food Saf 18:243–263. https://doi.org/10.1111/1541-4337.12406

Daliu P, Santini A, Novellino E (2019) From pharmaceuticals to nutraceuticals: bridging disease prevention and management. Expert Rev Clin Pharmacol 12:1–7

Das DR, Kumar D, Kumar P, Dash BP (2020) Molecular docking and its application in search of antisickling agent from Carica papaya. J Appl Biol Biotechnol 8:105–116. https://doi.org/10.7324/JABB.2020.80117

De Vries SJ, Van Dijk M, Bonvin AMJJ (2010) The HADDOCK web server for data-driven biomolecular docking. Nat Protoc 5:883–897

DeFelice SL (1995) The nutraceutical revolution: its impact on food industry R&D. Trends Food Sci Technol 6:59–61

Devi PB, Vijayabharathi R, Sathyabama S, Malleshi NG, Priyadarisini VB (2014) Health benefits of finger millet (Eleusine coracana L.) polyphenols and dietary fiber: a review. J Food Sci Technol 51:1021–1040. https://doi.org/10.1007/s13197-011-0584-9

Dudeja P, Gupta RK (2017) Nutraceuticals. Elsevier Inc.

Eberhardt J, Santos-Martins D, Tillack AF, Forli S (2021) AutoDock Vina 1.2. 0: new docking methods, expanded force field, and python bindings. J Chem Inf Model 61:3891–3898

Edeoga HO, Okwu DE, Mbaebie BO (2005) Phytochemical constituents of some Nigerian medicinal plants. African J Biotechnol 4:685–688

Grosdidier A, Zoete V, Michielin O (2011) SwissDock, a protein-small molecule docking web service based on EADock DSS. Nucleic Acids Res 39:W270–W277

Guan L, Yang H (2018) ADMET-score—a comprehensive scoring function for evaluation of chemical drug-likeness. https://doi.org/10.1039/c8md00472b

Halgren TA, Murphy RB, Friesner RA, Beard HS, Frye LL, Pollard WT, Banks JL (2004) Glide: a new approach for rapid, accurate docking and scoring. 2. Enrichment factors in database screening. J Med Chem 47:1750–1759

Hollingsworth SA, Dror RO (2018) Molecular dynamics simulation for all. Neuron 99:1129–1143. https://doi.org/10.1016/j.neuron.2018.08.011

Kitchen DB, Decornez H, Furr JR, Bajorath J (2004) Docking and scoring in virtual screening for drug discovery: methods and applications. Nat Rev Drug Discov 3:935–949. https://doi.org/10.1038/nrd1549

Kozakov D, Hall DR, Xia B, Porter KA, Padhorny D, Yueh C, Beglov D, Vajda S (2017) The ClusPro web server for protein–protein docking. Nat Protoc 12:255–278

Kumar R, Singh AK, Gupta A, Pandey AK (2019) In-silico approaches to study therapeutic efficacy of nutraceuticals. Phytochem An in-silico in-vitro Updat:479–490. https://doi.org/10.1007/978-981-13-6920-9_25

Lakshmi Kumari P, Sumathi S (2002) Effect of consumption of finger millet on hyperglycemia in non-insulin dependent diabetes mellitus (NIDDM) subjects. Plant Foods Hum Nutr 57:205–213. https://doi.org/10.1023/A:1021805028738

Lakshmi Ranganatha V, Zameer F, Meghashri S, Rekha ND, Girish V, Gurupadaswamy HD, Khanum SA (2013) Design, synthesis, and anticancer properties of novel benzophenone-conjugated Coumarin analogs. Arch Pharm (Weinheim) 346:901–911

Nagre K, Singh N, Ghoshal C, Tandon G, Iquebal MA, Nain T, Bana RS, Meena A (2023) Probing the potential of bioactive compounds of millets as an inhibitor for lifestyle diseases: molecular docking and simulation-based approach. Front Nutr 10:1–19. https://doi.org/10.3389/fnut.2023.1228172

Nasri H, Baradaran A, Shirzad H, Rafieian-Kopaei M (2014) New concepts in nutraceuticals as alternative for pharmaceuticals. Int J Prev Med 5:1487

Pandey M, Verma RK, Saraf SA (2010) Nutraceuticals: new era of medicine and health. Asian J Pharm Clin Res 3:11–15

Qin L, Guo H, Wang Y, Hou X, Wu F, Du H, Li Z (2025) Quantification of 11 metabolites from quinoa seeds through HPLC-MS/MS and evaluation of their antidiabetic activities through molecular docking. J Food Compos Anal 140:107151

Rani V, Singh VK, Joshi DC, Singh R, Yadav D (2024) Molecular docking insights into nuclear factor Y (NF-Y) transcription factor and pyrabactin resistance 1 (PYL) receptor proteins reveal abiotic stress regulation in finger millet. Crop Design 3:100051. https://doi.org/10.1016/j.cropd.2023.100051

Schneidman-Duhovny D, Inbar Y, Nussinov R, Wolfson HJ (2005) PatchDock and SymmDock: servers for rigid and symmetric docking. Nucleic Acids Res 33:W363–W367

Singh J, Sinha S (2012) Classification, regulatory acts and applications of nutraceuticals for health. Int J Pharm Biol Sci 2:177–187

Tsiaka T, Kritsi E, Tsiantas K, Christodoulou P, Sinanoglou VJ, Zoumpoulakis P (2022) Design and development of novel nutraceuticals: current trends and methodologies. Nutraceuticals 2:71–90. https://doi.org/10.3390/nutraceuticals2020006

Verdonk ML, Cole JC, Hartshorn MJ, Murray CW, Taylor RD (2003) Improved protein–ligand docking using GOLD. Proteins Struct Funct Bioinforma 52:609–623

Vigneshwara RB, Chandra Mohan RCV, Sekhar AC, Reddy PCO, Srinivasulu K (2021) A new insights and novel targets for hyperglycemia from foxtail millet (Setariaitalica l.) using molecular docking studies. Curr Trends Biotechnol Pharm 15:213–219. https://doi.org/10.5530/ctbp.2021.2.23

Vilar S, Cozza G, Moro S (2008) Medicinal chemistry and the molecular operating environment (MOE): application of QSAR and molecular docking to drug discovery. Curr Top Med Chem 8:1555–1572

Weng G, Shen C, Cao D, Gao J, Dong X, He Q, Yang B, Li D, Wu J, Hou T (2021) PROTAC-DB: an online database of PROTACs. Nucleic Acids Res 49:D1381–D1387

Yi F, Li L, Xu L jia, Meng H, Dong Y mao, Liu H bo, Xiao P gen (2018) In silico approach in reveal traditional medicine plants pharmacological material basis. Chinese Med (United Kingdom) 13:1–20. https://doi.org/10.1186/s13020-018-0190-0

Yurdakok-Dikmen B, Filazi A (2021) Chapter 78—nutraceuticals: the Turkish perspective. In: Gupta RC, Lall R, Srivastava ABT-N (eds) , 2nd edn. Academic Press, pp 1309–1320

Tactical Nanogene Delivery for Plant Molecular Farming: Paving the Way for Sustainable Agriculture

Ayushi Malik, Khushboo Jain, Mayank Suthar,
Sunita Verma, Rajarshi Kumar Gaur, and Avinash Marwal

Abstract

Nanogene delivery in plants signifies a pioneering frontier in plant molecular farming, presenting revolutionary potential for crop enhancement and sustainability. This analysis delves into the most recent progressions in nanogene delivery systems, employing nanomaterials to elevate the precision and efficacy of genetic alterations in plants. Through harnessing the distinct characteristics of nanoparticles, scientists can realize targeted gene delivery, minimize off-target effects, and enhance the stability and expression of introduced genes. The fusion of nanotechnology with genetic engineering offers the prospect of cultivating crops with bolstered resistance to pests, diseases, and environmental pressures, as well as improved nutritional composition. This study also tackles the obstacles and future outlook of nanogene delivery, encompassing the necessity for scalable production techniques, regulatory considerations, and potential environmental ramifications. On the whole, nanogene delivery stands ready to transform plant molecular farming, paving the way for more resilient and fruitful agricultural systems.

A. Malik · M. Suthar · A. Marwal (✉)
Nanoparticle Synthesis and Bioinformatics Laboratory, Department of Biotechnology, Vigyan Bhawan, Block – B, New Campus, Mohanlal Sukhadia University, Udaipur, Rajasthan, India

K. Jain
Department of Biotechnology, Faculty of Science, MVSC, Janardhan Rai Nagar Rajasthan Vidyapeeth (Deemed to be University), Udaipur, Rajasthan, India

S. Verma
Manas Enclave Residential Society, Indirapuri Street, Bhuwana, Udaipur, Rajasthan, India

R. K. Gaur
Department of Biotechnology, Deen Dayal Upadhyaya Gorakhpur University, Gorakhpur, Uttar Pradesh, India

Uttar Pradesh Council of Agricultural Research, Uttar Pradesh, India

R. K. Gaur et al. (eds.), *Plant Molecular Farming*,
https://doi.org/10.1007/978-981-95-3823-2_14

Keywords

Target effect · Environment · Scalable production · Crop enhancement · Genetic alteration

1 Introduction

Nanomaterials have garnered significant attention from researchers in recent years due to their distinctive architecture and diminutive scale. Notable progress has been achieved in the fields of therapeutics, diagnostics, and the delivery of biomolecules in animal cells. However, the application of nanotechnology in plant science is still in its early stages. Nanotechnology presents substantial opportunities for enhancing crop yield and could lead to a considerable upsurge in agricultural productivity (Jat et al. 2020; Das et al. 2014a, 2015a). Nanotechnology enables broad advancements in agricultural research, encompassing areas such as reproductive science and technology, the conversion of agricultural and food wastes into energy, and other valuable byproducts through enzymatic nanobioprocessing, as well as the prevention and treatment of diseases in plants using various nanocides (Athanassiou et al. 2007; Tailor et al. 2021). Nanoscale devices with novel properties have the potential to render agricultural systems "smart," capable of autonomously responding to diverse conditions and initiating appropriate remedial actions. These intelligent systems facilitate the controlled and targeted delivery of chemicals, akin to the proposed application of nanodrug delivery in humans (Mewis et al. 2001, Liu et al. 2006). Smart delivery systems in agriculture should possess a combination of characteristics, including time control, specific targeting, high precision, remote regulation, preprogramming, self-regulation, and multifunctionality, to circumvent biological barriers and achieve successful targeting. Nanotechnological connections to agriculture and food science encompass agriculture and food system security, novel delivery systems for disease treatment, new tools for molecular and cellular biology, and new materials for detecting plant pathogens (Marwal and Gaur 2021). Moreover, plants or their extracts offer a biologically synthesized route for several metallic nanoparticles, which is environmentally friendly and allows for controlled synthesis with well-defined size and shape (Panáček et al. 2009; Solgi et al. 2009).

The purpose of this chapter is to critically assess and consolidate current advances in tactical nanogene delivery systems designed to enhance plant molecular farming. Our objective is to investigate the mechanisms and technologies utilized in nanogene delivery, evaluate their effectiveness and safety in the development of transgenic plants, and explore their potential to enhance crop yield, resilience, and sustainability. Additionally, this review aims to pinpoint gaps in existing literature and suggest future research directions that could facilitate the integration of nanotechnology within agricultural practices, ultimately contributing to the establishment of more sustainable agricultural systems.

1.1 Overview of Molecular Farming: Conventional Practices vs Modern Techniques

The inception of conventional molecular farming dates back to 1920, when Frederick Banting extracted insulin from animal tissues. However, limitations associated with this method prompted the exploration of alternative approaches. Consequently, plant molecular farming emerged as a prominent outcome, predominantly relying on transgenic plant cells as the primary source (da Cunha et al. 2014; Dynkevich et al. 2013). Within the realm of plant species, those with higher water content in their tissues, such as tomato plants, are found to be more conducive to molecular farming compared to plants with dry tissues, such as cereals. This observation is likely associated with the ease of extracting soft tissues (Horn et al. 2004). The utilization of higher plants offers numerous benefits, yet there exist limitations in acquiring products through plant molecular farming. One challenge is the unidentified mechanism leading to specific posttranslational irregularities in plant cells. The refinement of essential systems for maintaining the structural integrity and functionality of recombinant proteins in novel cellular environments is an ongoing debate in the field of plant molecular farming (Walsh and Jefferis 2006).

The adoption of advanced molecular techniques has significantly transformed the landscape of crop quality improvement, presenting a pivotal development in overcoming existing challenges (Draga et al. 2023; Marwal and Gaur 2020; Verma et al. 2024). Additionally, these techniques have facilitated the creation of crops resilient to environmental stressors, such as drought and extreme temperatures, thereby enhancing their adaptability to the impacts of climate change (Maheshwari et al. 2023; Mishra et al. 2024). With the sustained progression and integration of advanced molecular techniques, it is foreseeable that the enhancement of both crop quality and yield will persist, thereby ensuring global food security for the welfare of future generations (Ishtiaq et al. 2023; Iqbal et al. 2023).

1.2 Role of Nanotechnology in Agriculture

Nanomaterials have a significant impact on the fate, mobility, and toxicity of soil pollutants, playing a critical role in various biotic and abiotic remediation approaches. Their efficiency and behavior are greatly influenced by their properties and interactions with soil constituents (Usman et al. 2020; Das et al. 2015b). Nanosensors, as highly effective analyzers of nanomaterials, play a crucial role in assessing various stress factors, soil quality indicators, and heavy metals detection. They also facilitate advanced data collection, leading to precise decision-making and efficient resource management in agricultural settings. The process of nanoencapsulation, which involves encapsulating traditional chemical fertilizers into nanofertilizers and pesticides into nanopesticides, enables the gradual and sustained release of chemicals into the soil, ensuring precise dosages for plants. Furthermore, nano-based disease detection kits are extensively used for the early and rapid identification of viral diseases (Zain et al. 2023; Tailor et al. 2022). In the field of precision agriculture, various categories of nanobiosensors are employed. Optical nanosensors, which rely on the detection of changes in optical signals, are suitable for a variety of spectroscopic measurements, including absorption, phosphorescence, fluorescence, Raman effect, refraction, and Raman scattering. These sensors detect variations in wavelength, intensity, polarity, time, and phase of light (Dar et al. 2020). Additionally, optical biosensors make use of fluorescence spectroscopy, interferometry, surface plasmon resonance, and spectroscopy to operate effectively (Srivastava et al. 2018). Near-infrared fluorescent H_2O_2 nanosensors have been identified as a novel tool in precision agriculture for early detection of stress indicators (Wu et al. 2020). Furthermore, research by Ahmed et al. (2020) has demonstrated that red-colored selenium nanoparticles obtained from bioreduction can serve as optical biosensors for heavy-metal toxicity assays.

Electronic nose (e-Nose) and electronic tongue (e-Tongue) nanosensors, which mimic human sensory organs, are highly accurate, sensitive, redundant, selective, and reliable devices with diverse applications in precision agriculture (Srivastava et al. 2018). These sensors are utilized to detect a variety of gases, tastes, aromas, and their varying concentrations (Tan and Xu 2020). They play a critical role in regulating the quality and quantity of various materials, such as foods, beverages, agricultural products, pharmaceuticals, personal care products, and their manufacturing processes (Baldwin et al. 2011; Tan and Xu 2020). For example, an e-Nose can be employed to assess rice aroma, thereby preventing loss through timely intervention (Hu et al. 2020). Moreover, e-Noses are utilized for monitoring production processes and measuring plant diseases, insect infestations, and soil or water contaminants (Hu et al. 2019).

Wireless sensor networks are pivotal in driving automation in the precision agriculture domain. Wireless nanosensors installed in fields provide real-time data on crop, soil, and weather conditions, ensuring productivity and sustainability (Shafi et al. 2019). Common applications of wireless sensor networks in precision farming include smart irrigation, smart fertilization, smart pest control, and greenhouse monitoring. It is crucial to apply fertilizers in precise quantities to enhance productivity. An automated wireless sensor-based real-time fertilization system measures soil fertility. Additionally, electrochemical sensors are utilized for various purposes in the agricultural industry. These sensors gather chemical data on soil by detecting specific ions and providing information on pH and soil nutrient concentrations (Pudake et al. 2021). Apart from spectroscopic detection, electrochemical sensors are employed for the detection of nitrogen, phosphorus, and potassium (NPK), pesticide residues, pathogenic *Staphylococcus aureus*, chloramphenicol, and nitrogen species (Chauhan and Urooj 2019; Umapathi et al. 2022; Braiek et al. 2012; Yang and Zhao 2015; Ryu et al. 2020). Variable rate technology (VRT) plays a pivotal role in precision agriculture by facilitating the precise application of input resources such as herbicides, fertilizers, and pesticides based on soil classes and crop health. VRT empowers farmers to minimize cost and waste while maximizing crop growth and productivity, as resources are utilized only when necessary. Furthermore, VRT enables the appropriate seeding rate for specific field types (Sarauskis et al. 2022).

Nanosensors not only automate irrigation systems but also prevent overwatering, thereby mitigating negative impacts on crop growth. In addition to measuring soil moisture, nanosensors can monitor various climatic and environmental parameters such as humidity, temperature, and soil nutrient status (Mahmoud and Fawzy 2021). To advance the development of nanosensors and nanobiosensors with effective detection capabilities, researchers are integrating elements of biology, nanoscience, computing, and electronics to enhance crop management. The implementation of nanosensors or nanobiosensors is poised to significantly impact precision agricultural practices (Sekhon 2014). Nanobiosensors are capable of detecting physiological signals and converting them into well-defined signals that can be visually represented. A range of nanosensors has been designed and utilized for the detection of heavy metals. Among the sensors available for the detection of heavy-metal ions, the field-effect transistor sensor exhibits notable potential for rapid and real-time monitoring. Soil moisture content is another critical parameter that requires attention for effective crop yield management. Additionally, nanobiosensors are employed to assess nitrite and urease concentrations in soil and water through microfluidic impedimetric and colorimetric assays. The accurate detection of these nitrogenous compounds via nanobiosensors can provide valuable spatial and temporal variations in nutrient concentrations within the field, thus facilitating fertilizer analysis and applications in precision agriculture (Usman et al. 2020). Moreover, the current limitations of nanomaterial-based biosensors are such that they can only quantify total carbon, organic matter, sodium chloride, phosphate, and residual nitrate in soil through in vitro standards and water solutions.

1.3 Importance of Nanogene Delivery

Nanofiber arrays capable of efficiently and rapidly delivering genetic material to cells exhibit significant potential for applications in drug delivery, agricultural engineering, and environmental monitoring. Studies have reported the integration of surface-modified carbon nanofibers containing plasmid DNA into viable cells for controlled biochemical manipulations (McKnight et al. 2003, 2004). The successful delivery and incorporation of plasmid DNA were confirmed through gene expression analyses (as shown in Fig. 1). This approach exhibits similarities to the microinjection method of gene delivery (Segura and Shea 2001; Neuhaus and Spangerberg 1990; Bolik and Koop 1991), thus making it applicable to plant cells, where treated cells can regenerate into whole plants that express the introduced traits. It is feasible to tether DNA to carbon nanofibers without facilitating their integration into the host genome, while still permitting some transcription of the tethered genes. Consequently, this method does not transmit modified traits to subsequent generations, distinguishing it from conventional genetic engineering techniques and allowing for a one-time modification of the cells.

An application involving fluorescently labeled starch nanoparticles as transgenic vehicles for plants has been reported. In this case, the nanoparticle biomaterial was designed to bind and transport genes across the plant cell wall by inducing

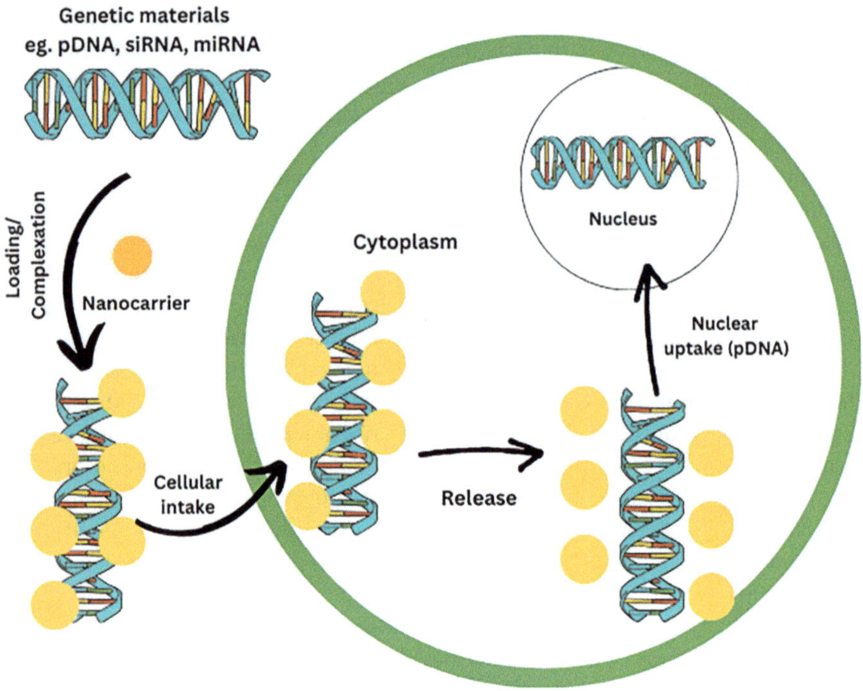

Fig. 1 Nanogene delivery inside the cell

instantaneous pore channels in the cell wall, cell membrane, and nuclear membrane using ultrasound (Liu et al. 2008). The integration of multiple genes onto the nanoparticle simultaneously is feasible, and fluorescence microscopy can be employed to image the fluorescent nanoparticles, thereby facilitating the study of gene movement and the expression of the transferred genes. Hence, the successful generation of pores in the cell wall and membrane by suitable agents enhances nanoparticle-mediated DNA transfer, demonstrating particular effectiveness in regenerative calli and soft tissues. Additionally, the capacity of surface-functionalized mesoporous silica nanoparticles (MSNs) to penetrate plant cell walls presents new avenues for the precise manipulation of gene expression at the single-cell level by delivering DNA and its activators in a controlled manner (Torney et al. 2007).

Reports indicate that a honeycomb MSN system with 3-nm pores can effectively transport DNA and other chemicals into both isolated plant cells and intact leaves. Mesoporous silica nanoparticles (MSNs) were integrated with a gene and its corresponding chemical inducer, subsequently capped with gold nanoparticles to prevent the leaching of molecules. The unblocking of gold nanoparticles facilitated the release of chemicals, thereby triggering gene expression in plants under controlled-release conditions. In preliminary experiments, protoplasts were incubated with MSNs that were fluorescently labeled. It was determined that the surface modification of MSNs with triethylene glycol was essential for cellular penetration. This modification also enabled the adsorption of plasmid DNA onto the MSN surfaces. Upon entering the protoplasts, the plasmid DNA was released from the MSNs, and the expression of the green fluorescent protein (GFP) marker encoded within the DNA was observed via microscopy. Notably, this method required a minimum amount of DNA that was 1000-fold less than that necessary for conventional delivery methods. This efficient delivery approach has significant implications for various protoplast-based gene expression studies. Currently, the gene gun or particle bombardment technique is among the prevalent tools utilized for delivering DNA into intact plant cells (Klein et al. 1989; Deng et al. 2001).

The particles employed in this bombardment are typically composed of gold, as they are capable of efficiently adsorbing DNA and are nontoxic to cellular structures. Given that MSNs possess a low weight, employing them to deliver foreign DNA by means of the gene gun method has been challenging. This issue was effectively resolved by capping the MSNs with gold nanoparticles, which enhanced their momentum when accelerated by the gene gun. Experimental results revealed that plasmid DNA transferred via the gene gun method using gold-capped MSNs was successfully expressed in intact tobacco and maize tissues. The primary advantage of this method lies in the simultaneous delivery of both DNA and effector molecules to specific sites, resulting in targeted delivery and expression of respective chemicals and genes. This process represents a departure from traditional genetic engineering techniques, such as electroporation and microinjection (Mahmood et al. 2017). Future directions include pore enlargement and multifunctionalization of MSNs, which may provide enhanced capabilities for target-specific delivery of proteins, nucleotides, and chemicals within the realm of plant biotechnology. Nanoparticles and nanocapsules present an efficient mechanism for the distribution of pesticides and fertilizers in a controlled manner, achieving high site specificity while reducing collateral damage. In the context of plant-pathogen interactions, the application of nanoparticle technology for the efficient transportation of substances, including systemic chemicals to designated sites, offers innovative solutions for the treatment of plants. If properly functionalized, nanoparticles can be distributed throughout the plant vascular system and directed to specific locations. These nanoparticles could be utilized to introduce chemicals (such as fungicides and insecticides) or other compounds (including plant hormones, elicitors, and nucleic acids) into localized regions of plant tissues. The aforementioned approach has the potential to advance various studies at physiological, biochemical, and genetic levels. The synthesis of metal nanoparticles mediated by plants presents a safe and effective method that allows for enhanced control over the morphology of the nanoparticles. The interaction between plant cells and nanoparticles leads to modifications in gene expression and associated biological pathways, which ultimately influence plant growth and development (Nair et al. 2010).

2 Fundamentals of Nanogene Delivery

Biosensors operate on various working principles, necessitating that each reader be conceived and constructed from the ground up. This aspect often represents the most significant cost in the development of sensors. As the sensor assesses the consumer's response to different stimuli, the transducer captures the information and converts the stimulus into an electrical impulse, which can be quantified as output data (Pollard et al. 2021). During fabrication, nanoparticles are integrated, leading to the creation of nanobiosensors. Nanomaterials are frequently the focus of investigation due to the extensive range of bioanalytical applications they facilitate in fields such as bioimaging, diagnostics, drug delivery, and therapeutic interventions (Munawar et al. 2019).

Amperometric devices have been utilized to evaluate enzyme-based reactions, while fluorescent quantum dot (QD) devices are employed to assess binding efficacy and immune-labeling applications that utilize conjugated nanoparticles for the analysis of biomolecular interactions. The intrinsic optical properties of nanoparticles, coupled with their ability to connect with fluorescent markers, render them promising candidates for biosensing applications. Electrochemiluminescent assays, fluorescent-based assays, and biological field-effect transistors (bio-FET) contribute to the development of carbon nanotubes (CNTs) and other relevant carbon-based nanostructures, including reduced graphene oxide (rGO), graphene, and graphene sheets. Fluorescent-based assays are typically detected through the quenching characteristics of graphene (Singh et al. 2016; Rasheed et al. 2019), and carbon nanomaterials can adopt diverse structures depending on the measurement methodology employed. Nanosensors, nanoprobes, and various nanoscale technologies have significantly enhanced the performance of chemical and biosensing systems. These nanodevices are designed with rapid reaction times and minimal power consumption in mind. The electrochemical indicators of biocatalytic processes at conductive junctions have seen marked improvements due to the incorporation of nanomaterials, including magnetic nanomaterials (Das et al. 2014b), metal and oxide nanoparticles, quantum dots, metallo-phthalocyanines, and carbon-based materials. Functionalized nanoparticles, which are bonded to organic molecules, have been engineered for potential applications in biosensing technologies.

A diverse array of methods exists for synthesizing nanostructured materials and devices, with the optimal selection largely dependent on the specific properties of the materials, the type of nanomaterial being produced (0D, 1D, 2D), and the requisite quantity, among other considerations. In particular, biosensing applications are increasingly utilizing various physical processes for the development of nanostructured materials, such as physical vapor deposition, high-energy ball milling, melt mixing, electric arc methods, laser ablation, and sputtering. Nanostructures are also synthesized through chemical methods, notably the sol–gel technique and inverse micelle formation. These materials can be produced via both bottom-up and top-down approaches. The bottom-up method involves the assembly of material building blocks through self-assembly processes to create nanostructures, yielding lower failure rates and more consistent elemental compositions, thereby making them a preferable alternative. Conversely, top-down methods commence with the predetermined nanostructure and involve meticulous processing of macroscopic materials. However, these techniques are often associated with a high incidence of surface flaws, which represents a significant disadvantage.

2.1 Types of Nanoparticles and Nanocarriers

Nanocarriers are nanoparticles or nanocapsules designed to carry active substances, facilitating their release through differences in chemical potential that drive mass transfer mechanisms, such as diffusion. The release may also be induced by specific stimuli, including enzymatic degradation of the nanocarriers (Fischer et al. 2019). Research has demonstrated that nanocarriers enhance the uptake efficiency of micronutrients (Yuvaraj and Subramanian 2018). Sustained-release formulations that utilize polymeric matrixes and shells offer significant advantages over conventional fertilizers and (bio)pesticides. These formulations reduce the risk of chemical runoff during irrigation and adverse weather conditions, while also decreasing degradation due to microbial action, photolysis, and hydrolysis. Furthermore, they mitigate soil contamination, biotic stress, and potential threats to human health (Tailor et al. 2024). Consequently, nanotechnology contributes to increased profitability, higher yields, and enhanced sustainability in the agricultural sector. In this context, biobased polymer nanocarriers (Babu et al. 2013; Wurm and Weiss 2014) exhibit notable characteristics, including biodegradability and biocompatibility. These materials, derived from renewable resources, are available in large quantities and are cost-effective, making them particularly well-suited for agricultural applications.

2.1.1 Chitosan

Chitosan, a naturally occurring polysaccharide composed of β-(1 → 4)-linked D-glucosamine and N-acetyl-D-glucosamine units, serves as a prominent biopolymer for the production of nanocarriers in agricultural applications (Kashyap et al. 2015; Sharma et al. 2018; Kumaraswamy et al. 2018; Kremer 2019; Kwak et al. 2019). The primary source of commercial chitosan is derived from the processing of sea crustaceans, particularly chitin extracted from seashells. According to Global Industry Analysts, the global market for chitin and its derivatives is anticipated to reach approximately 400 tons by 2024. Philibert et al. (2017) has identified over 2000 applications for chitin, chitosan, and their derivatives. Chitosan is recognized for its safety profile and has been approved for use as a food additive and dietary supplement in various countries, including Japan, the United States, and several European nations. Japan formally accepted chitosan as a functional food ingredient in 1992. The leading producing countries are Japan and the United States, with smaller-scale operations in India, Italy, and Poland (Achterberg 2014).

Historically, chitosan has been utilized in agricultural and horticultural practices, particularly as a biopesticide that enhances the innate immune response of plants (Chandra et al. 2015) and as a planting aid, such as seed coatings or growth promoters (Hadwiger 2013; Rinaudo 2006). Research has demonstrated that chitosan nanoparticles offer enhanced efficacy compared to bulk chitosan in these applications (Kumaraswamy et al. 2018; Divya et al. 2019; Chun and Chandrasekaran 2019; Sathiyabama and Manikandan 2016).

Chitosan nanocarriers have proven effective for the targeted delivery of fertilizers (Choudhary et al. 2019; Saharan et al. 2016), pesticides (Grillo et al. 2014; Cota-Arriola et al. 2013), and genetic material (Kwak et al. 2019). Additionally, they have been formulated with the herbicide paraquat, providing an environmentally responsible alternative for weed control (Grillo et al. 2014, 2015). Research conducted by Saharan et al. (2015) has focused on the application of Cu-chitosan hybrid nanoparticles to enhance both plant defense mechanisms and overall growth. The studies examined the release of copper from the chitosan matrix and its beneficial effects on plant health. Specifically, the research assessed antifungal and growth-promoting properties in tomato plants by utilizing an ionic gelation process that incorporated CuSO4 (0.01%) into chitosan formulations prior to cross-linking with tripolyphosphate (TPP) anions. Results indicated that Cu-loaded chitosan nanocarriers, with an average particle size of approximately 400 nm, significantly influenced the growth of tomato seedlings and exhibited powerful antifungal effects.

In subsequent investigations, Saharan's team explored the in vitro release profile of copper from the chitosan matrix across varying pH levels. Findings indicated a rapid increase in copper release as pH decreased, emphasizing the critical role of chitosan's amino groups in this process. The research demonstrated that at lower pH levels, the release of copper could achieve as much as 44.11%, while the release rate declined substantially at higher pH levels due to deprotonation. The Cu-chitosan nanoparticles not only improved release profiles but also positively impacted the activity of key antioxidant and defense enzymes, such as SOD (superoxide dismutase), POD (peroxidase), PAL (phenylalanine ammonia lyase), and PPO (polyphenol oxidase), revealing significant enhancements when compared to control groups and treatments with bulk chitosan.

Moreover, practical applications of Cu-chitosan nanoparticles resulted in notable improvements in plant growth metrics, including plant height, root length, and chlorophyll content, outperforming solutions of CuSO4 alone despite containing lower copper concentrations. These nanoparticles were also utilized to address challenges posed by diseases such as curvularia leaf spot (CLS) in tomato plants, demonstrating enhanced health outcomes and growth metrics compared to conventional treatment methods. Further studies involving zinc-loaded chitosan nanocarriers illustrated their effectiveness in controlling CLS disease while significantly increasing grain yield and enriching the grain with zinc micronutrients, showcasing the substantial potential of chitosan nanocarriers in enhancing agricultural productivity and sustainability (Choudhary et al. 2019).

2.1.2 Cellulose and Hemicellulose

Cellulose is recognized as the most abundant biopolymer, serving as the primary structural component of the cell wall in green plants. Additionally, certain bacterial strains are capable of producing cellulose. Its widespread availability, coupled with beneficial mechanical, optical, and thermal properties, as well as exceptional biocompatibility, has established cellulose as a biomaterial of increasing significance, particularly in tissue engineering. Reports indicate that cellulose derivatives are utilized in microcarrier formulations designed for the controlled release of pesticides. For instance, silica-epichlorohydrin-carboxymethylcellulose microcapsules demonstrated a sustained release of insecticides targeting *Myzus persicae* in the presence of cellulases. Genetic toxicity assays using *Allium cepa* revealed that these microcapsules exhibited significantly lower genotoxicity compared to undiluted emamectin benzoate. Hydroxypropyl cellulose (HPC) nanocapsules have been employed for the encapsulation of curcumin, an active component derived from turmeric (*Curcuma longa* Zingiberaceae), which can function as an insecticide. Nanocarriers with diameters ranging from 150 to 250 nm achieved encapsulation efficiencies of approximately 70%. Notably, a sustained in vitro release of curcumin, reaching up to 80% at room temperature over a period of 5 h, has been reported. Kumar et al. (2014) examined the delivery of clodinafop-propargyl herbicide to plants utilizing carboxymethyl cellulose nanocarriers. Furthermore, Machado et al. chemically modified cellulose with undec-10-enoic acid to facilitate the encapsulation of hydrophobic fungicides in situ via thiol-ene cross-linking within a miniemulsion. The encapsulation efficiencies for the fungicides captan and pyraclostrobin ranged from 80% to 100%. Moreover, cellulose nanocarriers have been evaluated in vitro for their antifungal efficacy against *Neonectria ditissima*.

2.1.3 Lignin

Lignin, a polyphenol, is produced in millions of tons annually as a by-product of the paper industry, particularly from the kraft wood pulping process. During this procedure, lignin is separated from cellulosic fibers under alkaline conditions in combination with sodium sulfide. A significant portion of the resulting kraft lignin is either incinerated for fuel or discarded as waste. Despite being categorized as an underutilized biobased compound, lignin holds considerable promise in biorefinery design and sustainability initiatives. The adage suggesting that one can produce anything from lignin except profit underscores the historical challenges in monetizing this resource. However, ongoing research endeavors aim to identify higher-value applications for lignin. Despite its status as a soluble and abundant renewable feedstock, lignin has been underexplored concerning the synthesis of nanostructures for various applications, including biomedicine, cosmetics, oil recovery, coatings, and agriculture. Lignin nanocarriers have been produced from diverse sources, such as kraft and sulfite wood pulping wastes and spent mushroom substrate. Research has further demonstrated the biodegradability of these nanocarriers in the presence of laccase-producing fungi. Consequently, lignin nanocarriers are being utilized to encapsulate antifungal agrochemicals, anticipating that release will be triggered by degradation facilitated by fungal laccases. Additionally, other waste streams, such as polyphenols derived from green tea and coffee extracts, have been employed as reducing agents in the production of nanomaterials. Fischer et al. developed pyraclostrobin-loaded lignin nanocarriers, with diameters ranging from 200 to 700 nm, utilizing aza-Michael addition within a miniemulsion process. These pyraclostrobin-loaded nanocarriers demonstrated selective inhibition of the growth of

Esca-related fungi, specifically *Phaeomoniella chlamydospora* and *Phaeoacremonium minimum,* in vitro (Fischer et al. 2019).

2.1.4 Polypeptides

Polypeptides, which are polymers constructed from amino acids, have garnered attention for their role as nanocarriers in precision agriculture; however, their application is not as widespread as that of carbohydrate or polyphenol-based nanocarriers. Certain studies have demonstrated the use of polypeptide nanocarriers due to their distinctive ability to traverse the plasma membrane of living cells, which selectively permits the entry of specific exogenous molecules (Chuah et al. 2016; Chugh et al. 2010). Research has shown that cross-linked nanoparticles derived from L-aspartic acid have been utilized for the sustained release of fertilizers within soil (Lü et al. 2016). Furthermore, these cross-linked polypeptide nanoparticles have exhibited superabsorbent properties, thereby enhancing the water-holding and water-retention capacities of the soil. The results indicate the significant potential of nano-γPGA/CSGA3 for agricultural applications (Pereira et al. 2017). Recent investigations have focused on polysuccinimide (PSI) nanoparticles conjugated with glycine methyl ester for the targeted delivery of fludioxonil, a fungicide, to the phloem tissue of banana trees (Wu et al. 2021). This strategy leverages the pH-responsive characteristics of this polymer; in vitro studies have demonstrated that PSI nanoparticles can undergo hydrolysis at a pH of approximately 8, characteristic of the plant phloem, thereby facilitating the release of encapsulated fludioxonil. The decision to explore the conjugation of glycine ester was informed by prior research conducted by the authors (Wang et al. 2018), which revealed that glycine conjugation to polysuccinimide (PGA) enhanced the bioactivity and absorption of avermectin in rice. PGA nanocarriers containing approximately 29% fludioxonil underwent in vitro drug release assays, which confirmed their pH-responsiveness. Notably, nanoparticles at pH 8 (the physiological pH in the phloem) released double the amount of fludioxonil compared to those at pH 5. In vivo experiments conducted with potted banana plants displayed effective downward transport of nanocarriers to the rhizomes and roots following foliar application, resulting in a 50% reduction in disease severity. Furthermore, studies concerning phloem transport and uptake revealed that the phloem loading of nanocarriers was facilitated by an active transport mechanism at the organ level in castor bean seedlings, and that these nanocarriers could be absorbed by mesophyll cells and subsequently loaded into vascular tissues via the symplastic pathway. Additionally, transmembrane uptake at the cellular level was significantly enhanced. Polypeptide-based nanocarriers present promising opportunities in agriculture due to their capability to degrade into metabolizable units and to change their secondary conformation in response to external stimuli. Unlike synthetic copolymers, which predominantly exhibit a coil or globular structure, polypeptides can assume ordered conformations, such as α-helices or β-strands (Zhao et al. 2014).

2.1.5 Alginate

The anionic polysaccharide alginate is extensively utilized in the development of nanocarriers for a variety of pharmaceutical agents due to its favorable biocompatibility and nontoxic characteristics. Alginate has been employed to formulate nanocarriers that encapsulate pesticides such as imidacloprid (Kumar et al. 2014), cypermethrin (Patel et al. 2018), and phloxine B (Yin et al. 2020), all of which are intended for the management of insect pests. These formulations are created through methods such as the self-assembly of polymeric micelles or emulsion cross-linking (Patel et al. 2018; Kumar et al. 2014), or by conjugation with the polymer to form a polymeric prodrug. Notably, the development of a prodrug nanocarrier for phloxine B effectively prevents the premature photoactivation of the pesticide prior to it reaching the intended target. In cross-linked alginate delivery systems, the principal mechanism governing the release of active compounds within physiological fluids is the sodium–calcium exchange process. When calcium alginate is introduced into an environment rich in monovalent salts, the insoluble calcium alginate is converted into soluble alginate. This transformation facilitates the solubilization of the nanoformulation and the subsequent release of its encapsulated cargo. Conversely, in prodrug systems, the release mechanism is typically activated by a specific stimulus, such as enzymatic degradation or metabolic compounds, including glutathione (GHS), which are essential for the regulation of oxidative stress in cells (Yin et al. 2020).

2.1.6 Polylactic Acid and Copolymers

Poly(lactic acid) (PLA) and poly(lactic acid-co-glycolic acid) (PLGA) are members of the aliphatic polyester family, which are typically synthesized from α-hydroxy acids and are recognized for their biodegradable and compostable properties. PLA has been extensively utilized as a biodegradable polymer across a variety of applications, particularly within the biomedical sector. Recent years have seen a marked increase in interest in the utilization of PLA for agricultural purposes, including the advancement of nanoformulations (Yu et al. 2017; Knoch et al. 2020; Sun et al. 2020; Tong et al. 2017). PLGA has been employed in conjunction with polyethylene glycol (PEG) to develop polymeric nanocarriers (ranging from 90 to 125 nm) through a dialysis method, successfully encapsulating the hydrophobic herbicide metolachlor with an encapsulation efficiency exceeding 85%. Researchers have investigated the uptake of these carriers by rice seedlings and evaluated the efficacy of the nanopesticide in controlling Gramineae weeds (Tong et al. 2017). The objective was to enhance the water solubility of metolachlor, thereby eliminating the necessity for organic solvents and surfactants, which are significant contributors to pesticide pollution. Absorption studies involving Cy5-labeled nanoparticles indicated a potential pathway for the transmission of this metolachlor formulation into rice roots, which enhanced the overall utilization of metolachlor. Additionally, bioassay tests revealed that the nanopesticide exhibited a significantly stronger impact than traditional bulk formulations on Gramineae weeds, including *Oryza sativa, Digitaria sanguinalis*, and *Arabidopsis thaliana* seedlings. The exploration of PLA-based nanocarriers has also been extended to foliar applications. Researchers synthesized functionalized abamectin poly(lactic acid) (Abam-PLA) nanoparticles (Yu et al. 2017) to increase foliar adhesion and the effective utilization rate of pesticides; three distinct variations of Abam-PLA nanoparticles—CH3CO-PLA-NS, HOOC-PLA-NS, and H2N-PLA-NS—were developed, each with unique adhesion properties to cucumber leaves. These PLA-based nanoparticles were produced via a miniemulsion/solvent evaporation technique, resulting in diameters of approximately 450 nm and an abamectin loading efficiency of up to 50%. This nanoformulation facilitated continuous release of abamectin while demonstrating superior photostability when compared to conventional abamectin. The authors noted a slightly enhanced deposition of Abam-PLA nanocarriers on cucumber leaves, influenced by the modulation of the interaction between the nanoparticles and the leaf surface. Increased polar groups enhanced adhesion to the leaf surface, despite its generally hydrophobic nature. This observation can be rationalized by the presence of carbohydrates, fatty acids, fatty alcohols, and fatty aldehydes. It is plausible that the interactions of these polar groups on cucumber leaves with the functional groups present on the surface of Abam-PLA nanoparticles contributed to improved adhesion.

2.2 Design Considerations and Mechanisms of Nanogene Delivery

To establish an efficient gene delivery system utilizing nanomaterials, it is essential for the nanocarrier to effectively transport macromolecules into cells. This involves traversing the cell wall to reach the plasma membrane and subsequently targeting the nucleus and cytosol. Research has indicated that positively charged nanoparticles are particularly advantageous for the genetic transformation of plants, as the naturally negative charge of the plant cell wall facilitates the binding and transfer of genes (Torney et al. 2007). There are seven significant categories of nanoparticles, including nanocomposites, nanofibers, nanocapsules, metal-based, carbon-based, magnetic, and dendrimer nanoparticles, all of which serve as vital tools for gene manipulation (Kanwar et al. 2019, Sekhon 2014, Panpatte et al. 2016, Wang et al. 2016). The four primary groups most commonly utilized for gene delivery are metallic/magnetic nanoparticles, carbon-based nanoparticles, silicon-based nanoparticles, and polymeric-based nanoparticles (Sandhu et al. 2002; Zhao et al. 2017a).

The careful selection of a suitable delivery system is critical, as it directly impacts not only the controlled release of the cargo but also the biocompatibility of the carrier. Mesoporous silica nanoparticles (MSNs) possess unique characteristics, such as optimized surface properties, which may significantly enhance targeted gene delivery in plants. Several mechanisms enable nanoparticles to enter plant cells, including endocytosis, pore formation, transfer mediated by carrier proteins, and pathways through plasmodesmata. To maximize the potential of MSNs, it is crucial to consider both pore size and morphology during the synthesis process. The particle diameter can be adjusted from a few micrometers down to 10 nm, depending on the specific conditions employed (Wu et al. 2013). The interactions between Si–O–Si and other species in the synthesis mixture influence the pore characteristics, shape, order, and overall morphology, which are affected by surfactant properties and their respective ratios.

Common methods for synthesizing MSNs include sol–gel, microwave-assisted, hydrothermal, modified aerogel, soft and hard templating, and fast self-assembly techniques. Frequently used silica precursors for MSN synthesis comprise tetraalkoxy silanes (such as tetramethyl orthosilicate (TMOS) or tetraethyl orthosilicate (TEOS)) and sodium silicate (Na2SiO3) (Ernawati et al. 2016; Zhao et al. 2017a). Cargo loading onto MSNs is primarily a surface phenomenon, where interactions between the host and guest materials significantly impact the process. The surface chemistry of MSNs allows for precise tuning of surface properties through the functionalization of silanol groups, facilitating the engineering of versatile nanocarriers through covalent attachment (Hoffmann et al. 2006). Upon exposure to aqueous and biological environments, MSNs can undergo hydrolytic breakdown to form water-soluble and biocompatible orthosilicic acid, a compound recognized by the FDA as safe and excreted through urine. The biodegradability of MSNs is influenced by the dissolution mechanism of SiO2 particles into silicic acid in biological media.

Key parameters affecting the biodegradability of MSNs include surface area, pore size, particle morphology, degree of condensation, surface functionalization, and medium characteristics such as pH, temperature, and concentration (Manzano and Vallet-Regi 2020). Additionally, in plant systems, nanoparticle uptake not only involves the traditional xylem and phloem pathways but also supplementary cellular mechanisms such as signaling, recycling, and plasma membrane regulation, indicating an active transport process (Banerjee et al. 2019). In vitro studies have demonstrated that plant cells utilize endocytic pathways to uptake engineered nanoparticles (Iversen et al. 2011).

Recent advancements in specialized analytical techniques—including quantitative analysis, speciation, location analysis, and stable isotope-based methods—are essential for understanding the uptake, translocation, and transformation of nanoparticles in plants. Techniques such as inductively coupled plasma mass spectrometry (ICP-MS), laser ablation ICP-MS (LA-ICP-MS), and fluorescence microscopy (FM) have been employed to analyze the behavior of metal-based nanoparticles in plant systems). Furthermore, the use of gene fluorescence-labeled nanoparticle conjugates facilitates the tracking of transferred genes following nanoparticle-mediated genetic transformation methods (Mohamed and Abd-Elsalam 2019). Fluorescent quantum dots also show considerable promise as biolabels for monitoring the dynamic intracellular trafficking and fate of carrier-DNA nano-complexes post-transfection and transformation in plants (Wang et al. 2011).

3 Applications of Nanogene Delivery

3.1 Enhanced Crop Yield and Quality

Nanogene delivery has shown promise in improving crop yield and quality by introducing specific genetic traits that promote growth and enhance stress resistance. The use of nanocarriers, such as carbon nanotubes, silica nanoparticles, and lipid-based nanoparticles, facilitates the transfer of growth-promoting genes directly into plant cells. This method can enhance nutrient uptake, improve photosynthetic efficiency, and boost resistance to abiotic stressors like drought and salinity (Cunningham et al. 2018). For instance, studies have demonstrated that carbon nanotube-mediated delivery of genes related to growth regulation can significantly improve the growth rate of various crops (Torney et al. 2007). Additionally, nanoparticle-mediated gene delivery can be more precise than traditional methods, minimizing the risk of unintended genetic modifications and promoting sustainable agricultural practices (Giraldo et al. 2019).

3.2 Biopharmaceutical Production in Plants

Nanogene delivery is pivotal in molecular farming, where plants are engineered to produce biopharmaceuticals, including vaccines, antibodies, and therapeutic proteins. This approach utilizes plants as biofactories for the production of valuable pharmaceuticals, providing a cost-effective and scalable alternative to conventional manufacturing systems. Nanoparticle-based gene delivery systems, such as mesoporous silica nanoparticles and dendrimers, have been employed to introduce foreign genes into plant cells efficiently (Mohammadinejad et al. 2019). This technology ensures stable gene expression, increasing the yield and quality of the produced biopharmaceuticals. For example, plant-based production of antibodies using nanogene delivery has shown potential in creating affordable treatments for various diseases, such as cancer and infectious diseases (Gleba et al. 2007).

3.3 Disease Resistance

Nanogene delivery also offers solutions for improving disease resistance in plants by introducing genes that enhance the plant's innate defense mechanisms. This application is particularly valuable in developing resistance against viral, bacterial, and fungal pathogens. For instance, genes encoding antimicrobial peptides or RNA interference (RNAi) constructs can be delivered to target-specific pathogens. Nanocarriers such as chitosan nanoparticles have been used to deliver these genes, resulting in the activation of systemic acquired resistance in plants (Singh et al. 2021a, b). This method can reduce the need for chemical pesticides, contributing to more sustainable and environmentally friendly agricultural practices. Recent studies have demonstrated the use of RNAi-mediated gene silencing to control insect pests and plant viruses effectively, highlighting the potential of nanogene delivery in integrated pest management (Das et al. 2011; Sahu et al. 2014; Marwal and Gaur 2019; Koeppe et al. 2023).

3.4 Nutrient Biofortification

Addressing micronutrient deficiencies through the biofortification of staple crops is another significant application of nanogene delivery. By introducing genes that enhance the biosynthesis or accumulation of essential nutrients, such as iron, zinc, and vitamins, the nutritional content of crops can be improved. This approach is particularly crucial in developing countries where dietary deficiencies are prevalent. For example, nanocarriers have been used to deliver genes encoding enzymes involved in the synthesis of vitamin A precursors, leading to increased carotenoid content in crops like rice and maize (Tomar et al. 2020). Similarly, the use of nanoparticles to deliver genes related to iron transport has resulted in enhanced iron accumulation in edible plant tissues, addressing iron deficiency anemia (Low et al. 2017).

4 Challenges of Nanogene Delivery in Plants

4.1 Regulatory and Safety Considerations

While nanogene delivery offers numerous benefits, it also raises significant regulatory and safety concerns. The novel nature of nanomaterials used in gene delivery requires careful risk assessments to ensure their safety for human health and the environment. Potential risks include the toxicity of certain nanoparticles to plant cells, soil microbes, and non-target organisms. For example, metal-based nanoparticles like silver and zinc oxide can exhibit cytotoxic effects on plants and beneficial soil bacteria at high concentrations (Rico et al. 2011). There are also concerns about the long-term persistence of nanoparticles in the environment and the possibility of gene flow to wild relatives, which could impact biodiversity. As the regulatory framework for nanotechnology in agriculture is still evolving, there is a need for standardized safety protocols and guidelines (Massawe et al. 2016). Regulatory agencies need to set specific guidelines for evaluating the safety and environmental effects of nanogene delivery systems.

4.2 Limitations of Nanogene Delivery

Despite its potential, nanogene delivery in plants faces technical and biological challenges. The efficiency of nanoparticle uptake by plant cells is influenced by factors such as particle size, shape, surface charge, and coating material. These parameters affect the ability of nanoparticles to penetrate the plant cell wall, which is a major barrier due to its complex structure and composition. For example, cellulose nanofibers can hinder the uptake of larger nanoparticles, limiting their effectiveness in delivering genetic material (Deshmukh et al. 2024). Additionally, the intracellular fate of nanoparticles is not fully understood, and their stability within plant cells can vary, leading to inconsistent gene expression levels. There is also a risk of gene silencing or off-target effects, which could affect the desired genetic modifications. Addressing these limitations will require advancements in nanomaterial engineering to optimize nanoparticle properties for specific plant species and delivery contexts (Rodríguez et al. 2022). Moreover, new strategies such as the use of biomimetic nanocarriers and combined delivery systems are being explored to overcome these challenges and enhance the efficiency of gene delivery. Biomimetic nanocarriers, like nanoparticles coated with cell membranes, have been designed to improve gene delivery by imitating natural cell membranes. This approach helps avoid detection by the immune system and enhances biocompatibility. For instance, nanoparticles coated with red blood cell membranes have demonstrated potential in delivering CRISPR/Cas9 components for in vivo genome editing (Hu et al. 2012). Similarly, combined delivery systems, such as lipid-polymer hybrid nanoparticles, effectively co-deliver therapeutic genes and siRNA. By leveraging the stability of polymers alongside the transfection efficiency of lipids, these systems can achieve gene expression and silencing simultaneously, enhancing overall therapeutic results (Shi et al. 2014).

5 Emerging Technologies in Nanogene Delivery Systems

To date, the development of nanotechnology-based delivery systems for transferring DNA into plant cells remains limited, with each approach encountering notable challenges. However, recent advancements in nanomaterial-mediated DNA delivery have opened new possibilities for the targeted introduction of biomolecules into plant cells. The unique physicochemical properties of nanomaterials make them highly suitable for uptake by plant cells. For instance, nanoparticles, which are widely used as pesticides and fertilizers, have been shown through transmission electron microscopy to successfully penetrate plant cells. Despite these observations, the precise mechanisms governing their entry remain only partially understood. One proposed mechanism is the lipid exchange envelope penetration (LEEP) model, which explains how nanotubes traverse the double lipid layer of plant cells. Traditionally, the primary pathways for nanomaterial entry into plant cells include cuticular and stomatal uptake through foliar surfaces. Additionally, the unique micro- and nanostructures present on the leaf surfaces of different plant species significantly influence nanoparticle uptake. Particle size has emerged as a critical determinant for the entry of nanomaterials into plant cells. Yong et al. (2021) hypothesized that nanoparticles approximately 50 nm in size can freely penetrate plant pollen cells. Other studies have suggested that nanoparticles must have at least one dimension under 20 nm to facilitate unrestricted entry into plant cells. Beyond particle size, other characteristics such as shape and zeta potential also play crucial

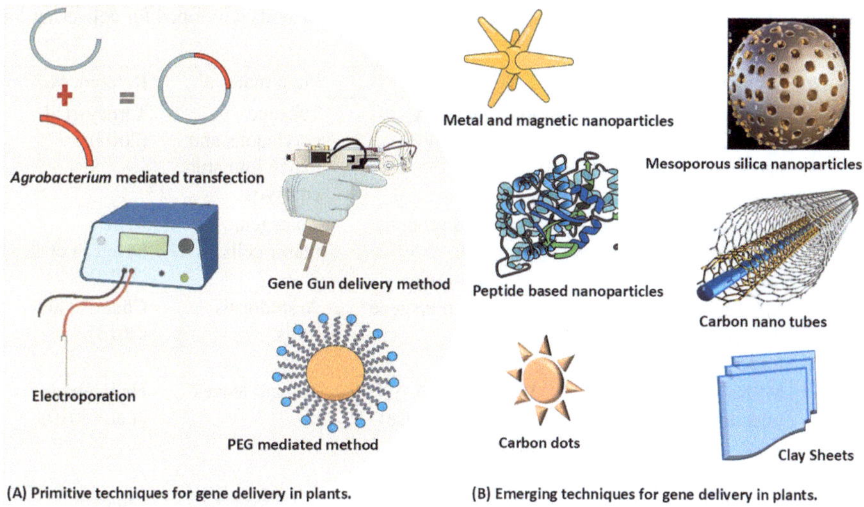

Fig. 2 Various techniques used to deliver genetic material into the plants

roles in nanoparticle uptake. Nano-delivery systems demonstrate improved transfer efficiency when the net zeta potential exceeds 30 mV, as low zeta potential values lead to particle aggregation, while higher values enhance system stability (Wang et al. 2016; Keasberry et al. 2017; Jat et al. 2020).

Nanomaterials have already been employed successfully to mediate the expression of foreign genes and to facilitate CRISPR/Cas-based genome editing in human cells for therapeutic purposes. However, in plants, the presence of the cell wall represents a significant barrier that limits the entry of foreign materials into the cytoplasm and organelles. Despite this challenge, several nanomaterials have emerged as promising vectors for gene delivery into plant cells. Examples include mesoporous silica nanoparticles (MSNs), liposomes (Das et al. 2015c), modified magnetic metal particles, layered double hydroxides, and carbon nanotubes (Fig. 2).

A common feature of these nanomaterials is that they typically possess at least one dimension under 100 nm, a critical factor for achieving efficient intracellular delivery. The use of nanoparticles for cargo delivery into plant cells is a rapidly advancing field with immense potential to transform plant genetic engineering and genome editing. This chapter provides an overview of various nanomaterials currently being investigated for gene delivery in plant systems, as well as the progress achieved in this exciting domain (Table 1).

5.1 Mesoporous Silica Nanoparticles (MSNs)

Mesoporous silica nanoparticles (MSNs) have emerged as versatile nanocarriers due to their unique solid framework, extensive surface area, thermal stability, and ability to undergo surface functionalization with various groups. Their inert and

Table 1 Describing the advanced genetic material delivery methods developed for delivering the genes to various locations in plants

S. No.	Technique used	Cargo	Plant material	References
1	Functionalized mesoporous silica nanoparticles (MSNs) with gold nanoparticles	Plasmid DNA encoding GFP	Tobacco cotyledons and maize immature embryos	Torney et al. (2007)
2	Functionalized au-MSNs with large pore diameters (~10 nm)	DNA and proteins (e.g., Cre recombinase)	Transgenic maize cells	Martin-Ortigosa et al. (2014)
3	MSN-mediated DNA delivery without external force	Reporter transgene	Arabidopsis roots	Chang et al. (2013)
4	MSNs by spraying and injection	Plasmid DNA (GUS (β-glucuronidase) gene and cry1Ab gene)	Tomato leaves	Hajiahmadi et al. (2019)
1	Oxidized single-walled carbon nanotubes (SWCNTs)	DNA and fluorescein isothiocyanate (FITC)	Tobacco BY2 cells	Liu et al. (2019)
2	CNT-cellulase	Plasmid DNA (GFP gene)	Arabidopsis cells	Fouad et al. (2008)
3	SWCNTs and MWCNTs	DNA encoding YFP (Yellow fluorescent protein) reporter gene	Diverse plant tissues and protoplasts	Burlaka et al. (2015)
4	Chitosan-coated SWCNTs	Plasmid DNA (YFP gene)	Chloroplasts of tobacco, spinach, watercress, and *Eruca sativa*	Kwak et al. (2019)
5	Arginine-functionalized SWCNTs (Arg-SWCNTs)	Plasmid DNA (GFP gene)	Tobacco root cells	Golestanipour et al. (2018)
6	Polyethyleneimine-modified CNTs	ssDNA-FITC	German chamomile cells	Ghaghelestany et al. (2020)
8	Poly-L-lysine (PLL) coated ZnS (zinc sulphide) nanoparticles with ultrasound assistance	Plasmid DNA	Tobacco cells	Yu-qin et al. (2012)
9	Magnetic gold nanoparticles (mGNPs) with magnetic field	Plasmid DNA, fluorescein isothiocyanate (FITC)	Canola protoplasts	Hao et al. (2013)
10	Pollen magnetofection using MNP-DNA complexes	Exogenous DNA	Cotton pollen	Zhao et al. (2017a, b)
11	Layered double hydroxide (LDH) clay nanosheets (BioClay)	dsRNA	Arabidopsis leaves	Mitter et al. (2017)

(continued)

Table 1 (continued)

S. No.	Technique used	Cargo	Plant material	References
12	Fluorescence-tagged starch nanoparticles with ultrasound	DNA	*Dioscorea* sp. suspension cells	Liu et al. (2008)
13	Calcium phosphate nanoparticles	DNA	Various plant cells	Naqvi et al. (2012)
14	Silicon carbide whiskers (SiC)	DNA (e.g., chitinase gene)	Peanut and wheat embryos	Wang et al. (2011)

biocompatible nature, facilitated by the presence of silanol groups, enables facile functionalization for enhanced stability and suitability in biomedical applications. MSNs are widely recognized for their efficacy in intracellular delivery of nucleic acids in animal cells and tissues and have been employed to transport therapeutic agents for disease treatment. While their use in plant systems is relatively nascent, research is advancing rapidly in this domain (Martin-Ortigosa et al. 2014).

One limitation in plant gene delivery is the small pore size of MSNs, which may hinder the accommodation of large DNA fragments. Polymer coatings such as polyethyleneimine (PEI), poly (allylamine hydrochloride) (PAH), and polyamidoamine dendrimers are commonly applied to modify MSN surfaces, imparting positive charges that facilitate biomolecule loading and delivery. Additionally, the morphology and size of nanoparticles play a critical role in overcoming plant cell wall barriers. A recent study by Zhang et al. compared spherical (5–20 nm) and rod-shaped (13 × 68 nm) gold nanoparticles functionalized with thiol-modified single-stranded DNA or RNA. The study revealed that although spherical nanoparticles could not penetrate the cell wall, they bound to its surface, while the rod-shaped nanoparticles traversed it effectively. Both types successfully delivered siRNA for gene silencing of a GFP construct, with spherical nanoparticles demonstrating higher delivery efficiency despite their limited cell wall penetration. These findings highlight the influence of nanoparticle morphology on plant gene delivery and provide valuable insights for optimizing nanocarrier designs for plant applications (Law et al. 2023).

5.2 Carbon Nanotubes (CNTs)

Carbon nanotubes (CNTs) are cylindrical nanostructures formed by rolling graphene sheets into tubes and are classified as single-walled (SWCNTs) or multi-walled (MWCNTs) based on the number of graphene layers. SWCNTs typically have diameters below 2 nm, while MWCNTs range from 5 to 100 nm, with lengths extending from micrometers to millimeters. Their remarkable mechanical and physical properties—such as high tensile strength, flexibility, lightweight nature, and excellent thermal conductivity—make CNTs valuable for diverse applications, including electronics, biosensors, transistors, and advanced energy storage systems. CNTs also feature a reactive surface that supports functionalization for cargo attachment and a hollow structure capable of encapsulating biomolecules like drugs, peptides, or nucleic acids. This functionalized surface not only enhances the delivery efficiency but also protects the cargo from degradation under adverse environmental conditions. In mammalian systems, CNTs have been widely utilized for delivering DNA, siRNAs, and other therapeutic agents (Cunningham et al. 2018).

Recent research has extended the application of CNTs to plant systems, demonstrating their ability to deliver DNA and biomolecules into plant cells and protoplasts. CNTs can penetrate plant cell membranes and localize to specific organelles such as the nucleus, plastids, and vacuoles. These findings highlight the potential of CNTs as versatile nanocarriers for genetic materials and bioactive compounds in plant biotechnology.

5.3 Use of Metal Nanoparticles (MNPs)

Metal nanoparticles (MNPs), such as gold, silver, platinum, and palladium, possess unique optical absorption and scattering properties driven by localized surface plasmon resonance (SPR). These attributes make them ideal for bioimaging and various biomedical applications. Among these, gold and silver nanoparticles are particularly prominent due to their tunable size, shape, straightforward synthesis, flexible surface modification, bioconjugation potential, and customizable optical characteristics.

MNPs have been extensively explored for the intracellular delivery of biomolecules, including DNA, drugs, and peptides, in animal and mammalian cells. Although their applications in plant sciences are primarily focused on roles as nanofertilizers and micronutrients to enhance crop productivity, their use as carriers for biomolecule delivery in plant cells remains underdeveloped. In many instances, MNPs function as capping agents alongside other nanomaterials or serve as triggers for controlled cargo release. For example, a study by Vijayakumar et al. demonstrated the potential of carbon "bullets" functionalized with gold nanoparticles for efficient gene delivery, employing a gene gun to enhance delivery efficiency (Fig. 3). These findings suggest that MNP-based systems hold significant promise for advancing gene delivery platforms in both biomedical and plant biotechnology.

5.4 Use of Magnetic Nanoparticles (MNPs)

It combines the general attributes of metal nanoparticles with unique magnetic properties, making them highly versatile for various applications. These nanoparticles typically consist of metals (e.g., nickel, cobalt), metal oxides (e.g., iron oxides), or metal alloys (e.g., FeCo, FePt) (Das et al. 2014c). They exhibit biocompatibility and superparamagnetic behavior, displaying magnetism only in the presence of an external magnetic field. Positively charged MNPs have shown substantial potential as autonomous carriers for nucleic acid delivery in both animal and plant cells. Their superparamagnetic properties provide several advantages, including enhanced

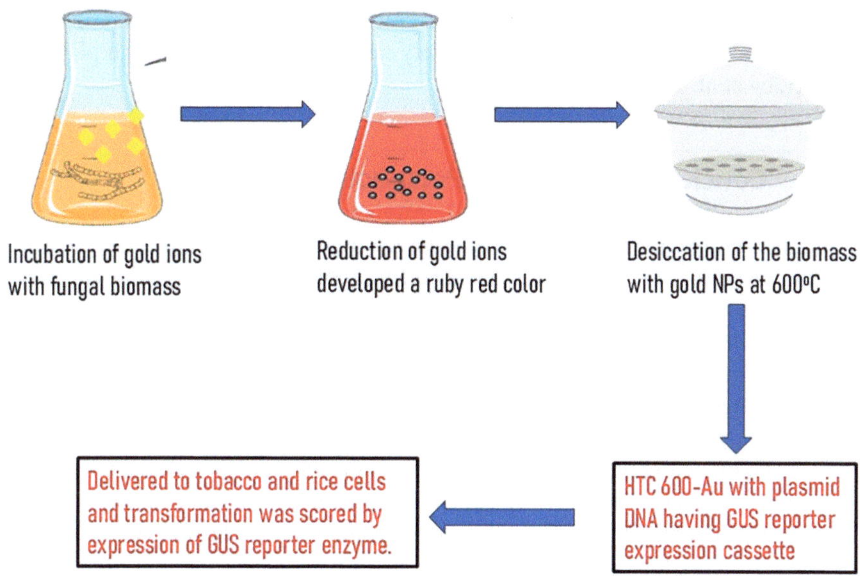

Fig. 3 Synthesis and application of metal nanoparticles

protection of nucleic acids from enzymatic degradation, increased cargo loading capacity, improved cellular penetration, cost-efficiency, and targeted delivery via external magnetic fields. A prominent application of MNPs is in "magnetofection," a technique that employs magnetic forces to direct MNPs loaded with biomolecules into cells. Magnetofection is a simple and efficient method requiring minimal instrumentation, making it a practical and effective approach for gene delivery in various biological systems (Zhi et al. 2022).

5.5 Other Nanomaterials for Gene Delivery

Emerging nanomaterials, such as clay nanosheets and carbon dots (CDs), have shown promise as innovative platforms for gene delivery in plants. Clay nanosheets are sheet-like nanoparticles with diameters of 30–80 nm, composed of clay and inorganic layers. They have demonstrated potential for delivering RNA to induce gene silencing through siRNA pathways.Carbon dots (CDs), on the other hand, are small (~20 nm), carbon-based nanoparticles distinguished by their fluorescence and unique optical properties, making them suitable for imaging and photosensitive applications. Functionalization with polyethyleneimine (PEI) enables CDs to bind double-stranded RNA (dsRNA) and protect it from enzymatic degradation. CDs have successfully penetrated intact leaves of *Nicotiana benthamiana* and tomato plants via adaxial spray application. In one study, CDs carrying siRNA targeting the GFP gene effectively silenced GFP expression in transgenic plants. Furthermore, CDs demonstrated their ability to silence endogenous genes by targeting magnesium chelatase, a key enzyme in chlorophyll synthesis, illustrating their potential as versatile tools for gene delivery and functional genomics in plant systems (Law et al. 2023).

6 Conclusion and Future Perspectives

Nanomaterial (NM)-mediated organelle-targeted gene delivery has undergone significant progress over the last decade, enabling researchers to overcome the previously insurmountable barrier posed by the plant cell wall. This advancement has also alleviated many of the limitations associated with earlier transgene delivery systems. Notable improvements have been achieved in the development of various NMs designed to target-specific organelles within plant cells, all while maintaining high biocompatibility and minimizing entry barriers. However, several critical challenges remain before the widespread commercialization of NMs in agricultural and industrial applications can be realized. One of the primary obstacles is the long-term toxicity of NMs, including concerns over bioaccumulation, as well as the scalability of these systems for large-scale genetic transformation. Furthermore, the precise mechanisms governing NM interactions with plant cells are still not fully understood, which remains a significant bottleneck for the rational design of NMs tailored for specific, targeted delivery. In the context of organelle-specific transformations in plants, the most significant challenge lies in achieving selectivity and overcoming the inherent difficulty of delivery. This is further complicated by the lack of effective selection systems for obtaining transformants and the difficulty in achieving homoplasmic mitochondrial or chloroplast transformants. Strict regulations surrounding genetically modified organisms (GMOs) have spurred the development of plant genetic engineering techniques that enable instantaneous gene expression without integrating foreign DNA into the host plant's genome.

However, technical barriers persist in genetic engineering, which conventional methods—such as gene gun technology—struggle to overcome. These methods often lead to high rates of random insertions, and the bombardment process can cause genomic damage in the host plant, triggering multiple repair mechanisms. Similarly, *Agrobacterium*-mediated transformation, though widely used in plant genetic engineering, can also result in random insertions, making it difficult to obtain transgene-free plants (Lakra et al. 2019). The polyethylene glycol (PEG)-mediated method, which involves harvesting protoplasts from leaf tissue lacking cell walls, allows for the delivery of various foreign gene cargoes into the nucleus or organelles via the lipid bilayer. While this approach offers distinct advantages, such as bypassing the cell wall barrier, it is still a time-consuming process and presents challenges in plant regeneration from protoplasts. In contrast, functional nanomaterials have shown superior capabilities in delivering a range of biomolecules—including DNA, RNA, and ribonucleoproteins (RNPs)—into animal cells. These nanomaterials hold great promise for efficient, transgene-free genetic engineering in plants, particularly in achieving high-efficiency homology-directed repair (HDR). To enhance the effectiveness of these systems, the design of vectors and the incorporation of exogenous genes that facilitate differentiation and regeneration are crucial. Looking to the future, NM delivery systems stand out as a more biocompatible, nontoxic, and cost-effective alternative to traditional genetic engineering methods. Nanovectors, with their remarkable potential, are poised to revolutionize advanced crop breeding techniques.

A transformative, though achievable, milestone for NM-based delivery in plants would be the enablement of large-scale, parallel genetic studies that can uncover key factors influencing tissue regeneration, as well as facilitate the direct manipulation of germline cells in plants. The application of machine learning (ML) and artificial intelligence (AI) tools holds significant promise in advancing the field of nanogene delivery for plant molecular farming. These technologies can accelerate the design and optimization of nanomaterials by predicting and modeling their interactions with plant cells at a molecular level. ML algorithms can analyze large datasets from experimental studies to identify patterns in NM behavior, such as their ability to penetrate cell walls or target-specific organelles. Additionally, AI-powered tools can be employed to design highly efficient nanovectors tailored for specific genetic modifications, minimizing trial-and-error in the experimental process. Moreover, AI techniques such as deep learning could assist in the prediction of toxicity profiles and bioaccumulation risks associated with different nanomaterials, thereby helping to ensure the safety and sustainability of nanogene delivery systems. With the integration of AI and ML, the potential for rational and efficient design of nanomaterial-based delivery systems is greatly enhanced, making these technologies indispensable for the future of plant molecular farming.

Acknowledgments The authors are thankful to Mohanlal Sukhadia University, Udaipur, for providing the necessary facilities during the course of study.

Declaration of Competing Interest The authors declare that there are no conflicts of interest. They agree to be accountable for all aspects of the work and ensure that questions related to the accuracy or integrity of any part of the work are appropriately investigated and resolved..

Funding This research did not receive any specific grant from funding agencies in the public, commercial, or not-for-profit sectors.

References

Achterberg EP (2014) Grand challenges in marine biogeochemistry [editorial material]. Front Mar Sci 1:7

Ahmed F, Dwivedi S, Shaalan NM, Kumar S, Arshi N, Alshoaibi A, Husain FM (2020) Development of selenium nanoparticle-based agriculture sensor for heavy metal toxicity detection. Agriculture 10(5):610

Athanassiou CG, Kavallieratos NG, Meletsis CM (2007) Insecticidal effect of three diatomaceous earth formulations, applied alone or in combination, against three stored-product beetle species on wheat and maize. J Stored Prod Res 43(4):330–334

Babu RP, O'Connor K, Seeram R (2013) Current progress on bio-based polymers and their future trends. Progress in Biomaterials 2(8):8

Baldwin EA, Hagenmaier R, Bai J (2011) Edible coatings and films to improve food quality. CRC Press

Banerjee K, Pramanik P, Maity A, Joshi DC, Wani SH, Krishnan P (2019) Methods of using nanomaterials to plant systems and their delivery to plants (mode of entry, uptake, translocation, accumulation, biotransformation and barriers). In: Ghorbanpour M, Wani SH (eds) Advances in phytonanotechnology, from synthesis to application. Academic Press, pp 123–152

Bolik M, Koop HU (1991) Identification of embryogenic microspores of barley (Hordeum vulgare) by individual selection and culture and their potential for transformation by microinjection. Protoplasma 162:61–68

Braiek M, Rokbani KB, Chrouda A, Mrabet B, Bakhrouf A, Maaref A, Jaffrezic-Renault NJB (2012) An electrochemical immunosensor for detection of Staphylococcus aureus bacteria based on immobilization of antibodies on self-assembled monolayers-functionalized gold electrode. Biosensors 2(2):417–426

Burlaka OM, Pirko YV, Yemets AI, Blume YB (2015) Cytol Genet 49:349–357

Chandra S, Chakraborty N, Dasgupta A, Sarkar J, Panda K, Acharya K (2015) Chitosan nanoparticles: a positive modulator of innate immune responses in plants. Sci Rep 5(1):1–14

Chang FP, Kuang LY, Huang CA, Jane WN, Hung Y, Hsing Y-IC, Mou CY (2013) J Mater Chem B 1

Chauhan N, Urooj S (2019) The effect of *Azadirachta indica* extract on the soil nutrients and the NPK value determination by electrochemical sensor. International journal of. Biosens Bioelectron 5

Choudhary RC, Kumaraswamy RV, Kumari S, Sharma SS, Pal A, Raliya R, Rai MK (2019) Zinc encapsulated chitosan nanoparticle to promote maize crop yield. Int J Biol Macromol 127:126–135

Chuah JA, Horii Y, Numata K (2016) Peptide-derived method to transport genes and proteins across cellular and organellar barriers in plants. J Vis Exp 113:e54972

Chugh A, Eudes F, Shim YS (2010) Cell-penetrating peptides: Nanocarrier for macromolecule delivery in living cells. IUBMB Life 62(3):183–193

Chun SC, Chandrasekaran M (2019) Chitosan and chitosan nanoparticles induced expression of pathogenesis-related proteins genes enhances biotic stress tolerance in tomato. Int J Biol Macromol 125:948–954

Cota-Arriola O, Onofre Cortez-Rocha M, Burgos-Hernández A, Marina Ezquerra-Brauer J, Plascencia-Jatomea M (2013) Controlled release matrices and micro/nanoparticles of chitosan with antimicrobial potential: development of new strategies for microbial control in agriculture. J Sci Food Agric 93(7):1525–1536

Cunningham FJ, Goh NS, Demirer GS, Matos JL, Landry MP (2018) Nanoparticle-mediated delivery towards advancing plant genetic engineering. Trends Biotechnol 36(9):882–897

da Cunha NB, Vianna GR, da Almeida Lima T, Rech EL (2014) Molecular farming of human cytokines and blood products from plants: challenges in biosynthesis and detection of plant produced recombinant proteins. Biotechnol J 9(1):39–50

Dar FA, Qazi GR, Pirzadah TB (2020) Nano-biosensors: NextGen diagnostic tools in agriculture. In: Nanobiotechnology in agriculture: An approach towards sustainability. Springer, Singapore, pp 129–144

Das S, Marwal A, Choudhary DK, Gupta VK, Gaur RK (2011) Mechanism of RNA interference (RNAi): current concept. Int Proc Chem Biol Environ Eng 9:240–245

Das AK, Marwal A, Sain D (2014a) One-step green synthesis and characterization of flower extract-mediated mercuric oxide (HgO) nanoparticles from Callistemon viminalis. Res Rev 2(2):25–28

Das AK, Marwal A, Verma R (2014b) Bio-reductive synthesis and characterization of plant protein coated magnetite nanoparticles. Nano Hybrids 7:69–86

Das AK, Marwal A, Verma R (2014c) *Datura inoxia* leaf extract mediated one step green synthesis and characterization of magnetite (Fe3O4) nanoparticles. Res Rev 2(2):21–24

Das AK, Marwal A, Verma R (2015a) Preparation and characterization of nano-bio hybrid based magneto liposome. Int J Pharm Sci Res 6(1):367

Das AK, Marwal A, Sain D, Pareek V (2015b) One-step green synthesis and characterization of plant protein-coated mercuric oxide (HgO) nanoparticles: antimicrobial studies. Int Nano Lett 5:125–132

Das AK, Marwal A, Pareek V (2015c) Nanoparticles-protein hybrid based magnetic liposome. IJCMMME 9(2):230–233
Deng XY, Wei ZM, An HL (2001) Transgenic peanut plants obtained by particle bombardment via somatic embryogenesis regeneration system. Cell Res 11(2):156–160
Deshmukh R, Sethi P, Singh B, Shiekmydeen J, Salave S, Patel RJ, Kumar A (2024) Recent review on biological barriers and host–material interfaces in precision drug delivery: advancement in biomaterial engineering for better treatment therapies. Pharmaceutics 16(8):1076
Divya K, Vijayan S, Nair SJ, Jisha MS (2019) Optimization of chitosan nanoparticle synthesis and its potential application as germination elicitor of Oryza sativa L. Int J Biol Macromol 124:1053–1059
Draga S, Gabelli G, Palumbo F, Barcaccia G (2023) Genome-wide datasets of chicories (Cichorium intybus L.) for marker-assisted crop breeding applications: a systematic review and meta-analysis. Int J Mol Sci 24:663
Dynkevich Y, Rother KI, Whitford I, Qureshi S, Galiveeti S, Szulc AL, Danoff A, Breen TL, Kaviani N, Shanik MH (2013) Tumors, IGF-2, and hypoglycemia: insights from the clinic, the laboratory, and the historical archive. Endocr Rev 34:798–826
Ernawati L, Ogi T, Balgis R, Okuyama K, Stucki M, Hess SC, Stark WJ (2016) Hollow silica as an optically transparent and thermally insulating polymer additive. Langmuir 32(1):338–345
Fischer J, Beckers SJ, Yiamsawas D, Thines E, Landfester K, Wurm FR (2019) Targeted drug delivery in plants: enzyme-responsive lignin nanocarriers for the curative treatment of the worldwide grapevine trunk disease esca. Advanced Science 6:1802315
Ghaghelestany AB, Jahanbakhshi A, Taghinezhad E (2020) Sci Hortic 263:109106
Giraldo JP, Wu H, Newkirk GM, Kruss S (2019) Nanobiotechnology approaches for engineering smart plant sensors. Nat Nanotechnol 14(6):541–553
Gleba Y, Klimyuk V, Marillonnet S (2007) Viral vectors for the expression of proteins in plants. Curr Opin Biotechnol 18(2):134–141
Golestanipour M, Nikkhah A, Aalami S, Hosseinkhani S (2018) Mol Biotechnol 60:863–878
Grillo R, Pereira AES, Nishisaka CS, De Lima R, Oehlke K, Greiner R et al (2014) Chitosan/tripolyphosphate nanoparticles loaded with paraquat herbicide: An environmentally safer alternative for weed control. J Hazard Mater 278:163–171
Grillo R, Clemente Z, De Oliveira JL, Campos EVR, Chalupe VC, Jonsson CM et al (2015) Chitosan nanoparticles loaded with the herbicide paraquat: the influence of aquatic humic substances on colloidal stability and toxicity. J Hazard Mater 286:562–572
Hadwiger LA (2013) Plant science review: multiple effects of chitosan on plant systems: solid science or hype. Plant Sci 208:42–49
Hajiahmadi Z, Shirzadian-Khorramabad R, Kazemzad M, Sohani MM (2019) Sci Hortic 243:367–375
Hao Y, Yang X, Shi Y, Song S, Xing J, Marowitch J, Chen J, Chen J (2013) Botany 91:457–466
Hoffmann F, Cornelius M, Morell J, Froba M (2006) Cover picture: silica-based mesoporous organic–inorganic hybrid materials (Angew. Chem. Int. Ed. 20/2006). Angewandte Chemie International Edition, 45(20), 3187
Horn M, Woodard S, Howard J (2004) Plant molecular farming: systems and products. Plant Cell Rep 22:711–720
Hu CMJ, Fang RH, Zhang L (2012) Erythrocyte-inspired delivery systems. Adv Healthc Mater 1(5):537–547
Hu W, Wan L, Jian Y, Ren C, Jin K, Su X, Bai X, Haick H, Yao M, Wu W (2019) Electronic noses: from advanced materials to sensors aided with data processing. Adv Mater Technol 4:1800488
Hu X, Lu L, Guo Z, Zhu Z (2020) Volatile compounds, affecting factors and evaluation methods for rice aroma: a review. Trends Food Sci Technol 97:136–146
Iqbal B, Li G, Alabbosh KF, Hussain H, Khan I, Tariq M, Javed Q, Naeem M, Ahmad N (2023) Advancing environmental sustainability through microbial reprogramming in growth improvement, stress alleviation, and phytoremediation. Plant Stress 10:100283
Ishtiaq M, Mazhar M, Maqbool M, Muzamil M (2023) Genetic engineering of horticultural crops. In: Genetic engineering: Volume 2: Applications, Bioethics, and Biosafety

Iversen TG, Skotland T, Sandvig K (2011) Endocytosis and intracellular transport of nanoparticles: present knowledge and need for future studies. Nano Today 6(2):176–185

Jat SK, Bhattacharya J, Sharma MK (2020) Nanomaterial-based gene delivery: a promising method for plant genome engineering. J Mater Chem B 8(19):4165–4175

Kanwar MK, Sun S, Chu X, Zhou J (2019) Impacts of metal and metal oxide nanoparticles on plant growth and productivity. In: Husen A, Iqbal M (eds) Nanomaterials and plant potential. Springer, Cham, pp 1–13

Kashyap PL, Xiang X, Heiden P (2015) Chitosan nanoparticle-based delivery systems for sustainable agriculture. Int J Biol Macromol 77:36–51

Keasberry NA, Yapp CW, Idris A (2017) Biochem Mosc 82(5):655–666

Klein TM, Kornstein L, Stanford JC, Fromm ME (1989) Genetic transformation of maize cells by particle bombardment. Plant Physiol 91(1):440–444

Knoch S, Pelletier F, Larose M, Chouinard G, Dumont MJ, Tavares JR (2020) Surface modification of PLA nets intended for agricultural applications. Colloids Surf A Physicochem Eng Asp 598:124787

Koeppe S, Kawchuk L, Kalischuk M (2023) RNA interference past and future applications in plants. Int J Mol Sci 24(11):9755

Kremer RJ (2019) Bioherbicides and nanotechnology: current status and future trends. Elsevier Inc.

Kumar S, Bhanjana G, Sharma A, Sidhu MC, Dilbaghi N (2014) Synthesis, characterization, and on-field evaluation of pesticide-loaded sodium alginate nanoparticles. Carbohydr Polym 101:1061–1067

Kumaraswamy RV, Kumari S, Choudhary RC, Pal A, Raliya R, Biswas P et al (2018) Engineered chitosan-based nanomaterials: bioactivities, mechanisms, and perspectives in plant protection and growth. Int J Biol Macromol 113:494–506

Kwak S-Y, Lew TTS, Sweeney CJ, Koman VB, Wong MH, Bohmert-Tatarev K, Snell KD, Seo JS, Chua N-H, Strano MS (2019) Nat Nanotechnol 14:447–455

Lakra N, Mahmood S, Marwal A, Sudheep NM, Anwar K (2019) Bioengineered plants can be an alternative source of omega-3 fatty acids for human health. Plant Human Health 2:361–382. Phytochemistry and Molecular Aspects

Law SSY, Miyamoto T, Numata K (2023) Organelle-targeted gene delivery in plants by nanomaterials. Chem Commun 59(47):7166–7181

Liu XM, Feng ZB, Zhang FD, Zhang SQ, He XS (2006) Preparation and testing of cementing and coating nano-subnanocomposites of slow/controlled-release fertilizer. Agric Sci China 5(9):700–706

Liu J, Wang FH, Wang LL, Xiao SY, Tong CY, Tang DY, Liu XM (2008) Preparation of fluorescence starch-nanoparticle and its application as a plant transgenic vehicle. J Cent S Univ Technol 15:768–773

Liu Q, Chen B, Wang Q, Shi X, Xiao Z, Lin J, Fang X (2019) Nano Lett 9:1007–1010

Low JW, Mwanga RO, Andrade M, Carey E, Ball AM (2017) Tackling vitamin, a deficiency with biofortified sweet potato in sub-Saharan Africa. Glob Food Sec 14:23–30

Lü S, Feng C, Gao C, Wang X, Xu X, Bai X et al (2016) Multifunctional environmental smart fertilizer based on L-aspartic acid for sustained nutrient release. J Agric Food Chem 64:4965–4974

Maged Fouad N K, Jabasini M, Tokeshi M, Baba Y (2008). Twelfth International Conference on Miniaturized Systems for Chemistry and Life Sciences October 12–16, 2008, San Diego

Maheshwari Y, Tailor S, Marwal A, Mishra A (2023) Plant-microbe interactions to reduce salinity stress in plants for the improvement of the agricultural system. In: Plant-microbe interaction-recent advances in molecular and biochemical approaches. Academic Press, pp 297–309

Mahmood S, Lakra N, Marwal A, Sudheep NÁ, Anwar K (2017) Crop genetic engineering: An approach to improve fungal resistance in plant system. Plant-microbe interactions in agro-ecological perspectives. volume 2: Microbial Interactions and Agro-Ecological Impacts, pp 581–591

Mahmoud AED, Fawzy M (2021) Nanosensors and nanobiosensors for monitoring environmental pollutants. In: Waste Recycling Technologies and Nanomaterials Manufacturing, pp 229–246

Manzano M, Vallet-Regí M (2020) Mesoporous silica nanoparticles for drug delivery. Adv Funct Mater 30(2):1902634

Martin-Ortigosa S, Peterson DJ, Valenstein JS, Lin VSY, Trewyn BG, Lyznik LA, Wang K (2014) Plant Physiol 164:537–547

Marwal A, Gaur RK (2019) Nanophytovirology: an emerging field for disease management. Plant Diseases-Current Threats and Management Trends. https://doi.org/10.5772/intechopen.86653

Marwal A, Gaur RK (2020) Molecular markers: tool for genetic analysis. In: Animal biotechnology. Academic Press, pp 353–372

Marwal A, Gaur RK (2021) Plant viruses as an engineered nanovehicle (PVENVs). In: Plant Virus-Host Interaction, pp 525–536

Massawe F, Mayes S, Cheng A (2016) Crop diversity: An unexploited treasure trove for food security. Trends Plant Sci 21(5):365–368

McKnight TE, Melechko AV, Griffin GD, Guillorn MA, Merkulov VI, Serna F, Hensley DK, Doktycz MJ, Lowndes DH, Simpson ML (2003) Intracellular integration of synthetic nanostructures with viable cells for controlled biochemical manipulation. Nanotechnology 14(5):551–556

McKnight TE, Melechko AV, Hensley DK, Mann DGJ, Griffin GD, Simpson ML (2004) Tracking gene expression after DNA delivery using spatially indexed nanofiber arrays. Nano Lett 4(7):1213–1219

Mewis I, Ulrichs C (2001) Action of amorphous diatomaceous earth against different stages of the stored product pests *Tribolium confusum, Tenebrio molitor, Sitophilus granarius,* and *Plodia interpunctella*. J Stored Prod Res 37(2):153–164

Mishra A, Marwal A, Tailor S, Jain K, Malik A, Suthar M, Meena M (2024) Overview of cell signaling response under plant stress. In: Molecular dynamics of plant stress and its management. Springer Nature, Singapore, pp 121–158

Mitter N, Worrall EA, Robinson KE, Li P, Jain RG, Taochy C, Fletcher SJ, Carroll BJ, Lu GQ, Xu ZP (2017) Nat Plants 3:16207

Mohamed MA, Abd-Elsalam KA (2019) Magnetic nanoparticles: a unique gene delivery system in plant science. In: Abd-Elsalam K, Mohamed M, Prasad R (eds) Magnetic nanostructures: nanotechnology in the life sciences. Springer, Cham, pp 1–12

Mohammadinejad R, Shavandi A, Raie DS, Sangeetha J, Soleimani M, Hajibehzad SS, Varma RS (2019) Plant molecular farming: production of metallic nanoparticles and therapeutic proteins using green factories. Green Chem 21(8):1845–1865

Munawar A, Ong Y, Schirhagl R, Tahir MA, Khan WS, Bajwa SZ (2019) Nanosensors for diagnosis with optical, electric, and mechanical transducers. RSC Adv 9:6793–6803

Nair R, Varghese SH, Nair BG, Maekawa T, Yoshida Y, Kumar DS (2010) Nanoparticulate material delivery to plants. Plant Sci 179(3):154–163

Naqvi S, Maitra AN, Abdin MZ, Akmal M, Arora I, Samim M (2012) J Mater Chem 22:3500–3507

Neuhaus G, Spangerberg G (1990) Plant transformation by microinjection techniques. Physiol Plant 79:213–217

Panáček A, Kolář M, Večeřová R, Prucek R, Soukupová J, Kryštof V, Kvítek L (2009) Antifungal activity of silver nanoparticles against Candida spp. Biomaterials 30(31):6333–6340

Panpatte DG, Jhala YK, Shelat HN, Vyas RV (2016) Nanoparticles—the next generation technology for sustainable agriculture. In: Singh DP, Singh HB, Prabha R (eds) Microbial inoculants in sustainable agricultural productivity volume 2: functional applications. Springer, New Delhi, pp 289–300

Patel S, Bajpai J, Saini R, Bajpai AK, Acharya S (2018) Sustained release of pesticide (Cypermethrin) from nanocarriers: An effective technique for environmental and crop protection. Process Saf Environ Prot 117:315–325

Pereira AES, Sandoval-Herrera IE, Zavala-Betancourt SA, Oliveira HC, Ledezma-Pérez AS, Romero J et al (2017) γ-Polyglutamic acid/chitosan nanoparticles for the plant growth regulator gibberellic acid: characterization and evaluation of biological activity. Carbohydr Polym 157:1862–1873

Philibert T, Lee BH, Fabien N (2017) Current status and new perspectives on chitin and chitosan as functional biopolymers. Appl Biochem Biotechnol 181:1314–1337

Pollard TD, Ong JJ, Goyanes A, Orlu M, Gaisford S, Elbadawi M, Basit AW (2021) Electrochemical biosensors: a nexus for precision medicine. Drug Discov Today 26:69–79

Pudake RN, Jain U, Kole C (2021) Biosensors in agriculture: recent trends and future perspectives. Springer

Rasheed T, Nabeel F, Adeel M, Rizwan K, Bilal M, Iqbal HM (2019) Carbon nanotubes-based cues: a pathway to future sensing and detection of hazardous pollutants. J Mol Liq 292:111425

Rico CM, Majumdar S, Duarte-Gardea M, Peralta-Videa JR, Gardea-Torresdey JL (2011) Interaction of nanoparticles with edible plants and their possible implications in the food chain. J Agric Food Chem 59(8):3485–3498

Rinaudo M (2006) Chitin and chitosan: properties and applications. Prog Polym Sci 31:603–632

Rodríguez F, Caruana P, De la Fuente N, Español P, Gámez M, Balart J, Céspedes MV (2022) Nano-based approved pharmaceuticals for cancer treatment: present and future challenges. Biomolecules 12(6):784

Ryu H, Thompson D, Huang Y, Li B, Lei Y J S (2020). Electrochemical sensors for nitrogen species: a review. Sensors and Actuators Reports, 2, 100022

Saharan V, Sharma G, Yadav M, Choudhary MK, Sharma SS, Pal A et al (2015) Synthesis and in vitro antifungal efficacy of Cu-chitosan nanoparticles against pathogenic fungi of tomato. Int J Biol Macromol 75:346–353

Saharan V, Kumaraswamy RV, Choudhary RC, Kumari S, Pal A, Raliya R et al (2016) Cu-chitosan nanoparticle mediated sustainable approach to enhance seedling growth in maize by mobilizing reserved food. J Agric Food Chem 64:6148–6155

Sahu AK, Marwal A, Nehra C, Choudhary DK, Sharma P, Gaur RK (2014) RNAi mediated gene silencing against betasatellite associated with croton yellow vein mosaic begomovirus. Mol Biol Rep 41:7631–7638

Sandhu KK, McIntosh CM, Simard JM, Smith SW, Rotello VM (2002) Gold nanoparticle-mediated transfection of mammalian cells. Bioconjug Chem 13:3–6

Sarauskis E, Kazlauskas M, Naujokiene V, Bručienė I, Steponavičius D, Romaneckas K, Jasinskas A (2022) Variable rate seeding in precision agriculture: recent advances and future perspectives. Agriculture (Basel) 12:305

Sathiyabama M, Manikandan A (2016) Chitosan nanoparticle induced defense responses in fingermillet plants against blast disease caused by *Pyricularia grisea* (Cke.) Sacc. Carbohydr Polym 154:241–246

Segura T, Shea LD (2001) Materials for non-viral gene delivery. Annu Rev Mater Res 31(1):25–46

Sekhon BS (2014) Nanotechnology in Agri-food production: an overview. Nanotechnol Sci Appl 7:31

Shafi U, Mumtaz R, García-Nieto J, Hassan SA, Zaidi SAR, Iqbal N (2019) Precision agriculture techniques and practices: from considerations to applications. Sensors 19:3796

Sharma N, Singh D, Rani R, Sharma D, Pandey H, Agarwal V (2018) Chitosan and its nanocarriers: applications and opportunities. In: Kumar D, Tripathi A, Ahmad P, Sharma S, Chauhan D, Dubey NK (eds) Nanomaterials in plants, algae and microorganisms, 1st edn. Academic Press, pp 267–286

Shi J, Xu Y, Xu X, Zhu X, Pridgen E, Wu J, Farokhzad OC (2014) Hybrid lipid–polymer nanoparticles for sustained siRNA delivery and gene silencing. Nanomedicine 10(5):e897–e900

Singh P, Pandey SK, Singh J, Srivastava S, Sachan S, Singh SK (2016) Biomedical perspective of electrochemical nanobiosensor. Nano-Micro Letters 8(3):193–203

Singh S, Sangwan S, Sharma P, Devi P, Moond M (2021a) Nanotechnology for sustainable agriculture: an emerging perspective. J Nanosci Nanotechnol 21(6):3453–3465

Singh NA, Rai N, Marwal A (2021b) Nanosensors for the detection of chemical food adulterants. In: Nanotoxicology and Nanoecotoxicology, vol 2. Springer International Publishing, Cham, pp 25–53

Solgi M, Kafi M, Taghavi TS, Naderi R (2009) Essential oils and silver nanoparticles (SNP) as novel agents to extend vase-life of gerbera (*Gerbera jamesonii* cv. 'Dune') flowers. Postharvest Biol Technol 53(3):155–158

Srivastava A K, Dev A, Karmakar S (2018). Nanosensors and nanobiosensors in food and agriculture

Sun C, Yu M, Zeng Z, Francis F, Cui H, Verheggen F (2020) Biocidal activity of polylactic acid-based nano-formulated abamectin on *Acyrthosiphon pisum* (Hemiptera: Aphididae) and the aphid predator *Adalia bipunctata* (Coleoptera: Coccinellidae). PLoS One 15(1):e0228817

Tailor S, Marwal A, Meena M (2021) Application of nanotechnology in management of various plant diseases. In: Innovative approaches in diagnosis and management of crop diseases. Apple Academic Press, pp 1–17

Tailor S, Jain K, Marwal A, Meena M, Anbarasu K, Gaur RK (2022) Outlooks of nanotechnology in organic farming management. Def Life Sci J 7(1):52–60

Tailor S, Jain K, Malik A, Suthar M, Mishra A, Gaur R et al (2024) New insight of nanotechnology in combating plant stresses: scope and potential applications. In: Molecular dynamics of plant stress and its management. Springer Nature, Singapore, pp 475–490

Tan J, Xu J (2020) Applications of electronic nose (e-nose) and electronic tongue (e-tongue) in food quality-related properties determination: a review. Artif Intell Agr 4(1):104–115

Tomar R S, Jyoti A, Kaushik S (2020). Nanobiotechnology: concepts and applications in health, agriculture, and environment

Tong Y, Wu Y, Zhao C, Xu Y, Lu J, Xiang S (2017) Polymeric nanoparticles as a Metolachlor carrier: water-based formulation for hydrophobic pesticides and absorption by plants. J Agric Food Chem 65(35):7371–7378

Torney F, Trewyn BG, Lin VSY, Wang K (2007) Mesoporous silica nanoparticles deliver DNA and chemicals into plants. Nat Nanotechnol 2(5):295–300

Umapathi R, Ghoreishian SM, Sonwal S, Rani GM, Huh YS (2022) Portable electrochemical sensing methodologies for on-site detection of pesticide residues in fruits and vegetables. Coord Chem Rev 453:214305

Usman M, Farooq M, Wakeel A, Nawaz A, Cheema SA, ur Rehman H et al (2020) Nanotechnology in agriculture: current status, challenges and future opportunities. Sci Total Environ 721:137778

Verma RK, Mishra M, Marwal A, Akhtar A, Manchev N, Petrov RK CRISPR/Cas9 protection system for chilli leaf curl virus. Acta Microbiologica Bulgarica 40(1):112–115

Walsh G, Jefferis R (2006) Post-translational modifications in the context of therapeutic proteins. Nat Biotechnol 24(10):1241–1252

Wang Q, Chen J, Zhang H, Lu M, Qiu D, Wen Y, Kong Q (2011) Synthesis of water-soluble quantum dots for monitoring carrier-DNA nanoparticles in plant cells. J Nanosci Nanotechnol 11(3):2208–2214

Wang P, Lombi E, Zhao FJ, Kopittke PM (2016) Nanotechnology: a new opportunity in plant sciences. Trends Plant Sci 21(8):699–712

Wang G, Xiao Y, Xu H, Hu P, Liang W, Xie L et al (2018) Development of multifunctional avermectin poly(succinimide) nanoparticles to improve bioactivity and transportation in rice. J Agric Food Chem

Wu SH, Mou CY, Lin HP (2013) Synthesis of mesoporous silica nanoparticles. Chem Soc Rev 42(7):3862–3875

Wu H, Nißler R, Morris V, Herrmann N, Hu P, Jeon SJ, Kruss S, Giraldo JP (2020) Monitoring plant health with near-infrared fluorescent H2O2 nanosensors. Nano Lett 20(4):2432–2442

Wu H, Hu P, Xu Y, Xiao C, Chen Z, Liu X et al (2021) Phloem delivery of fludioxonil by plant amino acid transporter-mediated polysuccinimide nanocarriers for controlling Fusarium wilt in banana. J Agric Food Chem

Wurm FR, Weiss CK (2014) Nanoparticles from renewable polymers. Front Chem 2:1–13

Yang G, Zhao FJB (2015) Electrochemical sensor for chloramphenicol based on novel multiwalled carbon nanotubes@ molecularly imprinted polymer. Biosens Bioelectron 64:416–422

Yin Y, Yang M, Xi J, Cai W, Yi Y, He G, Zhao F (2020) A sodium alginate-based nanopesticide delivery system for enhanced in vitro photostability and insecticidal efficacy of phloxine B. Carbohydr Polym 116677

Yong J, Zhang R, Bi S, Li P, Sun L, Neena M, Carroll BJ, Xu ZP (2021) Sheet-like clay nanoparticles deliver RNA into developing pollen to efficiently silence a target gene. Plant Physiol 2:886–899

Yu M, Yao J, Liang J, Zeng Z, Cui B, Zhao X et al (2017) Development of functionalized abamectin poly(lactic acid) nanoparticles with regulatable adhesion to enhance foliar retention. RSC Adv 7(9):11271–11280

Yu-qin F, Lu-hua L, Pi-wu W, Jing Q, Yong-ping F, Hui W, Jing-ran S, Chang-li L (2012) Chem Res Chinese Universities 28:672–678

Yuvaraj M, Subramanian KS (2018) Development of slow-release Zn fertilizer using nano-zeolite as carrier. J Plant Nutr 41(3):311–320

Zain M, Ma H, Chaudhary S, Nuruzzaman M, Azeem I, Mehmood F et al (2023) Nanotechnology in precision agriculture: advancing towards sustainable crop production. Plant Physiol Biochem 108244

Zhao L, Li N, Wang K, Shi C, Zhang L, Luan Y (2014) A review of polypeptide-based polymersomes. Biomaterials 35(4):1284–1301

Zhao P, Liu MC, Lin HC, Sun XY, Li YY, Yan SQ (2017a) Synthesis and drug delivery applications for mesoporous silica nanoparticles. J Med Biotechnol 1(3):1–8

Zhao X, Meng Z, Wang Y, Chen W, Sun C, Cui B et al (2017b) Nat Plants 3:956–964

Zhi H, Zhou S, Pan W, Shang Y, Zeng Z, Zhang H (2022) The promising nanovectors for gene delivery in plant genome engineering. Int J Mol Sci 23:8501

Advancements in Artificial Intelligence and Machine Learning to Optimize Plant Cell and Tissue Culture

Khushbu Kumari, Suman Gusain, Jhilmil Nath, Anita Kumari, Shubham Joshi, Meenakshi Rawat, and Rohit Joshi

Abstract

Micropropagation techniques provide an opportunity to induce in vitro organogenesis and proliferation, enabling the maintenance and comprehension of the molecular mechanisms of plant responses in precisely controlled conditions, free from external factors. This forms the foundation for numerous applications in plant biotechnology and breeding new crop varieties. Unfortunately, current strategies to optimize in vitro organogenesis are cumbersome and time-consuming because of the enormous variety of parameters that influence plant cell and tissue culture. Therefore, there is a significant potential to employ novel computational approaches for data analysis in order to produce more efficient protocols. Artificial intelligence (AI) and machine learning (ML) are widely used in many scientific and technological domains, and they are currently employed to enhance various stages of micropropagation. ML technology stands out as an innovative approach to characterize the combined role of different plant growth regulators, classify somatic embryos, predict and optimize the length and number of in vitro raised shoots or roots, and optimize environmental conditions to maximize productivity. Despite its potential, due to complex computational algorithms, the implementation of AI and ML in this field has yet to achieve its maximum potential. Therefore, this chapter concentrates on the current advancements in plant tissue culture using AI–ML. A number of modelling principles, as well as different algorithms, have been discussed and highlighted. This chapter also offers an innovative viewpoint for future study objectives.

K. Kumari · S. Gusain · J. Nath · A. Kumari · S. Joshi · M. Rawat · R. Joshi (✉)
Division of Biotechnology, CSIR-Institute of Himalayan Bioresource Technology, Palampur, Himachal Pradesh, India

Academy of Scientific and Innovative Research (AcSIR), Ghaziabad, India
e-mail: rohitjoshi@ihbt.res.in

R. K. Gaur et al. (eds.), *Plant Molecular Farming*,
https://doi.org/10.1007/978-981-95-3823-2_15

Keywords

Micropropagation · In vitro organogenesis · Somatic embryogenesis · Tissue culture · Plant growth regulators · Artificial intelligence · Algorithms · Machine learning · Computational algorithms · Data analysis

1 Introduction

Micropropagation is a method that allows a single cell, tissue, or organ to be used to regenerate an entire plant. While pluripotency entails the differentiation of a certain set of tissues or organs, totipotency refers to a cell's capacity to differentiate into an entire individual (Long et al. 2022). A century ago, when the plant tissue culture technique was initially proposed, the goal was to regenerate entire plants in vitro using somatic cells (Haberlandt 1902). In Steward et al. 1958, Steward and colleagues used isolated phloem cells from carrot roots to regenerate new somatic embryos, which they successfully developed into roots and shoots, proving plant cell totipotency. Since then, tissue culture techniques, driven by this regenerative capacity, have become essential in basic research and transgenic breeding. Numerous techniques, including in vitro propagation, in vitro shoot and root regeneration, the synthesis of metabolites obtained from plants, and somatic embryogenesis, would not be possible without in vitro culture (Ozyigit et al. 2023). Optimal in vitro culture conditions are crucial for various growth stages, like callogenesis, somatic embryogenesis, shooting, and rooting. These conditions vary by species and, in some cases, by genotype within a species (Gray and Trigiano 2018). For example, as multiple studies with various plants have shown, the composition and concentrations of macro- and micronutrients, vitamins, and amino acids have a significant impact on organogenesis.

While MS (Murashige and Skoog 1962) salts have been employed in many studies as a basal medium for organogenesis in many plants, the composition of MS medium is based on the examination of tobacco tissue (Akin et al. 2020). Because different plant species and plant tissue culture methods have variable nutrient requirements, it is vital to create medium formulations that are optimized for certain species and developmental stages in order to achieve maximum efficiency. According to Hildebrandt et al. (1946), creating a novel culture medium required more than 16,000 distinct treatments. Additionally, Murashige and Skoog (1962) took around 5 years to create and refine the culture medium employing 81 distinct vitamin and macro- and microelement combinations. However, because there are so many different media components, designing and altering a medium for a particular purpose requires a high level of expertise and takes a lot of time (Phillips and Garda 2019). Artificial intelligence (AI) and other computer technologies could assist in alleviating this issue by shortening the time-consuming and laborious process.

Thus, machine learning (ML) algorithms can serve as effective and dependable computational tools for interpreting and predicting various unpredictable datasets (Jamshidi et al. 2019). Recently, the multilayer perceptron (MLP), a type of artificial neural network (ANN), has been extensively used for modelling and predicting the results of in vitro experiments. These applications include in vitro sterilization (Hesami et al. 2019; Ivashchuk et al. 2018), callogenesis (Mansouri et al. 2016; Munasinghe et al. 2020), cell growth and protoplast culture (Albiol et al. 1995; Shiotani et al. 1994), somatic embryogenesis (Niazian et al. 2018a, b), shoot regeneration (Barone 2019; Arab et al. 2016), androgenesis (Niazian et al. 2019), hairy root culture (Osama et al. 2013), and in vitro rooting and acclimatization (Arab et al. 2018). A form of nonlinear computing technique called an MLP can be applied to a number of tasks, including complex system classification, prediction, and clustering (Niazian et al. 2019; Araghinejad et al. 2018). Without depending on presumptions about the physical world, it may ascertain the relationship between input and output variables and extract the underlying knowledge from datasets (Sheikhi et al. 2020). The algorithm is made up of several networked processing units, or neurones, that operate in tandem to tackle particular issues. In order to prevent time wastage or imprecise model performance, MLP learns from examples, which need to be properly chosen (Silva et al. 2019). Rodríguez-Pérez and Bajorath (2022) introduced support vector machines (SVMs), which are another efficient and comprehensible class of machine learning algorithms. By examining data and identifying patterns, they are frequently utilized for regression analysis, classification, and clustering of nonlinear relationships (Su et al. 2017).

When it comes to network complexity, support vector machines (SVMs) outperform multilayer perceptrons (MLPs) in a number of ways. SVMs often use a large number of hidden units, whereas MLPs typically use a small number of hidden neurons. The way SVMs formulate the learning problem, which results in a quadratic optimization task, is one of their main advantages (Moravej et al. 2020). The regression equivalent of SVM is called support vector regression (SVR). In the recent past, multiple studies have looked at using SVM-based techniques to solve significant industrial and chemical issues (Sexton et al. 2018). In contrast to MLPs, SVR is not very well-known to scientists studying plant tissue culture. Furthermore, no thorough investigation has been conducted to compare MLP with alternative ML algorithms, such as SVR, in order to create a model that is appropriate for predicting characteristics related to in vitro culture, such as the rate of callogenesis and embryogenesis. A number of studies on the varied uses of AI models in various steps of micropropagation procedures have recently been published (Hesami et al. 2019). Every report has confirmed AI accuracy and reliability in predicting and improving growth and development processes in in vitro culture environments. Initially, the concepts of modelling and optimization were presented, along with several widely recognized algorithms. Next, we discussed how AI–optimization algorithm (OA) can be applied at different in vitro propagation stages.

Recently, several ML techniques have been used to model and predict various in vitro culture systems, such as decision trees, random forests, and artificial neural networks (ANNs). In vitro seed germination (Güney 2019), callogenesis (Salehi et al. 2020), androgenesis (Arab et al. 2016), shoot proliferation (Mridula et al. 2018), rhizogenesis (Niazian et al. 2018a, b), explant sterilization (Joung et al. 2020), in vitro secondary metabolite production (Sheikhi et al. 2020), and gene transformation (Arab et al. 2018) are some of these applications. Among these algorithms, a number of ANN types have been widely used for modelling and predicting in vitro culture processes, including multilayer perceptron (MLP), radial basis function (RBF) networks, and generalized regression neural networks (GRNN) (Hesami et al. 2019; Jamshidi et al. 2019). Typically, univariate to multivariate inputs are used in classical statistical methods to optimize plant tissue culture protocols in order to attain different intended outputs (traits). These traditional approaches, however, have drawbacks, including a high probability of mistakes and decreased efficacy when managing complicated, nonlinear data (Hesami and Jones 2020). The optimization process is extremely sensitive, with many unknowable and unpredictable variables influencing it. By using ML models for testing and optimization, this problem can be solved. While ML modelling is a relatively new approach, there are limited reports on its successful application in optimizing plant tissue culture processes, such as initiation of aseptic culture, callus formation, organogenesis, and production of secondary metabolites (Hesami and Jones 2020). Therefore, the recent advancements in AI–OA-based plant tissue culture are the primary focus of this chapter. This chapter also explains several ML techniques, including multilayer perceptron (MLP), decision trees (DT), random forest (RF), generalized regression neural network (GRNN), and probabilistic neural network (PNN).

2 Pathways of Plant Regeneration in Plant Tissue Culture

Plant tissue culture technique, like the regeneration of a whole plant from somatic cells under in vitro conditions, was proposed a century ago (Thorpe 2007). Tissue culture–based approaches were highlighted after the successful discovery of MS medium, which is required for the growth of plants and different concentrations of plant growth regulators (PGRs), including auxin and cytokinin, which lead to organogenesis (Nath et al. 2023). Plant regeneration includes a sequential process under which a whole plant can be developed from a tissue, an organ, or an unorganized mass of cells (Fig. 1). In this process of regeneration, physiological changes and repair occur in plant tissue (Pasternak and Steinmacher 2024). Tissue culture technique follows the principle based on totipotency and pluripotency. The totipotency of a cell depends upon the efficiency of a plant cell in regenerating a whole plant, whereas pluripotency is the ability of a cell to develop into an organ (McFarland and Kaeppler 2024). A successful regeneration of carrot roots and shoots from somatic embryos, which were developed from phloem cells of carrot roots, confirmed that plant cells are totipotent (Karami et al. 2024). Nowadays, numerous fields, including genetic engineering, transgenic breeding, and genome editing, have made substantial use of tissue culture techniques. Regeneration in plants is divided into two major processes, including somatic embryogenesis and de novo organogenesis.

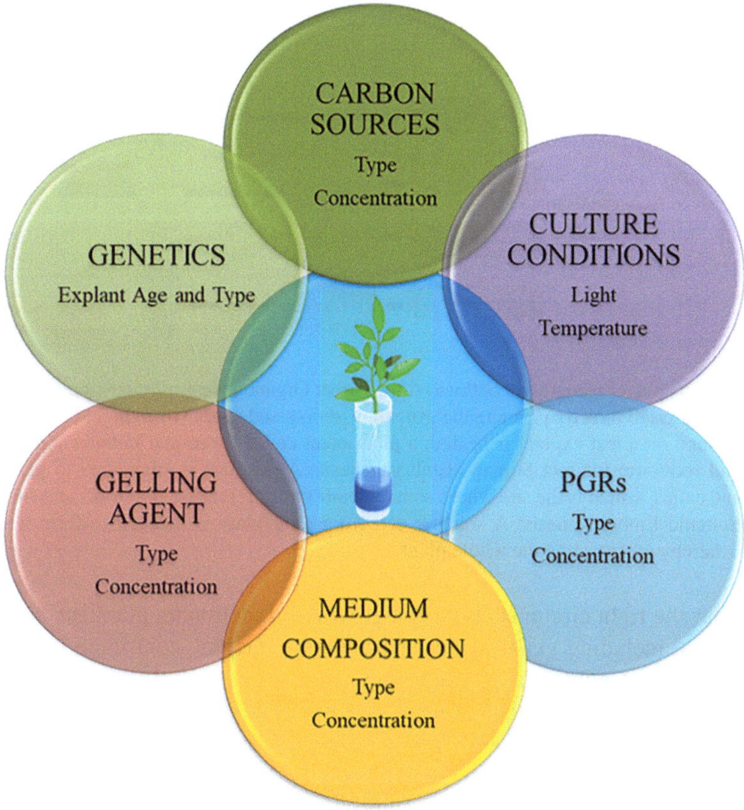

Fig. 1 Diagram illustrating various factors influencing plant cell and tissue culture

2.1 Somatic Embryogenesis

The process of plant regeneration through somatic embryogenesis is initiated through somatic cells, which undergo dedifferentiation and develop embryonic stem cells, which further convert into a whole plant. This process shows that plant cells exhibit totipotency (Mir et al. 2023). Conversion of somatic cell fate into an embryogenic cell leads to somatic embryogenesis. The phenomenon of plant development through this pathway takes place when the plant is treated with certain stress, hormonal treatment, or by modifying gene expression (Horstman et al. 2017). Somatic embryos can be produced indirectly from embryonic callus or directly from individual somatic cells (Yang et al. 2024) (Fig. 2). The most prevalent route, particularly in crop plants, is indirect somatic embryogenesis, which begins with the embryonic callus. Proembryonic masses occur on the surface or inside the callus mass after embryonic callus induction, from which individual cells or cell clusters develop into somatic embryos (Long et al. 2022). Organogenesis is a very complicated process that makes use of the totipotency of plant cells to create new organs

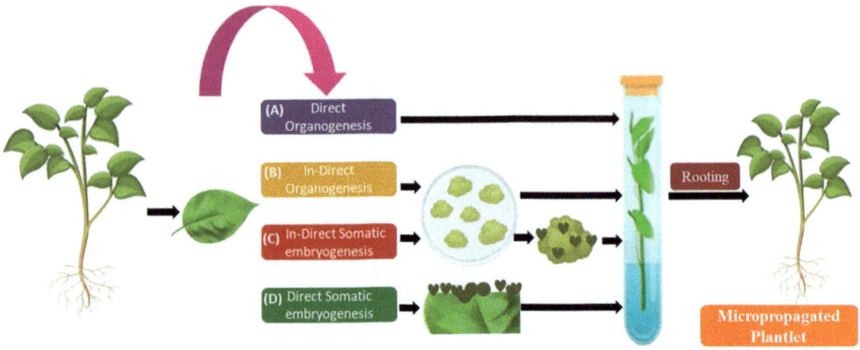

Fig. 2 Various plant regeneration pathways: (**a**) Direct Organogenesis: Preexisting meristems on the stem are used to directly promote the growth of shoots and roots. (**b**) Indirect Organogenesis: After a wound on a leaf explant is healed, a pluripotent callus forms and initiates the growth of shoots and roots. (**c**) Indirect Somatic Embryogenesis: An explant is stimulated to generate an embryonic callus, which is the site where somatic embryos differentiate into roots and shoots. (**d**) Direct Somatic Embryogenesis: A somatic cell from an explant is induced to form a somatic embryo, thereby regenerating the whole plant

that, under the right circumstances, can grow into functioning plantlets. A number of variables, including genetic and environmental factors, contribute to the complexity of understanding plant organogenesis.

Under favourable conditions, embryos can develop shoots and roots. Research on maize demonstrated the formation of embryogenic callus from immature maize embryos, which was later transferred to shoot induction media to promote shoot development. For root formation, the embryogenic callus was placed in root induction media containing auxin (Kumar et al. 2022). In direct somatic embryogenesis, callus formation is bypassed. Instead, compact calli with a low cell division rate formed, which led to the formation of new embryos under favourable conditions. These newly formed embryos have the potential to develop into whole plants. transcription factors are considered proteins, and therefore their names (e.g., WUSCHEL, BABY BOOM). This process leads to the development of somatic embryos on the scutella, which can grow into plants without the need for callus formation. This alternative pathway for plant regeneration streamlines the developmental process by eliminating the callus stage, making the system more efficient for plant propagation and research (Kumar et al. 2022). The use of WUS and BBM transcription factors thus presents a significant advancement in maize embryogenesis, offering a more direct route to plant development. Overall, these studies highlighted various approaches for plant tissue culture, focusing on both callus-mediated and direct embryogenesis, providing valuable insights for plant biotechnologists working on crop improvement and regeneration techniques (Kumar et al. 2022).

There are variations in the time required to produce regenerated plants, despite the fact that both direct and indirect somatic embryogenesis pathways can occur in the same explant. Plant regeneration via the indirect somatic embryogenesis pathway is slower than through the direct somatic embryogenesis pathway due to the callus induction process. Accordingly, the indirect somatic embryogenesis path is frequently associated with somaclonal variation (Bidabadi and Jain 2020).

2.2 De Novo Organogenesis

De novo organogenesis is a term used to describe a regenerative process that reflects the pluripotency of plant cells by differentiating the meristematic centre instead of using a somatic embryo (Raspor et al. 2021). Plants can regenerate adventitious roots and shoots in vitro or from damaged plant parts through mechanisms known as de novo organogenesis. This process is commonly observed in nature. This term, de novo organogenesis, is used to describe the process of adventitious shoot and root regeneration. De novo organogenesis, like somatic embryogenesis, can be classified as a direct or indirect regeneration pathway. Similar to somatic embryogenesis, under favourable conditions, shoots or roots are directly produced from preexisting meristems or wounded organs (Sang et al. 2018) (Fig. 2). De novo shoot organogenesis and pluripotency acquisition are the two steps in this process of plant regeneration. Explants are initially cultivated on auxin-rich media in order to generate a pluripotent callus. Additionally, pluripotent callus tissues express key regulators involved in the establishment of the root stem cell niche locally. These findings are consistent with the nature of pluripotency in relation to root stem cell identity. A subset of callus cells gains shoot identity and matures into a viable shoot when cultured on cytokinin-rich media, after demonstrating competence for organogenesis (Shin et al. 2020; Nath et al. 2024).

De novo organogenesis is crucial for various biological functions, including hormone interactions, cell cycle development, and reactive oxygen species (ROS) metabolism. Plant growth regulators (PGRs) enhance cell division during the process of indirect de novo organogenesis. As a result, cells remain in a dedifferentiated state for extended periods, especially with prolonged culture times, eventually forming a pluripotent callus. This callus undergoes numerous physiological and biochemical changes, which influence the orientation and positioning of cell division when placed in optimal conditions (Xu et al. 2024). The initiation of de novo organogenesis often involves using different combinations of auxin and cytokinin. These hormones play a significant role in regulating cell growth and division. While de novo organogenesis and somatic embryogenesis both facilitate plant regeneration, they differ in one key aspect, i.e. somatic embryogenesis involves the formation of somatic embryos, whereas de novo organogenesis does not (Rocha et al. 2015). In summary, de novo organogenesis is essential for various biological activities and is regulated by PGRs, which promote cell division. The process leads to the formation of a pluripotent callus, which undergoes significant physiological and biochemical changes. The main difference between this process and somatic embryogenesis lies in the absence of embryo development in de novo organogenesis.

3 Artificial Intelligence (AI) and Machine Learning (ML)

ML technology combines computational statistics and algorithms to create robust mathematical models from datasets with multiple independent variables (inputs) and dependent variables (outputs). It is commonly used in traditional and in vitro plant breeding programmes (Zheng et al. 2018). This combination enables machine learning techniques, such as ANNs, to deliver innovative and efficient insights into various mechanisms that drive plant organogenesis (Azevedo et al. 2015). Supervised learning, which is appropriate for life science issues, consists of algorithms for classification (predicting categorical outcomes) and regression (predicting numerical outcomes). ML algorithms may analyze formless information obtained by optical imaging or sequencing. CNNs (convolutional neural networks) are especially well-suited for classification research because of their capacity to extract features automatically (Wang et al. 2020). They are commonly used for image classification, object identification, object tracking, posture estimation, text recognition, visual saliency detection, action recognition, scene labelling, speech, and natural language processing.

However, neural networks, particularly CNNs, frequently lack interpretability for the retrieved characteristics. The combination of ANNs with fuzzy logic, known as neurofuzzy logic, improves data modelling by optimizing and predicting complicated processes while also producing clearly interpretable "IF-THEN" rules (García-Pérez et al. 2020a, b). This tool may categorize input and output values at various levels (low, mid, high, etc.) and assign membership degree values ranging from 0 to 1 (Niazian and Niedbała 2020). This AI-based technology has been effectively used in plant tissue culture, proving its efficacy in germination (Ayuso et al. 2017), micropropagation (Arteta et al. 2018), and phenolic compound production (García-Pérez et al. 2020a, b). This section seeks to give insights into the numerous applications of machine learning technologies in plant tissue culture by demonstrating their efficiency in in vitro experiments.

3.1 Applications of AI and ML in Plant Biotechnology

Biotechnology-based breeding techniques (BBBMs) have advanced traditional breeding methods, enabling quicker improvements in plant traits. A key component of these techniques is in vitro-based breeding, which relies on in vitro regeneration. This method plays a vital role in various aspects of plant breeding (Niazian 2019). It is widely used in both in situ and ex situ conservation efforts, as well as in micropropagation. Micropropagation, in particular, is essential for large-scale plant propagation, germplasm preservation, and the production of bioactive compounds. This is especially important for rare or endangered medicinal plant species (Ayuso et al. 2019). The success of in vitro plant regeneration is influenced by several factors. These include the plant genotype, PGRs, the composition of the nutrient medium, the type and age of the explant, and the presence of elicitors. These factors can be grouped into three main categories: initial regeneration triggers (environmental signals and physical stimuli), cellular responses to these signals, and chemical regulators that influence the establishment and development of new stem cell niches. The interactions among these elements create a complex, multifaceted process in in vitro plant regeneration (Niazian and Niedbała 2020). This intricate process is significant for its role in enhancing the efficiency of breeding programmes. By leveraging the advantages of in vitro techniques, breeders can achieve faster and more precise improvements in plant varieties. This is particularly useful for developing plants with specific desired traits, such as improved resistance to diseases or pests, enhanced yield, or better adaptability to environmental conditions. In vitro regeneration also supports the conservation of valuable plant genetic resources, ensuring that rare or endangered species can be preserved and propagated for future use. Overall, BBBMs and in vitro regeneration provide powerful tools for modern plant breeding, with applications in agriculture, conservation, and the pharmaceutical industry. These methods enable the rapid multiplication and improvement of plants, contributing to advancements in crop production and the sustainable management of plant resources.

MS medium, modified MS (MMS), woody plant Medium (WPM), gamborg's B5 medium, and driver and kuniyuki woody plant medium (DKW) are among the most often used basal culture mediums for in vitro research. Manipulating the basal medium can enhance the efficiency of in vitro assessments, but the intricacy of its constituents makes this difficult. Predicting the effects of components on the desired properties of cultures is an ideal option, and artificial ANNs have been used to indicate the appropriate culture media compounds for effective regeneration of diverse plants (Srivastava et al. 2020). The combination of auxin and cytokinin PGRs has a major influence on the growth of plant culture, directing them towards organogenesis or embryo formation. The ratio of cytokinin and auxin plays a very important role in this research. For example, Niazian et al. (2017a, 2017b) discovered that 2,4-dichlorophenoxyacetic acid (2,4-D) added with kinetin promoted indirect somatic embryogenesis in ajowan medicinal plants (*Carum copticum* L.), whereas another combination resulted in indirect shoot regeneration. In addition, Arab et al. (2017) used ANNs and genetic algorithms to anticipate and optimize the effects of different cytokinin-auxin combinations on micro-shoot proliferation in Garnem (GN15) rootstock, obtaining excellent prediction accuracy (Niazian and Niedbała 2020).

ANNs have also been used to model sterilization procedures using tissue culture techniques. Hesami et al. (2020a, b) used an MLP–ANN model and a genetic algorithm to optimize contamination frequency and explant survival in *chrysanthemum* culture as a function of multiple variables. The model accurately predicted the best parameters for obtaining the lowest contamination frequency and maximum explant viability. Sensitivity analysis demonstrated that immersion duration was the most important factor influencing these results. Beyond culture media components and PGR combinations, ANNs can simulate in vitro tissue growth, differentiate embryos from non-embryos, predict plantlet formation from embryos, estimate cell culture biomass, simulate temperature distribution in culture vessels, identify and estimate shoot length, and cluster regenerated plantlets (Hesami et al. 2017). These capabilities exhibit machine learning's flexibility and strength in optimizing in vitro plant biotechnology operations. Artificial polyploidy induction, twofold haploid creation, gene transformation, and genome editing are all multifactorial approaches that need advanced statistical interpretation (Alizadeh and Nikoo 2018).

Various chemical enhancers can increase the efficiency of doubled haploid generation, and ANN models have the ability to optimize these processes by predicting the optimal additive and other component combinations. For example, ANN models outperformed standard multiple linear regression models in accurately predicting callus induction percentages in tomato androgenesis. Artificial polyploidy induction, which is critical for increasing plant vigour and performance, entails using mitotic spindle poisons both in vivo and in vitro (Sadat Noori et al. 2017). The efficacy of this multifactorial technique is determined by a variety of elements, including culture medium components, PGR combinations, plant genotype, and antimitotic dosage particulars. Although no research has yet used ANNs to predict results in in vitro-induced artificial polyploidy, such applications might considerably improve efficiency by determining the best circumstances. *Agrobacterium*-mediated gene transformation is another important approach for plant genetic engineering. Optimizing numerous factors, such as *Agrobacterium* strain cell density, inoculation period, antibiotic concentrations, and acetosyringone levels, is critical for effective gene delivery (Niazian et al. 2017a, b). ML methods may be useful in predicting and optimizing these parameters, particularly for *Agrobacterium*-resistant plant species. In conclusion, ML, particularly ANNs, has shown enormous potential for improving in vitro-based plant biotechnology. These technologies offer powerful tools for optimizing complicated, multifactorial processes, hence boosting plant breeding and genetic engineering efforts (Niazian et al. 2019).

3.2 Application of AI in Plant Tissue Culture

Dedifferentiation, redifferentiation, and differentiation are examples of very complicated and nonlinear in vitro culture processes caused by a variety of genetic and environmental variables. Conventional mathematical approaches make it difficult to predict characteristics like callus induction, embryo formation, and the number of somatic embryos. Simple stepwise algorithms cannot represent biological processes like somatic embryogenesis, especially when the information is noisy and complicated. This is where ML techniques, namely AI, come into play, providing efficient and dependable computer methods for interpreting and predicting uncertain facts (Hesami et al. 2019). MLP, a form of ANN, is commonly used to model and predict in vitro culture systems. MLP applications in micropropagation include in vitro sterilization, callus formation, cell and protoplast culture, somatic embryogenesis, shoot regeneration, hairy root culture, and in vitro rooting and acclimatization. MLP, a nonlinear computational approach, is effective in clustering, predicting, and classifying complicated systems. It can identify relationships between input and output variables and uncover inherent knowledge within datasets without prior physical considerations. This algorithm comprises several highly linked processing neurons that operate in tandem to address certain challenges. However, MLP requires carefully chosen examples for learning to avoid inaccurate models or wasted time. Another potent ML approach for grouping, classification, and regression analysis of nonlinear connections is support vector machines (SVMs), created by Vapnik (Zheng et al. 2018). SVMs can handle intricate networks with a large number of hidden units and can convert learning issues into quadratic optimization challenges, which gives them an edge over MLPs. The regression variant of SVM, called support vector regression (SVR), has demonstrated promise in resolving important issues in the chemical and industrial domains. SVR is still not well understood in plant tissue culture, and there are not many thorough studies comparing MLP and SVR for predicting in vitro culture parameters (Hesami et al. 2017).

Genetic algorithms (GA), in particular, are evolutionary optimization techniques that have also been used to optimize several aspects of in vitro culture parameters (Arab et al. 2017). Compared to conventional optimization techniques, GA, a single-objective optimization algorithm, has benefits such as not requiring baseline estimations for decision variables. GA may only be used for single-objective functions, though. Srinivas and Deb (1994) created the non-dominated sorting genetic algorithm (NSGA), which handles multi-objective variables but lacks elitism and has a high computational cost. With the addition of elitism and an effective sorting algorithm, NSGA-II is an enhanced version of NSGA that may be used for multi-objective optimization involving several objective functions. It has been accomplished to concurrently optimize callus weight, shoot regeneration rate, and shoot count using NSGA-II. SVR was utilized in a recent work to predict factors related to somatic embryogenesis, such as callus formation, embryo formation, and the quantity of somatic embryos in *chrysanthemums*. To find the best model for simulating and predicting in vitro culture systems, the statistical performance of the SVR-based model was compared with that of MLP. Furthermore, the optimal model for maximizing PGR levels for somatic embryogenesis was associated with NSGA-II (Hesami et al. 2019). This work is the first to disclose the use of SVR in plant tissue culture, demonstrating the potential of cutting-edge AI methods to increase the precision and efficiency of in vitro plant biotechnology procedures (Table 1).

4 Background Required for Automated Plant Growth Detection

Automated technologies have been extensively utilized by researchers in plant tissue culture, agriculture, and plant production, as well as engineers, to enable more precise and efficient monitoring of plant health and development. In in vitro propagation, automated phenotyping can replace manual, destructive measurements with continuous and objective digital tracking of plant traits (Malabadi et al. 2023). Consequently, this can lead to a better understanding of crucial developmental processes in in vitro cultures. In the agriculture sector, these automated techniques are utilized in various crops such as potato, chilli, and tomato for the analysis and detection of disease infestation. The automated detection of plant growth using an

Table 1 Use of artificial intelligence (AI) models and optimization algorithms in plant tissue culture

Plant	Aim	Model	Inputs	Output	References
Passiflora caerulea	Predicting and optimizing the explant-specific effects of externally applied plant PGRs on callus formation	MLP	Types and concentrations of PGRs as well as explant types	Callogenesis rate	Jafari and Daneshvar (2024)
Lavandula	Optimization of in vitro organogenesis and in vitro rooting protocols for various lavender genotypes	Gaussian process (GP), MLP, XGBo-ost, and RBF	Various concentrations of BAP and IBA	Enhancing vitro organogenesis and in vitro rooting effectiveness in different lavender genotypes	Şimşek et al. (2024)
Strawberry cultivars	Identify and predict impact of water on strawberries	RF, MLP, and GP	MS media supplemented with PEG-6000 (0%, 2%, 4%, and 6%)	Strawberry responses to drought stress	Şimşek (2024)
Cannabis sativa L	Assessing and predicting the effects of chemical priming with H_2O_2 and LED exposure on industrial hemp cultivated in vitro	RF, MLP, and XGBoost	Chemical priming with H_2O_2 and LEDs	Chemical priming of seeds with 2.0% H_2O_2 yielded the best germination, development, and enzymatic activities	Aasim et al. (2024)
Bacopa monnieri	Optimizing in vitro regeneration using Ethyl Methane Sulfonate (EMS) and Sodium Azide (NaN_3) treatment	Support vector classifier (SVC), GP, XGBoost, and RF, and MLP	NaN_3 and EMS were applied at varying concentrations (0.025, 0.05, 0.1, and 0.2 mg l^{-1}), and for different durations of treatment	Maximum shoot length, node numbers, and leaf numbers	Mirza et al. (2023)

Hemianthus callitrichoides and Riccia fluitans	Prediction and confirmation of in vitro organogenesis in two aquatic plants, *Hemianthus callitrichoides* (Cuba) and *Riccia fluitans*	MLP, RF, and XGBoost	Basal medium, agar, sucrose, and PGRs concentration	Clump diameter, rooting (%)	Özcan et al. (2023)
Cotinus coggygria	Optimizing macro-elements and balance of cytokinin-auxin for micropropagation	MLP, RF, SVR, and XGBoost	Basal medium and PGRs concentrations	Regeneration (%) and Shoot count	Aasim et al. (2023)
Cannabis sativa	Optimizing in vitro seed sprouting and development indices	SVC, GP, XGBoost, RF, and MLP	Variable concentrations of H_2O_2	Seed germination	Aasim et al. (2022)
Cicer arietinum	Modelling the in vitro regeneration	SVR, GPR,, XGBoost, RF, and MLP	PGRs type and concentration	Shoot counts and shoot length	Kirtis et al. (2022)
Triticum aestivum	Predict and optimize shoot proliferation	GRNN-GA	Types of explants, genotype and various concentrations of PGRs (BAP, IAA, KIN, zeatin, IBA, 2,4-D, NAA) and $CuSO_4$	Frequency of shoot induction	Hesami et al. (2020a)
Cucumis melo	Predict and optimize indirect shoot induction	MLP-GA	Agar concentration, light exposure time, culture temperature, and humidity levels	Percentage of shoot induction	Zhang et al. (2020)

(continued)

Table 1 (continued)

Plant	Aim	Model	Inputs	Output	References
Solanum lycopersicum	Predict and optimize androgenesis	MLP	Length of flower, genotype, KIN, 2,4-D, cold pretreatment period, and gum arabic	Rate of callogenesis and quantity of calluses that regenerate	Niazian et al. (2019)
Wrightia tinctoria	Predict and optimize in vitro rooting	GP	Different concentration of charcoal and NAA	The maximum root length, number of roots, rhizogenesis rate, basal callus size, and lateral root number	Mridula et al. (2018)
Pistacia vera	Predict and optimize shoot proliferation	MLP-GA and Neurofuzzy logic	IBA, BAP, glycine, pyridoxine-HCl, nicotinic acid, thiamine-HCl, NO_3^-, NH_4^+, K^+, PO_4^{2-}, cl^-, Ca^{2+}, Mg^{2+}, SO_4^{2-}, Fe^{2+}, BO_3-, Mn^{2+}, Zn^{2+}, Cu^{2+}, MoO_4^{2-}, Na^+, $Co^{2+,}$ I^-, and EDTA	Rate of proliferation, length of shoots, overall weight, and healthy fresh weight	Nezami-Alanagh et al. (2017)

6-Benzylaminopurine (BAP), (Extreme gradient boosting (XGBoost), gaussian process (GP), generalized regression neural network (GRNN), genetic algorithm (GA), Indole-3-butyric acid (IBA), Kinetin (KN), Light Emitting Diode (LED), multilayer perceptron (MLP), Murashige and Skoog (MS), Polyethylene glycol (PEG), Plant growth regulator (PGR), radial basis function (RBF), random forest (RF) models, support vector classifier (SVC), support vector regression (SVR).

AI-based detection system occurs in predetermined steps, such as image acquisition and image processing. Automated plant growth detection is an emerging technology that leverages advanced imaging techniques and image processing algorithms to monitor and analyze the stages of development and disease infestation in plants.

4.1 Plant Image Acquisition

Image acquisition is the first step in the process of automated plant growth detection. It involves capturing images of plants at various growth stages using different types of imaging devices. The quality and type of images acquired significantly affect the accuracy and reliability of subsequent analysis. High-resolution digital cameras are used in this section to take pictures (Camargo and Smith 2009). Smartphones are also effective in recording images in multiple formats like JPG, PNG, TIFF, and others (Wani et al. 2022). Time-lapse cameras are used to capture a series of images over time, enabling the observation of growth dynamics and temporal changes. Nowadays, CMOS (complementary metal oxide semiconductor) and CCD (charge-coupled device) fused with TDI (time delay integration) technologies are being used in sensors for image acquisition. Bethge et al. (2023) acquired phenotypic data of in vitro plants in their study. They utilized a large-distance sensor for depth imaging of in vitro plants to determine maximum plant height, average canopy height, and culture media volume. Proper lighting is also crucial for obtaining high-quality images. Natural sunlight is often used, but controlled artificial lighting can provide more consistent results. A uniform background can help in segmenting the plant from the background during image processing. High-resolution images allow for the detection of fine details but require more storage and processing power. There are seven major types of image acquisition systems used for phenotyping of plants, such as stereo vision, mono-RGB vision, ToF (Time-of-flight) cameras, multi- and hyperspectral cameras, LIDAR, thermography, and fluorescence and tomography imaging. Following the capture of all required photos, preprocessing is applied to make any necessary alterations before usage. Image-enhancement techniques are used if the photos do not match the necessary requirements (Basavaiah and Arlene Anthony 2020). Since the obtained images are used to train ML models, their quality has a significant impact on the system's overall efficiency (Camargo and Smith 2009).

4.2 Image Processing

Image processing is essential for enhancing image contrast and removing background disturbances such as unwanted elements, noise, uneven lighting, and blurriness. As a result, processing raw data is crucial to make it suitable for monitoring plant developmental stages using automated methods. The unprocessed data is converted into a certain format by filtering out background noise. After that, images undergo essential segmentation and feature extraction. Image processing enables researchers to optimize the use of their computing resources and ensure consistent image resolutions according to a set standard. Image preprocessing includes grayscale adjustment, image binarization, distortion removal, and the utilization of filters to detect edges. Several studies have contributed to advancements in image processing for plant phenotyping. For example, Wang et al. (2017) proposed a comprehensive pipeline that includes grayscale conversion, the use of a new filter, binarization, and edge detection to improve plant phenotyping accuracy. Similarly, Thangadurai and Padmavathi (2014) developed an image processing pipeline tailored for identifying plant species under varying illumination conditions, demonstrating the adaptability of these techniques. Militante et al. (2019) highlighted the benefits of using a white background for images, as it enhances clarity and makes interpretation easier. In the field of agriculture, the hue, saturation, and value (HSV) colour space is widely utilized for image analysis because it aligns closely with human colour perception, providing more intuitive and accurate colour analysis (Jadhav and Patil 2015). Overall, image processing is an indispensable tool that improves image quality and analysis, enabling more effective monitoring and research in plant development and other fields.

4.3 Image Segmentation

In agricultural research, segmentation is a basic approach that involves carefully breaking an image up into its constituent parts. The main objective is to conduct a more thorough analysis of every object, identifying useful characteristics that could improve our comprehension and knowledge (Jafar et al. 2022). Based on the collected traits, it is possible to distinguish between unaffected and diseased locations in a leaf (Makandar and Halalli 2015). It is essential to segment the preprocessed pictures in order to classify diseased leaves and extract a variety of potentially useful attributes. The process of segmenting a picture into multiple subregions is called image segmentation. Conventional methods like region-based, edge detection, thresholding, and clustering depend on mathematical and image processing expertise to segment the provided images. One of the best methods for image segmentation is thresholding, which divides images according to pixel intensity levels. A classic technique for segmenting images, image threshold segmentation has the advantages of being easy to use, requiring little computation, and producing consistent results. It works particularly well with images where the background and target regions' pixel grey values are not in the same range (Huixian 2020). As a result, image segmentation technology has been widely used. It is extensively employed in many diverse applications, including remote sensing, detection, and classification.

Global, variable, and adaptive thresholding segmentation have three varieties. Every category has its own techniques for segmenting images; for instance, Otsu's thresholding, mean, and median are included in the global thresholding approaches (Makandar and Halalli 2015). Edge detection divides an image into segments according to its edges, which are usually referred to as the image's borders. The edge segmentation method divides an image into segments by identifying the areas where abrupt changes occur in the texture structure or in the grey value of a pixel. In image segmentation, the Roberts operator, Prewitt operator, Laplacian operator, and Krisch operator are frequently employed. The primary idea behind the threshold segmentation technique is the variation in grayscale characteristics between the several target areas in the picture. The Sobel operator, Laplacian of Gaussian (LoG) filter, and Canny edge detector are a few well-known edge detection techniques (Singh and Misra 2017). With region-based segmentation, the image is divided into several areas according to how similar the pixels are to each other in terms of form, colour, and intensity value.

Region growing and region splitting are two popular region-based segmentation techniques (Aubry et al. 2014). The two approaches for segmenting the image work in reverse order, with one expanding the region by including neighbouring pixels' seed pixels. Another method for segmenting images is clustering, which puts pixels in groups according to how similar they are in terms of texture, colour, or other necessary characteristics. Fuzzy-C means and K-means are well-known clustering techniques for picture segmentation that are extensively utilised in a variety of applications (Jafar et al. 2024). The performance of automatic segmentation techniques based on deep learning has surpassed that of conventional techniques in recent years. Semantic segmentation and instance segmentation are two popular DL-based segmentation techniques. Every pixel in an image is given a category label via semantic segmentation, which divides the image into sets that are mutually exclusive and each represents a substantial portion of the original image (Jafar et al. 2022). Convolutional neural network (CNN) is one of the DL models that outperforms and improves higher-level segmentation accuracy. An upgraded version of semantic segmentation that is intended to handle difficult or complex jobs is called instance segmentation. Many methods have been created, and each method makes use of well-known deep learning architectures, such as Tensor Mask, YOLO (You Look Only Once), RCNN (Region-based Convolutional Neural Network), and instance cuts (Jafar et al. 2024).

4.4 Feature Extraction

Feature extraction is the method of obtaining features from unprocessed data in agriculture. The attributes of form, colour, and texture are the feature descriptors of the input image. It is a crucial component of categorization tasks. Feature engineering, as used in ML, is a basic method that comprises turning unprocessed data into a set of pertinent and significant characteristics (Basavaiah and Arlene Anthony 2020). It describes the process of abstracting an image into a certain mathematical representation or vector description and retrieving the invariant image information through zooming, translation, scale change, or lighting. Based on the various extraction ranges, there are two types of image feature extraction: global and local features. The representation of every feature in a picture that describes its overall information, including its colour, shape, and texture, is referred to as its global feature (Huixian 2020). The incomplete aspects of an image inside a certain location are referred to as local features. These are limited to a region's local features inside the image. In local detection techniques, corner detection and speckle detection are frequently employed. This process determines whether or not plants are healthy by using the dataset as input. The colour, texture, morphology, and other associated aspects are an image's fundamental elements. Morphological aspects are more useful than other characteristics in recognising the damaged area on a leaf (Khirade and Patil 2015).

Colour moments and Gabor texture are two examples of often utilised colour properties. The colour histogram, colour correlogram, colour R moment, and other techniques can be used to acquire these features (Jafar et al. 2024). The texture can be enhanced by contrast, homogeneity, variation, and entropy, among others. It has been found that using textural features in the context of plant disease detection difficulties produces more positive results (Kaur et al. 2019). The grey-level co-occurrence matrix (GLCM) approach can be used to determine the area's energy, entropy, contrast, homogeneity, moment of inertia, and other textural features (Kaya et al. 2021). Wavelet packet decomposition and FT can be used to isolate texture properties (Kaur et al. 2019). Higher efficacy has been demonstrated by additional features, such as the Pyramid Histogram of Visual Words (PHOW), the Histogram of Oriented Gradients, and robust features with improved computational speed (Kaur et al. 2019).

5 Innovative Applications of AI and ML for Optimization of Plant Tissue Culture

According to Ali and Aasim (2024), AI modelling is significantly more efficient compared to other traditional methods and has proven to be a promising tool for plant tissue culture. In recent years, both AI and optimization algorithms (OA) have been employed to enhance various stages of micropropagation. These stages involve complex biological processes that require precise modelling and optimization to achieve the desired outcomes. AI models are created using data from experimental results and observations, which allows for the development of accurate and efficient solutions to complex problems. In the realm of AI, several types of ANNs have been widely used to support plant tissue culture. Some of the key types discussed in this context include multilayer perceptron (MLP), random forest (RF), decision tree (DT), data fusion models, PNN, and GRNN (Fig. 3). Each of these models is recognized for its simplicity in design and efficiency in solving complex biological problems, making them ideal for plant tissue culture applications. Before delving deeper into the specifics of these ANNs, it is important to clarify a few fundamental concepts. In the context of ANNs, a "neuron" refers to the basic unit that processes a particular input variable and applies a transfer function to generate a corresponding output or response. The "architecture" of a neural network is the overall structure, which includes the arrangement of input, hidden, and output layers, the number of neurons in each layer, the connections between neurons, the direction of data flow (whether recurrent or feed forward), and the transfer functions that are employed.

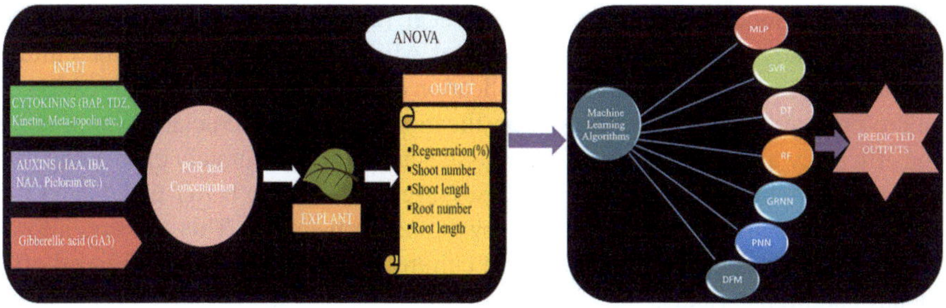

Fig. 3 The schematic diagram illustrating optimization of in vitro organogenesis via machine learning algorithm including a multilayer perceptron (MLP), support vector regression (SVR), decision tree (DT), random forest (RF), generalized regression neural network (GRNN), probabilistic neural network (PNN), and design for manufacturing (DFM)

Training a neural network involves adjusting the model's parameters based on input and output data pairs, allowing it to improve its performance through an iterative learning process.

AI employs a combination of mathematical, statistical, and engineering principles along with expert judgement to tackle complex problems in plant tissue culture. One of the most valuable aspects of AI is its ability to integrate optimization algorithms (OA), which can be used to select the best possible conditions for various plant tissue culture processes. This approach, particularly when employing genetic algorithms, is highly advantageous for plant tissue culture because it reduces both the cost and time associated with traditional methods. For instance, the stages of plant tissue culture that are particularly well-suited for AI and OA optimization include somatic embryogenesis, callogenesis, bud formation, shoot, and root development. Several studies have demonstrated the effectiveness of AI and OA in micropropagation. Hesami et al. (2017) applied these methods to predict and optimize the conditions necessary for successful embryogenesis, while Niazian et al. (2019) explored their utility in callus formation. Additionally, Nezami-Alanagh et al. (2017) used AI to optimize bud formation, and Mridula et al. (2018) applied these techniques to improve root development (Table 1). These studies underscore the importance of AI and OA in addressing the many variables involved in plant tissue culture, including plant genotypes, growth media composition, sterilization techniques, and the types and concentrations of PGRs.

By accurately predicting and optimizing these variables, AI models can significantly improve the success rates of in vitro organogenesis experiments. The length and number of micro-buds, root development, and biomass accumulation in plant cell cultures or hairy roots are other factors that have been successfully predicted and optimized using AI and OA. For example, Nezami-Alanagh et al. (2017) used these methods to predict the length and number of micro-buds, while Mridula et al. (2018) applied them to enhance root development. Similarly, Duan et al. (2015) demonstrated that AI could predict the biomass of plant cell cultures and hairy roots, which are crucial for biotechnological applications. Environmental conditions, such as temperature and sterilization techniques, have also been optimized through AI models, as shown by Ivashchuk et al. (2018). This wide range of applications illustrates the versatility and power of AI in optimizing plant tissue culture conditions. A particularly valuable application of AI and OA is the classification of micro-buds and somatic embryos. Classification systems are crucial for improving the efficiency and productivity of plant tissue culture. By accurately classifying different developmental stages of plant tissues, researchers can better understand the factors that contribute to successful growth and propagation. This, in turn, allows for more precise control over the tissue culture process, leading to enhanced efficiency and higher yields.

5.1 Multilayer Perception (MLP)

The MLP, a common type of artificial neural network (ANN), consists of multiple hidden layers in addition to input and output layers (Sheikhi et al. 2020). For designing an MLP, the number of hidden layers and neurons per layer must be specified (Niazian et al. 2018a). Hornik et al. (1989) showed that MLP functions as a universal approximator when it has a sigmoid activation function, meaning it can be trained to model any input–output relationship. An important factor influencing how the MLP is architecturally designed is the quantity of neurons in the hidden layer. Several studies, including Sheikhi et al. (2020), have provided guidelines for selecting the optimal number of neurons (m) based on factors such as data size (K) or the number of input variables (n). Tang and Fishwick (1993) and Wanas et al. (1998) suggested using "n," "2n", or "log(K)" as possible values for the number of neurons. Although these guidelines offer useful starting points, the ideal number of neurons is typically determined through trial and error. It is important to note that the quantity of neurons directly influences the complexity of the network. A high number of neurons adds complexity, while too few neurons can oversimplify the network. An overly simplified model may lead to underfitting, where it misses the underlying patterns of data. Conversely, an excessively complex network may lead to overfitting, where the model becomes too tailored to the training data and performs poorly on new data (Domingues et al. 2020). Therefore, balancing the number of neurons is essential for achieving optimal performance in MLPs.

5.2 Decision Trees (DT)

For regression and classification applications, decision trees (DT) are popular ML models (Thomas et al. 2021). A root node, internal nodes, and leaf nodes make up their structure. Class probability distributions and labels are contained in leaf nodes, whereas internal nodes represent attribute values. Several algorithms are available for creating DT, including exhaustive CHAID, chi-squared automatic interaction detector (CHAID), and classification and regression tree (CART). These models provide data in an understandable manner (Akin et al. 2018). The key distinctions between these algorithms are in how they build their trees—CART: uses binary splits (each node splits into two)— CHAID and exhaustive CHAID: use multiway splits (each node can split into more than two) by default (Akin et al. 2020). All methods prevent overfitting by limiting tree size—CART: produces the entire tree, then performs backward pruning (post-pruning)—CHAID and exhaustive CHAID: restrict growth at the optimal tree size using a stopping criterion (Thomas et al. 2021).

Tree construction process: (1) All descriptive variables are evaluated for optimal splits; (2) Trial splits are used to determine the best split point; (3) The best split maximizes between-group variance and minimizes within-group variance. CART's backward pruning (post-pruning) uses *V*-fold cross-validation: (1) The entire tree is randomly divided into V groups; (2) One group validates the model, while the rest construct it; (3) This process is repeated V times, creating *V* models for confirmation (Thomas et al. 2021). CHAID and exhaustive CHAID tree-producing methods involve: (1) Merging (2) Splitting (3) Stopping. For continuous output variables (regression-type), CHAID uses the F test for the best-next split. For categorical outputs, it employs chi-squared. Continuous variables are grouped with similar sample sizes. The response variables are divided according to the smallest Bonferroni-adjusted p-value, and the significance (typically 0.05) of the descriptive variable pairs with the largest p-values is examined. Pairs with p-values >0.05 are merged into a single node. To determine which split is most significant, groups consisting of three or more classes are analysed. If the node size of the mother plant and the new plant is less than the predetermined minimum value, the tree will not develop further; otherwise, it will grow (Rashidi et al. 2014). Exhaustive CHAID, an enhanced version of the CHAID method, takes longer to process because it performs more tests and merges for the dependent variable (Akin et al. 2020).

5.3 Random Forest (RF)

Breiman (2001) introduced the ensemble learning method known as random forest (RF). It is an ensemble of unpruned decision trees, and its simplicity and effectiveness have made it widely successful in both classification and regression tasks. RF offers several key advantages, such as preventing overfitting, managing noisy data, and efficiently handling a large number of features, as reported by researchers like Silva et al. (2019) and Biau et al. (2019). One of the core features of the RF model is that it combines the outputs of individual decision trees to produce the final result. This approach relies on maintaining a low correlation between the trees while ensuring each individual tree is strong enough to provide valuable insights. This ensemble method helps to avoid the pitfalls of traditional decision trees, like overfitting, and contributes to the overall stability of the model. Two important factors contribute to the accuracy of RF in classification tasks: (1) The bootstrap replica method, and (2) Random feature selection. The bootstrap method involves training each tree on a randomly drawn sample set with replacement. This introduces variability into the trees, which helps prevent them from becoming too similar to one another. The second factor, random feature selection, refers to the process where each tree makes splits based on a small, randomly chosen subset of variables. This ensures that no single feature dominates the decision-making process across all trees, further reducing overfitting and increasing the robustness of the model.

RF trees are grown fully, allowing for high variance in outputs with low bias. This means that while individual trees may be prone to high error, the overall forest compensates for this by averaging the outputs of all the trees, leading to more reliable predictions. In essence, RF benefits from the diversity of its trees, leveraging the collective strength of the ensemble to outperform individual decision trees. When applied to regression problems, RF uses mean squared error (MSE) as a key metric to guide the decision-making process at each node. MSE calculates the gap between expected results and the real outcomes, helping the model to optimize splits in the data. This allows RF to accurately predict continuous variables, making it versatile for a wide range of tasks. In conclusion, RF has the ability to prevent overfitting, handle noisy datasets, and efficiently manage large feature sets, making it a highly effective ML method. Its combination of bootstrap sampling and random feature selection enables it to maintain low correlation among trees while preserving the strength of individual trees. Whether in classification or regression tasks, RF provides a powerful and reliable approach, with key metrics like MSE helping it adapt effectively to different types of problems.

5.4 Generalized Regression Neural Network

GRNN (Generalized Regression Neural Network), introduced by Specht (1991), is a type of statistical ANN characterized by a rapid training process. Key features of GRNN include: (1) The hidden layer contains as many neurons as there are data points; (2) The input layer and the pattern layer are entirely connected; (3) The summation layer includes two different kinds of neurons: D-summation and S-summation; (4) Both D-summation and S-summation neurons are connected to outputs from each pattern layer neuron; (5) D-summation neurons calculate the unweighted sum of pattern layer outputs, with unity connection weights; (6) S-summation neurons calculate the weighted sum of pattern layer outputs, with connection weights equal to the target output; (7) The output layer determines the unknown output value by dividing the output of each S-summation neuron by the output of each D-summation neuron (Lan et al. 2020). This structure allows GRNN to efficiently process input data and generate output predictions, making it suitable for various regression and function approximation tasks.

The GRNN is a particular type of RBF network that excels in interpolation tasks, rendering it particularly suitable for pattern recognition and regression applications. Its structure comprises four essential layers: input, pattern, summation, and output, each serving a specific function in data processing and result generation. The efficacy and accuracy of this ML approach in simulating and predicting various in vitro cultivation systems have been confirmed across multiple plant species. These include chickpeas (Hesami et al. 2021), ajowan (Niazian et al. 2018a, b), *Prunus* rootstock (Arab et al. 2018), *Chrysanthemum* (Hesami et al. 2020a, b), pear rootstock (Jamshidi et al. 2019), *Passiflora caerulea* (Jafari and Daneshvar 2024) (Table

1), wheat (Hesami et al. 2020a, b), wallflower (Fakhrzad et al. 2022), walnut (Sadat-Hosseini et al. 2022), and tomato (Niazian et al. 2019).

5.5 Data Fusion Model

The rising demand for higher accuracy and precision in AI models has led researchers to develop advanced techniques like ensemble methods, also known as multi-model fusion. These techniques combine outputs from multiple models to improve prediction accuracy, offering significant advantages over using individual models alone. Alizadeh and Nikoo (2018) noted that ensemble models can integrate data from various sources, which results in more precise predictions than standalone models. Ensemble modelling has found applications across many fields of study, including environmental science, healthcare, and industrial processes. Meng et al. (2020) observed that data fusion, a key component of ensemble models, has been used to merge information from multiple datasets to produce more accurate and reliable outcomes. This approach is particularly effective in managing complex systems, where numerous variables or stages need to be considered. For instance, ensemble models have been used in different steps of micropropagation, where accurate predictions are crucial for optimizing growth conditions and outcomes. Various techniques are available for fusing individual models into an ensemble. Among the most commonly mentioned are methods like bagging, the K-nearest neighbours (KNN) algorithm, and the ordered weighted averaging (OWA) method. Each of these approaches has distinct features and advantages based on the data type and the desired outcome. For instance, KNN works well for classification problems, while OWA relies on specific weighting methods such as ORLIKE and ORNESS to aggregate information, balancing between full agreement and maximal disagreement among models.

Among these approaches, bagging emerges as one of the most efficient and uncomplicated techniques for combining data. Bootstrap aggregating, also known as bagging, is a method that involves training several models on distinct subsets of the dataset, which are generated through random sampling with replacement. The outcomes from these individual models are then averaged to produce a final prediction. The strength of bagging lies in its ability to reduce variance without increasing bias, making it particularly useful for preventing overfitting in complex datasets. Additionally, the simplicity of bagging makes it a favoured approach for researchers and practitioners who need reliable results with minimal computational complexity. The growing importance of ensemble models in AI research is evident, as they provide more accurate, reliable, and robust predictions for a variety of applications. Their ability to handle complex systems, merge multiple sources of data, and offer more precision than individual models has made them a critical tool in AI development. Whether in the context of environmental monitoring, medical diagnosis, or industrial optimization, ensemble techniques like bagging are becoming integral to advancing AI's predictive capabilities. In conclusion, ensemble models, particularly bagging, have emerged as essential tools for increasing the precision of AI systems, with applications ranging from scientific research to real-world problem solving. These models represent a powerful approach to handling the complexities of modern datasets and optimizing prediction accuracy.

5.6 Probabilistic Neural Network (PNN)

Probabilistic neural networks (PNN) are a specialized form of ANN specifically created for classification purposes. Their architecture shares similarities with radial basis function (RBF) networks. The PNN operates as follows: (1) Input Layer: Receives the input vector; (2) Pattern Layer: Calculates distances between the input vector and calibration input vectors, generating a probability vector f(Xr, Xb); (3) Output Layer: Applies a transfer function to select the maximum probability, determining the classification. Key Characteristics of PNN: Suitable for qualitative prediction—not applicable for continuous data predicting—based on kernel density estimator (KDE). The choice of kernel function affects the estimation quality by approximately 4%. The cauchy kernel is selected for its analytical properties and suitability for derivation operations. The product form of KDE with cauchy realization is employed for PNN implementation in this research.

6 Conclusion

Plant tissue culture is a fundamental technique in the field of plant biotechnology and vital to the multiplication and breeding of many plant species. It plays a crucial function in a number of processes, including micropropagation, in vitro shoot and root organogenesis, secondary metabolite production, and direct and indirect somatic embryogenesis. To achieve success in in vitro culture, optimization of conditions is crucial for several phases of plant growth and development, like callogenesis (the formation of callus), embryogenesis (the formation of embryos), in vitro shooting (the development of shoots), and in vitro rooting (the formation of roots). Given the complexity of these processes, ML has emerged as a valuable tool, based on AI, to address the challenges associated with plant tissue culture. By utilizing multiple models, ML can provide effective solutions for optimizing plant tissue culture protocols. A key advancement in this area is the application of AI–optimization algorithms (AI–OA). Now it is possible to apply these AI-driven techniques with fewer experimental treatments, which significantly reduces the time and expense required to design plant tissue culture protocols, particularly on an industrial scale. AI–OA methods offer great potential in analyzing in vitro culture data, interpreting the results, and providing a deep understanding of the biological systems involved.

Researchers can enhance their understanding of AI–OA applications by expanding the database of the generated model to include more input and output parameters. This iterative procedure improves the effectiveness of data interpretation while offering new perspectives on the physiological, developmental, and regulatory aspects of in vitro plant growth. Looking towards the future, combining multiple AI models, also known as a data fusion strategy, can further refine the predictive capabilities of these models. By integrating various data sources, more precise models can be developed that are capable of predicting and optimizing the outcomes of micropropagation protocols and in vitro biological mechanisms. Such advancements could lead to more reliable and efficient plant tissue culture systems, thus accelerating the development of plant propagation techniques. Moreover, AI–OA methods hold significant promise for automating and mechanizing genetic engineering, genome editing technologies, and in vitro plant breeding. One notable application of this is technologies such as CRISPR (clustered regularly interspaced short palindromic repeats) and CRISPR-associated protein 9 (Cas9), which are used for gene editing. Gene transformation is a complex process influenced by multiple variables. These variables include in vitro regeneration parameters, PGRs, sources of carbon, nutrient media composition, light and temperature conditions, density of bacterial cells, concentrations of antibiotics and chemical stimulants, and the duration of inoculation. Each of these factors can have a significant impact on the efficacy of genetic transformation.

Agrobacterium-mediated gene transformation, a commonly used method in plant genetic engineering, presents a particularly complex system. To optimize this process, it is necessary to understand the effects of various factors that influence T-DNA delivery into different explants (the plant tissues or cells used for culture). Following the transformation process, further analysis is required to verify T-DNA integration and stability within the plant genome, as well as to assess the efficiency of the gene transformation. Optimization of gene transformation protocols is critical to achieving success in genetic engineering. This is where AI–OA plays a crucial role, offering a powerful and reliable approach to developing novel computational methods for both genetic engineering and genome editing in plants. By applying ML and AI–OA techniques, researchers can streamline the process of genetic transformation, making it more efficient and predictable. The chapter discussed here provides an overview of the applications of ML in plant tissue culture. It highlights how adopting appropriate data analysis methods in future research could enhance the outcomes of studies in this field. By leveraging AI and machine learning, researchers can improve the precision and efficiency of in vitro plant breeding, plant tissue culture, and genetic engineering techniques. The ability to analyze complex datasets, optimize experimental protocols, and predict outcomes with greater accuracy will undoubtedly propel advancements in plant biotechnology.

In conclusion, the integration of ML and AI–OA into plant tissue culture and genetic engineering opens new doors for innovation and efficiency. By refining plant tissue culture protocols and automating complex procedures like gene transformation, AI-driven methods can significantly reduce both time and cost in research and industrial applications. As AI models continue to evolve, their predictive power and utility in biological systems will become even more significant, leading to more precise and efficient plant breeding and propagation techniques. The future of plant biotechnology will likely see an increasing reliance on AI to solve complex problems and to drive advancements in the field.

Acknowledgements The authors thank the Director, CSIR-Institute of Himalayan Bioresource Technology, Palampur, India, for infrastructure and support. SJ thanks the Department of Biotechnology, Government of India (File No. DBT/JRF/BET-19/1/2019/AL/212) for providing the Junior Research Fellowship. AK, SG, JN, and KK thank CSIR-UGC (NTA Ref. No. 191620063807; UGC Ref. No. 574/CSIR-UGC NET JUNE 2019; File No. 1029/CSIR-UGC NET JUNE 2019; NTA Ref. No. 191620158661) for providing Senior Research Fellowships. MR thanks the Department of Science and Technology, Government of India (No. DST/INSPIRE Fellowship [IF200333]) for providing the INSPIRE fellowship. The authors also acknowledge the financial support (No. BT/PR45280/NER/95/1918/2022) from DBT and (Grant number: No. Agr. H (Project cell) F(7)-KSY-01/2020 110032) from the Agriculture Department, Himachal Pradesh. This manuscript represents CSIR-IHBT Communication No. 5730.

Author Contributions KK, SG, JN, AK, SJ, and MR participated in the writing of the manuscript; RJ conceptualized the idea of the study, reviewed, and edited the final draft of the manuscript. All authors have read and agreed to the final version of the manuscript.

Conflicts of Interest The authors declare no conflict of interest.

References

Aasim M, Katırcı R, Akgur O, Yildirim B, Mustafa Z, Nadeem MA, Baloch FS, Karakoy T, Yılmaz G (2022) Machine learning (ML) algorithms and artificial neural network for optimizing in vitro germination and growth indices of industrial hemp (Cannabis sativa L.). Ind Crops Prod 181:114801

Aasim M, Ayhan A, Katırcı R, Acar AŞ, Ali SA (2023) Computing artificial neural network and genetic algorithm for the feature optimization of basal salts and cytokinin-auxin for in vitro organogenesis of royal purple (Cotinus coggygria Scop). Ind Crop Prod 199:116718

Aasim M, Yıldırım B, Say A, Ali SA, Aytaç S, Nadeem MA (2024) Artificial intelligence models for validating and predicting the impact of chemical priming of hydrogen peroxide (H2O2) and light emitting diodes on in vitro grown industrial hemp (Cannabis sativa L.). Plant Mol Biol 114(2):33

Akin M, Hand C, Eyduran E, Reed BM (2018) Predicting minor nutrient requirements of hazelnut shoot cultures using regression trees. Plant Cell Tissue Organ Cult 132(3):545-59

Akin M, Eyduran SP, Eyduran E, Reed BM (2020) Analysis of macro nutrient related growth responses using multivariate adaptive regression splines. Plant Cell Tissue Organ Cult 140:661–670

Ali SA, Aasim M (2024) Response surface methodology and artificial intelligence modeling for in vitro regeneration of Brazilian micro sword (Lilaeopsis brasiliensis). Plant Cell Tissue Organ Cult 157(1):10

Alizadeh MR, Nikoo MR (2018) A fusion-based methodology for meteorological drought estimation using remotesensing data. Remote Sens Environ 211:229-47

Arab MM, Yadollahi A, Shojaeiyan A, Ahmadi H (2016) Artificial neural network genetic algorithm as powerful tool to predict and optimize in vitro proliferation mineral medium for G× N15 rootstock. Front Plant Sci 7:1526

Arab MM, Yadollahi A, Ahmadi H, Eftekhari M, Maleki M (2017) Mathematical modeling and optimizing of in vitro hormonal combination for G× N15 vegetative rootstock proliferation using Artificial Neural Network-Genetic Algorithm (ANN-GA). Front Plant Sci 8:1853

Arab MM, Yadollahi A, Eftekhari M, Ahmadi H, Akbari M, Khorami SS (2018) Modeling and optimizing a new culture medium for in vitro rooting of G× N15 Prunus rootstock using artificial neural network-genetic algorithm. Sci Rep 8(1):1–18

Araghinejad S, Fayaz N, Hosseini-Moghari SM (2018) Development of a hybrid data driven model for hydrological estimation. Water Resour Manag 32:3737–3750

Arteta TA, Hameg R, Landin M, Gallego PP, Barreal ME (2018) Neural networks models as decision-making tool for in vitro proliferation of hardy kiwi. Eur J Hortic Sci 83:259–265
Aubry M, Paris S, Hasinoff SW, Kautz J, Durand F (2014) Fast local laplacian filters: theory and applications. ACM Trans Graph 33(5):1–14
Ayuso M, Ramil-Rego P, Landin M, Gallego PP, Barreal ME (2017) Computer-assisted recovery of threatened plants: keys for breaking seed dormancy of Eryngium viviparum. Front Plant Sci 8:2092
Ayuso M, García-Pérez P, Ramil-Rego P, Gallego PP, Barreal ME (2019) In vitro culture of the endangered plant Eryngium viviparum as dual strategy for its ex situ conservation and source of bioactive compounds. Plant Cell Tissue Organ Cult 138:427–435
Azevedo AM, Andrade VCD, Pedrosa CE, Oliveira CMD, Dornas MFS, Cruz CD, Valadares NR (2015) Application of artificial neural networks in indirect selection: a case study on the breeding of lettuce. Bragantia 74(4):387–393
Barone JO (2019) Use of multiple regression analysis and artificial neural networks to model the effect of nitrogen in the organogenesis of Pinus taeda L. Plant Cell Tissue Organ Cult 137(3):455–464
Basavaiah J, Arlene Anthony A (2020) Tomato leaf disease classification using multiple feature extraction techniques. Wirel Pers Commun 115(1):633–651
Bethge H, Winkelmann T, Lüdeke P, Rath T (2023) Low-cost and automated phenotyping system "Phenomenon" for multi-sensor in situ monitoring in plant in vitro culture. Plant Methods 19(1):42
Biau G, Cadre B, Rouvìère L (2019) Accelerated gradient boosting. Machine learning 108(6):971-92
Bidabadi SS, Jain SM (2020) Cellular, molecular, and physiological aspects of in vitro plant regeneration. Plants 9(6):702
Breiman L (2001) Random forests. Machine learning 45(1):5-32
Camargo A, Smith JS (2009) Image pattern classification for the identification of disease causing agents in plants.Comput Electron Agric 66(2):121-5
Domingues GF, Soares VP, Leite HG, Ferraz AS, Ribeiro CAAS, Lorenzon AS, Marcatti GE, Teixeira TR, de Castro NLM, Mota PHS, de Souza GSA (2020) Artificial neural networks on integrated multispectral and SAR data for high-performance prediction of eucalyptus biomass. Comput Electron Agric 168:105089
Duan Y, Jiang Y, Ye S, Karim A, Ling Z, He Y, Yang S, Luo K (2015) PtrWRKY73, a salicylic acid-inducible poplar WRKY transcription factor, is involved in disease resistance in Arabidopsis thaliana. Plant Cell Rep 34:831–841
Fakhrzad F, Jowkar A, Hosseinzadeh J (2022) Mathematical modeling and optimizing the in vitro shootproliferation of wallflower using multilayer perceptron non-dominated sorting genetic algorithm-II (MLP-NSGAII).PLoS One 17(9):e0273009
García-Pérez P, Lozano-Milo E, Landín M, Gallego PP (2020a) Combining medicinal plant in vitro culture with machine learning technologies for maximizing the production of phenolic compounds. Antioxidants 9(3):210
García-Pérez P, Lozano-Milo E, Landín M, Gallego PP (2020b) Machine learning technology reveals the concealed interactions of phytohormones on medicinal plant in vitro organogenesis. Biomolecules 10(5):746
Gray DJ, Trigiano RN (2018) Introduction to plant tissue culture. In: Plant tissue culture concepts and laboratory exercises. Routledge, New york, pp 3–7
Güney M (2019) Development of an in vitro micropropagation protocol for Myrobalan 29C rootstock. Turk J Agric For 43(6):569–575
Haberlandt G (1902) Cultur Versuche Mit Isolierten Pflanzenzellen Sitz. Ber Mat Nat ki Kais Akad Wiss, Wien 111:69–92
Hesami M, Jones AMP (2020) Application of artificial intelligence models and optimization algorithms in plant cell and tissue culture. Appl Microbiol Biotechnol 104(22):9449–9485
Hesami M, Naderi R, Yoosefzadeh-Najafabadi M, Rahmati M (2017) Data-driven modeling in plant tissue culture. J Appl Environ Biol Sci 7(8):37–44

Hesami M, Naderi R, Tohidfar M (2019) Modeling and optimizing in vitro sterilization of chrysanthemum via multilayer perceptron-non-dominated sorting genetic algorithm-II (MLP-NSGAII). Front Plant Sci 10:282

Hesami M, Condori-Apfata JA, Valderrama Valencia M, Mohammadi M (2020a) Application of artificial neural network for modeling and studying in vitro genotype-independent shoot regeneration in wheat. Appl Sci 10(15):5370

Hesami M, Naderi R, Tohidfar M, Yoosefzadeh-Najafabadi M (2020b) Development of support vector machine-based model and comparative analysis with artificial neural network for modeling the plant tissue culture procedures: effect of plant growth regulators on somatic embryogenesis of chrysanthemum, as a case study. Plant Methods 16:1–15

Hesami M, Jones AMP (2021) Modeling and optimizing callus growth and development in Cannabis sativa usingrandom forest and support vector machine in combination with a genetic algorithm. Appl Microbiol Biotechnol2021:1–12

Hildebrandt AC, Riker AJ, Duggar BM (1946) The influence of the composition of the medium on growth in vitro of excised tobacco and sunflower tissue cultures. Am J Bot 33:591–597

Hornik K, Stinchcombe M, White H (1989) Multilayer feedforward networks are universal approximators. Neural Netw 2(5):359–366

Horstman A, Bemer M, Boutilier K (2017) A transcriptional view on somatic embryogenesis. Regeneration 4(4):201-216

Huixian J (2020) The analysis of plants image recognition based on deep learning and artificial neural network. IEEE Access 8:68828–68841

Ivashchuk OA, Fedorova VI, Shcherbinina NV, Maslova EV, Shamraeva EO (2018) Microclonal propagation of plant process modeling and optimization of its parameters based on neural network

Jadhav SB, Patil SB (2015) Grading of soybean leaf disease based on segmented image using k-means clustering. Int J Adv Res Electr Commun Eng 4(6):1816–1822

Jafar A, Hameed MT, Akram N, Waqas U, Kim HS, Naqvi RA (2022) CardioNet: automatic semantic segmentation to calculate the cardiothoracic ratio for cardiomegaly and other chest diseases. J Pers Med 12(6):988

Jafar A, Bibi N, Naqvi RA, Sadeghi-Niaraki A, Jeong D (2024) Revolutionizing agriculture with artificial intelligence: plant disease detection methods, applications, and their limitations. Front Plant Sci 15:1356260

Jafari M, Daneshvar MH (2024) Machine learning-mediated Passiflora caerulea callogenesis optimization. PLoS One 19(1):e0292359

Jamshidi S, Yadollahi A, Arab MM, Soltani M, Eftekhari M, Sabzalipoor H, Sheikhi A, Shiri J (2019) Combining gene expression programming and genetic algorithm as a powerful hybrid modeling approach for pear rootstocks tissue culture media formulation. Plant Methods 15:1–18

Joung YH, Wu X, Roh MS (2020) Production of high-quality Ornithogalum thyrsoides cut flowers in one year from in vitro propagated plantlets influenced by plant growth regulators. Sci Hortic 269:109395

Karami O, Khadem A, Rahimi A, Zagari N, Aigner S, Offringa R (2024) Transient efflux inhibition improves plant regeneration by natural auxins. Plant J 118(2):295–303

Kaur S, Pandey S, Goel S (2019) Plants disease identification and classification through leaf images: a survey. Arch Comput Methods Eng 26:507–530

Kaya Y, Kuncan M, Kaplan K, Minaz MR, Ertunç HM (2021) A new feature extraction approach based on one dimensional gray level co-occurrence matrices for bearing fault classification. J Exper Theor Artif Intell 33(1):161–178

Khirade SD, Patil AB (2015) Plant disease detection using image processing. In: 2015 International conference on computing communication control and automation (ICCUBEA). IEEE, Pune, pp 768–771

Kirtis A, Aasim M, Katırcı R (2022) Application of artificial neural network and machine learning algorithms for modeling the in vitro regeneration of chickpea (Cicer arietinum L.). Plant Cell, Tissue Organ Cult 150(1):141–152

Kumar K, Jha AK, Kumar B, Karjagi CG, Abhishek A, Gambhir G, Aggarwal C, Tyagi A, Sharma P, Pandey P, Rakshit S (2022) Development of an efficient and reproducible in vitro regeneration and transformation protocol for tropical maize (Zea mays L.) using mature seed-derived nodal explants. Plant Cell Tissue and Organ Cult 148(3):557–571

Lan T, Tong C, Yu H, Shi X, Luo L (2020) Nonlinear process monitoring based on decentralized generalizedregression neural networks. Expert Syst Appl 150:113273

Long Y, Yang Y, Pan G, Shen Y (2022) New insights into tissue culture plant-regeneration mechanisms. Front Plant Sci 13:926752

Makandar A, Halalli B (2015) Image enhancement techniques using highpass and lowpass filters. Int J Comp Appl 109(14):12–15

Malabadi RB, Nethravathi TL, Kolkar KP, Chalannavar RK, Mudigoudra BS, Lavanya L, Abdi G, Baijnath H (2023) Cannabis sativa: applications of artificial intelligence and plant tissue culture for micropropagation. Int J Res Innov Appl Sci 8(6):117–142

Mansouri A, Fadavi A, Mortazavian SMM (2016) An artificial intelligence approach for modeling volume and fresh weight of callus–a case study of cumin (Cuminum cyminum L.). J Theor Biol 397:199–205

McFarland FL, Kaeppler HF (2024) History and current status of embryogenic culture-based tissue culture, transformation and gene editing of maize (Zea mays L.). Plant Genome:e20451

Meng T, Jing X, Yan Z, Pedrycz W (2020) A survey on machine learning for data fusion. Inform Fusion 57:115-29

Militante SV, Gerardo BD, Dionisio NV (2019) Plant leaf detection and disease recognition using deep learning. In: 2019 IEEE Eurasia conference on IOT, communication and engineering (ECICE). IEEE, Yunlin, pp 579–582

Mir R, Calabuig-serna A, Segui-simarro JM (2023) Plant somatic embryogenesis-A focus on the role of calcium in embryo induction and development. Sci Bull Ser F Biotechnol 27(1)

Mirza K, Aasim M, Katırcı R, Karataş M, Ali SA (2023) Machine learning and artificial neural networks-based approach to model and optimize ethyl methanesulfonate and sodium azide induced in vitro regeneration and morphogenic traits of water hyssops (Bacopa monnieri L.). J Plant Growth Regul 42(6):3471–3485

Moravej M, Amani P, Hosseini-Moghari SM (2020) Groundwater level simulation and forecasting using interior search algorithm-least square support vector regression (ISA-LSSVR). Groundw Sustain Dev 11:100447

Mridula MR, Nair AS, Kumar KS (2018) Genetic programming based models in plant tissue culture: an addendum to traditional statistical approach. PLoS Comput Biol 14(2):e1005976

Munasinghe SP, Somaratne S, Weerakoon SR, Ranasinghe C (2020) Prediction of chemical composition for callus production in Gyrinops walla Gaetner through machine learning. Inform Process Agric 7(4):511–522

Murashige T, Skoog F (1962) A revised medium for rapid growth and bio assays with tobacco tissue cultures. Physiol Plant 15(3):473

Nath J, Devi K, Kumar V, Sharma P, Sharma RK, Joshi R (2023) In vitro flower induction and cyto-genetic fidelity assessment of Chlorophytum comosum (Thunb.) Jacques var. comosum. S Afr J Bot 159:678–685

Nath J, Joshi S, Gupta S, Kesarwani V, Shankar R, Joshi R (2024) Genome-wide identification of WUSHEL-related homeobox genes reveals their differential regulation during cold stress and in vitro organogenesis in Picrorhiza kurrooa Royle ex Benth. In Vitro Cell Dev Biol 60(4):439–455

Nezami-Alanagh E, Garoosi GA, Maleki S, Landín M, Gallego PP (2017) Predicting optimal in vitro culture medium for Pistacia vera micropropagation using neural networks models. Plant Cell Tissue Organ Cult 129:19–33

Niazian M (2019) Application of genetics and biotechnology for improving medicinal plants. Planta 249:953–973

Niazian M, Niedbała G (2020) Machine learning for plant breeding and biotechnology. Agriculture 10(10):436

Niazian M, Noori SAS, Galuszka P, Mortazavian SMM (2017a) Tissue culture-based agrobacterium-mediated and in Planta transformation methods. Czech J Genet Plant Breed 4:133–143
Niazian M, Sadat Noori SA, Tohidfar M, Mortazavian SMM (2017b) Essential oil yield and agromorphological traits in some Iranian ecotypes of ajowan (Carum copticum L.). J Essent Oil Bear Plants 20(4):1151–1156
Niazian M, Sadat-Noori SA, Abdipour M (2018a) Modeling the seed yield of Ajowan (Trachyspermum ammi L.) using artificial neural network and multiple linear regression models. Ind Crops Prod 117:224–234
Niazian M, Sadat-Noori SA, Abdipour M, Tohidfar M, Mortazavian SMM (2018b) Image processing and artificial neural network-based models to measure and predict physical properties of embryogenic callus and number of somatic embryos in ajowan (Trachyspermum ammi (L.) Sprague). In Vitro Cell Dev Biol 54:54–68
Niazian M, Shariatpanahi ME, Abdipour M, Oroojloo M (2019) Modeling callus induction and regeneration in an anther culture of tomato (Lycopersicon esculentum L.) using image processing and artificial neural network method. Protoplasma 256:1317–1332
Osama K, Somvanshi P, Pandey AK, Mishra BN (2013) Modelling of nutrient mist reactor for hairy root growth using artificial neural network. Eur J Sci Res 97(4):516–526
Özcan E, Atar HH, Ali SA, Aasim M (2023) Artificial neural network and decision tree–based models for prediction and validation of in vitro organogenesis of two hydrophytes—Hemianthus callitrichoides and Riccia fluitans. In Vitro Cell Dev Biol 59(5):547–562
Ozyigit II, Dogan I, Hocaoglu-Ozyigit A, Yalcin B, Erdogan A, Yalcin IE, Cabi E, Kaya Y (2023) Production of secondary metabolites using tissue culture-based biotechnological applications. Front Plant Sci 14:1132555
Pasternak TP, Steinmacher D (2024) Plant growth regulation in cell and tissue culture in vitro. Plants 13(2):327
Phillips GC, Garda M (2019) Plant tissue culture media and practices: an overview. In Vitro Cell Dev Biol 55:242–257
Rashidi S, Ranjitkar P, Hadas Y (2014) Modeling bus dwell time with decision tree-based methods. Transp ResRec 2418(1):74-83
Raspor M, Motyka V, Kaleri AR, Ninković S, Tubić L, Cingel A, Ćosić T (2021) Integrating the roles for cytokinin and auxin in de novo shoot organogenesis: from hormone uptake to signaling outputs. Int J Mol Sci 22(16):8554
Rocha DI, Monte-Bello CC, Dornelas MC (2015) Alternative induction of de novo shoot organogenesis or somatic embryogenesis from in vitro cultures of mature zygotic embryos of passion fruit (Passiflora edulis Sims) is modulated by the ratio between auxin and cytokinin in the medium. Plant Cell Tissue Organ Cult 120:1087–1098
Rodríguez-Pérez R, Bajorath J (2022) Evolution of support vector machine and regression modeling in chemoinformatics and drug discovery. J Comput Aided Mol Des 36(5):355–362
Sadat Noori SA, Norouzi M, Karimzadeh G, Shirkool K, Niazian M (2017) Effect of colchicine-induced polyploidy on morphological characteristics and essential oil composition of ajowan (Trachyspermum ammi L.). Plant Cell Tissue Organ Cult 130:543–551
Sadat-Hosseini M, Arab MM, Soltani M, Eftekhari M, Soleimani A, Vahdati K (2022) Predictive modeling of Persianwalnut (Juglans regia L.) in vitro proliferation media using machine learning approaches: a comparative study ofANN, KNN and GEP models. Plant Methods 18(1):48
Salehi M, Farhadi S, Moieni A, Safaie N, Ahmadi H (2020) Mathematical modeling of growth and paclitaxel biosynthesis in Corylus avellana cell culture responding to fungal elicitors using multilayer perceptron-genetic algorithm. Front Plant Sci 11:1148
Sang YL, Cheng ZJ, Zhang XS (2018) Plant stem cells and de novo organogenesis. New Phytol 218(4):1334–1339
Sexton J, Everingham Y, Donald D, Staunton S, White R (2018) A comparison of non-linear regression methods for improved on-line near infrared spectroscopic analysis of a sugarcane quality measure. J Near Infrared Spectrosc 26(5):297–310
Sheikhi A, Mirdehghan SH, Arab MM, Eftekhari M, Ahmadi H, Jamshidi S, Gheysarbigi S (2020) Novel organic-based postharvest sanitizer formulation using Box Behnken design and mathe-

matical modeling approach: a case study of fresh pistachio storage under modified atmosphere packaging. Postharvest Biol Technol 160:111047
Shin J, Bae S, Seo PJ (2020) De novo shoot organogenesis during plant regeneration. J Exp Bot 71(1):63–72
Shiotani S, Fukuda T, Arai F, Takeuchi N, Sasaki K, Kinosita T (1994) Cell recognition by image processing: recognition of dead or living plant cells by neural network. JSME 37(1):202–208
Silva JCF, Teixeira RM, Silva FF, Brommonschenkel SH, Fontes EP (2019) Machine learning approaches and their current application in plant molecular biology: a systematic review. Plant Sci 284:37–47
Şimşek Ö (2024) Machine learning offers insights into the impact of in vitro drought stress on strawberry cultivars. Agriculture 14(2):294
Şimşek Ö, Dalda Şekerci A, Isak MA, Bulut F, İzgü T, Tütüncü M, Dönmez D (2024) Optimizing micropropagation and rooting protocols for diverse lavender genotypes: a synergistic approach integrating machine learning techniques. Horticulturae 10(1):52
Singh V, Misra AK (2017) Detection of plant leaf diseases using image segmentation and soft computing techniques. Inform Process Agric 4(1):41–49
Specht DF (1991) A general regression neural network. IEEE Trans Neural Netw 2(6):568-76
Srinivas N, Deb K (1994) Muiltiobjective optimization using nondominated sorting in genetic algorithms. Evol Comput 2(3):221–248
Srivastava A, Gupta S, Shanker K, Gupta N, Gupta AK, Lal RK (2020) Genetic diversity in Indian poppy (P. somniferum L.) germplasm using multivariate and SCoT marker analyses. Ind Crops Prod 144:112050
Steward FC, Mapes MO, Smith J (1958) Growth and organized development of cultured cells. I. Growth and division of freely suspended cells. Am J Bot 45:693–703
Su Q, Lu W, Du D, Chen F, Niu B, Chou KC (2017) Prediction of the aquatic toxicity of aromatic compounds to tetrahymena pyriformis through support vector regression. Oncotarget 8(30):49359
Thomas J, McDonald S, Noel-Storr A, Shemilt I, Elliott J, Mavergames C, Marshall IJ (2021) Machine learningreduced workload with minimal risk of missing studies: development and evaluation of a randomized controlled trialclassifier for Cochrane Reviews. J Clin Epidemiol 133:140-51
Tang Z, Fishwick PA (1993) Feedforward neural nets as models for time series forecasting. ORSA J Comput 5(4):374–385
Thangadurai K, Padmavathi K (2014) February. Computer visionimage enhancement for plant leaves disease detection. In: 2014 world congress on computing and communication technologies. IEEE, pp 173–175
Thorpe TA (2007) History of plant tissue culture. Mol Biotechnol 37:169–180
Wanas N, Auda G, Kamel MS, Karray FAKF (1998) On the optimal number of hidden nodes in a neural network. In: Conference proceedings. IEEE Canadian conference on electrical and computer engineering (Cat. No. 98TH8341), vol 2. IEEE, pp 918–921
Wang Z, Li H, Zhu Y, Xu T (2017) Review of plant identification based on image processing. Arch Comput Methods Eng 24:637–654
Wang H, Cimen E, Singh N, Buckler E (2020) Deep learning for plant genomics and crop improvement. Curr Opin Plant Biol 54:34–41
Wani JA, Sharma S, Muzamil M, Ahmed S, Sharma S, Singh S (2022) Machine learning and deep learning based computational techniques in automatic agricultural diseases detection: methodologies, applications, and challenges. Arch Comput Methods Eng 29(1):641–677
Xu P, Zhong Y, Xu A, Liu B, Zhang Y, Zhao A, Yang X, Ming M, Cao F, Fu F (2024) Application of developmental regulators for enhancing plant regeneration and genetic transformation. Plants 13(9):1272
Yang G, Huang LJ, Jiang D, Huang J, Cui C, Li N (2024) Development of indirect somatic embryogenesis and plant regeneration system with immature embryos of the cultivated traditional Chinese medicinal herb Polygonatum cyrtonema. Ind Crop Prod 214:118557

Zhang Q, Deng D, Dai W, Li J, Jin X (2020) Optimization of culture conditions for differentiation of melon based on artificial neural network and genetic algorithm. Sci Rep 10(1):3524

Zheng H, Li W, Jiang J, Liu Y, Cheng T, Tian Y, Zhu Y, Cao W, Zhang Y, Yao X (2018) A comparative assessment of different modeling algorithms for estimating leaf nitrogen content in winter wheat using multispectral images from an unmanned aerial vehicle. Remote Sens 10(12):2026

Artificial Intelligence and Machine Learning Models in Transformation of Biorefineries: Current Perspectives

Khushboo Jain, Pranhita Nimbalkar, Ayushi Malik, Mayank Suthar, Sunita Verma, and Avinash Marwal

Abstract

The use of artificial intelligence (AI), machine learning (ML), and real-time data capture to digitally alter biorefineries has drawn a lot of interest recently. The artificial intelligence (AI) and machine learning (ML) technologies analyze operational dynamics, process control to boost productivity, and high-dimensional output data. The focus of this chapter is on the categorization, past and present state of the biorefinery, and biorefineries that have achieved commercial success. There has been discussion on other uses of ML models and algorithms in supply chain analysis, biomass characterization, pretreatment, and physicochemical conversion processes. The strengths and limitations of AI and ML in biorefinery systems are comprehensively analyzed. This review finds that we can decrease the amount of time spent on experiments and increase productivity with the use of artificial intelligence and machine learning. When compared to biological processes, the use of AI and ML-based conversion models in biorefineries is simple, straightforward, and economically viable. In order to have a better knowledge of the biorefinery field's future, this chapter also presents the difficulties encountered in building and implementing such models.

K. Jain
Department of Biotechnology, Faculty of Science, MVSC, Janardhan Rai Nagar Rajasthan Vidyapeeth (Deemed to be University), Udaipur, Rajasthan, India

P. Nimbalkar
Research Scientist, Tojo Vikas Biotech Pvt Ltd, Bengaluru, Karnataka, India

A. Malik · M. Suthar · A. Marwal (✉)
Nanoparticle Synthesis and Bioinformatics Laboratory, Department of Biotechnology, Vigyan Bhawan, Block – B, New Campus, Mohanlal Sukhadia University, Udaipur, Rajasthan, India

S. Verma
Manas Enclave Residential Society, Indirapuri Street, Bhuwana, Udaipur, Rajasthan, India

R. K. Gaur et al. (eds.), *Plant Molecular Farming*,
https://doi.org/10.1007/978-981-95-3823-2_16

Keywords

Artificial intelligence · Machine learning · Biorefinery · Biological processes

1 Introduction

Biorefineries embody a sustainable framework for producing diverse bioenergy products by utilizing various biomass feedstocks and incorporating advanced conversion technologies. This multidisciplinary approach integrates hybrid methodologies from fields such as bioengineering, polymer chemistry, food science, and agriculture. Analogous to petroleum refineries that process crude oil into numerous petrochemical products, biorefineries are engineered to produce a wide range of bio-based outputs from biomass (Ohara 1993). Increasingly, the sustainability of bio-based products is being evaluated through certification systems designed to measure the environmental and societal impacts of such production processes (Ladu and Blind 2017). Moreover, incorporating sustainability principles during the early design stages of biorefineries enables the exploration of alternative configurations, significantly enhancing performance. This perspective is well-documented in academic research, spanning sustainability evaluations of electrolysis-enhanced biomass-to-liquid fuel technologies (Bernical et al. 2013) to the use of evolutionary algorithms in optimizing bioenergy supply chains (Ayoub et al. 2009). The methodologies and insights from these studies demonstrate substantial diversity, often revealing conflicting interpretations across various sources.

The growing global population has driven an exponential increase in demand for critical resources such as energy, chemicals, and consumer goods. This surge has exacerbated the depletion of finite, nonrenewable resources, which remain central to the global economy (Abid et al. 2023; Boretti and Rosa 2019; Henckens 2021; Xiao et al. 2022; Xu and Zhao 2023). A significant proportion of modern products relies on fossil-based resources, which are being exploited at a rate threatening their long-term availability. Therefore, reducing dependence on these resources is imperative—not only due to concerns over resource exhaustion but also to mitigate their profound environmental impacts, which degrade ecological quality and disrupt the Earth's natural cycles (Boccia et al. 2019; Gani 2021; Kühne et al. 2022; Megura and Gunderson 2022). These challenges have prompted a shift toward sustainable and circular production models centered on bio-based resources. However, transitioning to bio-based systems entails numerous obstacles, including limited raw material availability, immature technologies, scale-up challenges, and inconsistent regulatory frameworks (Dixon et al. 2015; Gerassimidou et al. 2021; Hina et al. 2022). Furthermore, the economic feasibility of biorefineries remains a critical issue, as they typically involve higher production costs, lower output capacities, and significant capital investments compared to conventional fossil fuel-based processes (Dragone et al. 2020; Patel and Shah 2021; Reshmy et al. 2022; Singh et al. 2022). Promoting large-scale biorefinery systems that are sustainable, economically viable, and socially acceptable is, therefore, essential.

The challenges associated with biorefineries can be effectively addressed through the integration of artificial intelligence (AI), a transformative tool that enhances their efficiency and facilitates market adoption. AI integration promotes the development of circular and sustainable production systems (Liao and Yao 2021; Nayak et al. 2018; Oruganti et al. 2023; Sharma et al. 2023). Implementing AI in biorefinery operations—commonly referred to as "smart biorefineries"—establishes a foundation for environmentally sustainable, efficient, and industrially viable production models. By leveraging AI technologies, biorefineries can process vast datasets, detect patterns, and deliver accurate insights to optimize bio-based production processes, ensuring the delivery of high-quality products that align with market requirements (Fig. 1) (Culaba et al. 2022; Ding and Grundmann 2022; Oldfrey et al. 2021; Solarte-Toro and Cardona Alzate 2021). AI's ability to optimize resource utilization is key to its transformative potential in biorefineries. By identifying optimal production stages and operational parameters, AI can improve product quality, minimize waste generation, and enhance supply chain management, encompassing production capacity, transportation logistics, and market integration (Franke et al. 2022; Haefner et al. 2021; Ramirez-Asis et al. 2022). This review chapter focuses on demonstrating how artificial intelligence and machine learning can be employed to optimize biorefinery operations and enhance the efficiency of bio-based product outputs.

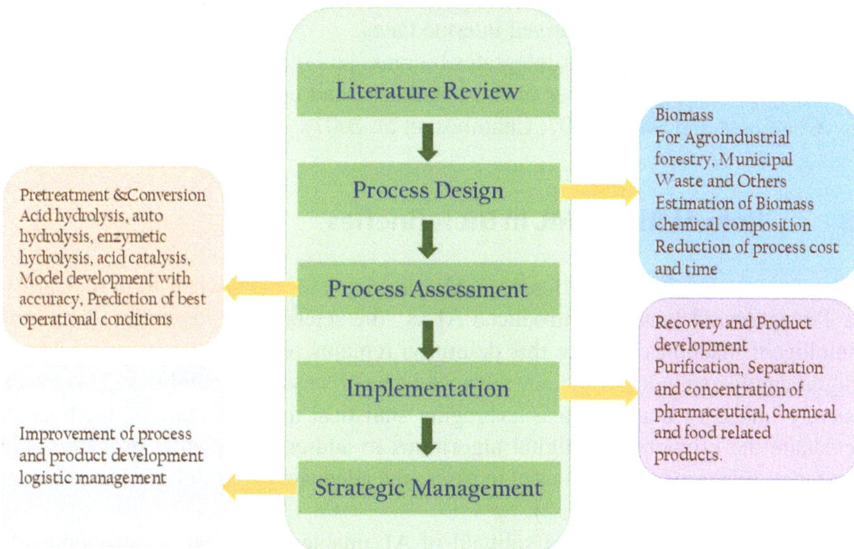

Fig. 1 Role of AI and ML in process designing of biorefineries

2 Types of Biorefineries

The classification of biorefinery systems has been extensively discussed in the literature, with two primary classification frameworks recognized (Kamm and Kamm 2005; Ree van and Annevelink 2007). Additional classification schemes have been proposed for specific biorefinery setups, such as the "liquid-phase catalytic processing biorefinery" concept (NSF 2008) and the "forest-based biorefinery." The major categories of biorefineries include:

1. *Lignocellulosic feedstock biorefinery*: Utilizes naturally dry raw materials like cellulose-rich biomass and waste.
2. *Whole crop biorefinery*: Processes raw materials such as cereals and maize.
3. *Green biorefinery*: Focuses on naturally wet biomass, including green grass, alfalfa, clover, or immature cereals.
4. *Two-platform concept biorefinery*: Involves sugar and syngas platforms for production.
5. *Conventional biorefinery*: Operates based on existing industries, such as sugar and starch production.
6. *Thermochemical biorefinery*: Employs a combination of various technological approaches.
7. *Marine biorefinery*: Relies on marine biomass as feedstock.
8. *Liquid-phase catalytic processing biorefinery*: Produces functionalized hydrocarbons from biomass-derived intermediates.
9. *Forest-based biorefinery*: Integrates biomass and other feedstocks to simultaneously produce pulp, paper fibers, chemicals, and energy (Axegård et al. 2007; Chambost and Stuart 2007; Chambost et al. 2007).

3 Role of AI and ML in Biorefineries

The origins of artificial intelligence (AI) trace back to 1956, when John McCarthy, a Turing Award laureate, introduced AI as "the science and ingenuity of making intelligent machines." While this definition remains relevant, the scope of AI has significantly expanded to include advanced data processing capabilities. AI is intrinsically linked to mathematics, leveraging statistical and probabilistic methods to evaluate the efficiency of digital algorithms in addressing specific problems. For instance, linear regression is used to assess the adequacy of model-generated values against real-world data (Fig. 2).

Machine learning (ML), a subfield of AI, enables machines to independently solve problems by analyzing data from various sources, including historical datasets and statistical trends (Jha et al. 2019). ML relies on inductive reasoning to establish relationships between inputs and outputs, which can then guide decision-making in new scenarios (Andrade Cruz et al. 2022). In recent years, ML has been increasingly employed in environmental applications, including air pollution control, wastewater management, and solid waste treatment. AI and ML algorithms provide

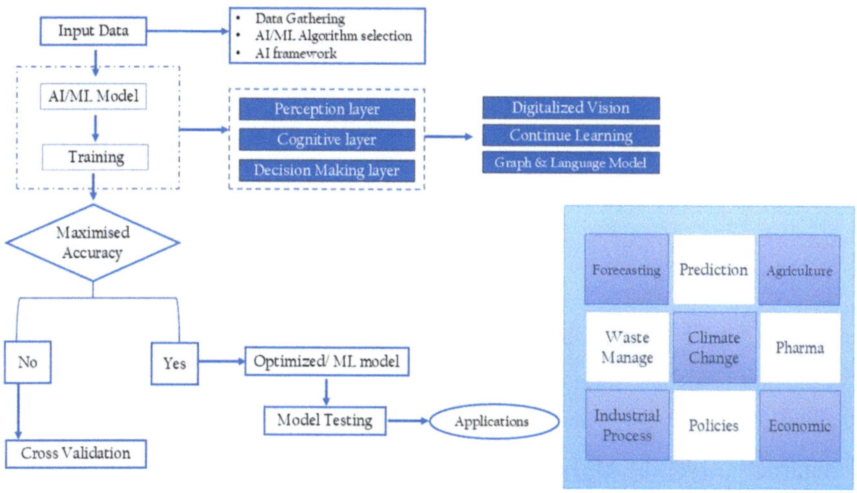

Fig. 2 General principle and working of AL and ML with applications

significant advantages in optimizing biorefinery systems. Unlike traditional kinetic or phenomenological models, ML approaches offer deeper insights into the uncertainties inherent in biological processes (Sundui et al. 2021). These algorithms can seamlessly integrate to monitor, optimize, predict, and detect faults in real time, addressing complex environmental systems and process parameters for resource recovery.

Interest in biorefineries' cascade production schemes has surged recently. However, the integration of biorefineries into value chains presents significant challenges due to the limited historical data available for processes still in the research and development phase. Critical considerations include the type of feedstock, its physicochemical properties, handling logistics, availability, and variability across temporal and geographical contexts. For waste streams, aspects such as transportation and pretreatment must also be effectively managed (Kumar et al. 2017). Moreover, emissions and waste streams generated by biorefineries must be carefully evaluated to minimize environmental impacts. The complexity of managing the vast array of variables in biorefineries necessitates the use of decision-support tools. These tools can evaluate interrelated variables to provide optimal solutions during the development of new production models. AI enables the analysis of large datasets through comprehensive algorithms, which are invaluable for forecasting, logistics optimization, and anomaly detection. These capabilities allow for immediate corrective actions, improving system reliability and performance (Jambrak et al. 2021; Mishra et al. 2021). Developing intelligent algorithm models capable of handling large datasets while accurately capturing interrelationships between parameters is essential. These models must be robust against external interferences to achieve optimal outcomes (Hatim et al. 2020; Wang et al. 2022).

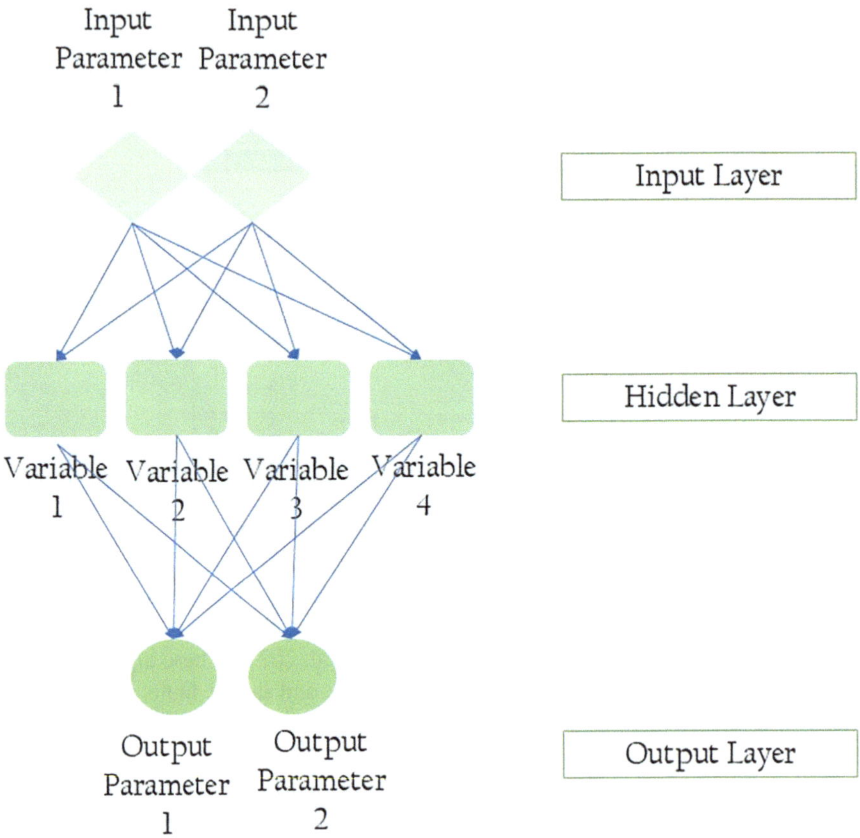

Fig. 3 Work flow chart of the ANN model in biorefinery

4 Main AI/ML Algorithms or Models

4.1 Artificial Neural Networks (ANNs)

Artificial Neural Networks (ANNs) are among the most widely used computational models for digitizing complex processes, particularly in biorefineries. ANNs are inspired by the nodal structures of the human brain and can process input parameters to generate output variables through interconnected layers (Olajuyin et al. 2022) (Fig. 3).

Structure and Functionality:

- ANNs comprise an *input layer*, one or more *hidden layers*, and an *output layer*.
- Input nodes correspond to biorefinery parameters and pass data to the first hidden layer, where nonrelevant patterns are filtered.
- Hidden layers process data by multiplying it with weights and applying functions (e.g., sigmoid and logistic) to improve efficiency.
- Output layers provide the final predicted values.

Learning Methods:

- Training of ANN models is achieved using methodologies such as *gradient descent*, *conjugate gradient*, and *quasi-Newton algorithms* (e.g., the Levenberg–Marquardt algorithm).

Applications:

ANNs have been used in environmental modeling, e.g., predicting microalgal growth in raceway ponds with multiple environmental factors such as pH, CO2, illumination, temperature, and hydraulic retention time (Supriyanto et al. 2018). These models have demonstrated high prediction accuracy ($R^2 > 0.93$).

Advantages:

- ANN models can "learn" by identifying relationships among variables, mimicking human brain functions (Nabavi-Pelesaraei et al. 2018).
- They are nonlinear, adaptive, and fault-tolerant (Guo et al. 2021).
- ANNs are capable of handling complex data, making them highly flexible and suitable for various predictive tasks.

Disadvantages:

- ANNs are "black-box" models that rely on empirical data and do not provide mechanistic insights into the processes (Guo et al. 2021).
- Overfitting can occur if the network size is too large, while too few neurons can lead to overgeneralization (Faris et al. 2019).
- They may converge slowly and tend to settle for local optima, making them less suitable for sparse datasets (Dong and Chen 2019).

4.2 Adaptive Neuro-Fuzzy Inference System (ANFIS)

ANFIS combines neural networks and fuzzy logic to evaluate process variables while accounting for epistemic and random uncertainties (Ayodeji et al. 2022).

Structure and functionality:

- ANFIS includes *membership functions (MFs)*, often referred to as "ANFIS rules," to create a Fuzzy Inference System (FIS).
- The model begins with a *knowledge-based layer* where MFs are defined, followed by fuzzy management rules.
- An *inference engine* adapts the rules to reduce inferences and optimize the solution.
- A final *transmutation stage* converts data into linguistic levels, generating output variables (Barati-Harooni et al. 2016) (Fig. 4).

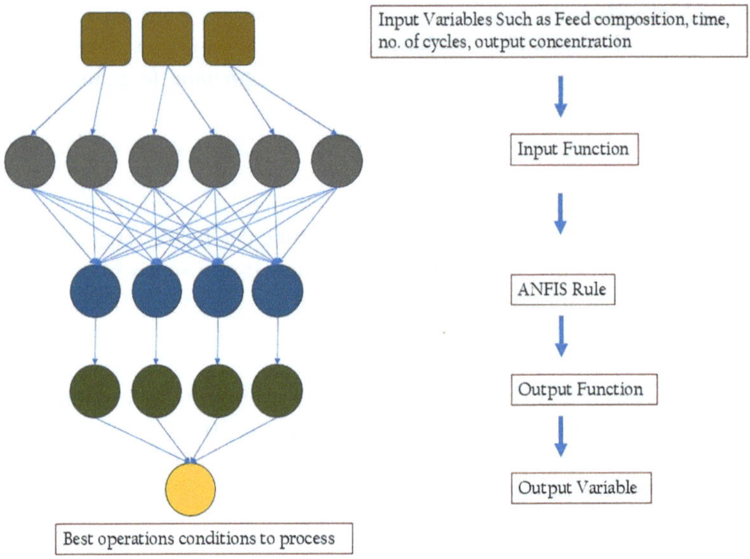

Fig. 4 Work flow chart of ANFIS model in biorefinery

Applications:
ANFIS has been applied to optimize process parameters in biorefineries and other environmental systems, particularly for scenarios involving uncertainty.

Advantages:
- ANFIS models are computationally efficient and faster to compute than ANNs.
- They are particularly effective for forecasting and performing time-period analyses.

Disadvantages:
- While ANFIS provides high accuracy in predictions, it may not be as statistically precise as ANNs for certain tasks that require high fidelity in the error between predicted and observed values.
- ANFIS models may not handle highly complex data relationships as effectively as ANNs.

4.3 Differential Evolution (DE)

Differential Evolution (DE) is a powerful optimization algorithm that generates initial random solutions and then refines them through mutation and recombination to find an optimal solution within defined boundaries (Magrassi et al. 2016). Although DE has proven useful in several fields, including biorefineries, there are some challenges associated with its application. DE can be affected by nonlinearity in input variables, overfitting, and lack of regularization. To overcome these challenges, it is recommended to construct the algorithm using predictive uncertainty, accounting for both epistemic and random uncertainties (Ayodeji et al. 2022). DE is often integrated into a Decision Support System (DSS) for comprehensive evaluation in environmental, economic, and social contexts. A DSS model helps in making transparent, sustainable decisions, particularly in early-stage design and implementation processes (Charlotte et al. 2016). The DSS interface is typically divided into four stages: input, intermediate, evaluation, and decision-making. These stages help optimize the model while ensuring accuracy and reliability in results. In cases requiring multiple criteria, Engineering Decision Support Systems (EDSS) can be used to assess environmental impacts while following life cycle assessment (LCA) principles (Buchert et al. 2019).

Advantages:

- DE is effective for complex optimization tasks and produces reliable results in diverse domains.
- The integration of DE with DSS models improves decision-making by providing a structured approach for sustainable solutions.

Disadvantages:

- DE's performance can be hindered by overfitting and nonlinearity in input data.
- It is sensitive to the lack of regularization, which can compromise optimization efficiency.

4.4 Support Vector Machine (SVM)

Support Vector Machine (SVM) is a supervised machine learning technique for regression and classification, initially introduced by Cortes and Vapnik (1995). The algorithm transforms data into a higher-dimensional feature space using a nonlinear mapping function and creates a decision boundary (hyperplane) that separates data points into different classes. SVM uses support vectors, which are the data points closest to the hyperplane, to maximize the margin between classes. The kernel technique and structural risk minimization (SRM) concept help SVM achieve better generalization compared to other algorithms like ANN, which uses empirical risk minimization (ERM). SVM is less prone to overfitting and has been successfully used in applications like microalgal classification and wastewater treatment (Hossain et al. 2022). However, SVM may experience reduced training efficiency when handling large datasets, and missing values can further complicate its performance (Fig. 5).

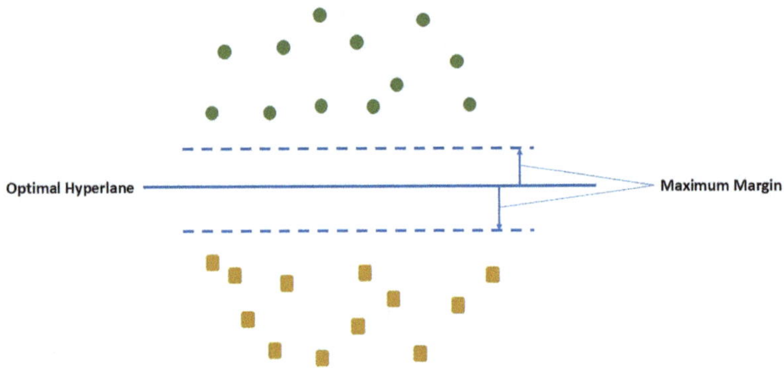

Fig. 5 Diagrammatic presentation of the SVM model in a biorefinery

Advantages:

- SVM is effective for large, complex datasets and can provide high accuracy in classification and regression tasks.
- It minimizes overfitting through the use of SRM, offering more robust performance compared to other models.

Disadvantages:

- SVM has slow training times with large datasets and struggles with missing data.
- Selection of an appropriate kernel function and handling of data size and missing values are critical for optimal performance.

4.5 Genetic Algorithm (GA)

Genetic Algorithm (GA) is an optimization model based on natural selection principles, such as selection, crossover, and mutation (Guo et al. 2021). In GA, the population undergoes iterative processes, starting with the selection of the fittest individuals, followed by recombination and mutation of genes. This approach continues until an optimal solution is found. GA is especially effective for combinatorial optimization problems and is widely used in optimizing microalgal growth and resource recovery parameters (Fig. 6).

Advantages:

- GA avoids local minima and is highly effective in solving complex optimization tasks.
- It can be integrated with other models, such as ANN and SVM, to enhance accuracy.

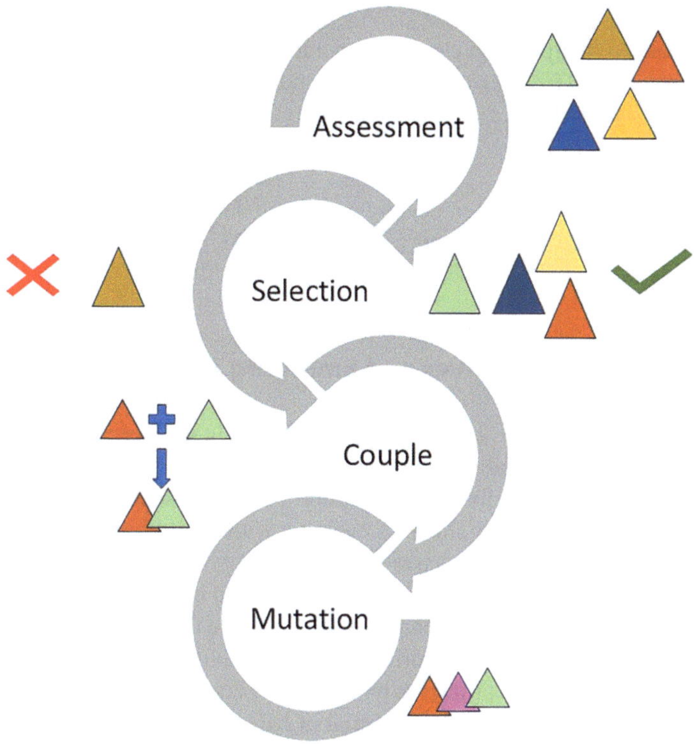

Fig. 6 Work flow chart of the GA model in biorefinery

Disadvantages:

- GA can suffer from premature convergence due to factors like selection operators and population size.
- The computational intensity of GA requires significant resources, making it more resource-demanding than other algorithms.

4.6 Decision Tree (DT) and Random Forest (RF) Algorithms

Decision Tree (DT) and Random Forest (RF) are popular supervised learning algorithms used for classification and regression. DT builds a tree structure to summarize training data, splitting it iteratively into smaller subsets based on input-output relationships. While DT is quick to train and interpret, it tends to overfit as the dataset grows larger. RF, an ensemble learning model, mitigates this issue by building multiple decision trees from randomly selected features and averaging their predictions.

Advantages:

- DT is easy to interpret and can model nonlinear relationships efficiently.
- RF overcomes DT's overfitting problem and is more accurate due to the use of multiple decision trees.

Disadvantages:

- DT models can overfit when dealing with large datasets.
- RF has a slower learning rate compared to DT, making it less efficient in some cases.

5 Future Perspectives and Emerging Trends

Machine learning (ML) has emerged as a crucial tool in the biorefinery sector, facilitating the analysis and prediction of complex processes with greater efficiency. The current landscape of biorefineries, including their classification and examples of commercially successful operations, is evolving with the integration of ML models. These models support various phases of the biorefinery lifecycle, including biomass characterization, pretreatment, lignin valorization, and both chemical and biochemical conversion processes. Additionally, they contribute to supply chain analysis and life cycle assessments. Each algorithm's benefits and limitations are discussed, highlighting their roles in optimizing biorefinery operations. The conclusion reflects on the current constraints and the potential of ML to drive future advancements in biorefinery technologies (Velidandi et al. 2023).

6 Advancements in AI and ML Technologies

The recent advancements in artificial intelligence (AI) and machine learning (ML) offer a comprehensive approach to data analysis, enabling real-time predictive systems for both microbes and feedstocks from laboratory to industrial scales. These data-driven techniques are particularly valuable in the design and development of bio-based fuels and chemicals. Initially applied to optimize algae growth for feedstock production, these methods now improve yield, product purity, analytics, and genetic and enzyme engineering strategies. While current models focus on optimizing specific product yields, there is significant potential for reducing resource intensity (in terms of time, capital, and operational costs) while creating adaptable product portfolios that can meet market demands (Decker et al. 2023). Machine learning (ML) plays an important role in the operation of biorefineries, especially in classification, regression, and optimization tasks. Various ML methodologies offer distinct advantages for analyzing lignocellulosic biorefineries. These models provide valuable insights across sectors such as high-value utilization transformation, chemical profiling, process emulation, and the development of functional materials from lignocellulose. The importance of correlated features, both numerical and categorical, is also highlighted, as these variables offer valuable information. Overall, ML tools hold significant promise for predicting the chemical composition of lignocellulosic feedstocks and improving understanding of pretreatment and conversion processes in biorefineries, as well as biofuel production (Jurinjak Tušek et al. 2024).

7 Integration of AI and ML in Biorefineries

The combination of AI and ML technologies in the biorefinery industry is enhancing biomass conversion processes, reducing production costs, and improving product yields. These technologies can process vast datasets generated during biorefinery operations, allowing for the prediction of optimal process parameters and enabling real-time decision-making for process control. For example, deep learning models have been utilized to predict biofuel yields from lignocellulosic feedstocks under various operational conditions, resulting in significant efficiency improvements (Garg et al. 2023). AI-powered modeling is also accelerating enzyme discovery and optimization, improving enzyme-substrate interactions and reaction pathways, which is essential for developing novel bioproducts (Singh et al. 2021). Another emerging trend involves AI-based predictive maintenance for biorefinery equipment, where machine learning models can forecast potential failures, minimizing downtime and maintenance costs (Carvalho et al. 2019). These innovations are expected to transform the biorefinery industry, making operations more sustainable and economically viable.

8 Impact on the Biorefinery Industry

AI and ML adoption in the biorefinery industry is poised to revolutionize traditional biomass conversion processes by optimizing feedstock utilization and minimizing waste. AI algorithms, for example, can predict biomass feedstock composition and its suitability for different conversion methods, guiding the selection of the most appropriate processing techniques. This approach can significantly mitigate the variability in feedstock quality, a challenge often encountered in biorefinery operations (Zhao et al. 2022). Additionally, AI-driven process optimization can lead to more efficient use of energy and resources in biorefineries. ML models can pinpoint the most energy-intensive stages of conversion and suggest adjustments to reduce energy consumption. This is especially crucial in the production of high-value chemicals and biofuels, where reducing operational costs can enhance the competitiveness of biorefined products relative to fossil-based alternatives (Arias et al. 2023). Moreover, AI can enable the development of integrated biorefineries that produce multiple products, such as biofuels, chemicals, and materials, thereby improving the economic sustainability of biorefineries (Shahid et al. 2021).

9 Role of Machine Learning in Biomass Valorization

Machine learning, particularly through artificial neural networks, has become an invaluable tool for predicting chemical behavior in biomass valorization processes relevant to biorefineries. These methods aim to optimize the production of desired products under specific reaction conditions. While neural networks have been predominantly applied in the petroleum industry, where extensive databases are available for effective algorithm training, they are now being adapted for biomass conversion. These systems excel in detecting patterns and assessing how different feedstock compositions affect complex conversion processes, without needing prior knowledge of reaction mechanisms or thermodynamics. This capability is crucial for dynamic, real-time control of biochemical processes such as fermentation and anaerobic digestion, enabling rapid adjustments to process parameters (Pomeroy et al. 2022).

10 Synergy Between Machine Learning and Biopolymers

The collaboration between machine learning (ML) and biopolymers offers significant potential to revolutionize sustainable energy, fuels, and biochemicals. Bio-based polymers, derived from renewable resources, are increasingly recognized for their role in sustainable energy and fuel industries. When integrated with ML techniques, these polymers can be optimized for energy systems, storage, and conversion processes. Case studies demonstrate the capabilities of bio-based polymers in energy applications, highlighting how ML can improve fuel efficiency and stimulate innovation. This integration also supports progress in biochemical production, particularly in drug delivery and medical device development. Leveraging the synergy between ML and bio-based polymers is crucial to advancing global sustainability goals in energy, fuels, and biochemicals, opening new avenues for innovation and sustainability (Bin Abu Sofian et al. 2024).

11 Conclusion, Research Gaps, and Recommendations

Despite the promising potential of AI and ML technologies in biorefineries, several research gaps must be addressed to fully leverage their capabilities. A major challenge is the scarcity of high-quality, large-scale datasets needed to train robust AI models. Biorefinery processes often involve complex and dynamic reactions, making it difficult to collect datasets that capture all relevant variables. Standardized data collection and sharing protocols are essential to support the development of more accurate predictive models (Arias et al. 2023). Another challenge lies in integrating AI and ML models with existing biorefinery infrastructures. While advancements in predictive modeling and process optimization have been made, incorporating AI into industrial-scale operations requires significant investment in digital infrastructure and personnel training. Collaborative efforts between academia, industry, and regulatory bodies are necessary to overcome these obstacles and create a roadmap for AI adoption in biorefineries (Shahid et al. 2021). Additionally, the ethical implications of AI use in biorefineries, such as data privacy and the potential for job displacement, must be proactively addressed. Establishing ethical guidelines and investing in workforce training will ensure that AI technologies are deployed responsibly and benefit all stakeholders (Zhao et al. 2022).

A systematic literature review is essential for understanding the current state of research in biorefinery design. Machine learning offers a powerful tool for enhancing the review process, particularly for topics where relevant information is spread across a wide range of publications. Traditional manual reviews of extensive literature can be time-consuming and inefficient, with many articles being irrelevant to the research questions. An ML-based approach can automate the screening of titles and abstracts, enabling researchers to conduct systematic reviews and meta-analyses with greater efficiency and accuracy. This automated process can be tailored to specific research queries and help identify knowledge gaps in biorefinery design and other related fields (Clauser et al. 2022).

Acknowledgments The authors are thankful to Mohanlal Sukhadia University, Udaipur, for providing the necessary facilities during the course of study.

Declaration of Competing Interest The authors declare that there are no conflicts of interest. The authors agree to be accountable for all aspects of the work to ensure that the questions related to the accuracy or integrity of any part of the work are appropriately investigated and resolved.

Funding This research did not receive any specific grant from funding agencies in the public, commercial, or not-for-profit sectors.

References

Abid N, Ceci F, Razzaq A (2023) Inclusivity of information and communication technology in ecological governance for sustainable resources management in G10 countries. Resour Policy 81:103378

Andrade Cruz I, Chuenchart W, Long F, Surendra KC, Renata Santos Andrade L, Bilal M, Liu H, Tavares Figueiredo R, Khanal SK, Fernando Romanholo Ferreira L (2022) Application of machine learning in anaerobic digestion: perspectives and challenges. Bioresour Technol 345:126433

Arias A, Feijoo G, Moreira MT (2023) How could Artificial Intelligence be used to increase the potential of biorefineries in the near future? A review. EnvironTechnol Innov 32:103277

Axegård P, Karlsson M, McKeogh P, Westenbroek A, Petit-Conil M, Eltrop N Niemela K (2007) A bio-solution to climate change. Final report of the Biorefinery Taskforce to the Forest-based sector Technology Platform. Available at: http://www.forestplatform.org/ [13 August 2009]

Ayodeji A, Amidu MA, Olatubosun SA, Addad Y, Ahmed H (2022) Deep learning for safety assessment of nuclear power reactors: reliability, explainability, and research opportunities. Prog Nucl Energy 151:104339

Ayoub N, Elmoshi E, Seki H, Naka Y (2009) Evolutionary algorithms approach for integrated bioenergy supply chains optimization. Energy Convers Manag 50(12):2944–2955

Barati-Harooni A, Najafi-Marghmaleki A, Mohammadi AH (2016) ANFIS modeling of ionic liquids densities. J Mol Liq 224:965–975

Bernical Q, Joulia X, Noirot-Le Borgne I, Floquet P, Baurens P, Boissonnet G (2013) Sustainability assessment of an integrated high temperature steam electrolysis-enhanced biomass to liquid fuel process. Ind Eng Chem Res 52(22):7189–7195

Bin Abu Sofian ADA, Sun X, Gupta VK, Berenjian A, Xia A, Ma Z, Show PL (2024) Advances, synergy, and perspectives of machine learning and biobased polymers for energy, fuels, and biochemicals for a sustainable future. Energy Fuels 38(3):1593–1617

Boccia F, Di Donato P, Covino D, Poli A (2019) Food waste and bio-economy: A scenario for the Italian tomato market. J Clean Prod 227:424–433

Boretti A, Rosa L (2019) Reassessing the projections of the World Water Development Report. Npj Clean Water 2:1–6

Buchert T, Ko N, Graf R, Vollmer T, Alkhayat M, Brandenburg E, Stark R, Klocke F, Leistner P, Schleifenbaum JH (2019) Increasing resource efficiency with an engineering decision support system for comparison of product design variants. J Clean Prod 210:1051–1062

Carvalho TP, Soares FA, Vita R, Francisco RDP, Basto JP, Alcalá SG (2019) A systematic literature review of machine learning methods applied to predictive maintenance. Comput Ind Eng 137:106024

Chambost V, Stuart PR (2007) Selecting the most appropriate products for the forest biorefinery. Ind Biotechnol 3:112–119

Chambost V, Eamer B, Stuart PR (2007) Forest biorefi nery: getting on with the job. Pulp Paper Canada 108:18–22

Charlotte LC, Abdellatif B, Jean-Pierre B, Patrice B, Guillaume B, Brigitte C, Stephane D, Sebastien D, Juliette D, Caroline S, Claire V (2016) A decision support system using multi-source scientific data, an ontological approach and soft computing- application to eco-efficient biorefinery. In: 2016 IEEE Int Conf Fuzzy Syst FUZZ-IEEE, pp 249–256

Clauser NM, Felissia FE, Area MC, Vallejos ME (2022) Integrating the new age of bioeconomy and industry 4.0 into biorefinery process design. Bioresources 17(3):5510

Cortes C, Vapnik V (1995) Support-vector networks. Mach Learn 20:273–297

Culaba AB, Mayol AP, San Juan JLG, Vinoya CL, Concepcion RS, Bandala AA, Vicerra RRP, Ubando AT, Chen WH, Chang JS (2022) Smart sustainable biorefineries for lignocellulosic biomass. Bioresour Technol 344:126215

Decker SR, Brunecky R, Yarbrough JM, Subramanian V (2023) Perspectives on biorefineries in microbial production of fuels and chemicals. Front Indus Microbiol 1:1202269

Ding Z, Grundmann P (2022) Development of biorefineries in the bioeconomy: A fuzzy-set qualitative comparative analysis among european countries. Sustainability 14:90

Dixon T, McCoy ST, Havercroft I (2015) Legal and regulatory developments on CCS. Int J Greenhouse Gas Control 40:431–448

Dong C, Chen J (2019) Optimization of process parameters for anaerobic fermentation of corn stalk based on least squares support vector machine. Bioresour Technol 271:174–181

Dragone G, Kerssemakers AAJ, Driessen JLSP, Yamakawa CK, Brumano LP, Mussatto SI (2020) Innovation and strategic orientations for the development of advanced biorefineries. Bioresour Technol 302:122847

Faris H, Mirjalili S, Aljarah I (2019) Automatic selection of hidden neurons and weights in neural networks using grey wolf optimizer based on a hybrid encoding scheme. Int J Mach Learn Cybern 10:2901–2920

Franke F, Franke S, Riedel R (2022) AI-based improvement of decision-makers' knowledge in production planning and control. IFAC-Papers Online 55:2240–2245

Gani A (2021) Fossil fuel energy and environmental performance in an extended STIRPAT model. J Clean Prod 297:126526

Garg D, Sridhar K, Stephen Inbaraj B, Chawla P, Tripathi M, Sharma M (2023) Nano-biofertilizer formulations for agriculture: A systematic review on recent advances and prospective applications. Bioengineering 10(9):1010

Gerassimidou S, Martin OV, Chapman SP, Hahladakis JN, Iacovidou E (2021) Development of an integrated sustainability matrix to depict challenges and trade-offs of introducing bio-based plastics in the food packaging value chain. J Clean Prod 286:125378

Guo HN, Wu SB, Tian YJ, Zhang J, Liu HT (2021) Application of machine learning methods for the prediction of organic solid waste treatment and recycling processes: a review. Bioresour Technol 319:124114

Haefner N, Wincent J, Parida V, Gassmann O (2021) Artificial intelligence and innovation management: A review, framework, and research agenda. Technol Forecast Soc Change 162:120392

Hatim QY, Saldana C, Shao G, Kim DB, Morris KC, Witherell P, Rachuri S, Kumara S (2020) A decision support methodology for integrated machining process and operation plans for sustainability and productivity assessment. Int J Adv Manuf Technol 107:3207–3230

Henckens T (2021) Scarce mineral resources: extraction, consumption and limits of sustainability. Resour Conserv Recycl 169:105511

Hina M, Chauhan C, Kaur P, Kraus S, Dhir A (2022) Drivers and barriers of circular economy business models: where we are now, and where we are heading. J Clean Prod 333:130049

Hossain SMZ, Sultana N, Jassim MS, Coskuner G, Hazin LM, Razzak SA, Hossain MM (2022) Soft-computing modeling and multiresponse optimization for nutrient removal process from municipal wastewater using microalgae. J Water Process Eng 45:102490

Jambrak AR, Nutrizio M, Djekić I, Pleslić S, Chemat F (2021) Internet of nonthermal food processing technologies (IoNTP): food industry 4.0 and sustainability. Appl Sci 11:686

Jha K, Doshi A, Patel P, Shah M (2019) A comprehensive review on automation in Agri culture using artificial intelligence. Artif Intell Agric 2:1–12

Jurinjak Tušek A, Petrus A, Weichselbraun A, Mundani R, Müller S, Barkow I et al (2024) Systematic review of machine-learning techniques to support development of lignocellulose biorefineries. Chem Biochem Eng Q 38(3):241–263

Kamm B, Kamm M (2005) Principles of biorefi neries. Appl Microbiol Biotechnol 64:137–145

Kühne K, Bartsch N, Tate RD, Higson J, Habet A (2022) Carbon bombs- mapping key fossil fuel projects. Energy Policy 166:112950

Kumar S, Srinivas GLK, Banerjee R (2017) Lignocellulosic biomass availability map: A GIS-based approach for assessing production statistics of lignocellulosics and its application in biorefinery. In: Lignocellul Prod Ind Appl, pp 197–214

Ladu L, Blind K (2017) Overview of policies, standards and certifications supporting the European bio-based economy. Curr Opin Green Sustain Chem 8:30–35

Liao M, Yao Y (2021) Applications of artificial intelligence-based modeling for bioenergy systems: A review. GCB Bioenergy 13:774–802

Magrassi F, Del Borghi A, Gallo M, Strazza C, Robba M (2016) Optimal planning of sustainable buildings: integration of life cycle assessment and optimization in a decision support system (DSS). Energies 9:490

Megura M, Gunderson R (2022) Better poison is the cure? Critically examining fossil fuel companies, climate change framing, and corporate sustainability reports. Energy Res Soc Sci 85:102388

Mishra A, Choudhary M, Das TR, Saren P, Bhattacherjee P, Thakur N, Tripathi SK, Upadhaya S, Kim HS, Murugan NA, Tiwari A, Patra S, Hussain CM, Mishra A, Shukla SK, Joshi GM (2021) Sustainable chemical preventive models in COVID-19: understanding, innovation, adaptations, and impact. J Indian Chem Soc 98:100164

Nabavi-Pelesaraei A, Rafiee S, Mohtasebi SS, Hosseinzadeh-Bandbafha H, Chau K w (2018) Integration of artificial intelligence methods and life cycle assessment to predict energy output and environmental impacts of paddy production. Sci Total Environ 631–632:1279–1294

Nayak M, Dhanarajan G, Dineshkumar R, Sen R (2018) Artificial intelligence driven process optimization for cleaner production of biomass with co-valorization of wastewater and flue gas in an algal biorefinery. J Clean Prod 201:1092–1100

NSF (2008) Breaking the chemical and engineering barriers to lignocellulosic biofuels: Next generation hydrocarbon biorefineries, edited by Huber GW. University of Massachusetts Amherst, National Science Foundation. Chemical, Bioengineering, Environmental, and Transport Systems Division, Washington. 180 pp. Available at: http://www.ecs.umass.edu/biofuels/Images/Roadmap2-08.pdf [13 August 2009]

Ohara H (1993) Kinetics and process development of l-lactic acid fermentation. PhD thesis, Osaka University, Osaka

Olajuyin EA, Olulope PK, Fasina ET (2022) An overview on reliability assessment in power systems using CI approaches. Arch Electr Eng 71:425–443

Oldfrey B, Barbareschi G, Morjaria P, Giltsoff T, Massie J, Miodownik M, Holloway C (2021) Could assistive technology provision models help pave the way for more environmentally sustainable models of product design, manufacture and service in a post-COVID world? Sustainability 13:10867

Oruganti RK, Biji AP, Lanuyanger T, Show PL, Sriariyanun M, Upadhyayula VKK, Gadhamshetty V, Bhattacharyya D (2023) Artificial intelligence and machine learning tools for high-performance microalgal wastewater treatment and algal biorefinery: A critical review. Sci Total Environ 876:162797

Patel A, Shah AR (2021) Integrated lignocellulosic biorefinery: gateway for production of second generation ethanol and value added products. J Bioresour Bioprod 6:108–128

Pomeroy B, Grilc M, Likozar B (2022) Artificial neural networks for bio-based chemical production or biorefining: a review. Renew Sust Energ Rev 153:111748

Ramirez-Asis E, Vilchez-Carcamo J, Thakar CM, Phasinam K, Kassanuk T, Naved M (2022) A review on role of artificial intelligence in food processing and manufacturing industry. Mater Today Proc 51:2462–2465

Ree van R, Annevelink B (2007) Status Report Biorefinery 2007. Report 847, SenterNovem, Wageningen

Reshmy R, Philip E, Madhavan A, Sirohi R, Pugazhendhi A, Binod P, Kumar Awasthi M, Vivek N, Kumar V, Sindhu R (2022) Lignocellulose in future biorefineries: strategies for cost-effective production of biomaterials and bioenergy. Bioresour Technol 344:126241

Shahid MK, Batool A, Kashif A, Nawaz MH, Aslam M, Iqbal N, Choi Y (2021) Biofuels and biorefineries: development, application and future perspectives emphasizing the environmental and economic aspects. J Environ Manag 297:113268

Sharma V, Tsai ML, Chen CW, Sun PP, Nargotra P, Di Dong C (2023) Advances in machine learning technology for sustainable biofuel production systems in lignocellulosic biorefineries. Sci Total Environ 886:163972

Singh N, Malik S, Gupta A, Srivastava KR (2021) Revolutionizing enzyme engineering through artificial intelligence and machine learning. Emerg Top Life Sci 5(1):113–125

Singh N, Singhania RR, Nigam PS, Di Dong C, Patel AK, Puri M (2022) Global status of lignocellulosic biorefinery: challenges and perspectives. Bioresour Technol 344:126415

Solarte-Toro JC, Cardona Alzate CA (2021) Biorefineries as the base for accomplishing the sustainable development goals (SDGs) and the transition to bioeconomy: technical aspects, challenges and perspectives. Bioresour Technol 340:125626

Sundui B, Ramirez Calderon OA, Abdeldayem OM, Lázaro-Gil J, Rene ER, Sambuu U (2021) Applications of machine learning algorithms for biological wastewater treat ment: updates and perspectives. Clean Techn Environ Policy 23:127–143

Supriyanto, Noguchi R, Ahamed T, Mikihide D, Watanabe MM (2018) A decision tree approach to estimate the microalgae production in open raceway pond. IOP Conf Ser Earth Environ Sci 209

Velidandi A, Gandam PK, Chinta ML, Konakanchi S, reddy Bhavanam A, Baadhe RR et al (2023) State-of-the-art and future directions of machine learning for biomass characterization and for sustainable biorefinery. J Energy Chem 81:42–63

Wang J, Zhang X, Zhang F, Wan J, Kou L, Ke W (2022) Review on evolution of intelligent algorithms for transformer condition assessment. Front Energy Res 10:618

Xiao C, Feng Z, You Z, Zheng F (2022) Population boom in the borderlands globally. J Clean Prod 371:133685

Xu Y, Zhao F (2023) Impact of energy depletion, human development, and income distribution on natural resource sustainability. Resour Policy 83:103531

Zhao H, Hillson N, Kleese van Dam K, Tanjore D (2022) Artificial intelligence and machine learning for bioenergy research: opportunities and challenges

Modern Approaches of Essential Oils in the Development of Botanical Pesticides

Shahzeen Fatma, Rajneesh Rao, and Ashok Kumar

Abstract

The increasing concerns over the environmental and health hazards associated with synthetic pesticides have driven the search for sustainable and eco-friendly alternatives. Higher plants are a great reservoir of various phytochemicals and secondary metabolites. Essential oils (EOs), derived from aromatic plants, have gained significant attention as an eco-friendly substitute for prevalent synthetic pesticides. Their diverse bioactive constituents, including terpenoids, phenolics, and other alkaloids, exhibit potent insecticidal, antifungal, bactericidal, and herbicidal properties. Advanced extraction methods, including supercritical fluid extraction and nanoencapsulation, are used in modern botanical pesticide development processes to improve stability, controlled release, and efficacy. Nanoformulation strategies, including nanoencapsulation, nanoliposomes, and polymeric nanoparticles, offer controlled release, enhanced bioavailability, and increased persistence in the field. Additionally, synergistic combinations with biopolymers, surfactants, and other botanical compounds further optimize the bioactivity, improve bioavailability, reduce environmental persistence, and minimize resistance development in target pests. Research advancements in nanotechnology and other omics technologies are further optimizing EO-based pesticides for targeted pest control with minimal nontarget effects. These developments establish EOs as viable options for sustainable agriculture, providing a safe and efficient substitute for conventional chemical pesticides.

S. Fatma · R. Rao · A. Kumar (✉)
Department of Botany, Deen Dayal Upadhyaya Gorakhpur University, Gorakhpur, India
e-mail: ashok.bot@ddugu.ac.in

R. K. Gaur et al. (eds.), *Plant Molecular Farming*,
https://doi.org/10.1007/978-981-95-3823-2_17

Keywords

Essential oils · Terpenoids · Botanical pesticides · Nanoencapsulation · Nanoemulsion · Integrated pest management · Sustainable agriculture

1 Introduction

In highly populated countries, crop production and protection are a burning issue due to limited arable land, pest attacks, and a critical level of pollution. In order to prevent losses in the field and during post-harvest storage, synthetic pesticides have been utilized extensively since the last several decades for pest management. The majority of prevalent synthetic pesticides are now evaluated to minimize their application due to resistance and negative effects on the environment and human health (Ahmed et al. 2024). Over the past few years, farmers and consumers have become more concerned about the residue and health impacts of pesticides due to their overuse and improper application. Nearly 3 million tonnes of synthetic pesticides are consumed globally every year, with an approximate expense of $100 billion (Kumar et al. 2022). The traces of chemical compounds that are left on or in crops, water, soil, and air, and that enter the environment either accidentally or as a result of application, are known as pesticide residues. Novel, highly selective, and readily degradable compounds are required to address the risks posed by persistent pesticides and residues. The exploration of natural pest control agents could potentially substitute synthetic chemical pesticides and reduce the adverse effects that synthetic chemicals cause to humans and the environment (Ayilara et al. 2023).

The growing interest of consumers toward green consumerism has resulted in the use of plant-derived natural products as eco-friendly biopesticides (Souto et al. 2021). These natural biopesticides, sometimes known as "green pesticides," might be essential oils (EOs) or crude extracts made from any plant parts. The term "green pesticides" refers to a variety of naturally occurring control materials that reduce the number of pests and ultimately improve quality and productivity (Mossa 2016). These are derived from various organic sources, which are thought to be less harmful to the environment, human health, animal health, habitats, and the ecosystem (Nollet and Rathore 2019; Abdelatti and Hartbauer 2020).

Essential oils (EOs) are volatile, hydrophobic substances with distinct aromatic fragrances that are derived from different plant parts. The average EO content in most plants ranges from 1% to 2%. The essence in EOs is chemically caused by monoterpenes, sesquiterpenes, and other higher terpenes. However, monoterpenes and sesquiterpenes are the main contributors, while higher terpenes are only present as minor constituents (Masyita et al. 2022). These EOs serve as secondary metabolites, playing key roles in plant defense against pests, pathogens, and herbivores, as well as in attracting pollinators (Al-Khayri, et al. 2023). Terpenoids, alcohols, phenylpropanoids, aldehydes, esters, and other organic compounds make up their complex composition and are valued for their therapeutic, antimicrobial, and aromatic properties (Masyita et al. 2022). EOs are widely used in industries such as cosmetics, pharmaceuticals, food flavoring, and aromatherapy, making them important both biologically and commercially (Bolouri et al. 2022).

Hence, EOs have received significant attention as sustainable alternatives to synthetic pesticides in modern agriculture and exhibited potent pesticidal properties, including insecticidal, antifungal, antiviral, and herbicidal activities. In recent decades, the effectiveness of EOs and their bioactive compounds as biopesticides has been extensively studied against various pathogens and their associated toxins (Kumar et al. 2013; Ben Miri 2025). The U.S. Code of Federal Regulations has already listed a number of EOs and their bioactive compounds as Generally Recognized As Safe (GRAS), including *Mentha piperita*, *Pimenta officinalis*, *Matricaria chamomilla*, *Cuminum cyminum*, *Coriandrum sativum*, *Ocimum basilicum*, carvacrol, carvone, cinnamaldehyde, citral, p-cymene, eugenol, limonene, menthol, linalool, vanillin, citral, thymol, etc. (Prakash et al. 2018; U.S. Code of Federal Regulations 2025). With growing concerns over the environmental and health impacts of chemical pesticides, EOs offer a promising solution for integrated pest management (IPM). This chapter explores the innovative strategies and technological advancements that have enabled the development of botanical pesticides from plant EOs.

2 Chemical Composition of Plant Essential Oils

Plant EOs are complex mixtures of volatile organic compounds produced as secondary metabolites in various plant tissues such as leaves, flowers, fruits, stems, and roots. The chemical profile of EOs varies significantly depending on the plant species, geographic location, growth conditions, harvesting time, and the method of extraction (e.g., steam distillation, cold pressing, or solvent extraction) (Tran et al. 2024). Advanced analytical techniques like gas chromatography-mass spectrometry (GC-MS) and high-performance liquid chromatography (HPLC) have facilitated the detailed characterization of these components, enabling a deeper understanding of their ecological roles and industrial applications (Rodilla et al. 2024). The main classes of compounds found in EOs are as follows:

2.1 Terpenes

These are the primary constituents of EOs, responsible for their characteristic aromas and biological activities. They play crucial roles in plant defense against herbivores and pathogens, as well as in attracting pollinators and seed dispersers. They are hydrocarbons derived from isoprene units (C_5H_8), i.e., 2-methyl-1,3-butadiene (Hillier and Lathe 2024), and are classified as follows based on the number of isoprene units.

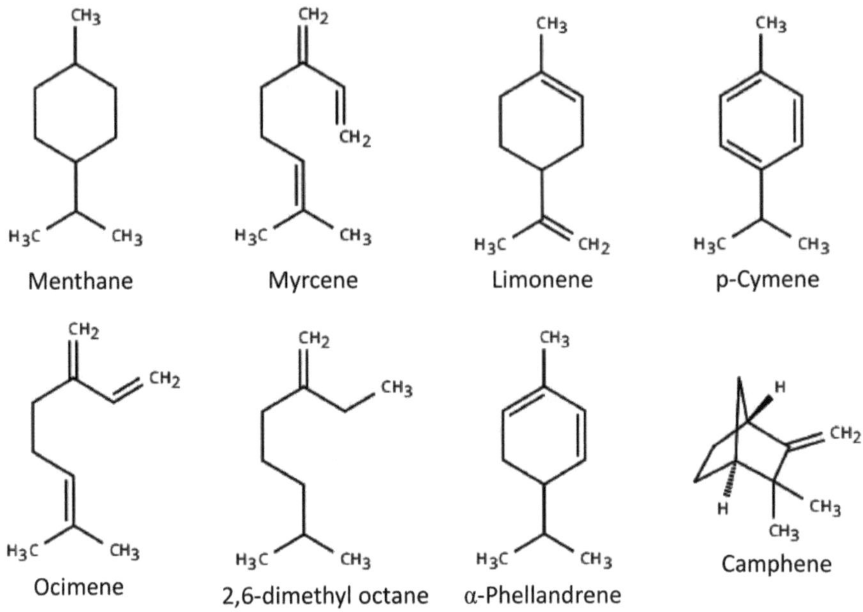

Fig. 1 Chemical structure of some common monoterpenes of EOs

2.1.1 Monoterpenes ($C_{10}H_{16}$)

Monoterpenes are a class of volatile organic compounds composed of two isoprene units ($C_{10}H_{16}$) and are the chief constituents of many EOs. They exhibit diverse biological activities, including insecticidal, antifungal, antimicrobial, and antioxidant properties, making them valuable in botanical pesticide formulations. Some monoterpenes, such as limonene, α-pinene, and linalool, disrupt insect neurotransmission, inhibit microbial growth, and interfere with pest metabolism (Qasim et al. 2024). Their high volatility and rapid degradation in the environment pose challenges, but modern nanoformulation techniques help to enhance their stability and efficacy for sustainable pest management (Fig. 1).

2.1.2 Sesquiterpenes ($C_{15}H_{24}$)

Sesquiterpenes are another class of terpenes that consist of three isoprene units ($C_{15}H_{24}$) and are commonly found in EOs of plants. They are characterized by diverse chemical structures, including acyclic, monocyclic, and bicyclic forms. Sesquiterpenes exhibited a wide range of biological activities, such as anti-inflammatory, antimicrobial, and antioxidant properties (Sun et al. 2024). Sesquiterpenes also play important roles in plant defense and ecological interactions (Chadwick et al. 2013). Some common sesquiterpenes, e.g., farnesene, caryophyllene, humulene, artemisinin, etc., are quite frequent in EOs (Fig. 2).

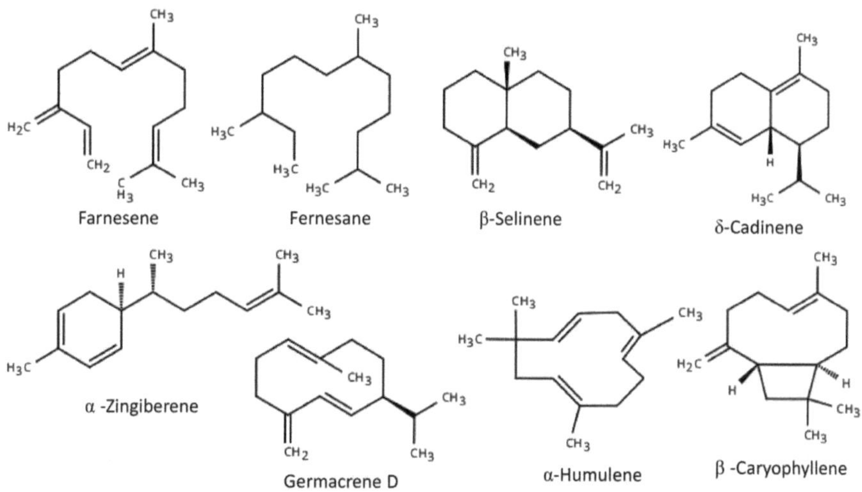

Fig. 2 Chemical structure of some sesquiterpenes

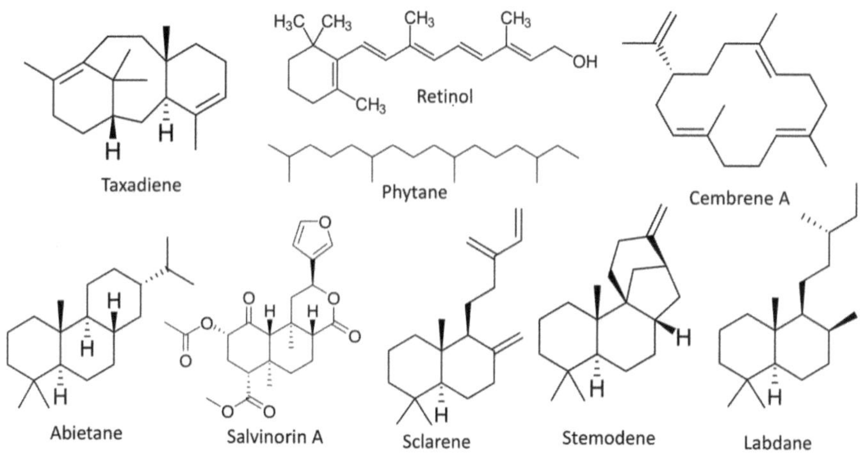

Fig. 3 Chemical structure of some diterpenes

2.1.3 Diterpenes ($C_{20}H_{32}$)

Diterpenes are also a class of terpenes composed of four isoprene units ($C_{20}H_{32}$). They occur in various plants, fungi, and marine organisms, often serving as defensive chemicals, growth regulators, or signaling molecules (Murphy and Zerbe 2020). Notable diterpenes include retinol (vitamin A), gibberellins (plant hormones), and taxol (an anticancer drug), highlighting their importance in medicine, agriculture, and ecology (Fig. 3).

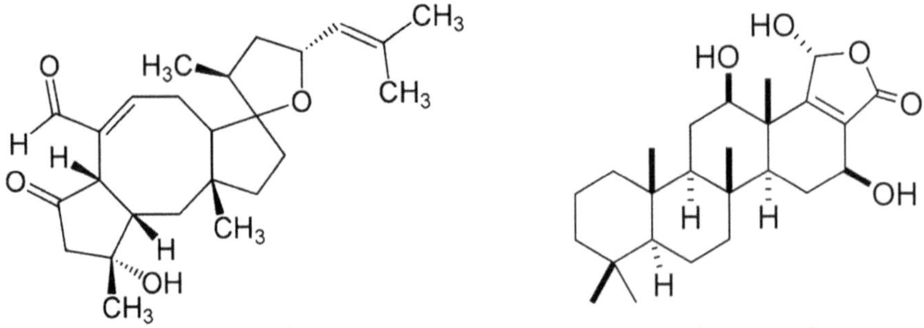

Ophiobolin Sesterstatin

Fig. 4 Chemical structure of ophiobolin and sesterstatin (sesterpenes)

2.1.4 Sesterpenes ($C_{25}H_{40}$)

Sesterpenes are composed of five isoprene units ($C_{25}H_{40}$), making them larger than simple terpenes. These terpenes are also rare and are mainly found in marine organisms and some plants. Sesterpenes also exhibit diverse biological activities, including antimicrobial, antifungal, and cytotoxic properties (Mabou and Yossa 2021). Ophiobolin and sesterstatin are notable examples that have potential in pharmaceutical applications (Fig. 4).

2.1.5 Triterpenes ($C_{30}H_{48}$)

Triterpenes are composed of six isoprene units ($C_{30}H_{48}$) and serve as precursors for steroids and other bioactive compounds (Liu et al. 2024). These compounds contribute to structural and defense roles and are widely found in plants, fungi, and marine organisms. Triterpenes exhibit various pharmacological properties, including anti-inflammatory, anticancer, and antiviral activities (Mantiniotou et al. 2025). Some common triterpenes, such as squalene, ursolic acid, and betulinic acid, have significant opportunities in medicinal and industrial applications (Fig. 5).

2.1.6 Tetraterpenes ($C_{40}H_{64}$)

Tetraterpenes are a special class of terpenes composed of eight isoprene units ($C_{40}H_{64}$) and are best known for their role as pigments in plants, algae, and some bacteria. They include carotenoids such as β-carotene, lutein, and lycopene, which are essential for photosynthesis and act as antioxidants. Many tetraterpenes serve as precursors to vitamin A and play crucial roles in vision, immune function, and cellular health (Bufka et al. 2024) (Fig. 6).

2.1.7 Polyterpenes ($(C_5H_8)n$)

Polyterpenes are long-chain terpenes composed of multiple isoprene units, often forming natural polymers found in several plants, fungi, and animals. Rubber, the most well-known example, is derived from polyisoprene. Polyterpenes have diverse applications, including in medicine, industry, and biotechnology, due to their elasticity, bioactivity, and chemical versatility (Sahu et al. 2020).

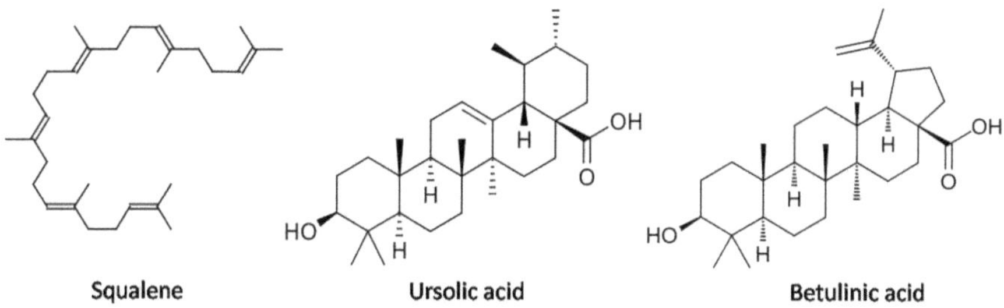

Fig. 5 Chemical structure of some triterpenes

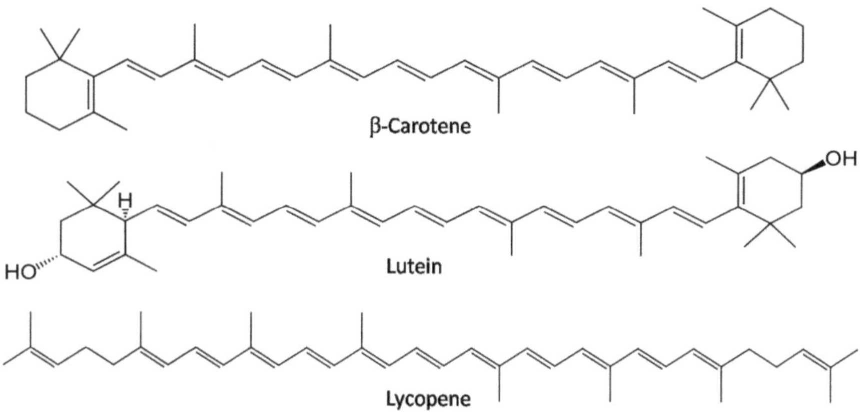

Fig. 6 Chemical structure of some common tetraterpenes

2.2 Phenolic Compounds

Phenolic compounds in EOs are bioactive molecules known for their strong antioxidant, antimicrobial, and anti-inflammatory properties. Their chemical structure, characterized by hydroxyl groups attached to an aromatic ring, enables them to neutralize free radicals and disrupt microbial cell membranes (Kruk et al. 2022), making them valuable in pharmaceuticals, cosmetics, and food preservation. Some phenolic compounds, including thymol, carvacrol, eugenol, etc., have greatly contributed to the therapeutic effects and characteristic aromas of many EOs (Fig. 7).

2.3 Aldehydes

Aldehydes in EOs are aromatic compounds known for their strong fragrances and biological activities. Some common aldehydes, such as citral, cinnamaldehyde, and vanillin, are found in lemongrass and lemon oil, cinnamon bark, and vanilla, respectively. Like other constituents of EOs, these compounds also contribute to the

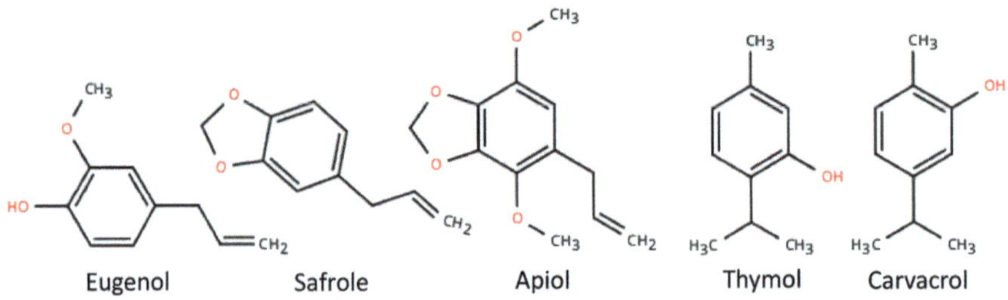

Fig. 7 Chemical structure of some common phenolic compounds of EOs

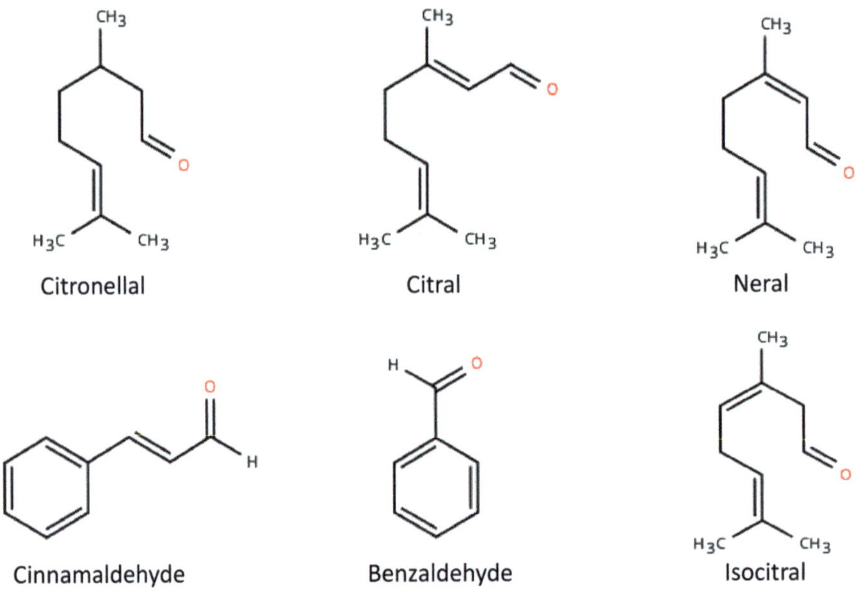

Fig. 8 Chemical structure of some common aldehydes found in EOs

characteristic aroma of EOs and exhibit potent antimicrobial, antifungal, and anti-inflammatory properties (Aljaafari et al. 2022). Their reactivity, due to the presence of a carbonyl group, makes them effective in therapeutic applications, food preservation, and perfumery. Furthermore, aldehydes also play a crucial role in plant defense mechanisms to protect against herbivores and pathogens (Fig. 8).

2.4 Esters

Esters in EOs are aromatic compounds known for their sweet, fruity, and floral fragrances. They are formed by the reaction of alcohols with carboxylic acids and contribute to the soothing and calming effects of many EOs. Some common esters of EOs include linalyl acetate (found in lavender), methyl salicylate (from wintergreen), and geranyl acetate (in roses and citronella). These compounds are valued for their anti-inflammatory, analgesic, and relaxing properties, making them widely used in aromatherapy, skincare, and perfumery (Sarmento-Neto et al. 2015) (Fig. 9).

2.5 Ketones

Ketones in EOs are bioactive compounds known for their distinctive aromas and therapeutic properties. They are characterized by a carbonyl group bonded to two hydrocarbon chains, making them chemically stable and effective in various applications. Common ketones, including menthone, camphor, and verbenone, are found in peppermint, camphor trees, and rosemary, respectively. These compounds contribute to mucolytic, wound-healing, and cell-regenerating effects, making them valuable in respiratory treatments, skincare, and pain relief (Criollo-Mendoza et al. 2023) (Fig. 10).

Fig. 9 Chemical structure of some common ester compounds found in EOs

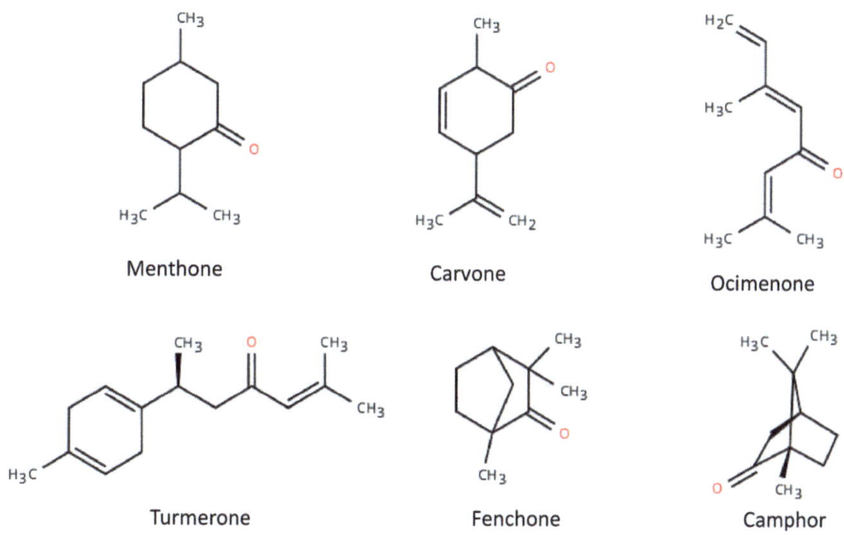

Fig. 10 Chemical structure of some common ketone compounds found in EOs

2.6 Alcohols

Alcohols in EOs are important aromatic compounds known for their antimicrobial and soothing properties (Kruk et al. 2022). They contain a hydroxyl (-OH) group, making them effective in therapeutic and cosmetic applications. Common alcohols found in EOs, including linalool, menthol, geraniol, etc., contribute to the pleasant fragrance, calming effects, and healing benefits and are highly demanded in aromatherapy, skincare, and natural medicine (Fig. 11).

2.7 Oxides

Oxides in EOs are oxygen-containing compounds known for their expectorant and respiratory benefits. The most common example is 1,8-cineole (eucalyptol), found in EOs of family Myrtaceae members like *Eucalyptus*, *Callistemon*, and tea tree oils. These compounds help clear congestion, support respiratory health, and have antimicrobial properties, making them valuable in aromatherapy and medicinal applications (Kumar et al. 2024) (Fig. 12).

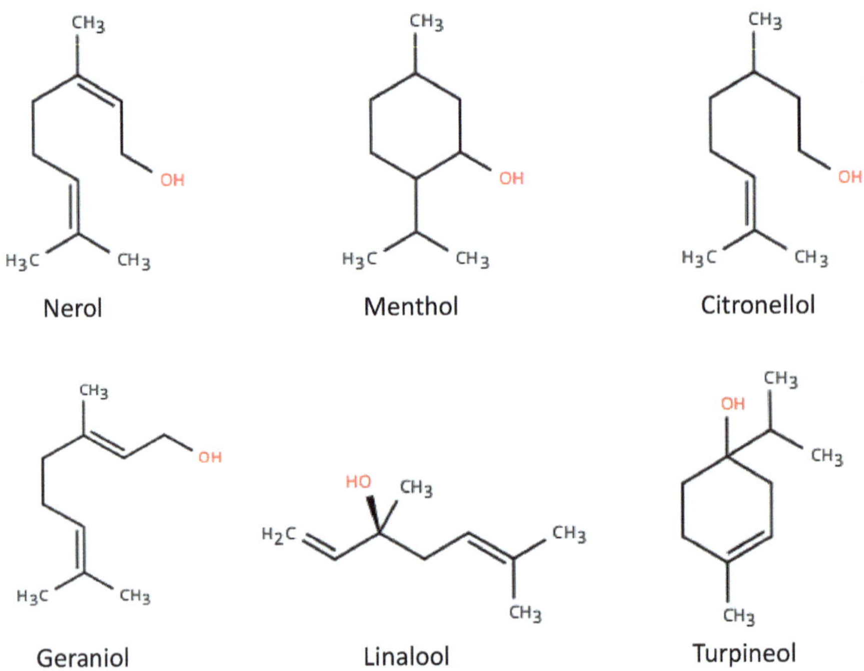

Fig. 11 Chemical structure of some common aromatic alcohols found in EOs

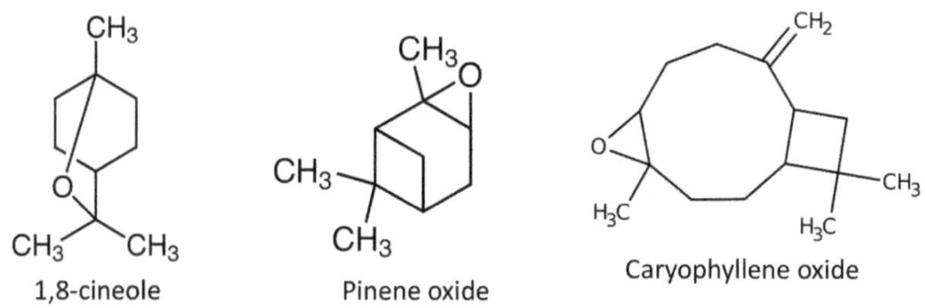

Fig. 12 Chemical structure of some common aromatic oxides found in EOs

3 Application of EOs in Pest Management

EOs are natural alternatives to synthetic pesticides, offering insecticidal, repellent, and antimicrobial properties for pest management (Das et al. 2021). Well-known oils like neem, citronella, and clove disrupt insect growth, deter feeding, and control fungal and bacterial pathogens (Tian et al. 2015; Takla et al. 2021; Zhang et al. 2022). They are applied as foliar sprays, soil treatments, and fumigants to protect crops and stored products.

3.1 Insect Pest Control

3.1.1 Agricultural Pests

EOs have been tested against a variety of agricultural pests, such as aphids, whiteflies, and stored-product insects. Oils from plants like neem (*Azadirachta indica*), citronella (*Cymbopogon nardus*), and peppermint (*Mentha piperita*) have shown strong insecticidal activity (Kumar et al. 2009; Takla et al. 2021; Zhang et al. 2022).

3.1.2 Household and Urban Pests

EOs are effective against common household pests like cockroaches, mosquitoes, and bed bugs. EOs from clove (*Syzygium aromaticum*) and eucalyptus (*Eucalyptus* sp.) are known to repel and kill these pests effectively (Tian et al. 2015).

3.1.3 Medical and Veterinary Pests

EOs are used to control vectors of diseases, such as mosquitoes transmitting malaria and dengue. EOs from citronella (*Cymbopogon nardus*) and lemongrass (*Cymbopogon citratus*) have been widely used as mosquito repellents (Aungtikun et al. 2021) (Table 1).

Table 1 Some commercially available EO-based insecticides

Commercial name	Plant oil used	Botanical name of plant	Target pests	References
EcoSMART®	Clove, rosemary, and thyme	*Syzygium aromaticum, Rosmarinus officinalis, and Thymus vulgaris*	Mosquitoes, ants, roaches, and other household pests	Isman (2006), Mossa (2016)
Neem Oil Products AzaMax®, Neemix®	Neem oil	*Azadirachta indica*	Aphids, whiteflies, thrips, and other soft-bodied insects	Mossa (2016)
Wondercide®	Cedarwood oil, lemongrass oil, and sesame oil	*Cedrus* sp., *Cymbopogon flexuosus, and Sesamum indicum*	Fleas, ticks, and mosquitoes	Machtinger and Li (2017)
Essentria IC3®	Rosemary oil, geranium oil, and peppermint oil	*Rosmarinus officinalis, Geranium* sp., and *Mentha piperita*	Mosquitoes, ticks, and other flying insects	Machtinger and Li (2017)
Bioganic®	Clove and thyme	*Syzygium aromaticum and Thymus vulgaris*	Ants, roaches, and flying insects	Copping and Duke (2007)
Mosquito Barrier®	Garlic oil	*Allium sativum*	Mosquitoes	Machtinger and Li (2017)
Cinnamite™	Cinnamon oil	*Cinnamomum zeylanicum*	Aphids, mites, and fungi	Dubey et al. (2010)
EcoTrol™	Rosemary oil, geranium oil, and peppermint oil	*Rosmarinus officinalis, Geranium* sp., and *Mentha piperita*	Aphids, thrips, whiteflies, mites, and plant bugs	Isman and Machial (2006)
Bioorganic™	Clove oil and sesame oil	*Syzygium aromaticum and Sesamum indicum*	Wide range of insects, including spider mites, flies, fleas, and beetles	Dayan et al. (2009)
Eco-exempt™	Clove oil, peppermint oil, and rosemary oil	*Syzygium aromaticum, Mentha piperita, and Rosmarinus officinalis*	Broad spectrum of insects	Dayan et al. (2009)

3.2 Fungal Disease Management

3.2.1 Agricultural Pathogens

EOs have been tested against a variety of fungal pathogens affecting crops, such as *Fusarium* sp., *Botrytis cinerea*, and *Aspergillus* sp. EOs from plants like thyme (*Thymus vulgaris*), clove (*Syzygium aromaticum*), and tea tree (*Melaleuca alternifolia*) have shown strong antifungal activity (Kumar et al. 2008; Tian et al. 2015).

3.2.2 Post-Harvest Pathogens

EOs are effective in controlling post-harvest diseases caused by fungi like *Penicillium* sp. and *Alternaria* sp. EOs from cinnamon (*Cinnamomum zeylanicum*) and oregano (*Origanum vulgare*) have exhibited significant antifungal efficacy in stored agricultural products (Yu 2025).

3.2.3 Medical and Veterinary Pathogens

EOs are also used to control fungal infections in humans and animals, including *Candida* sp. and other dermatophytes. EOs from eucalyptus and lemongrass have been widely studied for their antifungal properties (Rhimi et al. 2022) (Table 2).

Table 2 Some commercially available EO-based fungicides

Commercial name	Plant oil used	Botanical name of plant	Target pests	References
Timorex gold®	Tea tree oil	*Melaleuca alternifolia*	Powdery mildew, downy mildew, botrytis, black sigatoka, and rust	Reuveni et al. (2020)
Regalia®	Giant knotweed	*Reynoutria sachalinensis*	Powdery mildew, bacterial blight, downy mildew, and botrytis	Regnault-Roger et al. (2012)
Sporatec®	Rosemary oil, clove oil, and thyme oil	*Rosmarinus officinalis, Syzygium aromaticum, and Thymus vulgaris*	Powdery mildew, botrytis, anthracnose, and rust diseases	La Torre et al. (2014)
EcoSwing®	*Swinglea* extract and *citrus* oil	*Swinglea glutinosa* and *Citrus* sp.	Powdery mildew, botrytis, and brown rot	Deresa and Diriba (2023)
Fungastop®	Cinnamon oil, clove oil, and thyme oil	*Cinnamomum zeylanicum, Syzygium aromaticum, and Thymus vulgaris*	Powdery mildew, root rot, and other fungal pathogens	Martinez-Romero et al. (2008)
Sporan™	Rosemary oil, clove oil, thyme oil, and peppermint oil	*Rosmarinus officinalis Syzygium aromaticum, Thymus vulgaris, and Mentha piperita*	Black spot on roses, anthracnose, rust, and powdery mildew	Dayan et al. (2009)
Promax™	Thyme oil	*Thymus vulgaris*	Soilborne pathogens	Dayan et al. (2009)

3.3 Weed Control

EOs have emerged as a promising alternative to synthetic herbicides in weed management due to their eco-friendliness, biodegradability, and effectiveness against a wide range of weed species. This section highlights the application of EOs in weed control.

3.3.1 Preemergence Weed Control

EOs can be applied to the soil to inhibit seed germination and early seedling development. EOs from plants like cinnamon (*Cinnamomum zeylanicum*), clove (*Syzygium aromaticum*), and lemongrass (*Cymbopogon citratus*) have shown strong herbicidal activity (Verdeguer et al. 2020).

3.3.2 Postemergence Weed Control

EOs are effective in controlling established weeds by causing cell membrane disruption and desiccation. EOs from peppermint (*Mentha piperita*), thyme (*Thymus vulgaris*), and oregano (*Origanum vulgare*) have demonstrated significant contact herbicidal effects (Kitis 2023).

3.3.3 Integrated Weed Management

EOs can be combined with mechanical and cultural weed control strategies to enhance their effectiveness. Synergistic formulations with organic acids and other bioherbicides are being explored to improve efficacy and persistence. Cinnamon and clove EOs exhibit phytotoxic effects on weeds by inhibiting seed germination and root elongation to reduce competition for crops (Kaur et al. 2024) (Tables 3, 4, and 5).

3.4 Bacterial Disease Management

Bacterial pests pose significant challenges in agriculture, food safety, and human health. Traditional chemical pesticides, while effective, have led to antibiotic resistance, environmental contamination, and health hazards. EOs, derived from various plants, offer an eco-friendly and sustainable alternative due to their potent antimicrobial properties.

3.5 Nematode Disease Management

EOs have also gained significant attention as eco-friendly alternatives for managing nematode pests in agriculture. Their bioactive compounds exhibited nematicidal, repellent, and anti-feeding properties, making them effective against various plant-parasitic nematodes.

Table 3 Some commercially available EO-based herbicides/weedicides

Commercial name	Plant oil used	Botanical name of plant	Target pests	References
WeedZap®	Clove oil and cinnamon oil	*Syzygium aromaticum and Cinnamomum zeylanicum*	Broadleaf weeds and grasses	Dayan et al. (2009)
BurnOut®	Clove oil and *Citrus* oil	*Syzygium aromaticum and Citrus* sp.	Annual and perennial broadleaf weeds and grasses	Dayan et al. (2009)
Avenger® Weed Killer	d-Limonene (from *Citrus* oil)	*Citrus* sp.	Broadleaf and grassy weeds	U.S. Environmental Protection Agency (2025)
Green Gobbler® (Vinegar Weed Killer)	Clove oil, *Citrus* oil, thyme oil, cinnamon oil, and peppermint oil	*Syzygium aromaticum, Citrus* sp., *Thymus vulgaris, Cinnamomum zeylanicum, and Mentha piperita*	Dandelions, crabgrass, and other common weeds	Abouziena et al. (2009)
Nature's Avenger®	d-Limonene (from *Citrus* oil)	*Citrus* sp.	Annual and perennial weeds	U.S. Environmental Protection Agency (2006)
Matran® EC	Clove oil and cinnamon oil	*Syzygium aromaticum and Cinnamomum zeylanicum,*	Broadleaf weeds and grasses	Dayan et al. (2009)
Weed-A-Tak™	Clove oil, cinnamon oil, and *Citrus* oil	*Syzygium aromaticum, Cinnamomum zeylanicum, and Citrus* sp.	Broad spectrum of weeds and grasses	Dayan et al. (2009)
GreenMatch EX™	Lemongrass oil	*Cymbopogon citratus*	Wide range of broadleaf and grassy weeds	U.S. Environmental Protection Agency (2009)

4 Modern Approaches in EO Formulation Development

4.1 Nanoencapsulation

Nanoencapsulation of EOs is an advanced technique used to enhance their stability, bioavailability, and controlled release. EOs are highly volatile and sensitive to environmental factors such as light, heat, and oxygen, which can lead to rapid degradation and loss of efficacy. Nanoencapsulation involves enclosing these EOs within nanocarriers such as liposomes, polymeric nanoparticles, nanoemulsions, or solid lipid nanoparticles, providing a protective barrier against degradation (Liao et al. 2021). This technology improves the solubility of EOs in aqueous environments,

Table 4 Some commercially available EO-based bactericides

Commercial name	Plant oil used	Botanical name of plant	Target pests	References
Timorex Gold®	Tea tree oil	*Melaleuca alternifolia*	Bacterial leaf spot, bacterial canker, fire blight, and fungal diseases	U.S. Environmental Protection Agency (2016)
Regalia®	Giant knotweed	*Reynoutria sachalinensis*	Bacterial leaf spots, blights, and other bacterial infections	Regnault-Roger et al. (2012)
Sporatech®	Clove oil, thyme oil, and rosemary oil	*Syzygium aromaticum, Thymus vulgaris, and Rosmarinus officinalis*	Bacterial leaf blight diseases	Bhavaniramya et al. (2019)
EcoSwing®	*Swinglea* extract and *Citrus* oil	*Swinglea glutinosa* and *Citrus* sp.	Bacterial leaf blight, fire blight, and fungal diseases	Deresa and Diriba (2023)
OxiDate® 2.0	d-Limonene (from *Citrus* oil)	*Citrus* sp.	Bacterial spot, bacterial canker, and other plant diseases	U.S. Environmental Protection Agency (2011)
Ecotrol Plus	Rosemary oil, peppermint oil, and *geranium* oil	*Rosmarinus officinalis, Mentha piperita, and Geranium* sp.,	*Xanthomonas* and *Erwinia* sp.	Raveau et al. (2020)
Citrex	Lemongrass oil	*Cymbopogon citratus*	*Xanthomonas campestris and Clavibacter michiganensis*	Lobova and Cizek (2004)
Bioxeda™	Thyme oil	*Thymus vulgaris*	*Pseudomonas* sp., *Xanthomonas* sp.	Raveau et al. (2020)

enhances their therapeutic potential, and allows for targeted and sustained delivery. For instance, thymol-loaded chitosan nanoparticles have demonstrated prolonged insecticidal activity against agricultural pests. Applications of nanoencapsulated EOs extend to pharmaceuticals, food preservation, cosmetics, and agriculture, where they act as natural antimicrobials, antioxidants, and bioactive agents (Figs. 13 and 14).

4.2 Microencapsulation

Microencapsulation of EOs is a technique also used to protect these volatile compounds from environmental degradation and enhance their functional properties. This process involves enclosing EOs within microscopic capsules made of materials

Table 5 Some commercially available EO-based nematicides

Commercial name	Plant oil used	Botanical name of plant	Target pests	References
Nemastop®	Thyme oil, clove oil, and cinnamon oil	*Thymus vulgaris, Syzygium aromaticum*, and *Cinnamomum zeylanicum*	Root-knot nematodes and cyst nematodes	Metwaly and Zawam (2015)
Promax® 2.0	Thyme oil and other botanical extracts	*Thymus vulgaris*	Plant-parasitic nematodes and soilborne pathogens	Pérez and Lewis (2006)
Bionema	Thyme oil and Clove oil,	*Thymus vulgaris and Syzygium aromaticum*	Root-knot nematodes, cyst nematodes, and lesion nematodes	El-Sheikh et al. (2006)
Nema-Q®	*Quillaja saponaria* extract and oil	*Quillaja saponaria*	Root-knot nematodes and other plant-parasitic nematodes	Marais et al. (2010)
NEMGuard®	Garlic extract and oil-derived compounds (allicin)	*Allium sativum*	Root-knot nematodes (*Meloidogyne* sp.), cyst nematodes (*Heterodera* sp.), and lesion nematodes (*Pratylenchus* sp.)	Eder et al. (2021)

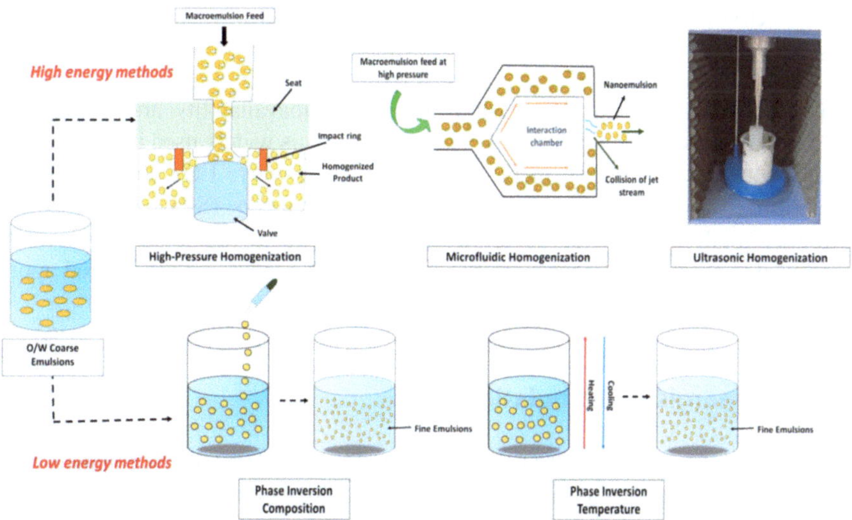

Fig. 13 Strategies for encapsulation of EOs. (Source: Soni et al. 2023)

such as polysaccharides, proteins, lipids, or synthetic polymers. The encapsulation matrix acts as a protective barrier, preventing oxidation, evaporation, and degradation due to light or heat (Sousa et al. 2022). Microencapsulation also enables

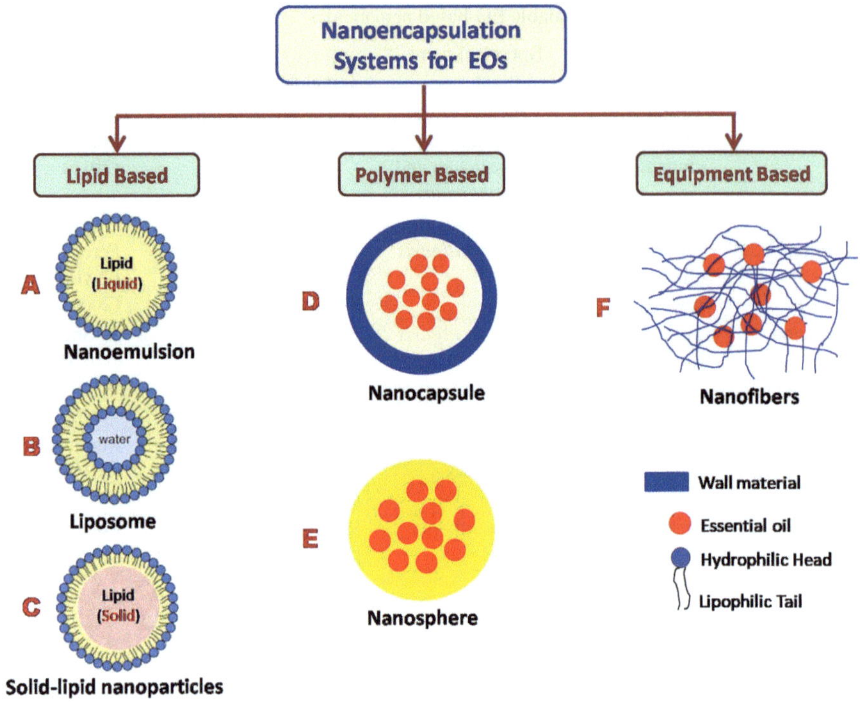

Fig. 14 Schematic representation of the different nanoencapsulation systems for EOs. (Source: Liao et al. 2021)

controlled and sustained release, improving the bioavailability and efficacy of essential oils in various applications. This technology is widely used in pharmaceuticals, food preservation, cosmetics, and agriculture, where EOs serve as natural preservatives, antimicrobial agents, and flavor enhancers (Sousa et al. 2022). Different methods, such as spray drying, coacervation, and freeze-drying, are employed to achieve efficient microencapsulation, depending on the desired application and release characteristics.

4.3 Synergistic Formulations

Synergistic formulations of EOs involve blending multiple oils to enhance their therapeutic efficacy through complementary or additive effects. Individual essential oils contain diverse bioactive compounds, and when combined strategically, they can exhibit stronger antimicrobial, antioxidant, anti-inflammatory, or medicinal properties than when used alone. This synergy can improve the effectiveness of EOs in pharmaceuticals, food preservation, and agriculture (Abu-Hussien et al. 2025). For instance, combining tea tree oil with eucalyptus oil enhances antimicrobial activity (Elangovan and Mudgil 2023), while a blend of lavender and peppermint oil provides improved analgesic and anti-inflammatory effects (Asih et al. 2021). Synergistic formulations also help in reducing the required dosage of each oil, minimizing potential toxicity or irritation while maintaining high efficacy. Advanced techniques such as nano- and microencapsulation further optimize the stability, controlled release, and targeted action of these formulations, making them highly effective in diverse applications.

4.4 Emulsification Technology

Emulsification technologies of EOs are now used to enhance their solubility, stability, and bioavailability by dispersing them in aqueous systems. EOs require emulsification due to their hydrophobic nature to effectively incorporate into water-based formulations for applications in various fields. This process involves the use of emulsifiers or surfactants to reduce interfacial tension and form stable oil-in-water or water-in-oil emulsions. Nanoemulsions and microemulsions, characterized by their small droplet size, offer improved stability, controlled release, and enhanced bioactivity of EOs (Kumar et al. 2025). Various methods such as high-energy homogenization, ultrasonication, and phase inversion are employed to achieve stable emulsions (Taha et al. 2020). Emulsification technology not only protects EOs from degradation but also enhances their targeted delivery, making them more effective in therapeutic and industrial applications.

4.5 Molecular Docking

The identification, optimization, and validation of plant-derived bioactive chemicals that interact with specific target proteins in pests, diseases, or weeds are made possible by molecular docking, which is essential to the development of botanical pesticides (Khursheed et al. 2022). This computational technique predicts the binding affinity, stability, and molecular interactions between phytochemicals and pest-specific enzymes, thereby allowing researchers to screen and design potent botanical pesticides with high selectivity and minimal nontarget effects. The process begins with choosing a target protein that is necessary for the pest's survival or reproduction, such as acetylcholinesterase (AChE) in insects (Johnson et al. 2021), acetolactate synthase (ALS) in weeds (Porri et al. 2024), or chitin deacetylase in fungi (Martinez-Cruz et al. 2021). To find possible inhibitors, molecular docking software like AutoDock, SwissDock, or Glide is used to screen large libraries of naturally occurring plant chemicals (Ivanova and Karelson 2022; Asiamah et al. 2023). To identify the best compounds, the binding energy, hydrogen bonding, and molecular interactions are examined. Potency, selectivity, and environmental safety are improved by further optimizing the chemical structures of lead compounds with high docking scores. This computational approach significantly reduces the time and cost required for experimental pesticide discovery while improving the likelihood of developing eco-friendly, target-specific botanical pesticides (Hou et al.

2023). Researchers can accelerate the development of innovative, sustainable substitutes for synthetic pesticides and support eco-friendly pest management techniques by integrating molecular docking with laboratory validation through in vitro and in vivo testing.

4.6 Integration with Smart Agriculture

Utilizing EO's inherent bioactive qualities, smart agriculture integrates them to improve soil health, crop protection, and sustainable farming methods. EOs, known for their antimicrobial, insecticidal, and growth-promoting effects, can be incorporated into precision agriculture systems using the aforementioned advanced technologies such as nanoencapsulation, emulsification, and controlled-release formulations. Automated sprayers, drones, and Internet of Things (IoT)-based sensors are examples of smart delivery systems that allow for the accurate distribution of EOs, reducing reliance on chemical pesticides and environmental impact (Song et al. 2015; Rejeb et al. 2022). These formulations can act as eco-friendly biopesticides, bioherbicides, and plant growth stimulants, contributing to sustainable and organic farming. By integrating EOs with smart agriculture techniques, farmers can optimize resource use, enhance crop yield, and promote eco-friendly pest and disease management strategies (Zhou et al. 2024; Kumar et al. 2025).

5 Advantages of EO-Based Pesticides

EO-based pesticides offer numerous advantages. Their natural composition and rapid biodegradability make them a sustainable choice in modern agriculture. As the demand for eco-friendly pest control practices continues to rise, EO-based phytopesticides provide a promising alternative to promote sustainable agricultural practices.

5.1 Eco-Friendly and Biodegradable

The safety of EO-based pesticides for the environment is one of their biggest benefits. EOs degrade quickly, in contrast to synthetic pesticides that can persist in the environment for longer durations. Their rapid biodegradability ensures minimal accumulation in soil and water, which reduces the risks of long-term contamination (Falleh et al. 2020).

5.2 Broad-Spectrum and Multifunctional Activity

The broad-spectrum action of EO-based pesticides against a variety of pests, such as insects, fungi, bacteria, and nematodes, is well-known. Many EOs, such as thyme oil (thymol), clove oil (eugenol), and lemongrass oil (citral), exhibit antimicrobial, antifungal, and insecticidal properties (Tourabi et al. 2023). This multifunctionality allows farmers to use a single product for multiple pest control applications, reducing the need for different chemical pesticides.

5.3 Low Toxicity to Humans and Beneficial Organisms

One of the primary concerns with conventional pesticides is their potential harm to humans, animals, and beneficial organisms. EO-based pesticides are generally low in toxicity to humans and pose minimal risks to pollinators such as bees and butterflies. Unlike synthetic chemicals, which can harm nontarget organisms, EOs selectively target pests, while beneficial insects like ladybugs, predatory mites, and earthworms are much less harmed. This selectivity supports biodiversity and promotes a healthier agricultural ecosystem (Giunti et al. 2022).

5.4 Reduced Risk of Pesticide Resistance

Pesticide resistance is a growing issue in modern agriculture, as pests evolve mechanisms to withstand synthetic chemicals. EO-based pesticides work through multiple modes of action, including disrupting cell membranes, interfering with neurotransmitters, and inhibiting enzyme functions (Devrnja et al. 2022). These diverse mechanisms make it difficult for pests to develop resistance, unlike conventional pesticides that rely on a single mode of action. As a result, EOs provide a more sustainable long-term solution for pest control.

5.5 Safe for Organic and Sustainable Farming

In integrated pest management (IPM) programs and organic farming, EO-based pesticides are preferred. A large number of EOs are organically certified and adhere to sustainable agriculture regulations. EO-based pesticides are suitable for farmers looking for environmentally friendly pest management techniques because of their natural origin and lack of harmful residues.

5.6 Compatibility with Other Biopesticides

Another advantage of EO-based pesticides is their compatibility with other microbial biopesticides, such as *Bacillus thuringiensis* and *Beauveria bassiana*. After the combination, resultant EO-based biopesticides enhanced pest control efficacy while reducing reliance on synthetic chemicals. This synergistic approach also provides an effective and sustainable alternative to conventional pest management (Assadpour et al. 2023).

6 Challenges and Future Perspectives

EO-based pesticides have emerged as promising substitutes for synthetic chemicals because of their biodegradability, efficacy against a variety of pests, and minimal toxicity to nontarget creatures. Derived from plant secondary metabolites, these botanical pesticides offer eco-friendly pest control solutions. However, a number of obstacles stand in the way of their general implementation. This section highlights the limitations of EO-based pesticides and discusses potential future directions to enhance their efficacy and applicability.

6.1 Challenges of EO-Based Pesticides

6.1.1 Limited Stability and Volatility

One of the primary challenges of EO-based pesticides is their high volatility and susceptibility to degradation. Exposure to light, heat, and oxygen can rapidly degrade their active components, reducing their effectiveness over time. This instability necessitates frequent reapplications, making them less practical compared to synthetic alternatives.

6.1.2 Inconsistent Efficacy

EOs exhibit variability in their chemical composition due to factors such as plant species, growing conditions, and extraction methods. This inconsistency can lead to fluctuation in efficacy, making it difficult to standardize formulations for large-scale agricultural use.

6.1.3 Mode of Action and Target Specificity

The precise mechanisms of action of EOs are frequently unclear, despite the fact that they typically function as neurotoxins, repellents, or growth inhibitors. This ambiguity makes it more difficult to create focused and effective pest management plans. Furthermore, EOs might have a wide range of effects, possibly affecting helpful insects like pollinators and natural predators.

6.1.4 Formulation and Delivery Challenges

EOs are difficult to dissolve in water due to their hydrophobic nature, leading to challenges in formulation and application. Enhancing their solubility, stability, and controlled release for extended efficacy requires the development of appropriate emulsifiers, encapsulating methods, or nano-formulations.

6.1.5 Regulatory and Commercial Constraints

EO-based pesticides, in contrast to synthetic ones, frequently encounter regulatory obstacles due to a lack of adequate toxicological and efficacious data. The cost and complexity of obtaining regulatory approval can hinder their market entry. Furthermore, their commercial viability is limited by their lower shelf life and comparatively higher production costs when compared to synthetic pesticides.

6.2 Future Perspectives

6.2.1 Enhancing Stability and Formulation Technologies

The stability and controlled release of pesticides based on EOs can be greatly enhanced by developments in encapsulation techniques, such as nano- and microencapsulation. Polymer-based formulations and emulsifiable concentrates can also enhance their application efficiency and longevity in the field.

6.2.2 Genetic and Biotechnological Approaches

The integration of biotechnology can help optimize the production of EOs with improved pest control properties. Genetic engineering of plants to produce higher concentrations of bioactive compounds or developing microbial-based biosynthesis of EO components could enhance their commercial viability.

6.2.3 Target-Specific Formulations

Future research should focus on identifying active components responsible for pest-specific toxicity while minimizing adverse effects on nontarget organisms. Combining EOs with synergistic natural compounds or selective botanical extracts may enhance their specificity and effectiveness. Artificial intelligence can accelerate the discovery of synergistic combinations and optimize formulations for specific pests.

6.2.4 Sustainable and Integrated Pest Management (IPM) Approaches

IPM programs should incorporate EO-based insecticides in addition to crop rotation, resistant crop types, and other biological control techniques. This all-encompassing strategy can encourage long-term pest management solutions and minimize reliance on artificial chemicals.

6.2.5 Regulatory and Policy Support

Governments and regulatory agencies should encourage small-scale producers, fund research, and establish clear rules to streamline the approval process for pesticides or preservatives based on EOs. Their integration into conventional agriculture should be facilitated by the establishment of global testing and formulation standards.

7 Conclusion

The global population has been exponentially increasing for many years, so food demand has to increase by 50–100% by 2050. There is an urgent need to find safe alternative strategies that may contribute to the provision of food while also protecting the environment and people. Plant EOs represent a transformative approach to sustainable pest management. EOs have emerged as a promising alternative to synthetic pesticides due to their biodegradable nature, diverse bioactivity, and minimal environmental impact. Modern approaches in the development of botanical pesticides have benefited from advancements in nanotechnology, encapsulation techniques, and formulation strategies to enhance the stability, efficacy, and controlled release of essential oil-based pesticides. Additionally, recent molecular docking and bioinformatics tools are revolutionizing the understanding of their mode of action, paving the way for targeted pest management strategies. Despite their potential, challenges such as variability in composition, limited field efficacy, and regulatory constraints must be addressed to facilitate large-scale adoption. Future research should focus on standardization, synergistic formulations, and sustainable production methods to improve their commercial viability. By integrating EO-based pesticides with conventional pest management practices, modern agriculture can orient toward a more sustainable, eco-friendly, and effective pest control system. Continued research and innovation will pave the way for their widespread acceptance and integration into global agricultural practices.

References

Abdelatti ZAS, Hartbauer M (2020) Plant oil mixtures as a novel botanical pesticide to control gregarious locusts. J Pest Sci 93:341–353. https://doi.org/10.1007/s10340-019-01169-7

Abouziena H, Omar AA, Sharma SD, Singh M (2009) Efficacy comparison of some new natural-product herbicides for weed control at two growth stages. Weed Technol 23(3):431–437. https://doi.org/10.1614/WT-08-185.1

Abu-Hussien SH, Nasry AR, Samy Z, El-Sayed SM, Bakry A, Ebeed N, Elhariry H, El-Noby T (2025) Synergistic antimicrobial activity of essential oils mixture of *Moringa oleifera*, *Cinnamomum verum* and *Nigella sativa* against *Staphylococcus aureus* using L-optimal mixture design. AMB Expr 15(1):15. https://doi.org/10.1186/s13568-024-01797-y

Ahmad MF, Ahmad FA, Alsayegh AA, Zeyaullah M, Al Shahrani AM, Muzammil K, Saati AA, Wahab S, Elbendary EY, Kambal N, Abdelrahman MH, Hussain S (2024) Pesticides impacts on human health and the environment with their mechanisms of action and possible countermeasures. Heliyon 10(7):e29128. https://doi.org/10.1016/j.heliyon.2024.e29128

Aljaafari MN, Alkhoori MA, Hag AM, Cheng WH, Lim SH, Loh JY, Lai KS (2022) Contribution of aldehydes and their derivatives to antimicrobial and immunomodulatory activities. Molecules 27(11):3589. https://doi.org/10.3390/molecules27113589

Al-Khayri JM, Rashmi R, Toppo V, Chole PB, Banadka A, Sudheer WN, Nagella P, Shehata WF, Mssallem MQ, Alessa FM, Almaghasla MI, Rezk AA (2023) Plant secondary metabolites: the weapons for biotic stress management. Meta 13(6):716. https://doi.org/10.3390/metabo13060716

Asiamah I, Obiri SA, Tamekloe W, Armah F, Borquaye LS (2023) Applications of molecular docking in natural products-based drug discovery. Sci Afr 20(8):e01593. https://doi.org/10.1016/j.sciaf.2023.e01593

Assadpour E, Can Karaça A, Fasamanesh M, Mahdavi SA, Shariat-Alavi M, Feng J, Kharazmi MS, Rehman A, Jafari SM (2023) Appication of essential oils as natural biopesticides; recent advances. Crit Rev Food Sci Nutr 64(19):6477–6497. https://doi.org/10.1080/10408398.2023.2170317

Aungtikun J, Soonwera M, Sittichok S (2021) Insecticidal synergy of essential oils from *Cymbopogon citratus* (Stapf.), *Myristica fragrans* (Houtt.), and *Illicium verum* Hook. f. and their major active constituents. Ind Crops PRO 164:113386. https://doi.org/10.1016/j.indcrop.2021.113386

Ayilara MS, Adeleke BS, Akinola SA, Fayose CA, Adeyemi UT, Gbadegesin LA, Omole RK, Johnson RM, Uthman QO, Babalola OO (2023) Biopesticides as a promising alternative to synthetic pesticides: a case for microbial pesticides, phytopesticides, and nanobiopesticides. Front Microbiol 14:1040901. https://doi.org/10.3389/fmicb.2023.1040901

Ben Miri Y (2025) Essential oils: chemical composition and diverse biological activities: a comprehensive review. Nat Prod Commun 20(1):1–29

Bhavaniramya S, Vishnupriya S, Al-Aboody MS, Vijayakumar R, Baskaran D (2019) Role of essential oils in food safety: antimicrobial and antioxidant applications. GOST 2(2):49–55. https://doi.org/10.1016/j.gaost.2019.03.001

Bolouri P, Salami R, Kouhi S, Kordi M, Lajayer BA, Hadian J, Astatkie T (2022) Applications of essential oils and plant extracts in different industries. Molecules 27(24):8999. https://doi.org/10.3390/molecules27248999

Bufka J, Vaňková L, Sýkora J, Křížková V (2024) Exploring carotenoids: metabolism, antioxidants, and impacts on human health. J Funct Foods 118:106284. https://doi.org/10.1016/j.jff.2024.106284

Chadwick M, Trewin H, Gawthrop F, Wagstaff C (2013) Sesquiterpenoids lactones: benefits to plants and people. Int J Mol Sci 14(6):12780–12805. https://doi.org/10.3390/ijms140612780

Copping LG, Duke SO (2007) Natural products that have been used commercially as crop protection agents. Pest Manag Sci 63:524–554

Criollo MMS, Contreras ALA, Leyva LN, Gutiérrez GEP, Jiménez OLA, Heredia JB (2023) Wound healing properties of natural products: mechanisms of action. Molecules 28(2):598. https://doi.org/10.3390/molecules28020598

Das S, Singh VK, Dwivedy A, Chaudhari AK, Dubey NK (2021) Insecticidal and fungicidal efficacy of essential oils and nanoencapsulation approaches for the development of next generation ecofriendly green preservatives for management of stored food commodities: an overview. Int J Pest Manag 70(2):1–32. https://doi.org/10.1080/09670874.2021.1969473

Dayan FE, Cantrell CL, Duke SO (2009) Natural products in crop protection. Bioorg Med Chem 17(12):4022–4034. https://doi.org/10.1016/j.bmc.2009.01.046

Deresa EM, Diriba TF (2023) Phytochemicals as alternative fungicides for controlling plant diseases: a comprehensive review of their efficacy, commercial representatives, advantages, challenges for adoption, and possible solutions. Heliyon 9(3):e13810. https://doi.org/10.1016/j.heliyon.2023.e13810

Devrnja N, Milutinović M, Savić J (2022) When scent becomes a weapon—plant essential oils as potent bioinsecticides. Sustainability 14(11):6847. https://doi.org/10.3390/su14116847

Dubey NK, Shukla R, Kumar A, Singh P, Prakash B (2010) Prospects of botanical pesticides in sustainable agriculture. Curr Sci 98(4):479–480

Eder R, Consoli E, Krauss J, Dahlin P (2021) Polysulfides applied as formulated garlic extract to protect tomato plants against the root-knot nematode *Meloidogyne incognita*. Plants 10(2):394. https://doi.org/10.3390/plants10020394

Elangovan S, Mudgil P (2023) Antibacterial properties of *Eucalyptus globulus* essential oil against MRSA: a systematic review. Antibiotics 12(3):474. https://doi.org/10.3390/antibiotics12030474

El-Sheikh MH, El-Motty EZA, Hasabo SAA (2006) Effect of two organic amendments, elemental sulphur, bionema and carbofuran soil application to control root-knot nematode on growth and productivity of thompson seedless grapes. AEJAES 1(3):191–197

Falleh H, Jemaa MB, Saada M, Ksouri R (2020) Essential oils: a promising eco-friendly food preservative. Food Chem 330:127268. https://doi.org/10.1016/j.foodchem.2020.127268

Fransisca Retno Asih FR, Husin F, Suwarsa O, Fidrianny I, Hilmanto D (2021) A randomized controlled trial of combination of peppermint, lavender, and turmeric oil for antipruritic agent in pregnant women. Med J Indones 30:39–44. https://doi.org/10.13181/mji.oa.204467

Giunti G, Benelli G, Palmeri V, Laudani F, Ricupero M, Ricciardi R, Maggi F, Lucchi A, Guedes RNC, Desneux N, Campolo O (2022) Non-target effects of essential oil-based biopesticides for crop protection: impact on natural enemies, pollinators, and soil invertebrates. BioControl 176:105071. https://doi.org/10.1016/j.biocontrol.2022.105071

Hillier SG, Lathe R (2024) Terpenes, hormones and life: isoprene rule revisited. J Endocrinol 242(02):R9–R22. https://doi.org/10.1530/JOE-19-0084
Hou Y, Bai Y, Lu C, Wang Q, Wang Z, Gao J, Xu H (2023) Applying molecular docking to pesticides. Pest Manag Sci 79:4140–4152. https://doi.org/10.1002/ps.7700
Isman MB, Machial CM (2006) Pesticides based on plant essential oils: from traditional practice to commercialization. In: Rai M, Carpinella MC (eds) Advances in phytomedicine, vol 3. Elsevier, Amsterdam, pp 29–44. https://doi.org/10.1016/S1572-557X(06)03002-9
Ivanova L, Karelson M (2022) The impact of software used and the type of target protein on molecular docking accuracy. Molecules 27:9041. https://doi.org/10.3390/molecules27249041
Johnson TO, Ojo OA, Ikiriko S, Ogunkua J, Akinyemi GO, Rotimi DE, Oche JR, Adegboyega AE (2021) Biochemical evaluation and molecular docking assessment of *Cymbopogon citratus* as a natural source of acetylcholine esterase (AChE)-targeting insecticides. Biochem Biophys Rep 28:101175. https://doi.org/10.1016/j.bbrep.2021.101175
Kaur A, Singh G, Menon S, Kumari K (2024) Integrated weed management: a comprehensive review of conventional, non-conventional, and emerging strategies for sustainable agriculture. J Adv Biol Biotechnol 27(8):156–167. https://doi.org/10.9734/jabb/2024/v27i81130
Khursheed A, Rather MA, Jain V, Wani AR, Rasool S, Nazir R, Malik NA, Majid SA (2022) Plant based natural products as potential ecofriendly and safer biopesticides: a comprehensive overview of their advantages over conventional pesticides, limitations and regulatory aspects. Microb Pathog 173:105854. https://doi.org/10.1016/j.micpath.2022.105854
Kitis YE (2023) Chemical composition and herbicidal activity of Oregano (*Origanum onites*) essential oil on weeds and wheat. Acta Agric Scand B 73(1):142–151. https://doi.org/10.1080/09064710.2023.2232680
Kruk J, Aboul-Enein BH, Duchnik E, Marchlewicz M (2022) Antioxidative properties of phenolic compounds and their effect on oxidative stress induced by severe physical exercise. J Physiol Sci 72:19. https://doi.org/10.1186/s12576-022-00845-1
Kumar A, Shukla R, Singh P, Prasad CS, Dubey NK (2008) Assessment of *Thymus vulgaris* L. essential oil a safe botanical preservative against post harvest fungal infestation of food commodities. Innov Food Sci Emerg Technol 9(4):575–580. https://doi.org/10.1016/j.ifset.2007.12.005
Kumar A, Shukla R, Singh P, Singh AK, Dubey NK (2009) Use of essential oil from *Mentha arvensis* L. to control storage moulds and insects in stored chickpea. J Sci Food Agric 89(15):2643–2649. https://doi.org/10.1002/jsfa.3768
Kumar A, Dubey NK, Srivastava S (2013) Antifungal evaluation of *Ocimum sanctum* essential oil against fungal deterioration of raw materials of *Rauvolfia serpentina* during storage. Ind Crop Prod 45:30–35. https://doi.org/10.1016/j.indcrop.2012.12.006
Kumar S, Mahapatro GK, Yadav DK, Tripathi K, Koli P, Kaushik P, Sharma K, Nebapure S (2022) Essential oils as green pesticides: an overview. Indian J Agric Sci 92(11):1298–1305. https://doi.org/10.56093/ijas.v92i11.122746
Kumar L, Vora K, Gholap AD, Hatvate NT, Naren P, Khan S, Chavda VP, Balar PC, Gandhi J, Khatri DK (2024) Essential oils for clinical aromatherapy: a comprehensive review. J Ethnopharmacol 330:118180. https://doi.org/10.1016/j.jep.2024.118180
Kumar A, Kanwar R, Mehta SK (2025) Nanoemulsion as an effective delivery vehicle for essential oils: properties, formulation methods, destabilizing mechanisms and applications in agri-food sector. Next Nanotechnol 7:100096. https://doi.org/10.1016/j.nxnano.2024.100096
La Torre A, Mandala C, Pezza L, Caradonia F, Battaglia V (2014) Evaluation of essential plant oils for the control of *Plasmopara viticola*. J Essent Oil Res 26(4):282–291. https://doi.org/10.1080/10412905.2014.889049
Liao W, Badri W, Dumas E, Ghnimi S, Elaissari A, Saurel R, Gharsallaoui A (2021) Nanoencapsulation of essential oils as natural food antimicrobial agents: an overview. Appl Sci 11(13):5778. https://doi.org/10.3390/app11135778
Liu J, Yin X, Kou C, Thimmappa R, Hua X, Xue Z (2024) Classification, biosynthesis, and biological functions of triterpene esters in plants. Plant Commun 5(4):100845. https://doi.org/10.1016/j.xplc.2024.100845

Lobova D, Cizek A (2004) Bactericidal efficacy of two disinfectants against *Brachyspira hyodysenteriae* and one feed supplement against *B. hyodysenteriae* and *B. pilosicoli*. Vet Med 49(5):156–160. https://doi.org/10.17221/5690-VETMED

Mabou FD, Yossa IBN (2021) Terpenes: structural classification and biological activities. IOSR J Pharm Biol Sci 16(3):25–40. https://doi.org/10.9790/3008-1603012540

Machtinger ET, Li AY (2017) Evaluation of four commercial natural products for repellency and toxicity against the lone star tick, *Amblyomma americanum* (Acari: Ixodidae). Exp Appl Acarol 73:451–460. https://doi.org/10.1007/s10493-017-0185-z

Mantiniotou M, Athanasiadis V, Kalompatsios D, Bozinou E, Lalas SI (2025) Therapeutic capabilities of triterpenes and triterpenoids in immune and inflammatory processes: a review. Compounds 5(1):2. https://doi.org/10.3390/compounds5010002

Marais L, Otero R, Riga E (2010) Control of plant-parasitic nematodes with the bionematicide Nema-Q (R) an extract of *Quillaja saponaria*. J Nematol 42(3):254–255

Martínez-Cruz JM, Polonio Á, Zanni R, Romero D, Gálvez J, Fernández-Ortuño D, Pérez-García A (2021) Chitin deacetylase, a novel target for the design of agricultural fungicides. J Fungi 7(12):1009. https://doi.org/10.3390/jof7121009

Martinez-Romero D, Serrano M, Bailén G, Guillén F, Zapata PJ, Valverde JM, Castillo S, Fuentes M, Valero D (2008) The use of a natural fungicide as an alternative to preharvest synthetic fungicide treatments to control lettuce deterioration during postharvest storage. Postharvest Biol Technol 47(1):54–60. https://doi.org/10.1016/j.postharvbio.2007.05.020

Masyita A, Sari RM, Astuti AD, Yasir B, Rumata NR, Emran TB, Nainu F, Simal-Gandara J (2022) Terpenes and terpenoids as main bioactive compounds of essential oils, their roles in human health and potential application as natural food preservatives. Food Chem 13:100217. https://doi.org/10.1016/j.fochx.2022.100217

Metwaly H, Zawam H (2015) Efficacy of some bioagents and nemastop compound in controlling root knot disease on peanut. J Plant Protoc Pathol 6(3):535–547. https://doi.org/10.21608/jppp.2015.53340

Mossa ATH (2016) Green pesticides: essential oils as biopesticides in insect-pest management. J Environ Sci Technol 9(5):354–378. https://doi.org/10.3923/jest.2016.354.378

Murphy KM, Zerbe P (2020) Specialized diterpenoid metabolism in monocot crops: biosynthesis and chemical diversity. Phytochemistry 172:112289. https://doi.org/10.1016/j.phytochem.2020.112289

Nollet LML, Rathore HS (2019) Green pesticides handbook: essential oils for pest control. CRC Press, pp 1–572

Pérez EE, Lewis EE (2006) Use of entomopathogenic nematodes and thyme oil to suppress plant-parasitic nematodes on English boxwood. Plant Dis 90:471–475. https://doi.org/10.1094/PD-90-0471

Porri A, Panozzo S, Sisay MT, Scarabel L, Lerchl J, Milani A (2024) 3D structure of acetolactate synthase explains why the Asp-376-Glu point mutation does not give the same resistance level to different imidazolinone herbicides. Pestic Biochem Physiol 204:106070. https://doi.org/10.1016/j.pestbp.2024.106070

Prakash B, Kujur A, Yadav A, Kumar A, Singh PP, Dubey NK (2018) Nanoencapsulation: an efficient technology to boost the antimicrobial potential of plant essential oils in food system. Food Control 89:1–11. https://doi.org/10.1016/j.foodcont.2018.01.018

Qasim M, Islam W, Rizwan M, Hussain D, Noman A, Khan KA, Ghramh HA, Han X (2024) Impact of plant monoterpenes on insect pest management and insect-associated microbes. Heliyon 10(20):39120. https://doi.org/10.1016/j.heliyon.2024.e39120

Raveau R, Fontaine J, Lounès-Hadj Sahraoui A (2020) Essential oils as potential alternative biocontrol products against plant pathogens and weeds: a review. Foods 9(3):365. https://doi.org/10.3390/foods9030365

Regnault-Roger C, Vincent C, Arnason JT (2012) Essential oils in insect control: low risk products in a high-stakes world. Annu Rev Entomol 57:405–424

Rejeb A, Abdollahi A, Rejeb K, Treiblmaier H (2022) Drones in agriculture: a review and bibliometric analysis. Comput Electron Agric 198:107017. https://doi.org/10.1016/j.compag.2022.107017

Reuveni M, Sanchez E, Barbier M (2020) Curative and suppressive activities of essential tea tree oil against fungal plant pathogens. Agron 10(4):609. https://doi.org/10.3390/agronomy10040609

Rhimi W, Mohammed MA, Zarea AAK, Greco G, Tempesta M, Otranto D, Cafarchia C (2022) Antifungal, antioxidant and antibiofilm activities of essential oils of *Cymbopogon* spp. Antibiotics 11(6):829. https://doi.org/10.3390/antibiotics11060829

Rodilla JM, Rosado T, Gallardo E (2024) Essential oils: chemistry and food applications. Foods 13(7):1074. https://doi.org/10.3390/foods13071074

Sahu P, Bhowmick AK, Kali G (2020) Terpene based elastomers: synthesis, properties, and applications. PRO 8(5):553. https://doi.org/10.3390/pr8050553

Sarmento-Neto JF, do Nascimento LG, Felipe CFB, de Sousa DP (2015) Analgesic potential of essential oils. Molecules 21(1):E20. https://doi.org/10.3390/molecules21010020

Song Y, Sun H, Li M, Zhang Q (2015) Technology application of smart spray in agriculture: a review. Intell Autom Soft Comput 21(3):319–333. https://doi.org/10.1080/10798587.2015.1015781

Soni M, Yadav A, Maurya A, Das S, Dubey NK, Dwivedy AK (2023) Advances in designing essential oil nanoformulations: an integrative approach to mathematical modeling with potential application in food preservation. Foods 12:4017. https://doi.org/10.3390/foods12214017

Sousa VI, Parente JF, Marques JF, Forte MA, Tavares CJ (2022) Microencapsulation of essential oils: a review. Polymers 14(9):1730. https://doi.org/10.3390/polym14091730

Souto AL, Sylvestre M, Tölke ED, Tavares JF, Barbosa-Filho JM, Cebrián-Torrejón G (2021) Plant-derived pesticides as an alternative to pest management and sustainable agricultural production: prospects, applications and challenges. Molecules 26(16):4835. https://doi.org/10.3390/molecules26164835

Sun Y, Xin J, Xu Y, Wang X, Zhao F, Niu C, Liu S (2024) Research progress on sesquiterpene compounds from artabotrys plants of Annonaceae. Molecules 29(7):1648. https://doi.org/10.3390/molecules29071648

Taha A, Ahmed E, Ismaiel A, Ashokkumar M, Xu X, Pan S, Hu H (2020) Ultrasonic emulsification: an overview on the preparation of different emulsifiers-stabilized emulsions. Trends Food Sci Technol 105:363–377. https://doi.org/10.1016/j.tifs.2020.09.024

Takla SS, El-Dars FMSE, Amien AS, Rizk MA (2021) Prospects of neem essential oil as bio-pesticide and determination of its residues in eggplant plants during crop production cycle. Egypt Acad J Biol Sci Toxicol Pest Control 13(2):1–15. https://doi.org/10.21608/eajbsf.2021.181340

Tian BL, Qi-zhi L, Liu ZL, Li P, Wang JW (2015) Insecticidal potential of clove essential oil and its constituents on *Cacopsylla chinensis* (Hemiptera: Psyllidae) in laboratory and field. J Econ Entomol 108(3):957–961. https://doi.org/10.1093/jee/tov075

Tourabi M, Nouioura G, Touijer H, Baghouz A, El Ghouizi A, Chebaibi M, Bakour M, Ousaaid D, Almaary KS, Nafidi HA, Bourhia M, Farid K, Lyoussi B, Derwich E (2023) Antioxidant, antimicrobial, and insecticidal properties of chemically characterized essential oils extracted from *Mentha longifolia*: *In vitro* and *in silico* analysis. Plants 12(21):3783. https://doi.org/10.3390/plants12213783

Tran ASN, Pham VH, Duong CD, Bui QTP (2024) Extraction conditions, chemical composition and biological activity of essential oil of *Allium schoenoprasum* L. bulb from Quang Tri province, Vietnam. Food Chem Adv 4:100574. https://doi.org/10.1016/j.focha.2023.100574

U.S. Code of Federal Regulations (2025) Title 21, Part 182, Section 182.20

U.S. Environmental Protection Agency (2006). https://www3.epa.gov/pesticides/chem_search/ppls/082052-00001-20060807.pdf

U.S. Environmental Protection Agency (2009). https://www3.epa.gov/pesticides/chem_search/ppls/082052-00004-20090115.pdf

U.S. Environmental Protection Agency (2011). https://www3.epa.gov/pesticides/chem_search/ppls/070299-00002-20110923.pdf

U.S. Environmental Protection Agency (2016). https://www3.epa.gov/pesticides/chem_search/ppls/086182-00001-20180531.pdf

U.S. Environmental Protection Agency (2025). https://www3.epa.gov/pesticides/chem_search/ppls/092967-00003-20250108.pdf

Verdeguer M, Sánchez-Moreiras AM, Araniti F (2020) Phytotoxic effects and mechanism of action of essential oils and terpenoids. Plants 9(11):1571. https://doi.org/10.3390/plants9111571

Yu J (2025) Chemical composition of essential oils and their potential applications in postharvest storage of cereal grains. Molecules 30(3):683. https://doi.org/10.3390/molecules30030683

Zhang X, Zhu H, Wang J, Li F, Wang J, Ma X, Li J, Huang Y, Liu Z, Zhang L, Li S (2022) Antimicrobial activity of citronella (*Cymbopogon citratus*) essential oil separation by ultrasound assisted ohmic heating hydrodistillation. Ind Crops Prod 176:114299. https://doi.org/10.1016/j.indcrop.2021.114299

Zhou W, Arcot Y, Medina RF, Bernal J, Cisneros-Zevallos L, Akbulut MES (2024) Integrated pest management: an update on the sustainability approach to crop protection. ACS Omega 9(40):41130–41147. https://doi.org/10.1021/acsomega.4c06628

MIX
Papier aus verantwortungsvollen Quellen
Paper from responsible sources
FSC® C105338

If you have any concerns about our products,
you can contact us on
ProductSafety@springernature.com

In case Publisher is established outside the EU,
the EU authorized representative is:
Springer Nature Customer Service Center GmbH
Europaplatz 3, 69115 Heidelberg, Germany

Printed by Libri Plureos GmbH
in Hamburg, Germany